Springer Series in Optical Sciences   Volume 71

Editor: Peter W. Hawkes

W0107236

Springer Series in Optical Sciences

Editorial Board: A. L. Schawlow A. E. Siegman T. Tamir

Managing Editor: H. K. V. Lotsch

Volumes 1-41 are listed at the end of the book

Ludwig Reimer (Ed.)

Energy-Filtering Transmission Electron Microscopy

With Contributions by
C. Deininger R. F. Egerton F. Hofer B. Jouffrey
D. Krahl R. D. Leapman J. Mayer L. Reimer H. Rose
P. Schattschneider J. C. H. Spence

With 199 Figures

 Springer

Professor Dr. Ludwig Reimer

Physikalisches Institut, Universität Münster
Wilhelm-Klemm-Straße 10
D-48149 Münster, Germany

Guest Editor: Dr. Peter W. Hawkes

Laboratoire d'Optique Electronique du CNRS
Boîte Postale No. 4347, F-31055 Toulouse Cedex, France

Editorial Board

Arthur L. Schawlow, Ph. D.

Department of Physics, Stanford University
Stanford, CA 94305-4060, USA

Professor Anthony E. Siegman, Ph. D.

Electrical Engineering
E. L. Ginzton Laboratory, Stanford University
Stanford, CA 94305-4085, USA

Theodor Tamir, Ph. D.

Department of Electrical Engineering
Polytechnic University, 333 Jay Street, Brooklyn, NY 11201, USA

Managing Editor: Dr.-Ing. Helmut K.V. Lotsch

Springer-Verlag, Tiergartenstrasse 17, D-69121 Heidelberg, Germany

ISBN 978-3-662-14055-0   ISBN 978-3-540-48995-5 (eBook)
DOI 10.1007/978-3-540-48995-5

Library of Congress Cataloging-in-Publication Data. Energy-filtering transmission electron microscopy/
Ludwig Reimer (ed.). p. cm. – (Springer series in optical sciences; 71) Includes bibliographical
references and index.
1. Transmission electron microscopy. I. Reimer, Ludwig, 1928- . II. Series: Springer series in optical
sciences; v. 71. QH212.T7E54   1995   94-43581

Typesetting: Camera-ready copy from the editor using a Springer T$_E$X macro package
SPIN 10069608   54/3144-5 4 3 2 1 0 - Printed on acid-free paper

Preface

The strength of transmission electron microscopy (TEM) lies in its ability to combine high-resolution imaging with analytical electron microscopy (AEM). The modern TEM can attain a resolution of about 0.1 nm. The use of electron probes of about 0.2–10 nm in the scanning transmission mode of a conventional TEM, or 0.1–0.2 nm in a dedicated scanning transmission electron microscope (STEM), therefore allows quantitative analysis to be performed by electron diffraction, x-ray spectroscopy, and electron energy-loss spectroscopy (EELS) in the nanometre range.

The development of energy-filtering transmission electron microscopy (EFTEM) using an imaging filter lens or spectrometer offers numerous new possibilities. The selection of zero-loss electrons in images and diffraction patterns not only allows a better comparison with computer simulations but also eliminates the inelastic background in diffraction patterns from thicker samples and avoids the blurring of images due to chromatic aberration. By taking images at different energy losses E (using an energy window $\Delta = 1$–10 eV) better contrast can be achieved, phases with different plasmon losses can be imaged selectively and element distribution images using the inner-shell ionisation edges can be formed. Additionally, angularly and spatially resolved EELS can be recorded, providing a local analysis from plasmon losses, the ionisation edges of inner-shell ionisation with energy-loss near edge structure (ELNES), or extended energy-loss fine structure (EXELFS).

This book summarizes the present state of the electron-optical design of imaging filter lenses and spectrometers, the physical background of plasmon losses and inner-shell ionisation, quantitative elemental analysis by EELS, the different operating and imaging modes of EFTEM for electron spectroscopic diffraction (ESD) and electron spectroscopic imaging (ESI), and the application of energy filtering to reflection electron microscopy (REM).

The editor and the contributors thank Dr. P.W. Hawkes for thorough correction of the manuscript and many helpful comments.

Münster,   L. Reimer
April 1994

Contents

Contributors

Ray F. Egerton
Physics Department
University of Alberta
Edmonton
T6G 2J1, Canada

Ferdinand Hofer
Forschungsinstitut für
Elektronenmikroskopie
Technische Universität Graz
A-8042 Graz
Austria

Bernard Jouffrey
Laboratoire de mécanique, sols,
structures et matériaux
École Centrale Paris
F-92295 Châtenay-Malabry
France

Dieter Krahl
Fritz-Haber-Institut der MPG
Abteilung Elektronenmikroskopie
Faradayweg 4-6
D-14195 Berlin
Germany

Richard D. Leapman
National Institutes of Health
Bethesda
MD 20892, U.S.A.

Joachim Mayer
Christine Deininger
Max-Planck-Institut für Metallforschung
Institut für Werkstoffwissenschaft
Seestraße 92
D-70174 Stuttgart
Germany

Ludwig Reimer
Physikalisches Institut
Universität Münster
Wilhelm-Klemm-Straße 10
D-48149 Münster
Germany

Harald Rose
Institut für Angewandte Physik
Technische Universität Darmstadt
Hochschulstraße 6
D-64289 Darmstadt
Germany

Peter Schattschneider
Institut für Angewandte und
Technische Physik
Technische Universität Wien
Wiedner Hauptstraße 8-10
A-1040 Wien
Austria

John C.H. Spence
Arizona State University
Department of Physics
Tempe
AZ 85287-1504, U.S.A.

1. Introduction

Ludwig Reimer

This introductory chapter summarizes the elastic and inelastic scattering processes and presents a short review of the instrumentation, imaging modes and analytical procedures in conventional transmission electron microscopy (CTEM). A short review of the historical development of electron energy-loss spectroscopy (EELS) and energy filtering transmission electron microscopy (EFTEM) is followed by a presentation of the new operation modes, which condiserably expand the possibilities of TEM. The acronym EELS is also used for "energy-loss spectrum".

1.1 Electron–Specimen Interactions

The information provided by electron microscopy and microanalysis results from the electron–specimen interactions. These are the elastic and inelastic scattering processes at atoms and in solids and the emission of secondary electrons, light quanta (cathodoluminescence), x-ray quanta and Auger electrons caused by the transfer of energy in inelastic scattering events.

1.1.1 Elastic Scattering and Cross-Sections

Elastic electron scattering is the most important interaction for obtaining contrast and high resolution in transmission electron microscopy (TEM). In a particle model, the electrons are attracted by the positively charged nucleus and pass the atom on a hyberbolic trajectory (Fig. 1.1a); the scattering through angle θ increases with decreasing impact parameter a. By analogy with the trajectories of alpha-particles, this is called Rutherford scattering. Elastic scattering preserves the total kinetic energy and momentum. The atomic electrons are not excited in an elastic collision and only screen the slowly decreasing Coulomb potential proportional to $1/r$ with increasing distance r from the nucleus.

The kinetic energy transferred to the nucleus can be neglected for small scattering angles θ owing to the low electron-to-nucleus mass ratio. Large energy losses E can only be observed at large scattering angles, e.g. about 1 eV for 80 keV electrons scattered through $\theta = 90°$ at copper atoms [1.1]. In high voltage electron microscopy where the electron energy reaches 0.3–1

Springer Series in Optical Sciences, Vol. 71
Energy-Filtering Transmission Electron Microscopy
Editor: Ludwig Reimer ©Springer-Verlag Berlin Heidelberg 1995

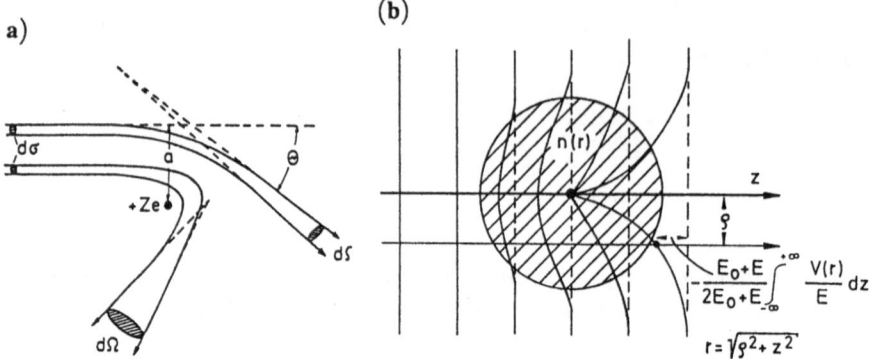

Fig. 1.1. (a) Elastic scattering and differential cross-section $d\sigma/d\Omega$, (b) phase shift of an electron wave when passing an atom.

MeV, the energy loss and energy transfer to the nucleus can increase at large $\theta > 90°$ to a few tens of eV; this in turn may cause a displacement of atoms in solids, though with a very small cross-section σ_d (Fig. 1.4). Consequently, those elastically scattered electrons with $\theta \leq 50$ mrad that contribute to an image or a diffraction pattern in TEM can in practice be treated as zero-loss electrons, given that the energy spread of electrons from thermionic sources is 1–2 eV and from Schottky or field-emission guns, 0.2–0.5 eV.

Figure 1.1a illustrates what a differential cross-section $d\sigma/d\Omega$ means. Electrons passing through a small area $d\sigma$ are scattered through angles θ into a solid angle $d\Omega$. However, for the calculation of $d\sigma_{el}/d\Omega$, a wave-optical model has to be used, in which the phase shift of plane incident waves (Fig. 1.1b) by the Coulomb potential of the nucleus, screened by the atomic jellium, is considered. This potential may be written

$$V(r) = -\frac{e^2 Z}{4\pi\epsilon_0 r} + \frac{e^2}{4\pi\epsilon_0}\sum_{j=1}^{Z}\int \frac{\rho(r_j)}{|r - r_j|}d^3 r_j \qquad (1.1)$$

where r and r_j are the coordinates of the incident electron relative to the nucleus and of the Z atomic electrons, respectively, and $\rho(r_j)$ is the probability density of the atomic electrons. If it is assumed to be rotationally symmetric, (1.1) simplifies to

$$V(r) = -\frac{e^2 Z_{eff}}{4\pi\epsilon_0 r} \quad \text{with} \quad Z_{eff} = Z - \sum_{j=1}^{Z}\int_0^r \rho(r_j)4\pi r_j^2 dr_j . \qquad (1.2)$$

The final term in Z_{eff} is a measure of the electron charge inside a sphere of radius r. Its contribution to the Coulomb potential $V(r)$ can be evaluated by concentrating this charge at the centre of the sphere, whereas charges outside the sphere with $r \geq r_j$ do not contribute (elementary law of electrodynamics). The following approximations for $V(r)$ are useful in calculations of the scattering amplitude $f(\theta)$ and of the differential elastic cross-section:

$$d\sigma_{el}/d\Omega = |f(\theta)|^2 \tag{1.3}$$

a) Wentzel atom model with one exponential screening term

$$V(r) = -\frac{e^2 Z}{4\pi\epsilon_0 r}\exp(-r/R) \quad \text{with} \quad R = a_H Z^{-1/3} \tag{1.4}$$

where $a_H = 0.0529$ nm is the Bohr radius.
b) A sum of $k = 2$ or 3 exponentials using Hartree or Hartree–Slater calculations of $\rho(r_j)$ [1.2,3]

$$V(r) = -\frac{e^2 Z}{4\pi\epsilon_0 r}\sum_{i=1}^{k} b_i \exp(-a_i r) \quad ; \quad \sum_{i=1}^{k} b_i = 1 \quad . \tag{1.5}$$

c) Overlap of potentials of neighbouring atoms to consider in a first-order approximation the dense packing of atoms in a solid (muffin-tin model)

$$V_{eff} = V(r) + V(2a - r) - 2V(a) \quad \text{for} \quad r \leq a \tag{1.6}$$
$$V_{eff} = 0 \quad \text{for} \quad r \geq a$$

where $2a$ is the distance between neighbouring atoms or the diameter of a sphere with the same volume as the Wigner–Seitz cell of the lattice [1.4].
$V(r)$ can be related to an electron optical refractive index

$$n(r) = 1 - \frac{1}{2}\frac{V(r)}{E_0}\frac{E_0 + m_0 c^2}{E_0/2 + m_0 c^2} \tag{1.7}$$

with $E_0 = eU$ and the electron rest energy $m_0 c^2 = 511$ keV; $n(r) \geq 1$ because $V(r)$ in matter is negative. A plane incident wave of amplitude ψ_o passing the nucleus at a distance r is shifted in phase by

$$\phi(r) = \frac{2\pi}{\lambda}\int_{-\infty}^{+\infty} n(\rho)dz \quad \text{with} \quad \rho^2 = r^2 + z^2 , \tag{1.8}$$

and the wave behind the atom can be written

$$\psi(r) = \psi_o + \psi_o\{\exp[i\phi(r)] - 1\} \simeq \psi_o + i\psi_o\phi(r) + \cdots \tag{1.9}$$

which becomes a superposition of the plane incident wave ψ_o and the 90° phase-shifted scattered wave [$\exp(i\pi/2)$=i]. The scattered wave $F(q)$ at a large distance is obtained by a Fourier transform

$$
\begin{aligned}
F(q) &= \int \psi(r)\exp(-2\pi i q \cdot r)d^2 r \tag{1.10}\\
&= \int \psi(r)\exp(2\pi iqr \cos\chi)\, rdrd\chi\\
&= \psi_o\left(\delta(0) + 2\pi\int\{\exp[i\phi(r)] - 1\}J_0(2\pi qr)rdr\right)
\end{aligned}
$$

where q denotes the spatial frequency, $|q| = \theta/\lambda$, and J_0 is the Bessel function of zero order. The Fourier transform of the first term in (1.9) results in a δ-function at $q = 0$, and represents the incident, unscattered wave. The scattering amplitude $f(\theta)$ is obtained from (1.10) by comparing it with the relation

$$F(q) = \psi_0\{\delta(0) + \mathrm{i}\,|f(\theta)|\exp[\mathrm{i}\eta(\theta)]\} \tag{1.11}$$

where $f(\theta) = |f(\theta)|\exp[\mathrm{i}\,\eta(\theta)]$ is the complex scattering amplitude. This is the so-called WKB method (Wentzel–Kramer–Brillouin) in the small angle approximation of *Molière* [1.5], also associated with the name of *Glauber* [1.6]. Figure 1.2 shows examples of calculations of $f(\theta)$ and the phase shift $\eta(\theta)$ for carbon and platinum atoms [1.7] using Hartree–Byatt potentials (1.5) and the muffin-tin model (1.6). The finite values of $f(\theta)$ at $\theta = 0$ are a consequence of screening and are affected sensitively by the choice of screening model.

Substituting the Taylor series of (1.9) (assumption of a weak-phase object) in (1.10) results in the Born approximation

$$f(q) = -\frac{\pi}{\lambda E_0}\frac{E_0 + m_0 c^2}{E_0/2 + m_0 c^2}\int V(r)\exp(2\pi\mathrm{i}\,q\cdot r)\mathrm{d}^3 r \tag{1.12}$$

where $f(\theta)$ becomes a real quantity and the additional phase shift $\eta(\theta)$ is zero. The Born approximation cannot, therefore, be used for atoms of high atomic number because these are never "weak-phase objects". The only advantage of the Born approximation is that an analytic solution is obtained when the Wentzel potential (1.4) is used for $V(r)$ [1.8]:

$$\mathrm{d}\sigma_{\mathrm{el}} = |f(\theta)|^2 = \frac{4Z^2 R^4(1 + E_0/m_0 c^2)^2}{a_{\mathrm{H}}^2[1 + (\theta/\theta_0)^2]^2} \quad \text{with} \quad \theta_0 = \lambda/2\pi R \tag{1.13}$$

where θ_0 is the characteristic angle of elastic scattering.

An exact solution of the elastic scattering problem is given by the partial wave method, in which $f(\theta)$ is expressed as an infinite series of Legendre polynomials. Unfortunately, even the calculation of $f(\theta)$ at low scattering angles for about 100 keV needs more than hundred partial waves.

The total elastic cross-section is defined by

$$\sigma_{\mathrm{el}} = \int_0^{\pi}(\mathrm{d}\sigma_{\mathrm{el}}/\mathrm{d}\Omega)2\pi\sin\theta\mathrm{d}\theta \simeq \int_0^{\infty}(\mathrm{d}\sigma_{\mathrm{el}}/\mathrm{d}\Omega)2\pi\theta\mathrm{d}\theta \tag{1.14}$$

and substitution of (1.13) in (1.14) results in

$$\sigma_{\mathrm{el}} = \frac{Z^2 R^2\lambda^2(1 + E_0/m_0 c^2)^2}{\pi a_{\mathrm{H}}^2} = \frac{c^2 h^2 Z^{4/3}}{\pi E_0^2\beta^2} = 1.87\times10^{-20}Z^{4/3}/\beta^2 \tag{1.15}$$

in units of cm², where $\beta = v/c$. The dash-dotted line in Fig. 1.3 shows this dependence of σ_{el} on the atomic number Z. The dashed line represents calculations using Hartree–Slater functions or the Thomas–Fermi model for high Z [1.9]. The latter have been approximated by [1.10]

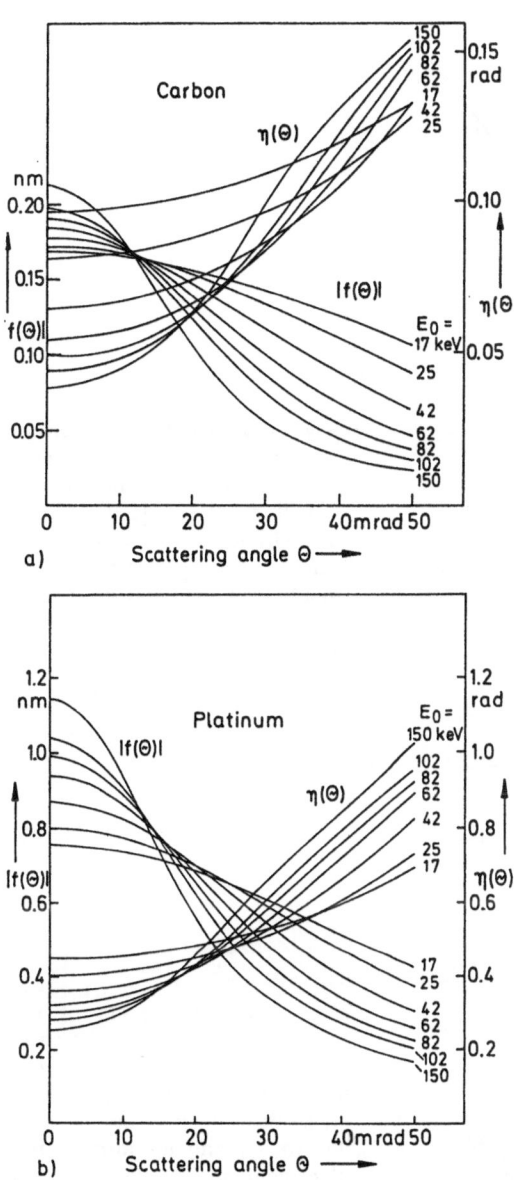

Fig. 1.2. Values of the scattering amplitude $f(\theta)$ and phase shift $\eta(\theta)$ of the complex scattering amplitude calculated by the WKB method using a muffin-tin model for (**a**) carbon and (**b**) platinum [1.7].

$$\sigma_{el} = \frac{1.5 \times 10^{-20} Z^{3/2}}{\beta^2} \left(1 - \frac{0.23 Z}{127 \beta}\right) \qquad (1.16)$$

which is represented by the full line in Fig. 1.3. Considering dense-packing by means of the muffin-tin model (1.6) and measurements of electron transmission result in lower values of σ_{el} [1.7]; the experimental values are shown in Fig. 1.3 as three circles for carbon, germanium and platinum. Figure 1.4 demonstrates the dependence of σ_{el} on electron energy [1.11]. The dashed lines are the cross-sections $\sigma(\pi/2)$ for backscattering through $\theta \geq 90°$ for carbon and platinum.

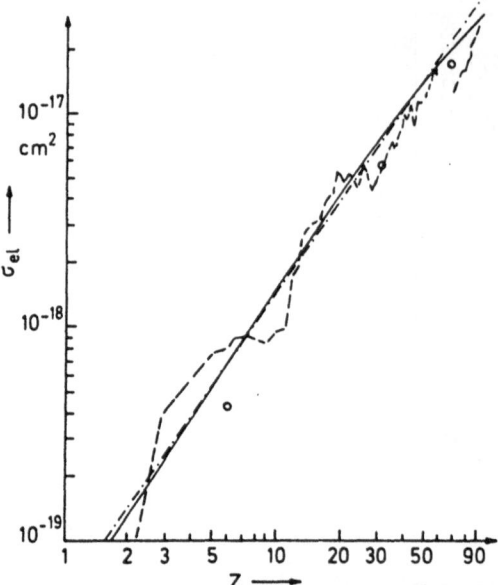

Fig. 1.3. Total elastic cross-sections σ_{el} for 100 keV electrons versus atomic number Z; full line: (1.10) [1.16]; dash-dotted line: (1.15); dashed line: calculated [1.9]; circles: experiments for C, Ge and Pt [1.7].

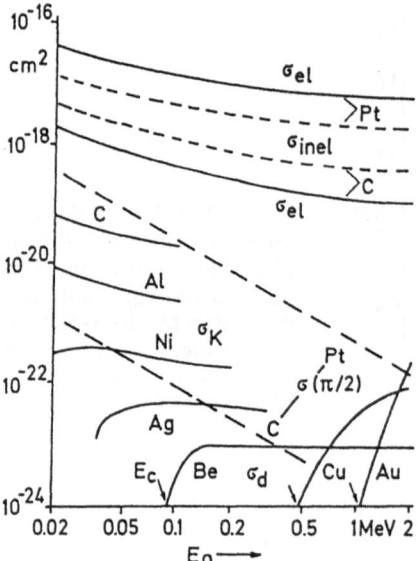

Fig. 1.4. Comparison of total cross-sections for elastic (σ_{el}) and inelastic (σ_{in}) scattering, for K-shell ionisation (σ_K), for backscattering into angles $\theta \geq 90°$ ($\sigma_{\pi/2}$) and for atomic displacement with a displacement energy of 20 eV (σ_d) as functions of electron energy.

Table 1.1. Experimental values of the mean-free-path x_{el} and the characteristic angle θ_0 of elastic scattering [1.7]

E_0	C		Ge		Pt	
	x_{el}	θ_0	x_{el}	θ_0	x_{el}	θ_0
[keV]	[μg/cm^2]	[mrad]	[μg/cm^2]	[mrad]	[μg/cm^2]	[mrad]
40	22	47	10	43	11	51
60	31	38	14	39	14	44
80	39	33	17	35	16	41
100	47	28	21	31	19	38
150	70	22	28	23	23	26
300	114	18	42	19	32	16

The mean-free-path x_{el} of elastic scattering in units of mass-thickness (g/cm^2) is given by

$$x_{el} = \rho \Lambda_{el} = 1/N\sigma_{el} \tag{1.17}$$

where $N = N_A\rho/A$ is the number of atoms per gram. N_A is Avogadro's number and A is the atomic weight. Experimental values of x_{el} and the characteristic angle θ_0 are listed in Table 1.1.

1.1.2 Inelastic Scattering and Electron Energy-Loss Spectrum

Inelastic scattering results from the excitation of inner-shell, valence or conduction electrons. The energy transferred can be observed as an energy loss E of the incident electron and reduces its kinetic energy to $E_0 - E$. Excitations are only possible from occupied states below the Fermi level to allowed states beyond it. An electron energy-loss spectrum (EELS) contains contributions and information from the following excitation processes (see also collections of EELS spectra [1.12,13]):

1) Excitation of oscillations in molecules and phonon excitations in solids below $E = 1$ eV. These energy losses are of the order of 20 meV – 1 eV. Because the energy spread of electrons is about 0.5–1 eV for thermionic and 0.2–0.5 eV for Schottky-emission or field-emission electron guns, the electron beam has to be monochromatized before reaching the specimen [1.14,15]. Figure 1.5 shows the EELS below 1 eV with a resolution of 4 meV for an evaporated Ge film with the excitation of phonons, oscillations of the Ge-O bonding and intraband transitions. This example demonstrates future possibilities of EFTEM, provided that an electron monochromator can be realized for high-energy electrons with sufficient electron beam intensity.

2) Excitation in the plasmon-loss region with $E = 0$–50 eV. The excitation of longitudinal oscillations of the plasma of valence or conduction electrons results in discrete volume and surface plasmon losses in the range $E = 0$–50 eV [1.16]. Sharp volume and surface plasmon losses in an aluminium foil are shown in Fig. 1.6a and broader plasmon losses in germanium in Fig. 1.6b. The energy loss is, in first order, proportional to the square root of the electron concentration. The dependence of plasmon-loss energy on the scattering angle

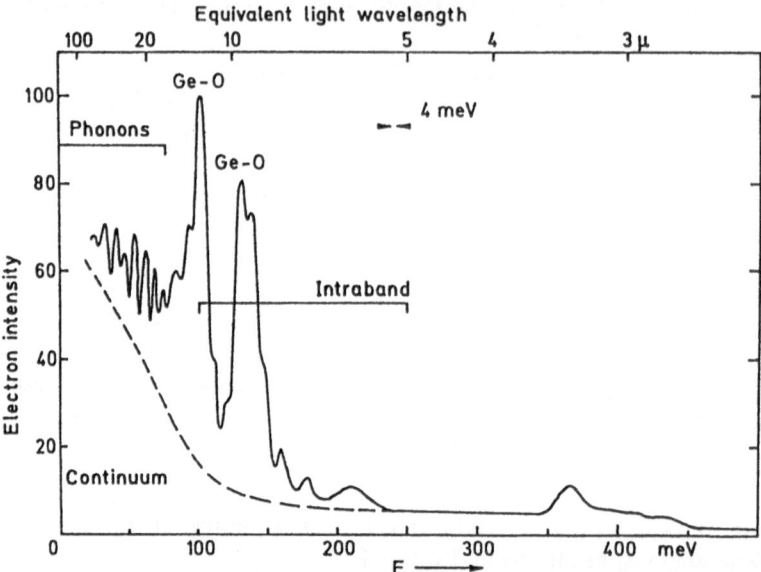

Fig. 1.5. Energy-loss spectrum of an evaporated germanium film due to phonon excitation, excitation of the GeO bonding and to interband transitions for energy losses $E \leq 500$ meV [1.14].

(dispersion) and the cut-off angle contain further information about the band structure of valence and conduction electrons (see Chap. 3 for further details). Not all metals show sharp plasmon losses, as demonstrated in Fig. 1.6c for gold. The low-loss region also contains energy losses from the excitation of intra- and interband transitions or excitation of Cerenkov radiation. This region can be overlapped by ionization edges of outer shells, such as the Ge M_{45} (Fig. 1.6b) or Au O_{23} ionization edges (Fig. 1.6c).

3) Excitation of quasi-free single electrons (Compton scattering). When the energy transferred is larger than the binding energy of excited electrons, the interaction can be treated as an electron–electron collision: The resulting most probable energy loss increases as the square root of the energy loss and, in an $E(\theta)$ diagram, forms the Bethe surface (Sect. 4.2, Figs. 1.23, 4.3, 6.31 and 6.32). Excitation with low scattering angles also contributes to the background of the EELS.

4) Inner-shell ionizations. When an electron in the K, L, M, N or O shell is excited to a free state or to the continuum of free electrons beyond the Fermi level [1.17,18], the intensity in the EELS increases at $E = E_{\mathrm{I}}$ and decreases in a long tail for $E > E_{\mathrm{I}}$, as shown for K-shell ionisation of oxygen in Fig. 1.7a. The edge can also show delayed maxima and the influence of sub-shells (Fig. 1.7b), or white lines (Fig. 1.7c), for example. The different shapes of ionization edges are discussed in detail in Sect. 4.4. Up to about 50 eV beyond the ionisation edge, an energy-loss near-edge structure (ELNES) can

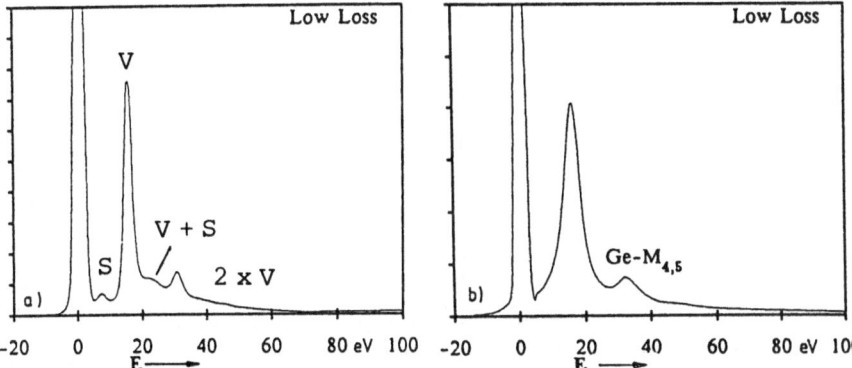

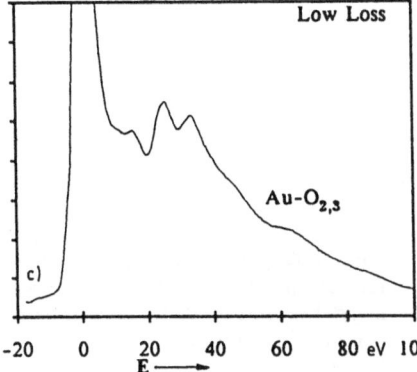

Fig. 1.6. (a) Surface (S) and volume (V) plasmon losses in an aluminium foil. (b) plasmon loss and Ge M$_{45}$ edge in amorphous germanium and (c) plasmon losses and Au O$_{45}$ edge in an evaporated gold film [1.13].

be observed, which depends sensitively on the binding state and coordination of atoms (Sect. 4.5.1). An extended energy-loss fine-structure (EXELFS) covering a few hundred electronvolts beyond the ionization edge results from interference between the electron wave backscattered from neighbouring atoms and the outgoing wave of the excited electrons. This yields information about the distance to neighbouring atoms (see Sect. 4.5.2 for further details).

Inelastically scattered electrons are more or less delocalized [1.19] and considerably reduce the high resolution contrast. Furthermore, the chromatic aberration of the objective lens reduces the resolution for thicker specimens. It is therefore of great interest to remove the inelastically scattered electrons using, the zero-loss mode of an energy-filtering microscope.

1.1.3 Inelastic Cross-Sections

Inelastic scattering results from the interaction of incident electrons with atomic electrons. The total kinetic energy is not conserved but reduced by the energy that is needed to excite an electron from the initial state with the wave function ψ_o to the final state with ψ_n. The cross-section of such an

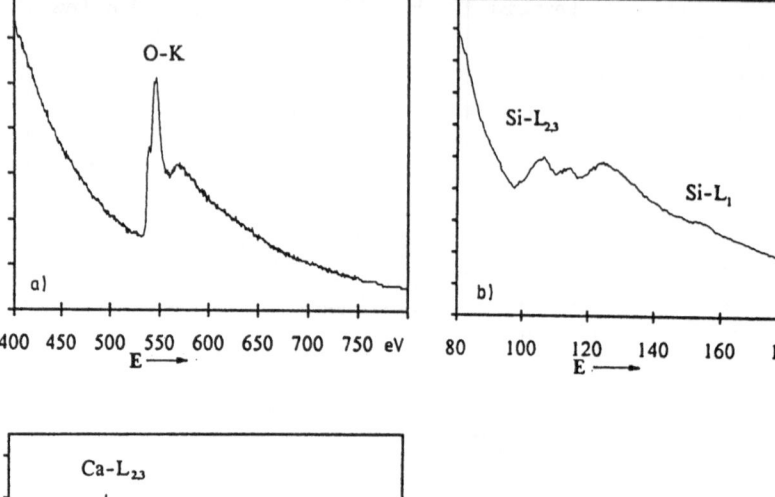

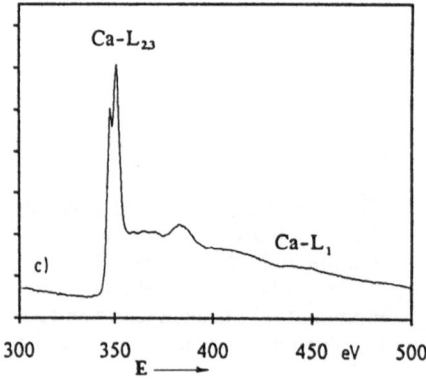

Fig. 1.7. (a) Saw-tooth shaped K ionisation edge of oxygen, **(b)** delayed L_{23} and weak L_1 edge of silicon and **(c)** white lines of Ca L edge [1.13].

inelastic excitation process with an energy loss E can be calculated using the golden role of quantum mechanics:

$$\frac{d\sigma_n}{d\Omega} = \frac{4\pi^2 m^2}{h^2} |\langle\psi_n|V(r)|\psi_o\rangle|^2 . \tag{1.18}$$

The wave functions ψ_o and ψ_n from the interaction of incident and scattered plane waves $\exp(-2\pi i\mathbf{k}\cdot\mathbf{r})$ with wave vectors $\mathbf{k}_o$ and $\mathbf{k}_n$ and wave functions u_o and u_n of the atomic electrons, respectively. Substitution of the potential $V(r)$ from (1.1) results in

$$\frac{d\sigma_n}{d\Omega} = \frac{e^4}{(4\pi\epsilon_0)^2 E_0 E} \frac{f_{on}(q')}{\theta^2 + \theta_E^2} \tag{1.19}$$

with the generalized oscillator strength (GOS)

$$f_{on}(q') = \frac{8\pi^2 mE}{h^2} |\langle u_n|\mathbf{u}\cdot\mathbf{r}|u_o\rangle|^2 . \tag{1.20}$$

The unit vector $\mathbf{u}$ is parallel to the scattering vector $\mathbf{q'}=\mathbf{k}_n - \mathbf{k}_o$ with $q'^2 = (\theta^2 + \theta_E^2)/\lambda^2$, where

$$\theta_E = \frac{E}{mv^2} = \frac{E}{2E_0}\frac{E_0 + m_0c^2}{E_0/2 + m_0c^2} \simeq \frac{E}{2E_0} \tag{1.21}$$

denotes a characteristic angle of inelastic scattering for an energy loss ; θ_E is much smaller than the characteristic angle θ_0 (1.13) of elastic scattering in Table 1.1.

In the case of ionization, the atomic electrons can be excited to a continuum of final states and a GOS per unit energy loss is introduced in the double-differential cross-section

$$\frac{d^2\sigma_{in}}{d\sigma dE} = \frac{e^4}{(4\pi\epsilon_0)^2 E_0 E}\frac{df_{on}(q', E)}{dE}. \tag{1.22}$$

This shows that an accurate knowledge of atomic eigenfunctions and the band structure in solids is necessary to calculate inelastic cross-sections (see Chaps. 3 and 4). The denominator $(\theta^2 + \theta_E^2)$ in (1.19) shows that a fraction of the inelastically scattered electrons is concentrated within scattering angles $\theta \leq \theta_E$ much lower than the characteristic angle θ_0 for elastic scattering. However, the cross-section shows also a long tail so that some of the inelastically scattered electrons are also distributed over angles $\theta \gg \theta_E$.

For the discussion of contrast effects, it is often sufficient to have an estimate of the fraction of inelastically scattered electrons as a function of atomic number without knowing the detailed EELS. The differential cross-section for inelastic scattering, analoguous to (1.13) for elastic scattering, can be written [1.8,20]

$$\frac{d\sigma_{in}}{d\Omega} = \frac{\lambda^4(1 + E_0/m_0c^2)^2}{4\pi^4 a_H^2}\frac{Z\{1 - [1 + (\theta/\theta_0)^2]^{-2}\}}{(\theta^2 + \theta_E^2)^2}. \tag{1.23}$$

Equations (1.13) and (1.23) for elastic and inelastic scattering can be used in the discussion of electron transmission through thin films in the unfiltered and zero-loss filtering modes of electron-spectroscopic imaging (Chap. 7).

Inelastic total cross-sections σ_{in} reported by *Inokuti* [1.21] and *Eusemann* et al. [1.22] are shown in Fig. 1.8 as dashed and dotted lines, respectively, and the approximate formula of *Wall* et al. [1.23]

$$\sigma_{in} = \frac{1.5 \times 10^{-20}}{\beta 2}Z^{1/2}\ln(2/\overline{\theta_E}) \tag{1.24}$$

as a full line, where $\overline{\theta_E} = J/mv^2$ and $J \simeq 13.5\ Z$ = mean ionization potential. These calculations for single atoms have to be interpreted with care because the wave functions and energy states of the outer electrons, which strongly contribute to σ_{in}, are quite different in a solid. Figure 1.8 also contains two calculations (triangles) for the excitation of an electron plasma [1.24]. Measurements of σ_{in} [1.25] are plotted as circles.

An important quantity for calculating the scattering contrast (Sect. 7.1.2) is the ratio

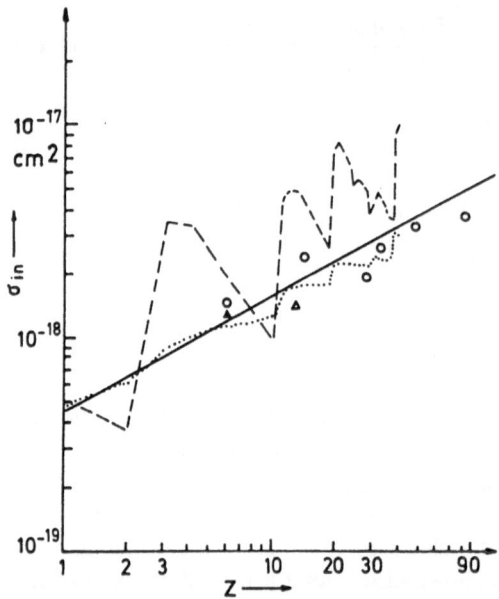

Fig. 1.8. Total inelastic cross-sections σ_{in} for 100 keV electrons versus atomic number Z; full line: calculated by (1.24) [1.23]; dashed line [1.21]; dotted line [1.22]: triangles: for plasmon losses [1.24]; circles: experimental values [1.25].

$$\nu = \sigma_{\rm in}/\sigma_{\rm el} \simeq 20/Z \qquad (1.25)$$

of the total inelastic to total elastic cross-sections [1.26,27]. Figure 1.9 shows experimental values of $\nu^{-1} \propto Z$. The ratio becomes too small when (1.16) and (1.24) are used. Figure 1.3 demonstrates that experiments (circles) result in lower $\sigma_{\rm el}$ and higher ν because of the dense packing of atoms in a solid. It is important that films used for the measurement of ν to be amorphous or polycrystalline with small crystallites. Otherwise, the dynamical theory of electron diffraction can result in higher values of ν (lower values of $1/\nu$, indicated by a value for crystalline Sb films in Fig. 1.9, for example). Equation (1.25) results in $\nu = 4.0$ for ice, 3.3 for carbon and 0.25 for platinum, for example. The cross-section $\sigma_{\rm in}$ is therefore much larger than $\sigma_{\rm el}$ for carbon and much lower for large Z elements (see also Fig. 1.4).

1.2 Conventional Transmission Electron Microscopy

1.2.1 Electron Guns and Specimen Illumination

Electrons are emitted from the Fermi level of a solid when their kinetic energy is increased so far that they can overcome the work function Φ, regarded as a potential barrier at the surface (Fig. 1.10). This can be achieved by heating (thermionic emission), applying in additional a strong electric field at a cathode tip of radius $r < 1\mu$m (Schottky emission) or by the quantum mechanical tunnelling effect (field emission, FE).

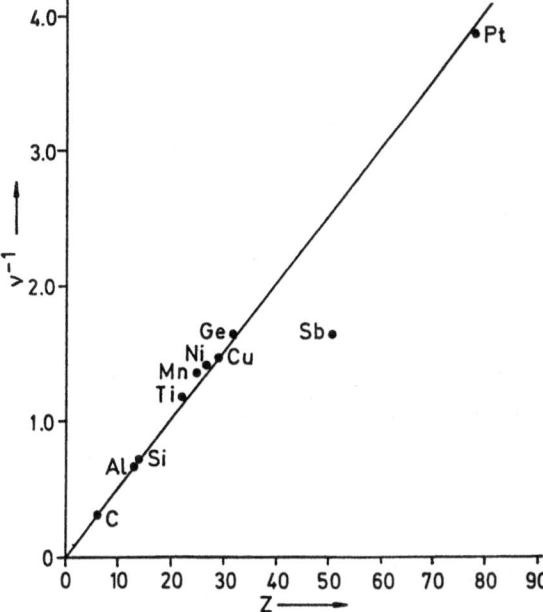

Fig. 1.9. Measurements of the ratio of total elastic-to-inelastic cross-section $1/\nu$ versus atomic number Z [1.27].

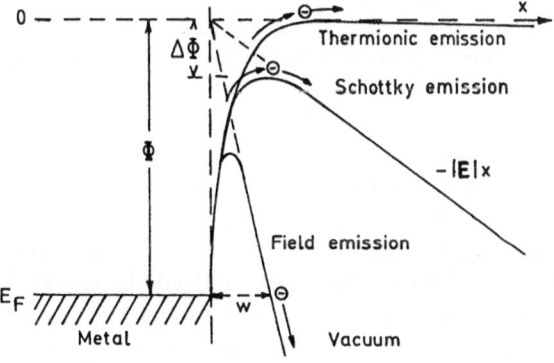

Fig. 1.10. Potential at cathode–vacuum interface for thermionic, Schottky and field-emission (FE) guns.

Thermionic emission from heated tungsten filaments ($\Phi = 4.5$ eV) or lanthanum hexaboride tips ($\Phi = 2.7$ eV) is the usual source of electrons for electron optical instruments. Such guns consist of a cathode, a Wehnelt electrode and an anode (Fig. 1.11a). The Wehnelt cup is negatively biased relative to the cathode and the electrostatic field of this gun forms a crossover of electron trajectories, which in turn acts as a virtual electron source of about 10–50 μm in diameter.

In Schottky emission from W/ZrO tips ($\Phi = 2.8$ eV) without a Wehnelt cup, the work function is further lowered by a strong electric field (Schottky

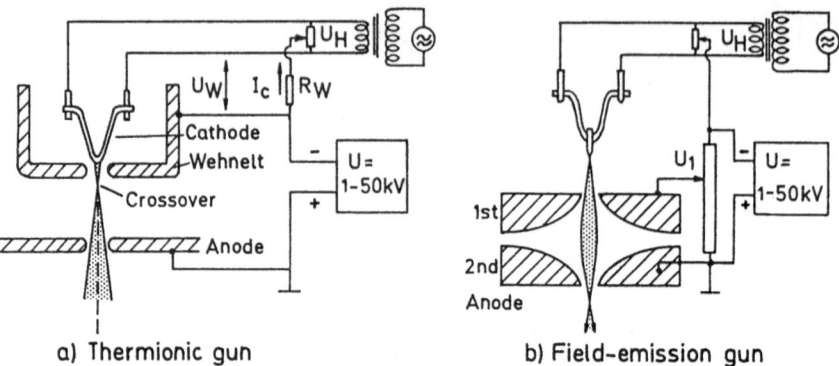

a) Thermionic gun   b) Field-emission gun

Fig. 1.11. (a) Thermionic electron gun consisting of cathode, Wehnelt cup, and anode, (b) one type of field-emission gun consisting of first extraction anode and second acceleration anode.

effect) but the electrons still have to overcome the potential barrier of the work function. The radius of the virtual electron source is about 0.5–1 μm.

In field emitters from tungsten tips of ≤ 0.1 μm in radius, the electrons can tunnel directly through the potential barrier, which is reduced in width to ≤ 10 nm. The diameter of the virtual source is further reduced to about 10-100 nm. The emission current is controlled by an extracting anode at a bias U_1 of about 1 kV and the electrons are accelerated by a second anode at a potential U (Fig. 1.11b).

The following parameters are characteristic of the different types of electron gun. The temperature T_c = 2500–3000 K at the tip of a thermionic tungsten cathode has to be near the melting point T_m = 3650 K of tungsten and the lifetime of the filament (10–100 h) is limited by tungsten evaporation. Lanthanum hexaboride (LaB$_6$) cathodes need a lower T_c = 1400–2000 K because of the lower value of Φ and have a lifetime of 150–300 h. The ZrO layer reduces the work function in Schottky emitters, and a lower value of T_c = 1800 K is sufficient. Field emission works with the tip at room temperature. However, tip temperatures of 1000–1800 K are often employed to avoid adsorption of gas molecules.

In thermionic cathodes, the Maxwellian distribution of the emitted electrons and stochastic Coulomb interactions of electrons in the crossover result in an energy spread of beam electrons of about δE = 1–2 eV (Boersch effect). This effect is strongly reduced in Schottky and field emitters (δE = 0.2–0.5 eV).

Thermionic tungsten cathodes can work in a vacuum of 10^{-3}–10^{-2} Pa, LaB$_6$ cathodes need 10^{-5}–10^{-4} Pa, and Schottky and field emitters only work under ultrahigh vacuum (10^{-8}–10^{-7} Pa).

An important quantity for characterizing a source is the gun brightness

$$\beta = \Delta I / \Delta A \Delta \Omega \tag{1.26}$$

where ΔI is the current passing an area ΔA with a current density $j = \Delta I/\Delta A$ and a solid angle $\Delta \Omega = \pi \alpha^2$ (α : aperture). Typical values of current density are $j = 2$, 20 and 500 A/cm^2 for W, LaB$_6$ and Schottky cathodes, respectively. The gun brightness of thermionic cathodes has a maximum value

$$\beta_{\max} = j_c E/\pi k T_c \qquad (1.27)$$

where j_c is the emission current density at the cathode. This means that the gun brightness is proportional to the electron energy E. Typical values for E = 100 keV are $\beta = (1\text{–}5)\times 10^5$ A/cm^{2}sr for thermionic tungsten; the value is 3–5 times larger for LaB$_6$ cathodes, whereas values of $2 \times 10^8 - 2 \times 10^9$ A/cm^{2}sr can be obtained by field emission, though the total electron current of the latter is lower.

The gun brightness remains constant on the axis of an electron optical system independent on any diaphragms and lenses. This means that current densities $j = \Delta I/\Delta A$ and apertures α ($\Delta \Omega = \pi \alpha^2$) cannot be changed independently but are related by the gun brightness (1.26). If we want to work with low illumination apertures $\alpha_i \leq 1$ mrad in high resolution TEM, the available current density will be limited. High current densities in an electron probe of small diameter are only possible with large apertures of about 10 mrad. Field-emission guns of high brightness will be preferred when high, coherent beam currents are required for high resolution or electron holography or when high currents must be concentrated in an electron probe of small diameter.

The condenser lens system between the electron gun and the specimen consists of one or two lenses (Fig. 1.12). The purpose of this system is

1) to focus the electron beam on the specimen, providing sufficient image intensity even at high magnifications,

2) to irradiate with uniform current density an area corresponding as closely as possible to the final screen to avoid unnecessary specimen irradiation, thereby reducing the specimen drift by heating and limiting radiation damage and contamination in non-irradiated areas,

3) to vary the illumination aperture α_i, which must be of the order of 1 mrad for medium magnification, less than 0.1 mrad for high resolution and phase contrast microscopy and below 0.01 mrad for small-angle electron diffraction and holography, and

4) to produce a small electron probe (1–100 nm in diameter) for x-ray microanalysis, electron microdiffraction and the scanning modes.

The illumination system of a modern TEM works with two or three condenser lenses and the prefield of an condenser-objective lens (Fig. 1.13). A first condenser lens demagnifies the crossover and produces a virtual electron source with a diameter of the order of one micrometre. A stigmator for correcting astigmatism is needed in this first lens.

In case of a conventional objective lens, the second condenser lens can either image this source onto the specimen to get the highest possible current

Source

Condenser diaphragm
Condenser lens

Specimen
Objective lens
Objective diaphragm
First diffraction pattern
First intermediate image
Selector diaphragm
Intermediate lens
Second diffraction pattern

Second intermediate image
Projector lens
Third diffraction pattern

Final image
Screen

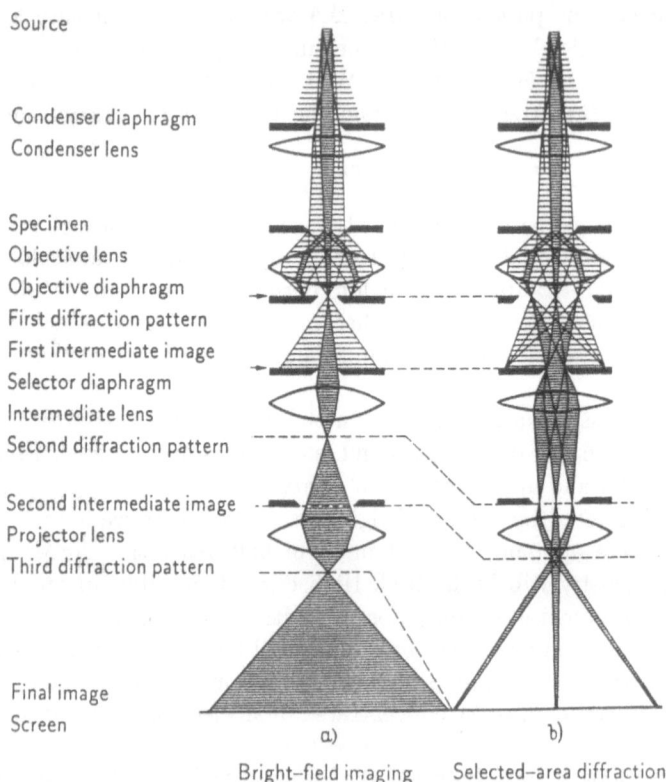

a)   b)

Bright–field imaging   Selected–area diffraction

Fig. 1.12. Ray diagram for a transmission electron microscope in (**a**) the bright-field mode and (**b**) selected-area electron diffraction (SAED) mode.

density and the smallest diameter of the irradiated area or be defocused in order to irradiate larger areas at low magnification. A diaphragm in the first condenser lens can vary the current density j_s, the diameter d_s of the irradiated area and the illumination aperture α_i.

In case of a condenser-objective lens, the second condenser lens at the top of Fig. 1.13 focuses the beam and produces an image of the electron source at the front focal plane (FFP) of the objective-prefield (condenser) lens (OPL). The diaphragm in the second condenser can vary the illuminated area at the specimen in the centre of the lens, and the diaphragm in the OPL can vary the illumination aperture (analogous to the Koehler illumination in light microscopy). In order to produce an electron probe of 2–5 nm diameter at the specimen, the second condenser is more strongly excited and produces a source image at the entrance image plane of the OPL, which is conjugate to the central specimen plane. The use of a zoom system of three condenser lenses [1.28] in front of the OPL and the correct position of scan coils allows fast switching from the illumination to the spot mode and tilting and shifting

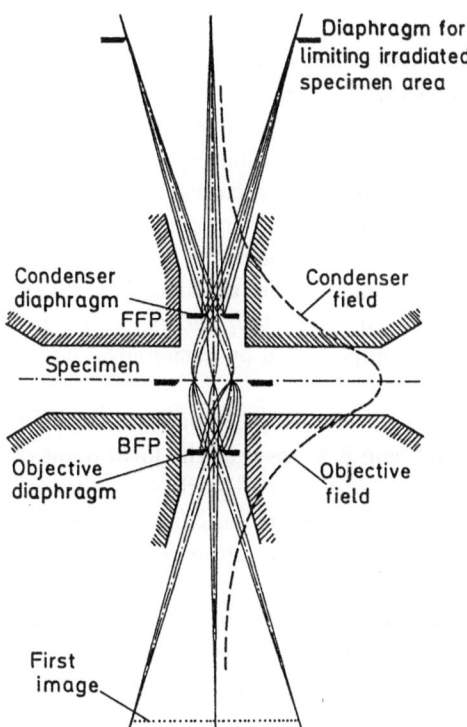

Diaphragm for
limiting irradiated
specimen area

Condenser
diaphragm

Condenser
field

FFP

Specimen

BFP

Objective
diaphragm

Objective
field

First
image

Fig. 1.13. Electron trajectories in a
single-field condenser-objective lens.

of the electron beam. By changing the angle of incidence (rocking) of the
electron beam in front of a condenser-objective lens, the electron probe will
scan across the specimen. By shifting the incident beam, the prefield lens
generates tilted illumination, which is important for the dark-field mode and
for hollow-cone illumination.

1.2.2 Imaging Modes

Electron lenses suffer from aberrations similar to those of glass lenses in
light optics. The most important aberrations are the spherical and chromatic
aberrations. The former causes off-axis rays to be focused more strongly, so
that a cone of rays does not converge to a sharp image point but to a circle
of least confusion, of diameter

$$d_{\mathrm{s}} = 0.5\,C_{\mathrm{s}}\alpha^3 \tag{1.28}$$

where C_{s} is the spherical aberration coefficient and α is the angle at the image
plane. In electron wave optics, spherical aberration is represented by a phase
shift (wave aberration, Sect. 7.4.2). The chromatic aberration is caused by
the energy spread δE of the beam electrons and results in a circle of least
confusion

$$d_{\mathrm{c}} = C_{\mathrm{c}}(\delta E/E_0)\alpha \tag{1.29}$$

where C_c is the chromatic aberration constant. The relatively large values of C_s and C_c of about 0.5–2 mm oblige us to work with apertures of only a few tens of milliradians and limit the resolution of a TEM to about 0.1–0.2 nm. The chromatic aberration strongly blurs the image of thick specimens owing to the broad spectrum of the inelastically scattered electrons.

Inhomogeneities in the polepiece material, ellipticities of the polepiece bore and charging of the specimen and aperture diaphragms result in astigmatism. Electron rays in sagittal and meridional bundles are focused at different focal distances, which results in crossed line foci at F_s and F_m a distance $\Delta f \simeq 10$–100 nm apart. This error can be compensated by a stigmator, which consists of a magnetic quadrupole lens.

The action of the objective lens is shown schematically in Fig. 1.12. The approximately parallel incident electron beam is focused at the focal point of this lens. Electrons scattered in the specimen through an angle θ are focused off-axis in the focal plane. Each point in the focal plane corresponds to a scattering angle θ. This plane thus contains the first diffraction pattern of the specimen and an objective diaphragm in this plane can select a cone of scattered electrons with $\theta \leq \alpha$ to increase the contrast. A first intermediate image of the specimen is formed in front of an intermediate lens which, together with the projector lens system, further magnifies the image onto the final viewing screen (Fig. 1.12a).

The normal imaging mode is the bright-field (BF) mode with a centred diaphragm in the focal plane of the objective lens (Fig. 1.12a) selecting the primary beam and a cone of scattered electrons with an objective aperture α. The interception of electrons with scattering angles $\theta \geq \alpha$ by the objective diaphragm causes scattering contrast (Sect. 7.3.2). In the case of crystalline specimens, the angles $\theta = 2\theta_B$, where θ_B is the Bragg angle, are normally larger than α and the diffraction spots are also intercepted by the diaphragm, resulting in Bragg contrast (Sects. 7.3.5 and 7.4.4). The dynamical theory of electron diffraction tells us that bright-field images will show the dependence of primary beam intensity on specimen thickness and tilt resulting in thickness fringes and bend contours.

The electrons passing through the diaphragm can produce phase contrast (Sect. 7.4.2) at high magnification when the scattered electron waves interfere with the primary wave. The transfer of spatial frequencies in the Fourier spectrum of specimen periodicities shows transfer gaps and cutoffs, which depend sensitively on defocusing and on the illumination aperture and energy spread of the electron beam (partial spatial and temporal coherence). The interference of the primary beam and a larger number of Bragg reflections results in the crystal-lattice imaging mode. Information about the crystal structure is obtained by comparing a defocus series with computer simulations. Because the inelastically scattered electrons result in a blurring of lattice images, zero-loss filtering becomes of increasing interest.

Dark-field (DF) images can be generated by tilting the illuminating beam so that the spot of the primary beam is intercepted by the objective diaphragm and only scattered electrons pass through the diaphragm. By selecting Bragg reflections, the contrast of dislocations and other crystal defects can be increased and the use of weak-beam reflections increases the resolution. The resulting asymmetry of the tilted-beam technique can be avoided either by introducing an annular diaphragm in the condenser lens or by sequentially changing the azimuth of a tilted beam by means of deflection coils.

1.2.3 Electron Diffraction Modes

Instead of the first image, the first diffraction pattern in the focal plane of the objective lens can be imaged onto the screen either by exciting an additional diffraction lens or by reducing the strength of the intermediate lens (Fig. 1.12b). The specimen area contributing to the diffraction pattern can be limited by inserting a selector diaphragm at the first image (selected area electron diffraction). Owing to the spherical aberration, the diameter of the selected area cannot be decreased below 0.2–1 μm. Smaller specimen areas have to be selected by reducing the size of the electron probe with 2–200 nm in diameter.

A diffraction pattern normally consists of the primary beam of small aperture ($\leq$1 mrad), the Bragg diffraction spots, and Kikuchi lines and bands caused by Bragg diffraction of thermal-diffusely and inelastically scattered electrons. Very small electron probes need a large probe aperture of about 20–100 mrad, which results in convergent beam electron diffraction. The circular primary beam and Bragg spots are modulated in intensity by the pendellösung of the dynamical theory and also contain dark lines from high-order Laue zone reflections (Chap. 6). A further increase of the electron probe aperture either externally by electron optics or internally by multiple scattering in a thick specimen results in a Kossel pattern with defect Kikuchi bands [1.29]. Inelastically scattered electrons contribute to the background, which increases with increasing thickness. Clearly, zero-loss filtering of electron diffraction patterns will increase the contrast and provide a better comparison with theoretical calculations. Records of diffraction pattern for different energy losses can be used to extract additional information and gain a better understanding of electron scattering in crystals.

1.2.4 Analytical Electron Microscopy

The most important modes of analytical TEM [1.30] are electron diffraction (Sects. 1.2.3 and Chap. 6), x-ray microanalysis (XRMA) and electron energy-loss spectroscopy (EELS, Sect. 1.5 and Chap. 5). For XRMA a TEM is equipped with an energy-dispersive x-ray spectrometer (EDS), which detects a cone of emitted x-rays that emerges through the polepiece gap. It is important that the solid angle of collection be large, about 10^{-2} sr, and the take-off angle between specimen plane and detector should be made as large as possible by using a cone-shaped objective lens polepiece. X-ray microanalysis in a TEM has the advantage over x-ray microprobe analysers that the lateral resolution can be improved by reducing the illumination area below 1 μm by means of a condenser prefield lens. The fraction of continuous x-ray quanta is lower than for bulk material because, in thin films, these are emitted preferentially in the forward direction. Care is necessary to prevent parasitic x-rays, excited on diaphragms and other parts of the microscope column that happen to be struck by electrons, from reaching the spectrometer.

An energy-dispersive spectrometer consists of a 3–5 mm thick intrinsic zone in a silicon crystal prepared by diffusion of Li atoms. X-ray quanta of energy $h\nu$ produce $N = h\nu/E_i$ electron-hole pairs where E_i is the mean energy for producing a pair. The pairs are separated by a voltage drop of about 1 kV at the crystal and result in a charge pulse proportional to the x-ray quantum energy. This pulse is amplified by a charge-sensitive preamplifier and then passes through a pulse-shaping unit so that the pulse can be recorded in a multichannel analyser (MCA); the number of counts in a channel that is proportional to the quantum energy is increased by one unit. This allows all quantum energies to be recorded simultaneously and the growth of a spectrum between quantum energies of 1–20 keV can be followed on the screen of the MCA. The resolution of x-ray energies by a EDS is of the order of 100–200 eV and about 10000 counts per second can be recorded. The vacuum inside an EDS is normally separated from the vacuum of the microscope by a 8–10 μm thick beryllium foil, which absorbes the characteristic radiation of elements below $Z = 11$ (Na), though windowless or ultrathin-window detectors can also record the lines of carbon, oxygen and nitrogen, for example.

X-ray microanalysis of thin specimens has the disadvantage that inner-shell ionisation may result in the emission not only of a characteristic x-ray quantum but also of Auger electrons. The fraction of x-ray emission, called the x-ray fluorescence yield, decreases with decreasing atomic number and falls below 1% for low Z elements. Furthermore, the characteristic x-rays are emitted isotropically in a solid angle $\Delta\Omega = 4\pi$ and only a fraction $\Delta\Omega = 10^{-3}$–10^{-2} is collected by the EDS. On the contrary, the inner-shell ionisation results in a characteristic edge in the energy-loss spectrum of transmitted electrons and a large fraction of inelastically scattered electrons, concentrated at very small scattering angles, can pass through the objective diaphragm. An electron spectrometer can therefore record a fraction of about 10% − 70% of the inner-shell ionisation processes. Furthermore, the ionisation edges of elements below Na can be analysed, which is not possible for an EDS with a beryllium window.

1.2.5 Scanning Transmission Electron Microscopy (STEM)

In a dedicated STEM [1.31] (Fig. 1.14) the virtual source of a field-emission gun is demagnified by an objective lens to an electron probe of about 0.1–0.2 nm at the specimen. An image is formed by scanning the probe in a raster across the specimen and recording various signals, which subsequently modulate the intensity on a TV tube scanned in synchronism. A prism spectrometer is used to create the EELS of the transmitted electrons; signals corresponding to unscattered, inelastically scattered and elastically scattered electrons into large angles can then be recorded simultaneously.

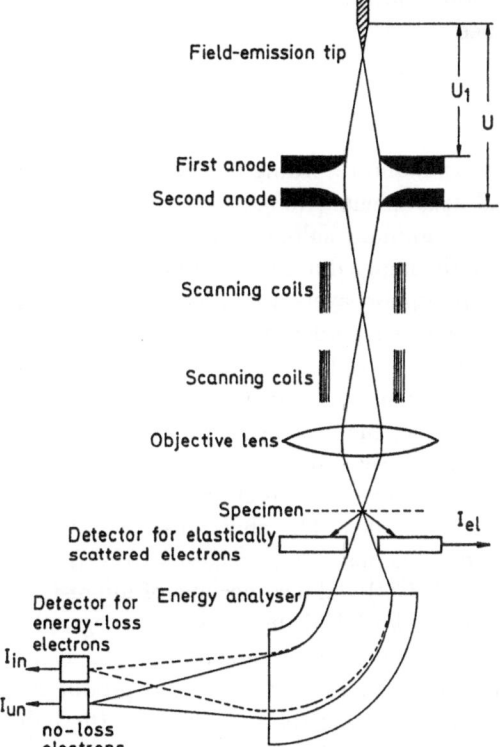

Fig. 1.14. Dedicated scanning transmission electron microscope (STEM).

The ray path of a STEM is the reciprocal of that of a TEM. In TEM, we work with small illumination apertures $\alpha_i \simeq 0.1$–1 mrad and large objective apertures $\alpha_o \simeq 5$–20 mrad; in a STEM with the large probe aperture $\alpha_p \simeq 10$–20 mrad necessary to obtain a small probe size, the same contrast as in TEM will be observed when a small detector aperture α_d is used. However, the signal-to-noise ratio will be better when $\alpha_d \simeq \alpha_p$.

The STEM has the advantage that the contrast for imaging single atoms can be optimized, by recording the ratio signal of elastically and inelastically scattered electrons for example. Thickness variations of the supporting film are recorded in both signals and are suppresssed in the ratio signal, whereas the inelastic signal of heavy single atoms is much broader (delocalized inter-action) than the elastic image resulting in a bright spot per atom. Such a ratio signal can also be used to generate Z-ratio contrast of biological sec-tions (Sect. 7.3.3). The signal coresponding to electrons scattered elastically through large angles can be recorded with an annular scintillation detector and crystal-lattice images can then be recorded with a signal that increases with increasing atomic number [1.32].

Modern TEMs are also equipped with a STEM imaging mode, which can be used for accurate positioning of a small electron probe for electron diffrac-tion, XRMA or EELS and also allows backscattered (BSE) and secondary (SE) electrons signals to be recorded.

1.2.6 Image Recording

The following methods for the detection and recording of electrons are com-mon in all types of electron-optical instruments [1.33].

Fluorescent screens of zinc and cadmium sulphide with resolutions of about 30–50 μm are used to observe the image during searching and focusing in direct-view instruments and the phosphor layers are used in the cathode-ray tubes used to display one of the signals recorded in a scanning mode. For high resolution TEM, the light intensity of a fluorescent screen may be in-adequate, especially when working with small illumination apertures, in the dark-field mode or in a low current-density mode to avoid radiation damage. In this case, an image intensifier can be used; this consists of a fluorescent screen observed by a low-light-level TV camera which even allows single 100 keV electrons to be recorded as a bright dot. The shot noise of the inci-dent electrons becomes the final limit for observing images. In the future, Peltier-cooled charge-coupled devices (CCD) will also become of interest.

The most efficient recording material for TEM images is a photographic emulsion, directly exposed to electrons inside the microscope vacuum. It has the advantage of a high sensitivity of about 10^{-11} C/cm^2 and produces a density $S = \log_{10} J_0/J = 1$ where J_0 and J are the intensities of the light transmitted through an unexposed and exposed area, respectively. Further advantages are the large dynamical range, which permits electron intensities of very different orders of magnitude to be recorded simultaneously and the high storage density. A resolution of about 20–50 μm results in 2×10^6–10^7 image points on a film area of 6×9 cm^2. The exposed emulsions can either be copied by the usual dark-room technique or scanned by a microdensitometer for digital image processing

Electron detectors for transmitted electrons in STEM modes or for sec-ondary and backscattered electrons in SEM modes need a low noise and a large bandwidth for recording fast scans up to TV rates. A combination of a scintillator, a light-pipe and a photomultiplier – also known as an Everhart–Thornley detector in SEM – is the most efficient detector system; it guarantees that each electron will generate a few light quanta in the scintillator material and a minimum of one photoelectron per incident electron, which is necessary for a low level of noise. Another alternative is to record the electron-beam-induced current in depletion layers of semiconductors, which separate the 1000-10000 electron-hole pairs generated per incident electron. A disadvantage of semiconductor detectors is the capacitance of the depletion layer, which limits the bandwidth especially for low incident currents.

For the simultaneous (parallel) recording of an EELS arrays of semiconductors diodes can be used. These are commercially available in sizes of a few micrometres in linear and square arrays of semiconductor diodes (CCD, charge coupled devices) and the signal can also be integrated over a longer exposure time. However, they cannot be exposed directly to the electron beam because of radiation damage. A transparent fluorescent screen therefore has to be used, coupled by a fibre-optic plate to the array.

1.3 The Development of EELS and EFTEM

1.3.1 EELS

The first energy-loss spectra of 40–900 eV electrons reflected at metal surfaces were published in 1930 by *Rudberg* [1.34]. He used a 180° magnetic spectrometer with a radius of 25 mm and achieved an energy resolution of one part in 200, adequate for low electron energies. These spectra already showed the maxima, later interpreted by *Bohm* and *Pines* [1.35] as plasmon losses, which are characteristic of the chemical composition of the specimen. Nowadays, the technique of EELS with low electron energies of 10–2000 eV can work with an energy resolution of a few meV and has become a powerful tool in surface physics [1.36,37].

The earliest recordings of energy spectra of transmitted 2–10 keV electrons, obtained by *Ruthemann* [1.38-40] in 1941–1948, showed the multiple plasmon losses of aluminium and the carbon, nitrogen and oxygen K edges of nitrocellulose films. The radius of the magnetic spectrometer was 175 mm and the energy resolution, one part in 2000. In 1944, *Hillier* and *Baker* [1.41] suggested that inner-shell ionisation edges could be used for elemental analysis. They built a 75 keV electron microscope with a 20–200 nm probe at the specimen and a 180° magnetic spectrometer.

An increase in energy resolution to about 1 eV for 20–60 keV electrons required the development of electron energy spectrometers with a resolution of about one part in 50000. In 1949, *Möllenstedt* and coworkers developed a cylindrical electrostatic lens for energy analysis [1.42,43], making use of the large chromatic aberration of electrostatic lenses. When the electrons are decelerated by the negative potential of the central electrode, the path of the trajectories depend very sensitively on electron energy and an energy spectrum with large dispersion can be obtained behind the lens. By placing a narrow slit parallel to the axis of the cylindrical electrodes, energy-loss spectra with both spatial and angular dispersion could be recorded (Sect. 1.5.4.). The action and performance of this Möllenstedt analyser is reviewed by *Metherell* [1.43]. It has been used in many laboratories and has also adapted for conventional transmission electron microscopes [1.44].

In the two decades between 1950 and 1970, the physics of plasmon losses were also investigated in many laboratories with home-made spectrometers of different types. Most of our knowledge about surface and volume plasmon losses and their dispersion and angular dependence was established during this period with instruments consisting only of an electron gun, condenser lenses and a spectrometer. The retarding potential of an electrostatic lens was most widely used [1.45–47]. The spectrum was recorded either with the retarded and reflected electrons or with the electrons that overcame the central potential barrier. This method worked as a high-pass filter and yielded integrated energy spectra. Superposition of a small ac potential on the central electrode voltage allowed an EELS to be recorded directly as an differentiated retarding curve [1.48]. One effective method of overcoming the potential drop at the lens centre was to use a long bore in the central retarding electrode and superimpose a longitudinal magnetic field by means of an external Helmholtz coil, thereby concentrating on the axis electrons that had overcome the central potential [1.1,49,50].

Other spectrometers decelerated the electrons to about 100 eV before they entered an electrostatic prism spectrometer [1.51,52]. The best resolution of 2 meV was achieved with a Wien filter of crossed magnetic and electric fields perpendicular to the axis, so that the action of Lorentz forces $e\boldsymbol{E}$ and $e\boldsymbol{v} \times \boldsymbol{B}$ was compensated for a particular electron velocity v [1.53,54]. By placing a second Wien filter in front of the specimen, the energy spread of the electron beam was reduced to 2 meV compared to 0.5–2 eV for electrons emitted by a thermionic cathode (applications see [1.14,15]). A monochromatizing Wien filter can also be incorporated in the column of a TEM [1.55].

All these spectrometers are limited to 60–80 keV for routine work due to electric breakdown at the retarding electrodes. A few scattered attempts have therefore been made to exploit the lower chromatic aberration of magnetic lenses in an analogue of the electrostatic Möllenstedt analyser [1.43,56–59]. However, the progress in double focusing in the lateral directions and in correcting the second-order aberrations of magnetic sector-field spectrometers by profiling the polepiece edges and using field clamps (mirror plates) (Fig. 1.15) has led to a regeneration of this classical type of spectrometer [1.31,61–63], which can be adapted to nearly any type of TEM at the end of the microscope column [1.64,65]. With the use of these 90° sector-field spectrometers, EELS has become an established technique of elemental analysis.

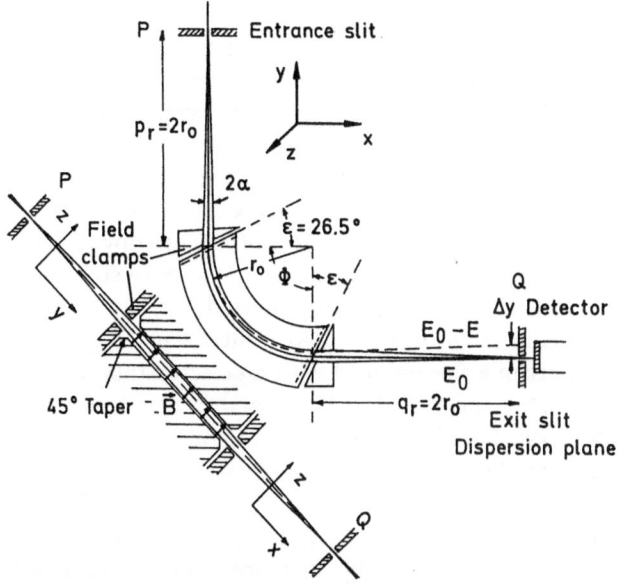

Fig. 1.15. Double focusing in a 90° electron prism with a tilt angle $\epsilon = 26.5°$ of the edges, tapered polepieces and field clamps (mirror plates). The deflection Δy of electrons with energy loss E is responsible for the dispersion $\Delta y/E$.

1.3.2 EFTEM

The first electron microscopists were already confronted with the problem of inelastic scattering, which influenced image quality, especially because the technique of preparing thin specimens has not yet been developed. They observed damage caused by specimen heating and blurring due to chromatic aberration [1.66]. A long road separated the first records of electron energy-loss spectra and attempts at energy-filtering of images and diffraction patterns from the first commercial energy-filtering transmission electron microscope, which became available few years ago.

Boersch [1.67,68] and *Möllenstedt* and *Rang* [1.69] first used zero-loss filtering of images and diffraction pattern (see example in Figs. 2.43 and 2.44) by means of a retarding grid or by increasing the potential at the central electrode of an electrostatic lens. But the high aberrations of such lenses limited their use. *Watanabe* and *Uyeda* [1.70,71] then used a Möllenstedt analyser as dispersive unit and selected an energy window by placing a slit in the dispersive plane.

Retarding electrostatic lenses were also used by *Wilska* [1.72] in low-voltage transmission electron microscopy with accelerating voltages of a few kilovolts to increase the contrast. Energy filtering of the zero-loss electrons was necessary to avoid the strong chromatic aberrations due to the more frequent energy losses.

In reflection electron microscopy of bulk specimens with grazing electron incidence, the reflected electrons show a broad energy distribution, which reduces the resolution to 20–50 nm owing to the chromatic aberration. Electrostatic filter lenses were therefore used to retard electrons with energy losses larger than 10 eV [1.73]. A resolution of 8 nm was reported when Bragg reflected spots, which contain a larger fraction of elastically scattered electrons were selected [1.74]. In the last decade, reflection electron microscopy (REM) has again become of interest for the direct investigation of surface steps and domains and the first results with an EFTEM are reported in Chap. 8.

In 1962 *Castaing* and *Henry* [1.75,76] presented a filter lens (Fig. 1.16) in which a double magnetic prism was combined with an electrostatic mirror (Chap. 2) which has been further perfected [1.77,78] and is now in routine use in the Zeiss EM 902 [1.79]. The limitation to accelerating voltages of the order of 80–100 keV due to the increased risk of electrical breakdown at higher voltages has been overcome by the use of purely magnetic filter lenses, first proposed in the thesis of *Sénoussi* (Paris-Orsay 1971), which can also be used in high-voltage electron microscopy [1.80]; they are discussed in detail in Chap. 2.

Prism spectrometers can also be used as imaging filters with an energy selecting slit in the energy-dispersive plane and the filtered image or diffraction pattern at a plane conjugate to the filter entrance plane. This plane can be magnified by an electron lens on a fluorescent screen coupled by a fibre-optic plate to a CCD array [1.81,82].

With these imaging filter lenses, a new era of analytical electron microscopy began. The different operating modes of EELS and EFTEM, which are described in the next section, are combined in one instrument and the operator can easily switch from electron spectroscopic imaging to diffraction and to the different modes of EELS.

1.4 Modes of Operation of EFTEM

1.4.1 Dispersive Coordinates and Selective Windows

An imaging filter lens in front of the last projector lens of a TEM offers the combination of electron spectroscopic imaging (ESI), electron spectroscopic diffraction (ESD) and electron energy-loss spectroscopy (EELS).

An electron microscope image represents a two-dimensional intensity distribution at the specimen; this distribution is a function of the space coordinates x and y and a projection of the third (z) coordinate parallel to the axis. A three-dimensional reconstruction of the specimen can be achieved by tilting the specimen or the incident electron beam. A stereometric reconstruction needs two images with tilts $\pm\gamma$ of about $1° - 5°$ when the specimen

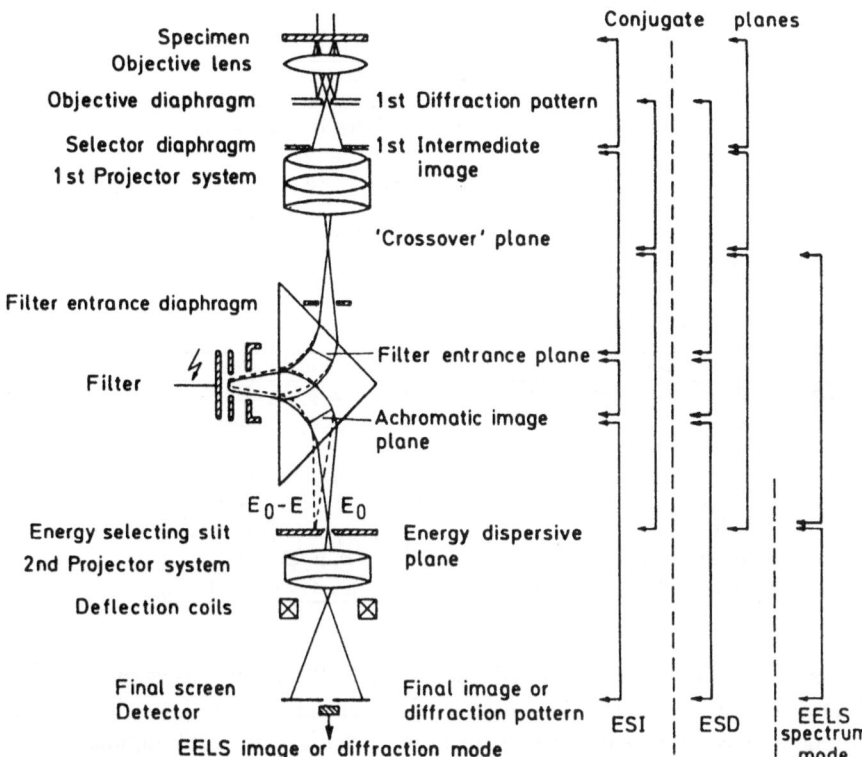

Fig. 1.16. Ray path diagram and important planes in a TEM with a Castaing–Henry imaging filter lens in the electron spectroscopic imaging (ESI) mode. Conjugate planes are also indicated at the right-hand side for the electron spectroscopic diffraction (ESD) mode and the "spectrum mode" of electron energy-loss spectroscopy (EELS).

contains sharp details that create a parallax effect. A tomographic reconstruction needs a series of 10–20 tilts in the range $\pm 60°$.

The electron diffraction pattern represents an intensity distribution with angular coordinates (spatial frequencies) q_x, q_y, where

$$| \boldsymbol{q} | = \theta/\lambda = | \boldsymbol{k} - \boldsymbol{k_o} | \tag{1.30}$$

in which θ denotes scattering angle, λ the electron wavelength and the $\boldsymbol{k_o}$ and $\boldsymbol{k}$ the wave vectors of the incident and scattered waves, respectively; $\boldsymbol{p} = h(\boldsymbol{k} - \boldsymbol{k_o})$ is the momentum transfer normal to the axis. The electron diffraction pattern is a projection of the reciprocal lattice. Information along the q_z coordinate can again be obtained by tilting the specimen or the incident beam, or by studying reflections from high-order Laue zones because of the curvature of the Ewald sphere.

The intensity distribution of an image or diffraction pattern can be parallel-recorded on a fluorescent screen, a photographic emulsion, or by

Table 1.2. Dispersive coordinates and selecting windows

	Dispersive coordinates		Selecting windows
x, y, z	Space coordinates	ΔA	Specimen area
		d	Diameter of selected area
		Δy	Width of slit parallel to z
q_x, q_y, q_z	Angular coordinates	$\Delta \Omega$	Solid angle (aperture)
	$q = \theta/\lambda$	α	Aperture of selected cone
		Δq_y	Width of angular slit
E	Energy loss	Δ	Width of energy window
t	Time		

Table 1.3. Operating modes of EFTEM

Mode	Coordinates
ESI (electron spectroscopic imaging)	$x, y/E, \Delta\Omega$
ESD (Electron spectroscopic diffraction)	$q_x, q_y/E, \Delta A$
EELS (Electron energy-loss spectroscopy)	
Spectrum mode	$E, \Delta a, \Delta\Omega$
Image mode	$x, y/E(t), \Delta A, \Delta\Omega$
Diffraction mode	$q_x, q_y/E(t), \Delta A, \Delta\Omega$
Spatially dispersive mode	$x, E/\Delta y, \Delta\Omega$
Angularly dispersive mode	$q_x, E/\Delta q_y, \Delta A$

means of a CCD (charge-coupled device) or SIT (silicon intensifed target) camera. In the scanning transmission (STEM) mode these intensity distributions are recorded sequentially pixel per pixel but different signals corresponding to elastically and inelastically scattered electrons and the EELS can be parallel-recorded.

The energy losses E in an electron energy-loss spectrum offers us a new dimension of electron microscopy and this is now available with imaging filter lenses or spectrometers in commercial instruments. The time t is, in case of radiation damage, environmental and dynamic experiments, for example, a further "dispersive" coordinate listed in Table 1.2.

Because only one or two coordinates of this 8-dimensional space can be parallel-recorded, fixed "selectinge windows" of the other coordinates have to be used (Table 1.2) and the various choices result in the different operating modes of EFTEM (Table 1.3) [1.84,85].

The selected specimen area is not necessarily circular with a diameter d, as is usual for selected-area electron diffraction (SAED) or for selected-area EELS, but can have other shapes better adapted to the problem in question. One example is a slit parallel to x and of a width Δy as used for spatially resolved EELS. The solid angle $\Delta\Omega$ is often conical and can be characterized by the aperture α measured in milliradians (mrad) and $\Delta\Omega = \pi\alpha^2$. This is realized by placing a circular diaphragm at the first diffraction pattern in the focal plane of the objective lens for the conventional bright- and dark-field modes. However, the angular selection can also be realized by placing a slit in the q_x direction with a width Δq_y, as used for the mode of angularly resolved

EELS, for example. The selected energy-loss window is characterized by the position E and the width Δ, so that the energy losses of the interval $E \pm \Delta/2$ are selected.

1.4.2 Electron Spectroscopic Imaging

For the discussion of the different EFTEM operating modes [1.84,55], we use the schematic ray-diagram of Fig. 1.17 for the region between specimen and final image plane (FIP); here, we consider the example of a Castaing–Henry imaging filter lens but the figure can also be used for imaging Ω filters and even for post-column imaging prism spectrometers if we treat the imaging filter as a black box (Figs. 1.17 and 2.1). The objective lens produces a first diffraction pattern in its focal plane and a first intermediate image in front of the next magnifying lens. (An additional intermediate lens between the 1st intermediate lens and the 1st projector lens is omitted so as not to overload Fig. 1.17.) A centred diaphragm in the focal plane selects a solid angle $\Delta\Omega = \pi\alpha^2$ with an objective aperture of about 4–40 mrad. In the bright-field mode (BF) the primary on-axis beam can pass through the aperture, as can electrons scattered elastically and inelastically through scattering angles $\theta \leq \alpha$. A dark-field mode (DF) can be realized by shifting the diaphragm off-axis or better, by tilting the primary beam, so that only the cone of scattered electrons passes through the objective diaphragm (OD) on axis. When Bragg diffracted beams can pass through a large diaphragm, a crystal-lattice image can be recorded.

In the imaging filter lens, a filter entrance (intermediate image) plane (FEP) is conjugate to the achromatic image plane (AIP) with a 1:1 "magnification", as discussed in Sect.2.1. A diaphragm in or near the FEP selects the region that contributes to the ESI and the EELS. In the achromatic image plane, electrons with different energy losses pass through the same image point but at different angles to the axis (see also Fig. 2.1). The EELS is produced in the energy dispersive plane (EDP), which is conjugate to the "source plane" (SP) at the focal plane of the first projector lens system. This means that the EELS is convolved with the intensity distribution in the source plane. In the case of electron spectroscopic imaging (ESI), this is a demagnified diffraction pattern limited by the shadow of the objective diaphragm. The decrease in its diameter is inversely proportional to the magnification of the intermediate image at the filter entrance plane. The diameter of the primary spot of unscattered electrons is proportional to the illumination aperture. The right-hand side of Fig. 1.16 indicates the conjugate planes in the ESI mode.

The second projector lens system magnifies the AIP on the final image plane (FIP), resulting in an unfiltered (global) image if there is no slit in the EDP plane or an electron spectroscopic image (ESI) if an energy window of width Δ has been selected by a slit. As an example of ESI, Fig. 1.17a shows the bright-field (BF) image of a 420 nm evaporated aluminium film unfiltered

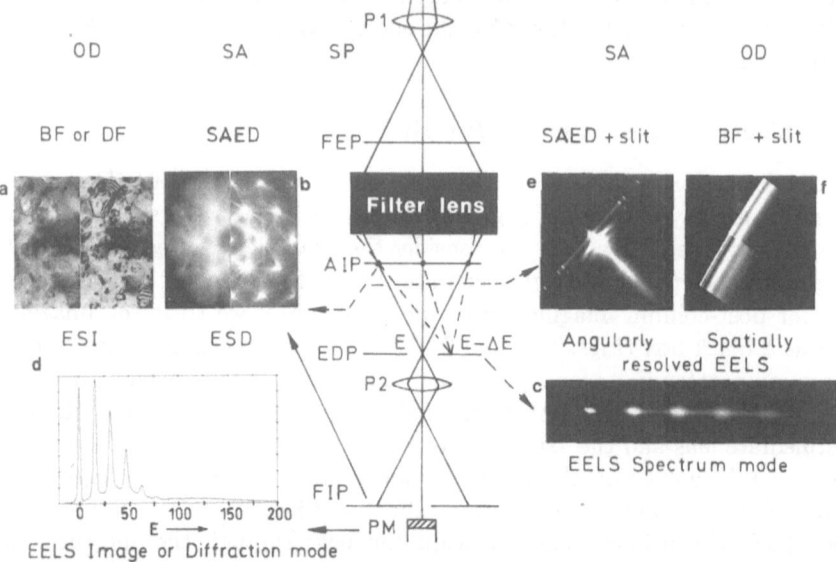

Fig. 1.17. Schematical ray path of an imaging filter lens in an EFTEM. P1, P2: Projector lenses; SP: source plane; FEP: filter entrance plane; AIP: achromatic image plane; EDP: energy dispersive plane; FIP: final image plane; PM: scintillator-photomultiplier. Examples of (**a**) electron spectroscopic imaging (ESI) of a 420 nm evaporated Al film unfiltered (left) and zero-loss filtered (right), (**b**) electron spectroscopic diffraction (ESD) of a [111]-oriented Si foil unfiltered (left) and zero-loss filtered (right), (**c**) EELS of a 40 nm Al foil recorded in the spectrum mode, (**d**) EELS image mode (150 nm Al with plasmon losses and convolved L edge, (**e**) angularly resolved EELS (40 nm Al) and (**f**) spatially-resolved EELS of graphite crystal (bottom) on carbon film (top).

(left) and zero-loss filtered (right). The ESI at some energy loss E is realized by increasing the accelerating voltage U at the cathode by $\Delta U = E/e$, which results in an increase of the electron energy $E_0 = eU$ by an ammount E in front of the specimen and hence in a shift of the EELS in the energy-dispersive plane. Electrons with an energy loss E in the specimen than have an energy E_0 behind the specimen and can pass through the energy selecting slit on-axis. This guarantees that the filtered image will not shift as the selected energy loss is increased. Any influence of the specimen illumination caused by the increase of electron energy can be compensated by variation of the condenser lens current.

1.4.3 Electron Spectroscopic Diffraction

A diffraction pattern can be recorded by magnifying the first diffraction pattern at the objective focal plane by focusing the intermediate lens on this plane and not on the first intermediate image. A selector diaphragm (SA) in the first intermediate image plane can select a specimen area ΔA of 0.1–1 μm in diameter, which is necessary for selected-area electron diffraction (SAED). The area selected cannot be reduced below 0.1–1 μm because of the spherical aberration. Electron diffraction from smaller areas can be achieved by using an electron probe of 0.2–10 nm in diameter. A large electron-probe aperture results in convergent-beam electron diffraction (CBED, [1.29] and Chap. 6).

The FEP and the AIP now contain a diffraction pattern, which can be magnified by the second projector lens system and energy-selected by a slit in the EDP. The source plane (SP) contains a demagnified image of the selector diaphragm, the decrease in diameter of which is inversely proportional to the magnification (camera length) of the diffraction pattern at the FEP. The EELS in the EDP is now convolved by this image of the selector diaphragm. The right-hand side of Fig. 1.16 indicates the conjugate planes in the ESD mode. As an example of ESD, Fig.1.17b shows the unfiltered (right) and zero-loss filtered diffraction pattern of a [111]-oriented Si foil.

1.5 EELS Modes in EFTEM

1.5.1 EELS Spectrum Mode

As discussed before, the EDP contains the EELS which can be magnified by the second projector lens system on the FIP. The right-hand side of Fig.1.16 indicates the conjugate planes in the EELS spectrum mode. The EELS is convolved with the demagnified diffraction pattern in the source plane (SP) in the case of the ESI mode (intermediate image at the FEP). This can be seen in Fig. 1.18a, which shows the plasmon losses of an aluminium film convolved with the first Debye–Scherrer diffraction rings. The diameter of the rings in the EELS decreases with increasing magnification of the intermediate image in the FEP. By using a 20 μm diaphragm at the objective focal plane, the rings at large scattering angles are intercepted and a better resolved EELSpectrum can be recorded (Fig. 1.18b). As the diameter of the aperture image in the source plane decreases inversely as the magnification M, smaller objective apertures are necessary at low magnifications; these should not exceed an acceptance angle [1.86]

$$\alpha_{\text{accept}} = kM\Delta/E_0. \tag{1.31}$$

The corresponding diameter of the objective diaphragm is

$$d \simeq 2\alpha_{\text{accept}}/f \tag{1.32}$$

where f denotes the focal length of the objective lens. The meaning of Δ depends on the actual acquisition condition. It can be the primary energy spread of the electron gun, the energy resolution of the spectrometer and detector or the selected window-width, whichever is larger. The ultimate

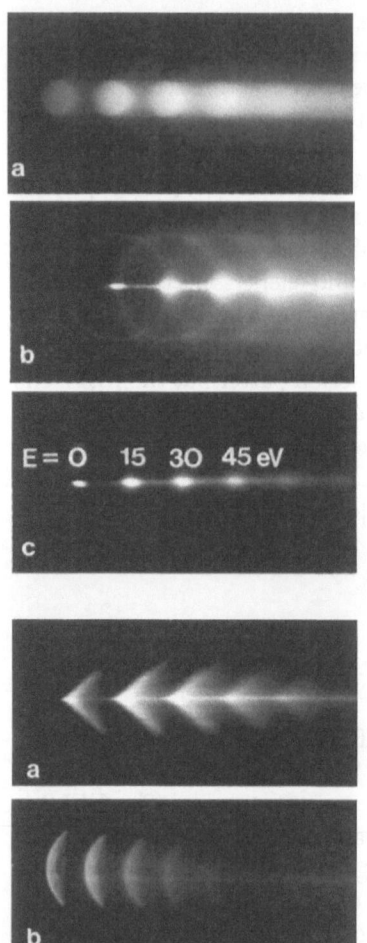

E = O   15   30   45 eV

Fig. 1.18. Magnified images of the EELS (spectrum mode) of a 40 nm aluminium film recorded in a Zeiss EM902. (a) Image ($M = 3000$) at the FEP with 50 μm selector diaphragm and no objective diaphragm. Each plasmon loss at multiples of 15 eV is surrounded by Debye–Scherrer rings. (b) as (a) with a 20 μm objective diaphragm. (c) Diffraction pattern (camera length $L = 0.8$ m) at the FEP with 50 μm selector and 20 μm objective diaphragm.

Fig. 1.19. Magnified images of the EELS. (a) Image ($M = 3000$) at the FEP with 100 μm selector and 20 μm objective diaphragm with the caustic due to second-order aberration recorded in a Zeiss EM902, (b) shows the caustic in an EM912 OMEGA.

limit of energy resolution for EELS recording is determined by the energy spread of about $\delta E = 0.5$–2 eV, which is increased by the Boersch effect (stochastic Coulomb interactions near the crossover) and can be minimized by not overheating the cathode. For element-distribution images, the window width can be $\Delta = 10$–20 eV. Typical values of k in (1.31) are $k = 13.5$ mrad for the Zeiss EM902 and $k = 23$ mrad for the EM912 OMEGA . Figure 1.20 demonstrates the relation between Δ or δE and the magnification M for different apertures. Magnifications larger 10000× will be needed for element-distribution images with $\Delta = 10$ eV and an aperture $\alpha = 20$ mrad, for example.

In the case of the ESD mode (diffraction pattern at the FEP), the EELS is convolved with the image of the selector diaphragm (Fig. 1.18c), which also decreases in diameter with increasing magnification (camera length) of the diffraction pattern. The EELS in Figs. 1.18a–c were recorded with a selector diaphragm of 50 μm, so that only a central part of the final image contributes to the EELS where the second-order aberration is small. When the diameter of this diaphragm is increased to 100 μm, the EELS is convolved with the caustic of the second-order aberration (Fig. 1.19a). Owing to this aberration, filtering of an image or diffraction pattern in the ESI or ESD modes with an energy loss E is correct only in the central area of the final image and the selected window drops parabolically to E–15 eV at the periphery of the final image. This second-order aberration can be totally compensated with the Ω filter lens of Krahl (Sect. 2.2.2), which means that the EELS has the same appearance as Fig. 1.18c even with a larger diaphragm in the FEP and the final image is filtered with the same energy loss E across its whole surface. Figure 1.19b shows the caustic of the Zeiss EM912 OMEGA with the partly corrected magnetic Ω filter lens. The central flat part of the caustic indicates that a well-resolved EELS can be recorded with a larger diameter in the FIP but the second-order aberration increases rapidly for excessively large diameters.

The EELS at the FIP can be observed visually on the fluorescent screen (Fig. 1.17c) with a dispersion of 0.5 mm/eV in the Zeiss EM 902 to adjust the position and the width of the energy selecting slit or can be parallel-recorded on a photographic emulsion or by means of a CCD or SIT camera

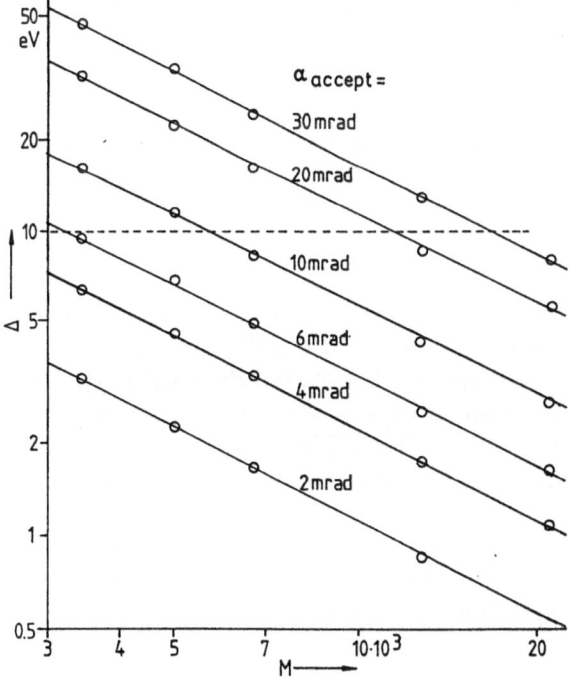

Fig. 1.20. Relation between the energy-selecting window width Δ, magnification M and acceptance angle (objective aperture) α_{accept} for 80 keV electrons in a ZEISS EM902.

after withdrawing the energy selecting slit. For sequential recording of an EELS as in Fig. 1.17c, a slit in the EDP or in front of a photomultiplier below the FIP limits the energy resolution and the spectrum is shifted across the slit by varying the accelerating voltage as described above.

1.5.2 EELS Image Mode

Besides the EELS spectrum mode discussed in Sect. 1.5.1 with either an image or diffraction pattern at the FEP, an EELS can be recorded sequentially in the EELS image mode with a magnified image (ESI mode) at the FIP. A diaphragm of a few millimetres in this final image selects an area ΔA, and the intensity passing through the diaphragm when the EELS is swept across the energy-selecting slit is recorded by an electron detector, which can either be a scintillator-photomultiplier combination for an analogue signal, or a semiconductor detector, which is preferable for single-electron counting. The latter avoids uncontrolled shifts of the zero level of operational amplifiers. This mode allows us to arrange that only very small specimen areas $\Delta A/M^2$ contribute to the EELS by increasing the magnification M of the final image, whereas in the spectrum mode only a much larger area can be selected by a diaphragm placed in the intermediate or filter entrance plane. For a diaphragm diameter of 3 mm, the diameter at the specimen plane is 30 nm with a 100000× magnification, for example. Because of the small selected area at the image centre, the triangularly shaped second-order aberration in the energy-loss spectrum (Fig. 1.19a) discussed above can be neglected. The loss of energy resolution caused by the convolution with the image of the objective diaphragm in the source plane (SP) (Fig. 1.18a) cannot be avoided but this blurring decreases with increasing magnification. Fig. 1.17d shows as an example the EELS of a 150 nm Al film with multiple plasmon losses, recorded in the EELS image mode.

1.5.3 Image-EELS Mode

When a series of ESI images is recorded sequentially with increasing energy loss E by an image recording system (CCD or SIT camera), a selected area on the image of arbitrary shape can be selected digitally. Integration of the intensity inside this structure of interest will produce its energy-loss spectrum with increasing energy loss E [1.87]. The contribution of any supporting film or embedding material not containing the structure can kept to a minimum, which is not the case when a circular diaphragm is employed, and the signal-to-noise ratio can be optimized because only the structure of interest contributes to the EELS. In practice, the energy resolution is reduced to 3–5 eV to get enough image intensity in the series of ESI.

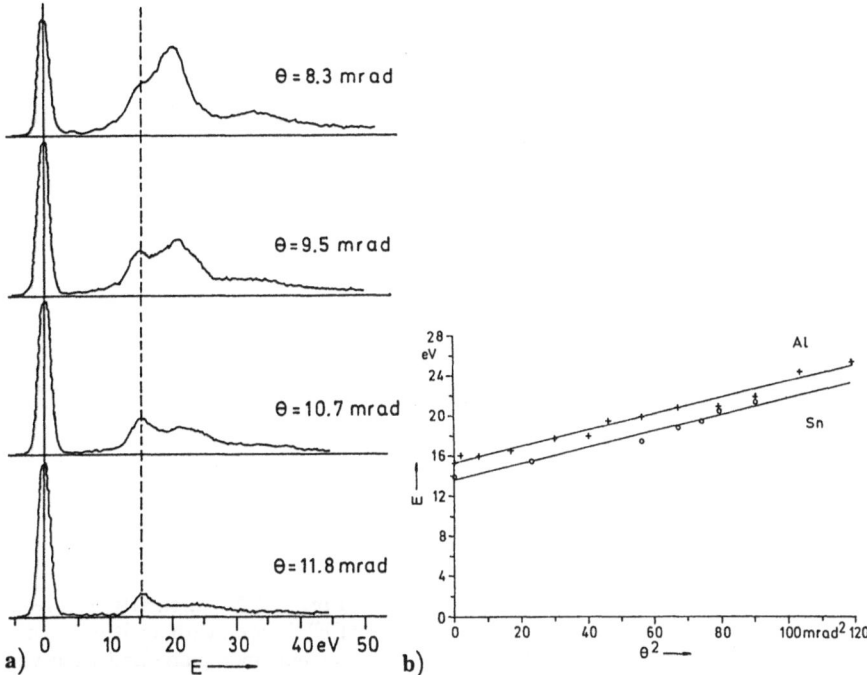

Fig. 1.21. (a) Example of EELS of an aluminium film recorded at increasing scattering angles θ and (b) plot of the shift of plasmon loss (dispersion) versus θ^2.

1.5.4 EELS Diffraction Mode

In the EELS diffraction mode, the FEP, AIP and FIP contain a selected-area electron diffraction (SAED) or a convergent-beam diffraction (CBED) pattern. When working with a large camera length (strong magnification of the SAED), a diaphragm in the final image can select much smaller solid angles than can a diaphragm in the focal plane of the objective lens, where the diameter of the first diffraction pattern is less than 1 mm. The energy-loss resolution of this mode is limited by the diameter of the image of the selector diaphragm or the irradiated area at the energy-dispersive plane, as discussed above.

In order to select a particular position inside the diffraction pattern, the latter can be shifted either by deflection coils below the final lens or by tilting (rocking) the primary beam by means of a deflection coil system in front of the specimen. In this way, the dispersion

$$E_{pl}(\theta) = E_{pl}(0) + a\theta^2 \tag{1.33}$$

of plasmon losses can be recorded, as demonstrated in Fig. 1.21a, for example (see also Sect. 3.1.3). The EELS at large scattering angles also contain the plasmon loss at scattering angle $\theta = 0$ because of elastic-inelastic scattering.

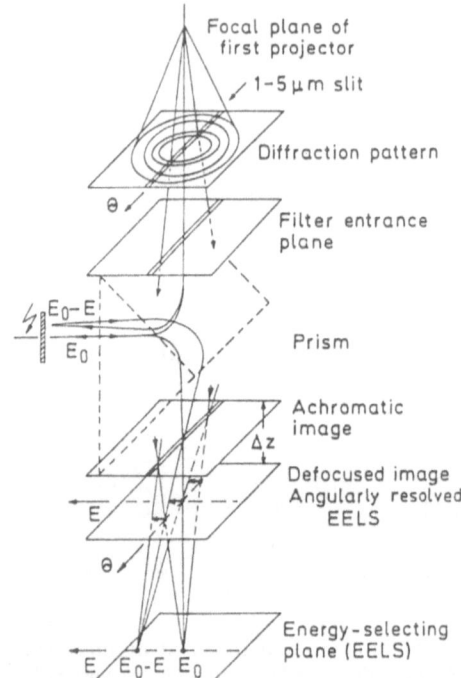

Focal plane of
first projector

1-5 μm slit

Diffraction pattern

θ

Filter entrance
plane

$E_0 - E$
E_0

Prism

Achromatic
image

Δz

Defocused image
Angularly resolved
EELS

E

θ

Energy-selecting
plane (EELS)

E $E_0 - E$ E_0

Fig. 1.22. Angularly resolved EELS mode with a slit in the filter entrance plane and observation of a defocused achromatic image plane.

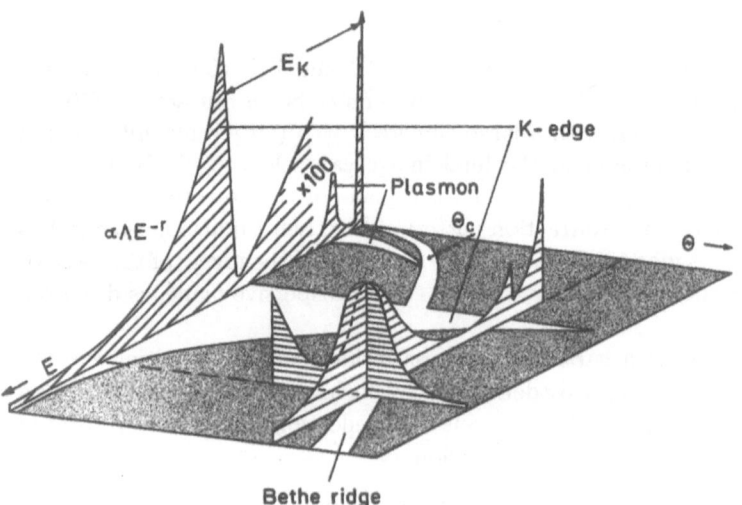

E_K

K- edge

×100

Plasmon

θ_c

$\alpha \Lambda E^{-r}$

θ →

E

Bethe ridge

Fig. 1.23. Schematical EELS intensity distributions as a function of energy loss E and scattering angle θ with profiles at $E =$ const and $\theta =$ const showing a Compton maximum when crossing the Bethe ridge.

Plotting the maximum of the shifted and broadened plasmon loss at large θ versus θ^2 results in a straight line of slope a (Fig. 1.21b).

1.5.5 Angularly and Spatially Dispersive EELS

In an angularly-dispersive EELS, a narrow slit of width Δq_y and parallel to q_x selects a line in the electron diffraction pattern with an energy dispersion normal to the slit in q_y. The selecting slit width of 1–5 μm is mounted close to the filter entrance plane (FEP) (Fig. 1.22) [1.88]. Although this plane does not exactly coincide with the FEP, a sharp shadow of the slit can be observed in the AIP. On defocusing the AIP, each point on the slit is extended to an EELS as shown for a 40 nm Al film in Fig. 1.17e; where the dispersion (curvature) of the plasmon loss at $E = 15$ eV, the decrease of plasmon loss intensity at the cut-off angle, the L-edge at 70 eV and the elastic parts of the Debye–Scherrer rings on both sides of the primary beam can be seen.

This technique of angularly-dispersive EELS has been used earlier with other types of spectrometers [1.43,88–92] for the investigation of the plasmon-loss region but can be used with an imaging filter lens for a range of energy losses from 0–500 eV and scattering angles up to 100 mrad. Such a diagram contains information in the θ–E plane, shown schematically in Fig. 1.23. The angular width of single and multiple plasmon losses (cut-off angle θ_c) and of inner-shell ionisations can be seen directly. The dispersion of plasmon losses causes the parabolic curvature and the broad maximum of the Bethe ridge (see also Fig. 4.3) at the most-probable scattering angle $\theta_C = (E/E_0)^{1/2}$ of electron-electron collisions can also be observed (Compton scattering, see also Sect. 6.5).

By using the ESI mode with a slit in the FEP to select a line through the image, we obtain a perpendicular EELS for each image point (spatially dispersive EELS), which can be used for parallel recording of EELS from different points of the specimen [1.43,88,93–96]. Figure 1.17f shows the spatially dispersive EELS of a graphite flake on a carbon supporting film and demonstrates the differences in the EELS in the plasmon-loss region.

References

1.1 H. Boersch, R. Wolter, H. Schoenebeck: Elastische Energieverluste kristall-gestreuter Elektronen. Z. Physik **199**, 124–134 (1967)

1.2 W.J. Byatt: Analytical representation of Hartree potentials and electron scattering. Phys. Rev. **104**, 1298–1300 (1956)

1.3 H.L. Cox, R.A. Bonham: Elastic electron scasttering amplitudes for neutral atoms calculated using the partial wave method at 10, 40, 70 and 100 kV for Z=1 to Z=54. J. Chem. Phys. **47**, 2599–2608 (1967)

1.4 H. Raith: Komplexe Atomstreuamplituden für die elastische Elektronen-streuung an Festkörperatomen. Acta Cryst. A **24**, 85–93 (1968)

1.5 G. Moliére: Theorie der Streuung schneller geladener Teilchen. Z. Natur-
forsch. A **2**, 133–145 (1947)

1.6 R.J. Glauber: *Lectures in Theoretical Physics*, ed. by W.E. Brittin, G. Dun-
ham (Interscience, New York 1959) p.315

1.7 L. Reimer, K.H. Sommer: Messungen und Berechnungen zum elektronenmikroskopischen Streukontrast für 17 bis 1200 keV-Elektronen. Z. Natur-
forsch. A **23**, 1569–1582 (1968)

1.8 F. Lenz: Zur Streuung mittelschneller Elektronen in kleinste Winkel. Z.
Naturforschg. A **9**, 185–204 (1954)

1.9 L. Schäfer, A.C. Yates, R.A. Bonham: New values for the partial wave electron scattering factor for the elements $1 < Z < 57$ and $72 < Z < 90$ for
incident electron energies of 10, 40, 70 and 100 keV. J. Chem.Phys. **55**,
3055–3056 (1971)

1.10 J.P. Langmore, J. Wall, M.S. Isaacson: The collection of scattered electrons
in dark field electron microscopy. I. Elastic scattering. Optik **38**, 335–350
(1973)

1.11 L. Reimer: *Transmission Electron Microscopy. Physics of Image Formation
and Microanalysis*, Springer Ser. in Opt. Sci. Vol. 36, 3rd ed. (Springer,
Berlin, Heidelberg 1993)

1.12 C.C. Ahn, O.L. Krivanek: *EELS Atlas* (Gatan Inc., Warendale, Pennsylvania
1983)

1.13 L. Reimer, U. Zepke, J. Moesch, St. Schulze-Hillert, M. Ross-Messemer, W.
Probst, E. Weimer: *EELSpectroscopy. A Reference Handbook of Standard
Data for Identification and Interpretation of Electron Energy Loss Spectra
and for Generation of Electron Spectroscopic Images* (Carl Zeiss, Oberkochen
1992)

1.14 B. Schröder, J. Geiger: Electron-spectrometric study of amorphous Ge and
Si in the two-phonon region. Phys. Rev. Lett. **28**, 301–303 (1972)

1.15 R.H. Jakobs, J. Geiger: A comparative study of phonons in polycrystalline
and single-crystalline LiF films by electron spectroscopy. Phys. stat. sol.(b)
95, 549–560 (1979)

1.16 H. Raether: *Excitation of Plasmons and Interband Transitions by Electrons*.
Springer Tracts in Modern Physics, Vol.88 (Springer, Berlin, Heidelberg,
1980)

1.17 C. Colliex: Electron energy loss spectroscopy in the electron microscope.
Adv. in Optical and Electron Microscopy, ed. by R. Barer and V.E. Cosslett
(Academic, New York 1984) Vol.9, 65–177 (1984)

1.18 R.F. Egerton: *Electron Energy-Loss Spectroscopy in the Electron Microscope*.
(Plenum, London 1986)

1.19 H. Kohl, H. Rose: Theory of image formation by inelastically scattered electrons in the electron microscope. Adv. Electr. Electron Phys. **65**, 173-200
(1965)

1.20 H. Koppe: Der Streuquerschnitt von Atomen für unelastische Streuung
schneller Elektronen. Z. Physik **124**, 658–664 (1948)

1.21 M. Inokuti: Electron-scattering cross sections pertinent to electron microscopy. Ultramicroscopy **3**, 423–427 (1979)

1.22 R. Eusemann, H. Rose, J. Dubochet: Electron scattering in ice and organic
materials. J. Micr. **128**, 239–249 (1982)

1.23 J. Wall, M. Isaacson, J.P. Lamgmore: The collection of scattered electrons
in dark field electron microscopy. II. Inelastic scattering. Optik **39**, 359–374
(1974)

1.24 J.C. Ashley, R.H. Ritchie: The mean free path of relativistic elctrons to
plasmon excitation. Phys. Stat. Sol. **40**, 623–630 (1970)

1.25 M.S. Isaacson: Specimen damage in the electron microscope. In *Principles and Techniques of Electron Microscopy*, ed. by M.A. Hayat (VanNostrand, New York 1977) pp.1–78

1.26 R.F. Egerton: Measurement of inelastic/elastic scattering ratio for fast electrons and its use in the study of radiation damage. Phys. Stat. Sol. A **37**, 663–668 (1976)

1.27 L. Reimer , M. Ross-Messemer: Contrast in the electron spectroscopic imaging mode of a TEM. II. Z-ratio, structure-sensitive and phase contrast. J. Micr. **159**, 143–160 (1990)

1.28 G. Benner, J. Bihr, M. Prinz: A new illumination system for an analytical TEM using a condenser objective lens. *Proc. XIIth Int'l Congr. on Electron Microscopy*, ed. by L.D. Peachey, P.B. Williams (San Francisco Press, San Francisco 1990) Vol.I, pp.138–139

1.29 L. Reimer: Electron diffraction methods in TEM, STEM and SEM. Scanning **2**, 3–19 (1979)

1.30 J.J. Hren, J.I. Goldstein, D.C. Joy (eds.): *Introduction to Analytical Electron Microscopy* (Plenum, New York 1979)

1.31 A.V. Crewe, M. Isaacson, D. Johnson: A high resolution electron spectrometer for use in transmission electron microscopy. Rec. Sci. Instr. **42**, 411–419 (1971)

1.32 S.J. Pennycook, D.E. Jesson: High resolution Z-contrast imaging of crystals. Ultramicroscopy **37**, 14–38 (1991)

1.33 K.H. Herrmann, H. Krahl: The detection quantum efficiency of electron image recording systems. J. Micr. **127**, 17–28 (1982)

1.34 E. Rudberg: Characteristic energy losses of electrons scattered from incandescent solid. Proc. Roy. Soc. (London) A **127**, 111–140 (1930)

1.35 D. Bohm, D. Pines: A collective description of electron interactions. Phys. Rev. **92**, 609–625 (1953)

1.36 H. Ibach, D.L. Mills: *Electron Energy-loss Spectroscopy and Surface Vibrations* (Academic, New York 1982)

1.37 J.M. Cowley: Surface energies and surface structure of small crystals studies by use of a STEM instrument. Surf. Sci. **114**, 587–606 (1982)

1.38 G. Ruthemann: Diskrete Energieverluste schneller Elektronen in Festkörpern. Naturwiss. **29**, 648 (1941)

1.39 G. Ruthemann: Elektronenbremsung an Röntgenniveaus. Naturwiss. **30**, 145 (1942)

1.40 G. Ruthemann: Diskrete Energieverluste mittelschneller Elektronen beim Durchgang durch dünne Folien. Ann. Physik **2**, 113–134 (1948)

1.41 J. Hillier, R.F. Baker: Microanalysis by means of electrons. J. Appl. Phys. **15**, 663–675 (1944)

1.42 G. Möllenstedt: Die elektrostatische Linse als hochauflösender Geschwindigkeitsanalysator. Optik **5**, 499–517 (1949)

1.43 A.J.F. Metherell: Energy analysing and energy selecting electron microscopes. *Adv. in Optical and Electron Microscopy*, ed. by R. Barer and V.E. Cosslett (Academic, London 1971) Vol.4, pp.263–361

1.44 S.L. Cundy, A.J.F. Metherell, M.J. Whelan: An energy analysing electron microscope. J. Sci. Instr. **43**, 712–715 (1966)

1.45 H. Boersch: Experimentelle Bestimmung der Energieverteilung in thermisch ausgelösten Elektronenstrahlen. Z. Physik **139**, 115–146 (1954)

1.46 H. Boersch, H. Miessner: Ein hochempfindlicher Gegenfeld-Energieanalysator für Elektronen. Z. Physik **168**, 298–304 (1962)

1.47 H. Boersch, S. Schweda: Eine inverse Gegenfeldmethode zur Energieanalyse von Elektronen und Ionenstrahlen. Z. Physik **167**, 1–10 (1962)

1.48 L.B. Leder, J.A. Simpson: Improved electric differentiation of retarding potential measurements. Rev. Sci. Instr. **29**, 571–574 (1958)

1.49 K. Brack: Über eine Anordnung zur Filterung von Elektroneninterferenzen. Z. Naturforschg. A **17**, 1066-1070 (1962)

1.50 M.T. Browne, S. Lockovic, R.E. Burge: Instrumentation and recording for the Vacuum Generators HB5 STEM instrument. In *Developments in Electron Microscopy and Analysis*, ed. by J.A. Venables (Academic, London 1976) pp.27–28

1.51 J. Lohff: Charakteristische Energieverluste bei der Streuung mittelschneller Elektronen an Al-Oberflächen. Z. Physik **171**, 442–448 (1963)

1.52 A.W. Blackstock, R.D. Birkhoff, M. Slater: Electron accelerator and high resolution analyzer. Rev. Sci. Instr. **26**, 274–275 (1955)

1.53 H. Boersch, J. Geiger, W. Stickel: Das Auflösungsvermögen des elektrostatisch-magnetischen Energieanalysators für schnelle Elektronen. Z. Physik **180**, 415–424 (1964)

1.54 G. H. Curtis, J. Silcox: A Wien filter for use as an energy analyzer with an electron microscope. Rev. Sci. Instr. **42**, 630–637 (1971)

1.55 M. Terauchi, R. Kuzuo, F. Satoh, M. Tanaka, K. Tsuno, J. Ohyama: Performance of a new high-resolution electron energy-loss spectroscopy microscope. Microsc. Microanal. Microstruct. **2**, 351–358 (1991)

1.56 F. Lenz: Über das chromatische Auflösungsvermögen von Elektronenlinsen bei der Geschwindigkeitsanalyse. Optik **10**, 439–446 (1953)

1.57 T. Ichinokawa: Electron energy analysis by cylindrical magnetic lens. Jpn. J. Appl. Phys. **7**, 799–813 (1968)

1.58 R. Shirota, T. Yanaka: An energy analyser with rotation symmetrical lenses. *Electron Microscopy 1974*, ed. by J.V. Sanders, D.J. Goodchild (Australian Acad. Sci., Canberra 1974) Vol.1, pp.368–369

1.59 L. Reimer, U. Riediger: Energieverlustspektroskopie mit einer modifizierten Kaustikmethode in einem 100 keV Transmissionselektronenmikroskop. Optik **46**, 67–73 (1976)

1.60 D.B. Wittry: An electron spectrometer for use with the TEM. J. Phys. D **3**, 1757–1766 (1969)

1.61 R.F. Egerton: A simple electron spectrometer for energy analysis in the trasnmission microscope. Ultramicroscopy **3**, 39–47 (1978)

1.62 R.F. Egerton: Design of an aberration-corrected electron spectrometer for the TEM. Optik **57**, 229–242 (1980)

1.63 H. Shuman: Correction of the second-order aberrations of uniform field magnetic sectors. Ultramicroscopy **5**, 45–63 (1980)

1.64 R.F. Egerton: The use of electron lenses betwen a TEM specimen and an electron spectrometer. Optik **56**, 363–376 (1980)

1.65 D.E. Johnson: Pre-spectrometer optics in CTEM/STEM. Ultramicroscopy **5**, 163–174 (1980)

1.66 B. von Borries: *Die Übermikroskopie. Einführung, Untersuchung ihrer Grenzen und Abriss ihrer Ergebnisse* (Saenger, Berlin 1949)

1.67 H. Boersch: Ein Elektronenfilter für Elektronenmikroskopie und Elektronenbeugung. Optik **5**, 436–450 (1949)

1.68 H. Boersch: Gegenfeldfilter für Elektronenbeugung und Elektronenmikroskopie. Z. Physik **134**, 156–164 (1953)

1.69 G. Möllenstedt, O. Rang: Die elektrostatische Linse als hochauflösendes Geschwindigkeitsfilter. Z. angew. Phys. **3**, 187–189 (1951)

1.70 H. Watanabe: Energy selecting microscope. Jpn. J. Appl. Phys. **3**, 480–485 (1964)

1.71 H. Watanabe, R. Uyeda: Energy-selecting electron microscope. J. Phys. Soc. Jpn. **17**, 569–570 (1962)

1.72 A.P. Wilska: Expectations and limitations of low voltage electron microscopy. Lab. Inv. **14**, 825–828 (1965).

1.73 K. Müller: Maßstabsgetreue und farbfehlerarme Rückstrahl-Mikroskopie. *5th Int'l. Congr. for Electron Microscopy*, ed. S. Breese (Academic, NewYork 1962) Vol.1, p.AA-15

1.74 J.S. Halliday, R.C. Newman: Reflection electron microscopy using diffracted beams. Brit. J. Appl. Phys. **11**, 158–165 (1960)

1.75 R. Castaing, L. Henry: Filtrage magnétique des vitesses en microscopie électronique. Compt. rend. Acad. Sci. (Paris) B **255**, 76–78 (1962)

1.76 R. Castaing, J. Hennequin, L. Henry, G. Slodzian: The magnetic prism as an optical system. In *Focussing of Charged Particles*, ed. by A. Septier (Academic, New York 1967) pp.265–293

1.77 R.F. Egerton, J.G. Philips, P.S. Turner, M.J. Whelan: The design and operation of an energy selecting electron microscope. Electron Microscopy 1974, ed. by J.V. Sanders, D.J. Goodchild (Australian Acad. of Sci., Canberra 1974) Vol.I, pp.376–377

1.78 J.W. Andrew, F.P. Ottensmeyer, E. Martell: An improved magnetic prism design for a TEM energy filter. *Electron Microscopy 1978*, ed by. J.M. Sturgess (Microscopical Soc. of Canada, Toronto 1978) Vol.I, pp.40–41

1.79 W. Egle, A. Rilk, F.P. Ottensmeyer: A new analytical TEM with imaging electron energy loss spectrometer. *Electron Microscopy 1984*, ed. by A. Csanády, P. Röhlich, D. Szabó (Progr. Comm. Eigth European Congr. on Electron Microscopy, Budapest 1984) Vol.1, pp.63–64

1.80 G. Zanchi, J.P. Perez, J. Sevely: Adaption of a magnetic filtering device on a one megavolt electron microscope. Optik **43**, 495–501 (1975)

1.81 O.L. Krivanek, A.J. Gubbens, N. Dellby: Developments in EELS instrumentation for spectroscopy and imaging. Microsc. Microanal. Microstruct. **2**, 315–332 (1991)

1.82 D. McMullan, J.M. Rodenburg, W.T. Pike: Post-spectrometer instrumentation for STEM. *Proc. XIIth Int'l. Congr. on Electron Microscopy*, ed. by L.D. Peachey, D.B. Williams (San Francisco Press, San Francisco 1990) Vol.2, pp.104–105

1.83 L. Reimer, I. Fromm, R. Rennekamp: Operation modes of electron spectroscopic imaging and electron energy-loss spectroscopy in a transmission electron microscope. Ultramicroscopy **24**, 339–354 (1988)

1.84 L. Reimer: Energy-filtering transmission electron microscopy. Adv. Electr. Electron Physics **81**, 43–126 (1991)

1.85 L. Reimer: Electron diffraction methods in TEM, STEM and SEM. Scanning **2**, 3–19 (1979)

1.86 J. Bihr, A. Rilk, G. Benner: Angle of acceptance and energy resolution in the imaging electron energy-loss spectrometer of the EM 902. *EUREM 88*, Inst. Phys. Conf. Ser. 93 (Inst. of Physics, Bristol 1988) pp.159–160

1.87 K.H. Körtje: Image-EELS: simultaneous recording of multiple electron energy-loss spectra from series of electron spectroscopic images. J. Micr. **174**, 149–159 (1994)

1.88 L. Reimer, R. Rennekamp: Imaging and recording of multiple scattering effects by angular resolved electron-energy-loss spectroscopy. Ultramicroscopy **28**, 258–265 (1989)

1.89 A.J.F. Metherell: Energy analysing and energy selecting electron microscopes. *Adv. in Optical and Electron Microscopy*, ed. by R. Barer and V.E. Cosslett (Academic, New York 1971) Vol.4, pp.263–360

1.90 C.H. Chen, J. Silcox, R. Vincent: Dispersion of solid-state excitations. In *Physical Aspects of Electron Microscopy and Microbeam Analysis*, ed. by B.M. Siegel and D.R. Beaman (Wiley, New York 1975) pp.303–313

1.91 F. Leonhard: Spektrometrie von Elektroneninterferenzen. Z. Naturforschg. A **9**, 727–734 and 1019–1031 (1954)

1.92 G.H. Curtis, J. Silcox: A Wien filter for use as an energy analyzer with an electron microscope. Rev. Sci. Instr. **42**, 630–637 (1971)

1.93 S.L. Cundy, A.J.F. Metherell, M.J. Whelan: Contrast preserved by elastic and quasi-elastic scattering of electrons near Bragg beams. Phil. Mag. **15**, 623–630 (1967)

1.94 S.L. Cundy, A.J.F. Metherell, M.J. Whelan: Microanalysis of Al-4wt%Cu by combined electron microscopy and energy analysis. Phil. Mag. **17**, 141–147 (1968)

1.95 S.L. Cundy, A. Howie, U. Valdre: Preservation of electron microscopic image contrast after inelastic scattering. Phil. Mag. **20**, 147–163 (1969)

1.96 L. Reimer, I. Fromm, P. Hirsch, U. Plate, R. Rennekamp: Combination of EELS modes and electron spectroscopic imaging and diffraction in an energy-filtering electron microscope. Ultramicroscopy **46**, 335–347 (1992)

2. Electron Optics of Imaging Energy Filters

Harald Rose and *Dieter Krahl*

2.1 Scattering and Filtering

Inelastic scattering always occurs when charged particles interact with the atoms of an object. Since the inelastic process is delocalized the image formed by the inelastically scattered electrons does not contain much information about the atomic structure of the specimen. Nevertheless, the inelastically scattered electrons do carry information about its elemental composition on a nanometre scale.

In a conventional transmission electron microscope (TEM) both the elastically scattered and the inelastically scattered electrons contribute to the image intensity. Accordingly, each micrograph can be considered as a superposition of a high–resolution "elastic image" and a blurred "inelastic image". The degradation of the inelastic image results from the unavoidable chromatic aberration of the round lenses. As a consequence, the contrast of the elastic image is reduced. This deterioration is relatively slight in the case of phase contrast images of thin objects because the inelastically scattered part of the electron wave is incoherent with respect to the unscattered part. The intensity of the inelastic image is hence of the same order of magnitude as the intensity of the "incoherent" elastic image resulting from the nonlinear terms of the projected object potential. Accordingly, for weak phase objects the contrast of these nonlinear images is negligibly small compared with the phase contrast.

A different situation arises in the case of relatively thick low-Z (organic) material when the object thickness is still somewhat smaller than the elastic mean-free-path length. For biological objects each elastic scattering event is accompanied on average by about three inelastic processes (Sect.1.1.3), with the result that the phase contrast is weakened while the intensity of the superimposed low-resolution image formed by the inelastically scattered electrons is strongly enhanced.

The chromatic aberration becomes extremely deleterious in the dark-field mode where the intensities of the elastic and inelastic images are comparable in magnitude. However, since the inelastic image is almost completely blurred, the contrast and specimen resolution of the total image are substantially diminished.

Springer Series in Optical Sciences, Vol. 71
Energy-Filtering Transmission Electron Microscopy
Editor: Ludwig Reimer   ©Springer-Verlag Berlin Heidelberg 1995

The harmful effect of the chromatic aberration can be reduced to a certain extent by increasing the voltage. However, a much better way of improving the image is to eliminate the inelastically scattered electrons by means of an imaging energy filter. Filtering completely eliminates the effect of inelastic scattering from the dark-field image. Surprisingly, this is not the case for the bright-field image. This unexpected behaviour is an immediate consequence of the optical theorem, according to which the elastic scattering amplitude contains all the information about the spatial distribution of the non-local potential.

The occurrence of an image of the non-local potential in the bright-field mode can be understood by considering an ideal microscope. Neglecting backscattering, no contrast will be visible in the Gaussian image plane because all elastically and inelastically scattered electrons originating from different object points are perfectly redirected by the aberration-free lenses into the corresponding image points. If the inelastically scattered electrons are removed by an imaging filter, a "negative" inelastic image will appear representing the nonlocal interaction potential. This shadow image is produced by the interference of the elastically scattered part with the unscattered part of the electron wave. The resulting contrast is proportional to the imaginary part of the complex elastic scattering amplitude and can be considered as a "negative" of the image, formed by the electrons removed by the filter.

Even in the case of a real lens, the shadow image is not affected by chromatic aberration since this image is entirely produced by the elastic part of the scattered wave. The resolution of the inelastic shadow image depends on the operating conditions of the microscope, as the corresponding contrast transfer function is proportional to the cosine of the phase shift $W(q)$ (Sect. 7.2.2) caused by the defocus and the spherical aberration of the objective lens. The inelastic shadow image would, therefore, vanish for an ideal phase shift $W(q) = \pi/2$. However, it is present in filtered phase contrast images of any current microscope because at Scherzer focus, $\cos W(q)$ remains close to unity for angles smaller than about a third of the optimum aperture angle.

The incorporation of an imaging energy filter in an electron microscope greatly increases the amount of information that can be obtained about the object. By considering the energy loss E as a new coordinate in addition to the two spatial x and y coordinates, the energy-filtering transmission electron microscope (EFTEM) is capable of yielding information about a distinct plane $E = const$ of the three-dimensional x, y, E space with a single exposure. By taking several exposures with different characteristic energy losses, "colour images" can be obtained, where each colour corresponds to a distinct energy loss representing a specific element. This technique, known as elemental mapping, yields information about the elemental composition of the object on a nanometre scale. It should be noted that elemental mapping requires background subtraction. In order to determine the background intensity, two additional energy-loss images must be taken, in front of each chosen ionization edge.

The complete removal of the inelastically scattered electrons increases the contrast and the resolution in both the imaging and the diffraction mode. In the latter mode, this improvement has been strikingly demonstrated by filtering convergent beam diffraction patterns [2.1]. Besides these applications the filter can also be used for recording energy-loss spectra of an arbitrary object area.

So far we have been considering an ideal filter. Such an ideal filter should act like a (thick) round lens with respect to the transmitted electrons with nominal energy E_0 and like a combination of a round lens and a prism for electrons whose energies differ from E_0. Moreover, the energy selection slit must (a) not affect the quality of the image formed by the transmitted electrons and (b) select the same energy for each object element ("obel"). The energy selection will then be "isochromatic" which means that the selected energy does not depend on the off-axial position of the "obels". To fulfil these conditions, all second-rank aberrations must be either eliminated or adequately suppressed.

Energy filtering is best performed at the diffraction plane located behind the filter, because the diffraction pattern represents the smallest waist of the beam. Moreover, selection at this plane does not cause vignetting. In order to demagnify the diffraction pattern of a given energy so that it is much smaller than the width of the energy selection slit and to reduce the resolution-limiting aberrations, the filter is usually placed between the last intermeditate lens and the projector lens of the microscope. As shown in Fig. 2.1 the filter images the polychromatic diffraction pattern located in front of the filter into a series of laterally displaced monochromatic spots in the energy selection plane. Simultaneously, the filter transfers the intermediate image of the object stigmatically into the achromatic (zero dispersion) image plane. These two consecutive intermediate images may be real or virtual depending on the optical properties of the filter. Most filters are symmetric with respect to their central plane. In this case both the diffraction plane and the intermediate image are imaged by the filter with unit magnification. If we regard the filter as a thick lens, the images of the object are then located at nodal planes which coincide with the principal planes, while the diffraction images are situated at conjugate planes, each located at a distance $2f$ from the corresponding principal plane, where f is the focal length of the stigmatically imaging filter.

In addition to the requirements mentioned above a suitable high-performance energy filter must fulfil the following supplementary conditions: first, the dispersion at the energy selection plane should be as high as possible to allow for small energy windows. Second, the incorporation of the straight-vision filter must not significantly enlarge the length of the column. Third, the filter must be precisely alignable in a systematic manner, such that each

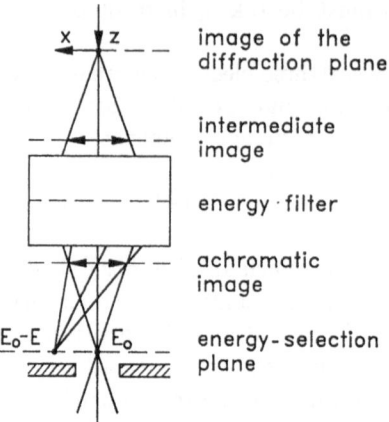

image of the
diffraction plane

intermediate
image

energy filter

achromatic
image

energy-selection
plane

Fig. 2.1. Action of an imaging energy filter showing the underlying principle of the energy selection at an image of the diffraction plane.

alignment step can be controlled and does not destroy the state of alignment obtained by the preceding adjustment steps. Fourth, wholly magnetic systems are preferable because they do not limit the accelerating voltage. Up to now, only the magnetic filter built in the Fritz–Haber–Institut in Berlin fulfils all the desiderata of a high-performance filter. We shall therefore concentrate most of our discussion on the optical properties and the alignment of this corrected magnetic imaging energy filter. Nevertheless, the treatment of the paraxial properties and the aberrations of imaging energy filters will be quite general, although only purely magnetic filters are considered in detail.

Unfortunately, no common notation exists in the field of charged particle optics. This situation dates from the early days of charged particle optics, when each group entering this new field of research introduced its own notation. As a result, even for the expert, it is sometimes very difficult to comprehend the results published by groups using different nomenclature. In the past, for example, little notice has been taken of results derived in electron optics by the nuclear instruments community and vice versa. Thus many results which are well established in a specific field of charged particle optics are often rediscovered by those working in another field. Since we shall restrict our discussion mainly to the optics of imaging energy filters used in electron microscopy, we employ the notation and terminology introduced by *Scherzer* [2.2].

In this nomenclature, we distinguish between "sections" and "planes". *Sections* are surfaces that contain the optic axis, while *planes* are plane surfaces perpendicular to the optic axis. Unlike a plane, a section can be curved, as happens in systems with curved optic axes. For most systems of interest, the torsion of the optic axis is zero. In this case all curved sections have an evolute. This behaviour facilitates the representation of the trajectories since they are generally visualized by their projections onto two perpendicular sections. For the description of the path of rays, it is advantageous to choose an

x, y, z coordinate system in which the z axis coincides with the optic axis. In deflection systems such as imaging energy filters this axis is curved. The surface $y = 0$ will be a plane section since we consider only systems for which the axis lies in a plane.

By the *order* of an aberration, we understand the so-called Seidel order, which is defined as the sum of the exponents of the geometrical ray parameters determining the ray. The exponent of the chromatic parameter describing the relative energy deviation is called the *degree* of the aberration. The sum of the order and degree will be referred to as the *rank*. For example, the familiar chromatic aberration is of first order and first degree and is hence a second-rank aberration. It should be noted that this terminology differs from that used in most of the literature, where unfortunately no such differentiation is made and the *order* always means the *rank* of an aberration. The rank is a measure of the magnitude of an aberration. The higher the rank the smaller is the influence of the corresponding aberration.

2.2 Types of Imaging Energy Filters

Before we outline in detail the optical properties of imaging energy filters, we discuss the different types of filters. Imaging energy filters are best characterized by the nature of their constituent elements. Accordingly, we differentiate between purely electric, purely magnetic and combined electric–magnetic filters. For producing a dispersion that spatially separates electrons according to their energy loss, dipole fields are mandatory for any energy filter or spectrometer. All such systems therefore have a curved optic axis with the exception of the Wien filter, in which the electric and magnetic deflections cancel out for electrons with nominal energy E_0.

Electric–magnetic imaging energy filters have the decisive disadvantage that they are limited to accelerating voltages below about 100 kV owing to the difficulties of handling large electric field strengths. It is for this reason that wholly electric filters have so far been proposed only as monochromators for reducing the energy width of the illuminating beam [2.3,4]. These monochromators are placed somewhere beyond the cathode, where the electrons have an energy of several keV.

Imaging energy filters can be incorporated in an electron microscope either within the column as part of the instrument or as an attachment beneath the viewing screen. In the latter case, we have a post-column filter, which need not be a straight-vision system. In the following sections, we shall discuss briefly the different types of filter that have been proposed. Although all of these systems are suitable as imaging energy filters only one of them fulfils all the conditions necessary for optimum performance.

2.2.1 Castaing–Henry Filter

The first imaging energy filter consisting of a triangular magnetic double prism and an electrostatic mirror was developed in 1964 by *Castaing* and *Henry* [2.5]. The schematic arrangement of the filter and the course of the corresponding fundamental rays are depicted in Fig. 2.2. Henry utilized earlier results of *Cotte* [2.6], who had shown that a specially designed 90° deflection magnet was capable of forming a real and a virtual stigmatic image for two distinct planes. By combining the triangular double prism with an electrostatic mirror Henry obtained a symmetric prism–mirror–prism system, which simply inverts the position of the two stigmatic planes in such a way that the entire filter acts like a thick round lens for electrons with nominal energy. By adjusting the intermediate lenses in front of the filter in such a way that the virtual stigmatic planes are conjugate to the object plane and the real stigmatic planes are conjugate to the diffraction plane of the objective lens, the filter yields an achromatic image and a high dispersion at the energy selection plane. Owing to the symmetry of the deflection fields and the fundamental rays with respect to the mirror plane, the filter introduces neither second-order axial aberrations nor second-order distortions at the final image. To our knowledge this fact has not been realized previously.

Unfortunately, Henry placed the mirror–prism filter directly behind the first intermediate lens. As a result, the remaining second-order aberrations (inclination of the image field and field astigmatism) strongly limited the field of view. This shortcoming was eliminated in the first commercial version of

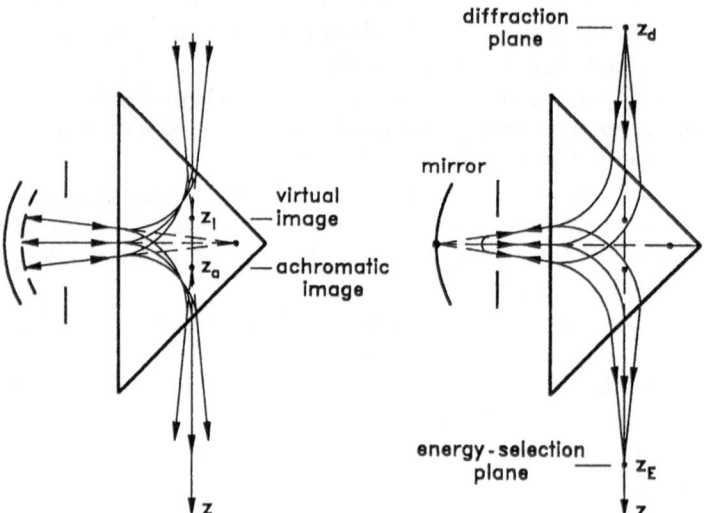

Fig. 2.2. Prism–mirror arrangement and path of the paraxial rays in the Castaing–Henry filter. Virtual images of the object plane and the conjugate achromatic image plane are located within the filter.

an energy filtering microscope, built by the French manufacturer Sopelem, in which the filter was placed behind the second intermediate lens. Later, in 1973, *Henkelman* and *Ottensmeyer* placed the filter in front of the projector lens [2.7]. Since the second-order field aberrations are inversely proportional to the magnification M_i of the intermediate image in front of the filter, the new position of the filter afforded a sufficiently large field of view. These convincing results initiated the second (successful) commercial version of an energy-filtering electron microscope, the Zeiss EM902.

It should be mentioned, however, that the decrease of the field aberrations is accompanied by a strong increase of the second-order axial aberrations at the energy-selection plane, since the latter increase with M_i^2. Consequently, isochromatic filtering is not possible, which implies that relatively large energy windows must be employed. Experimental tests have shown that the Zeiss EM902 yields satisfactory image quality over the whole image area as long as the energy window is not made smaller than about 8 eV. Even so, it should be noted that the allowable minimum energy width can always be diminished by reducing the size of the transferred object field.

2.2.2 Magnetic Filters

The need for higher accelerating voltages stimulated the development of entirely magnetic imaging energy filters. Magnetic straight-vision filters are usually named according to the characteristic course of their curved axis. If the sum of the deflection angles of the constituent magnets cancels out to zero, the system is called an "Omega" filter, while it is called an "alpha" filter if the sum adds up to 2π. *Senoussi*, in 1971, replaced the electric mirror of the Castaing–Henry filter by a single deflection magnet [2.8]. However, his asymmetric filter consisting of three differently shaped deflection magnets is of little use, because none of the second-order aberrations is eliminated, unlike the symmetric Castaing–Henry filter.

Symmetry principles for correcting the second-order aberrations were first introduced in 1974 by *Rose* and *Plies*, who proposed the first symmetric magnetic equivalent of the prism–mirror–prism system [2.9]. This magnetic Ω filter consists of four identical deflection magnets which are symmetrically arranged about the midplane of the system. Owing to the symmetry of the magnetic deflection field and the paraxial rays, this filter does not possess second-order axial aberrations and distortions, as is the case for the Senoussi filter. However, even this filter as well as the systems proposed later by *Zanchi* et al. [2.10] and *Perez* et al. [2.11] are of little practical use because the remaining field aberrations, which are significantly larger than those of the Castaing–Henry filter, strongly reduce the field of view.

Partly Corrected Ω and α Filters. For routine work, filter systems of simple structure are extremely desirable. An extensive search for optimum magnetic systems with small aberrations consisting of the least number of deflection magnets has been carried out by *Lanio* [2.12]. The remaining aberrations of a suitable filter must not affect appreciably the quality of the final high-resolution image. Two of the most promising systems that have been found are shown in Figs. 2.3 and 4. The first system (Fig. 2.3) is an Ω–type filter, while the second system is an α filter. Both systems consist of only three deflection magnets. The geometry of each system is defined by the tilt angles $\epsilon_{2\nu-1}$ and $\epsilon_{2\nu}$, the radii of curvature of the axis R_ν, the deflection angles Φ_ν, and the distances $\ell_{\nu,\nu+1}$ between the magnets ν and $\nu+1$. The diffraction plane and the intermediate image plane located in front of the filter are separated by the distance L. It should be noted that for the Ω filter (Fig. 2.3) the intermediate image is located within the first magnet.

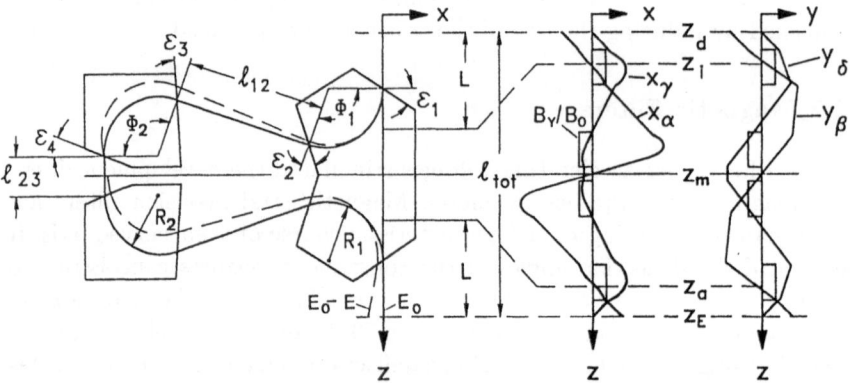

Fig. 2.3. Arrangement of the deflection magnets, course of the optic axis, and path of the paraxial fundamental trajectories along the straightened optic axis of an optimized magnetic Ω filter with simple structure.

The fundamental field-ray with components x_γ and y_δ intersects the centre of the diffraction plane, while the axial fundamental ray with components x_α and y_β intersects the centre of the intermediate image. The figures clearly demonstrate the symmetry properties. For both filters, the magnetic induction $B(z)$ along the optic z axis and the component x_γ of the field ray are symmetric with respect to the midplane z_m , while the component x_α of the axial fundamental ray is antisymmetric. The ray components y_β and y_δ in the vertical section behave differently for the two filters. For the Ω filter, y_β is symmetric and y_δ antisymmetic with respect to the midplane z_m; for the α filter these rays have the opposite symmetry.

The second-order aberrations of the two filters are comparable to those of the Castaing filter; their energy resolution is, however, improved by a factor of about four. By incorporating an additional sextupole centred about the

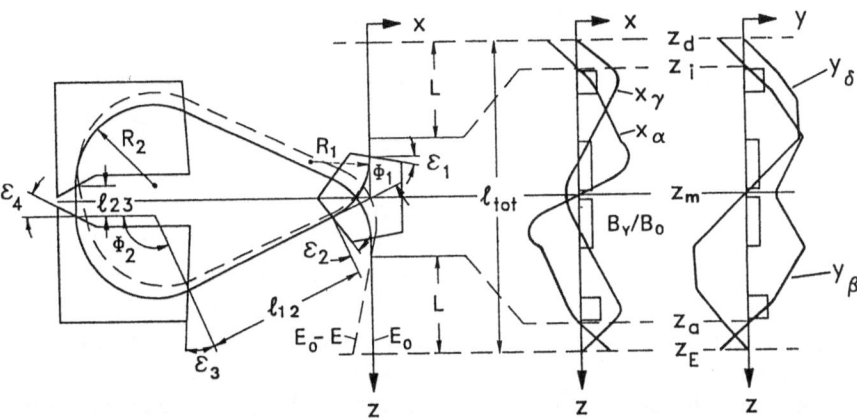

Fig. 2.4. Definition of the geometrical parameters and course of the optic axis and the fundamental paraxial rays of an optimized α filter composed of three homogeneous deflection magnets.

optic axis at the symmetry plane z_m, the energy resolution of the Ω filter can be further enhanced by a factor of 1.6. This version of the filter is corporated in the Zeiss EM912 Omega filter electron microscope.

Corrected Ω Filter. To allow high-resolution imaging of extended objects together with very small energy windows, a fully corrected Ω filter consisting of four magnets with curved pole-faces and three additional sextupoles has been designed by *Rose* and *Pejas* [2.13,14] and built and tested by *Krahl* et al. [2.15]. Experiment has revealed that precise alignment of the beam within the filter is not possible with this design. The reason for this failure is the curvature of the pole-faces. A small lateral displacement of the beam changes the angle ϵ between the optic axis and the normal to the pole-faces at the entrance and/or exit of the magnets. As a result, the quadrupole fields, which influence the paraxial rays, are changed because their strength is proportional to $\tan \epsilon$. Accordingly, the deviation between two initially parallel adjacent trajectories strongly increases with the number of elements. A preliminary investigation has revealed that the system shows chaotic behaviour, preventing a systematic adjustment. Another reason for the observed deviations stems from the fact that the influence of the fringing fields increases with increasing curvature of the pole-faces.

Owing to these negative results, the deflection magnets have been replaced by magnets with straight pole-faces, and separate sextupole elements were introduced instead. The upgraded system consists of four deflection magnets and seven sextupoles, six of which are excited in pairs [2.16]. The incorporation of the sextupole elements has the advantage that they can be adjusted electrically to compensate for the disturbing aberrations exactly. A detailed description of the correction procedure will be given in Chaps. 2.7–10.

2.2.3 Retarding Wien Filter

Another type of a combined electric–magnetic spectrometer is the well-known Wien filter. In order to use the Wien filter as an imaging energy filter, the electric and magnetic fields must be formed in such a way that the filter acts like a round lens for electrons with nominal energy $E_0 = mv_0^2/2$. Here m is the electron mass and v_0 the nominal velocity.

Unfortunately, the dispersion of the Wien filter is rather small. To avoid an unduly large filter length, it is therefore necessary to reduce the velocity of the electrons within the filter. For this purpose, the Wien filter must be placed between a retarding lens and an accelerating electrostatic immersion lens, as shown schematically in Fig. 2.5. The retarding Wien filter was first successfully employed by *Legler* [2.17] and *Boersch* et al. [2.18] as a monochromator.

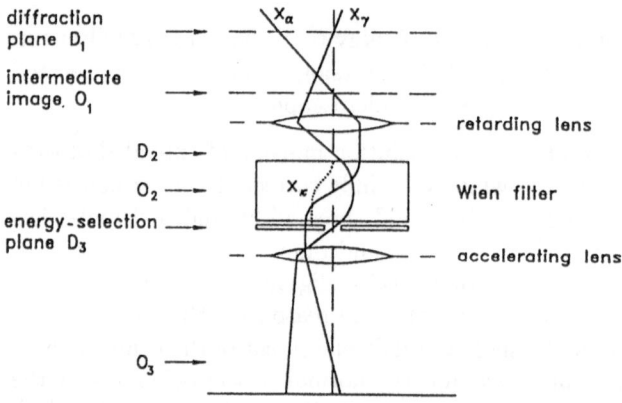

Fig. 2.5. Action of a retarding Wien filter and course of the dispersion ray x_κ and the paraxial rays demonstrating the filtering mechanism.

The possibility of using the retarding Wien filter as an imaging energy filter was first exploited experimentally by *Andersen* and *Kramer* [2.19] and later theoretically by *Rose* [2.20]. The standard Wien filter consists of crossed electric and magnetic dipole fields

$$E_x = -\Phi_{1c}(z) \ , \ \ B_y = -\Psi_{1s}(z) \tag{2.1}$$

which are perpendicular to the optic z–axis; $\Phi_{1c}(z)$ and $\Psi_{1s}(z)$ are the strengths of the dipole components of the electric potential φ and the magnetic potential ψ, respectively. In this case the Wien filter has the combined action of a cylindrical lens plus a straight-vision prism. The prism only deflects electrons with velocities that differ from the nominal velocity v_0 if the electric and magnetic dipole fields are excited such that their strengths satisfy the Wien condition

$$\Phi_{1c} = v_0 \Psi_{1s} \ . \tag{2.2}$$

Owing to the straight central trajectory for electrons that travel with velocity v_0 along the optic axis, the filter should be easier to align than systems with curved axes.

The cylindrical lens action of the standard Wien filter prevents its direct use as a stigmatically imaging filter. For this purpose, one must superimpose quadrupole fields which transform the cylindrical lens into a round lens. By superimposing crossed electric and magnetic quadrupoles with strengths $\Phi_{2c}(z)$ and Ψ_{2s} , respectively, one obtains an inhomogeneous Wien filter. The paraxial motion and the dispersion of the electrons within an inhomogeneous filter that satisfies the Wien condition (2.1) is governed by the Gaussian path equation

$$w'' + \frac{1}{8} \frac{\Phi_{1c}^2}{\Phi^2} w + \left(v_0 \frac{\Psi_{2s}}{\Phi} - \frac{\Phi_{2c}}{\Phi} + \frac{1}{8} \frac{\Phi_{1c}^2}{\Phi^2} \right) \bar{w} = -\frac{\kappa}{4} \frac{\Phi_{1c}}{\Phi} \ . \tag{2.3}$$

In this equation, relativistic effects are neglected because the electron velocity within the retarding filter is small compared with the velocity of light. The complex coordinates

$$w = x + iy \ , \quad \bar{w} = x - iy \tag{2.4}$$

characterize the off-axial position of the electron. The chromatic parameter

$$\kappa = \frac{E}{E_0} = \frac{\Delta\Phi}{\Phi} \tag{2.5}$$

denotes the relative energy deviation of an electron whose energy differs by the amount $E = e\Delta\Phi$ from the nominal energy $E_0 = e\Phi$.

For the Wien filter to act like a round lens in the paraxial approximation for electrons with nominal energy ($\kappa = 0$), the third term on the left-hand side of (2.3) must vanish. This happens if the quadrupole strengths Φ_{2c} and Ψ_{2s} fulfil the anastigmatism condition

$$\Phi_{2c} - v_0 \Psi_{2s} = \Phi_{1c}^2 / 8\Phi \ . \tag{2.6}$$

This relation has been satisfied approximately by tilting the magnets [2.21,22], by curving the electrodes [2.17] or by a combination of these methods [2.23,24]. The conditions (2.2) and (2.6) are most difficult to fulfil in the fringing field regions. The best approximation is obtained by using an electromagnetic dodecapole element, allowing independent excitation of the dipole, quadrupole and hexapole components of the electric and magnetic fields. Such an element has been built and successfully tested by *Haider* et al. [2.25].

A stigmatic Wien filter can only be used as a high-performance imaging filter if the second-order aberrations and the chromatic aberration of magnification are eliminated or sufficiently suppressed. It has been shown in [2.20] that these aberrations vanish if the quadrupole strengths satisfy the relation

$$\Phi_{2c} = \frac{4}{3} v_0 \Psi_{2s} \qquad (2.7)$$

in addition to the condition (2.6) and if hexapole components are introduced whose strengths Φ_{3c} and/or Ψ_{3s} are proportional to the third power of the dipole strength:

$$\Phi_{3c} - v_0 \Psi_{3s} = \frac{3}{32} \Phi_{1c}^3 / \Phi^2 . \qquad (2.8)$$

Although the correction conditions (2.6-8) cannot be met exactly in the region of the fringing fields, the residual aberrations will be small for a dodecapole Wien filter because the spacings between opposite electrodes and/or magnetic poles are identical in this case. Accordingly, the electric and magnetic strengths of a given multipole component have the same shape along the entire optic axis. The applicability of the retarding Wien filter is limited to accelerating voltages below about 100 kV, as is the case for the Castaing–Henry filter. However, compared to this filter the corrected retarding Wien filter has fewer and smaller aberrations and a significantly higher dispersion.

2.2.4 Post-Column Filter

Each of the filters discussed in the preceding subsections is an integral part of the filter electron microcope since the filters must be placed within the column somewhere between the first intermediate lens and the projector lens. To convert a conventional TEM retroactively into an EFTEM, the filter must be placed beneath the viewing screen.

For this purpose an attachable imaging energy filter has been developed by the manufacturer Gatan [2.26]. The Gatan filter consists of a single 90° sector magnet with curved pole-faces followed by a sequence of six quadrupoles and six or seven sextupoles. The sextupoles are necessary for eliminating the second-order distortions and the chromatic aberration of magnification. The Gatan filter can operate in either the imaging mode, yielding a filtered image, or in the spectrum mode, for parallel recording of the energy loss spectrum. Since the system is furnished with a CCD camera and a computer, it is possible to obtain background-subtracted chemical maps on line. The post-column filter usually operates at a large intermediate magnification. The diffraction pattern in front of the filter is then so small that the uncorrected second-order field aberrations do not appreciably affect the image quality. However, owing to the high magnification, the third-order distortions and the higher-order chromatic aberrations decisively limit the number of equally well resolved object elements. Reducing the magnification is of little help because this procedure increases the second-order field aberrations. Moreover, since the second-order axial aberrations at the energy selection plane are only partially corrected, isochromatic imaging is only possible for a limited field of view. Unfortunately, no exact quantitative statements can be made at present about the theoretical performance of the Gatan filter because no precise data about the arrangement of the filter and its electron-optical properties have yet been published.

2.3 Calculation Procedures

For determining the optical properties of beam guiding systems, three different methods are commonly employed:
  a) ray tracing
  b) matrix method
  c) eikonal method.
The last method is based on the ideas of Hamilton who showed that the properties of an optical system can be derived from a single characteristic function or eikonal. *Glaser* [2.27] applied the Hamiltonian formalism to electron optics for calculating the primary aberrations of electron lenses, while *Scherzer* [2.28] employed the "trajectory method", which starts from the equation of motion and derives the aberrations by means of a perturbation procedure using the method of variation of parameters. By combining the two methods, a very elegant and powerful formalism for the calculation of higher-order aberrations is obtained [2.29], which is similar to the treatment by *Sturrock* [2.30]. In the following we refer to this procedure as the eikonal method. Which of the three methods is best suited for determining the optical properties of a system is the subject of some on-going debate. Since each method has its benefits and shortcomings, we shall briefly discuss their specific features and emphasize the important differences.

2.3.1 Ray Tracing

This procedure is especially suited for determining the motion of charged particles in well-defined systems whose electromagnetic fields have been determined numerically, typically by the finite-element method or the finite-difference method. The calculation starts from the standard equation of motion

$$\frac{d}{dt}(m\boldsymbol{v}) = -e(\boldsymbol{E} + \boldsymbol{v} \times \boldsymbol{B}) \ . \tag{2.9}$$

In the first step the central trajectory is calculated. This ray defines the optic axis of the system. In the next step a ray originating from the centre of the object plane with a small angle of inclination is determined. The intersection of this ray with the optic axis defines the Gaussian image plane.

Information about the image quality is most conveniently obtained by calculating the endpoints of many trajectories, each emanating from a given point on the object, like the fine streams from a showerhead. The intersections of the rays with the Gaussian image plane form a so-called spot diagram. The diagrams of different object points are then used to determine the aberration coefficients numerically. Since these coefficients are obtained by numerical subtraction, only the constants of the primary aberrations can be determined with a high degree of accuracy.

Numerical ray tracing is best suited for calculating extremely non-paraxial rays, i.e. electrons with large off-axial distances and/or large inclination angles with respect to the optic axis, as is the case for cathodes. The method is less appropriate for small angles or for calculating higher-order aberrations because the numerical accuracy rapidly deteriorates with decreasing starting angle. Moreover, this procedure cannot give any clue about how best to optimize the system or to minimize disturbing aberrations. For example, 18 linearly independent coefficients exist for the primary aberrations of deflection systems with midsection symmetry. In addition, imaging filters are generally composed of several magnets. Hence a large number of free parameters must be varied in order to find the best system. Employment of a Newton approach will almost certainly end in a local minimum far from the absolute optimum. Owing to these deficiencies, ray tracing is usually employed for determining the performance of systems which have been optimized by other means.

2.3.2 Matrix Method

The matrix method was introduced into charged particle optics as a suitable tool for calculating the optical properties of systems composed of many elements such as accelerators and beam-guiding systems. Within the frame of Gaussian optics, each drift space and each optical element is characterized by a corresponding transfer matrix. The paraxial transfer matrix describes the transfer of the position and the momentum coordinates of a ray at a given plane into those at another plane. In the case of a thin lens, these planes coincide and the corresponding transfer matrix changes only the momentum, i.e. the direction of the particle trajectory.

The course of an electron with a particular initial energy E_0 is entirely defined by two position coordinates and two momentum or angular coordinates at an arbitrary plane. Accordingly, the paraxial transfer matrix is generally a 4×4 matrix. For orthogonal systems, this matrix is composed of two 2×2 matrices, one for the x and the other for the y component. In the case of a rotationally symmetric system these matrices are identical. For dispersive elements the relative energy deviation is another ray-defining parameter. Within the frame of Gaussian optics the dispersion is added to each trajectory of nominal energy E_0. In systems in which the axis lies in a plane, the dispersion has no component perpendicular to this plane. Accordingly, for such systems, the corresponding paraxial transfer matrix is a 5×5 matrix in the most general case.

The off-axial position coordinates x, y and the momentum components p_x and p_y of an electron moving within any non-trivial system are nonlinear functions of the initial parameters x_0, y_0, p_{x0}, p_{y0} and the chromatic parameter κ. By assuming that x_0 and y_0 are small with respect to the inner diameter of the optical elements and p_{x0} and p_{y0} are small with respect to the absolute value p_0 of the initial momentum, the position–momentum vector of the trajectory at a given reference plane z can be expanded as a power series with respect to the five initial parameters. Furthermore, each polynomial of degree n in these parameters can be expressed in terms of the matrix formalism. For example the polynomials of second degree can each be written as a vector–matrix–vector product. Since we have four deviations and each vector is composed of five initial parameters, four *symmetric* 5×5 matrices exist, resulting in 60 coefficients. This number can be reduced significantly by imposing symmetry conditions. If the system has a common plane section of symmetry, half of the 60 coefficients are zero, leaving a total of 30. For magnetic deflection systems with midsection symmetry, each of the two 5×5 matrices describing the positional deviations reduces to an array of a 3×3 matrix and a 2×2 matrix. Therefore six of the 15 coefficients of each position matrix are zero. The two matrices defining the second-rank deviations of the momentum components from their paraxial approximations reduce even further. The position and momentum matrices are complementary in a special way: all matrix elements that are zero for the position matrices are nonzero for the momentum matrices and vice versa. Accordingly, each momentum matrix possesses only six off-diagonal elements [2.31]. In addition 12 linear relations exist between the 30 elements of the four matrices. These relations do not emerge naturally in the matrix method but must be formed either by utilizing the eikonal method or by making use of the invariance of the Poisson or Lagrange brackets defined in Hamiltonian mechanics. The latter invariance is equivalent to the so-called symplectic conditions imposed on the momentum and position components. This requirement forms the basis for the modern Lie algebraic treatment of optical systems [2.32] which can be considered as equivalent to the eikonal approach [2.29].

Although the matrix method permits fast numerical calculation of the optical properties of a given system, it does not yield any insight into the structure of the coefficients of the aberration matrices. This shortcoming is a decisive disadvantage for designing systems with optimum performance. For this purpose a precise knowledge of the structure of each aberration coefficient is necessary in order to eliminate simultaneously all disturbing aberrations. Without such knowledge it is very difficult to find appropriate means for the simultaneous correction of several aberrations because the correction of one aberration usually causes all the other aberrations to increase. Even if several aberrations are eliminated, the induced higher-order aberrations often increase dramatically, thereby preventing an appreciable improvement of the optical performance.

The eikonal method avoids this shortcoming since it yields a detailed insight into the structure of the aberration coefficients and immediately reveals all interrelations existing between some of these coefficients.

2.3.3 Eikonal Method

In the past the usefulness of the eikonal method has been repeatedly questioned by many eminent light-optical scientists [2.33,34]. Owing to their scepticism, it is widely believed that the eikonal theory does not provide a satisfactory treatment of the imaging properties of electron and/or ion optical systems because the method will not give insight into the actual working of the constituent elements. The reasons for this negative opinion are manifold. First, the eikonal is defined by a nonlinear partial differential equation which can be solved analytically only in a few trivial cases. Second, the eikonal depends on four ray parameters, two of which belong to the initial plane and the others to the final reference plane. Usually the parameters are the off-axial coordinates of the trajectory at these planes. The initial slope components are given by the partial differentials of the eikonal with respect to the corresponding initial position coordinates. These two equations form a pair of implicit equations for the two unknown position coordinates of the ray at the final reference plane. Since the equations can generally not be solved by direct inversion, the eikonal method seemed to be of little use [2.33].

However, by employing sophisticated perturbation procedures, the eikonal approach has proven to be a very elegant and powerful tool for efficiently calculating and optimizing electron optical systems [2.29,30]. These techniques were presumably not appreciated in the field of light optics 40 years ago. The point eikonal

$$S(\boldsymbol{r}_0, \boldsymbol{r}_1) = \int_{P_0}^{P_1} n \, ds = \int_{z_0}^{z_1} \mu \, dz \qquad (2.10)$$

represents the optical path length between two points P_0 and P_1 defined by the position vectors $\boldsymbol{r}_0$ and $\boldsymbol{r}_1$. The integral has to be taken along the actual path of the charged particle. The index of refraction is given by the relation

$$n = \sqrt{\varphi^*/\Phi_0^*} - \frac{e}{p_0} \frac{\boldsymbol{A} \cdot \boldsymbol{v}}{v} \quad , \qquad (2.11)$$

where

$$\varphi^* = \varphi(1 + e\varphi/2m_0c^2) \, , \ \ \Phi_0^* = \Phi_0(1 + e\Phi_0/2m_0c^2) \qquad (2.12)$$

are the relativistically modified electric potentials at the plane z, and at the centre of the initial plane z_0, respectively,

$$p_0 = \sqrt{2em_0\Phi_0^*} = e/\eta \qquad (2.13)$$

is the axial kinetic momentum at this plane; m_0 is the electron rest mass and c the velocity of light. The expression (2.11) demonstrates that in the presence of a magnetic field with vector potential $\boldsymbol{A} = \boldsymbol{A}(x, y, z)$, the electromagnetic field forms an inhomogeneous and anisotropic medium for the electrons.

The function

$$\mu = \sqrt{\varphi^*/\varPhi_0^*} \, \frac{\mathrm{d}s}{\mathrm{d}z} - \eta \frac{\boldsymbol{A} \cdot \mathrm{d}s}{\mathrm{d}z} \tag{2.14}$$

is the so-called variational function. According to Fermat's principle, among all paths connecting the points P_0 and P_1 , the true path of the electron makes the integral (2.10) an extremum. This extremum is a minimum if the point P_1 is located in front of the caustic formed by all the rays emanating from P_0; the extremum is a maximum if P_1 is situated on the opposite side of the caustic. The conclusion of *Sturrock* [2.35], that the extremum is a true minimum in the case of rotationally symmetric systems, is erroneous.

As the eikonal must tend to an extremum, the path equations can be derived from the second relation of (2.10) by employing the calculus of variations. The resulting Euler–Lagrange equations

$$\frac{\partial\mu}{\partial x} - \frac{\mathrm{d}}{\mathrm{d}z}\left(\frac{\partial\mu}{\partial x'}\right) = 0 \ , \quad \frac{\partial\mu}{\partial y} - \frac{\mathrm{d}}{\mathrm{d}z}\left(\frac{\partial\mu}{\partial y'}\right) = 0 \tag{2.15}$$

describe the off-axial ray components $x = x(z)$, $y = y(z)$ along the optic z-axis. These equations can serve as the basis for the perturbation procedure yielding the paraxial rays and integral expressions for the aberration coefficients.

2.4 Scalar and Vector Potential of Magnetic Deflection Systems with Midsection Symmetry

Magnetic deflection systems can only be used as high-performance imaging energy filters if the systems are designed in such a way that the number of aberrations is kept as small as possible. Symmetry properties are a very effective means of reducing the number of aberrations [2.36-38]. For this purpose one aims for systems which possess as many internal symmetries as possible. The imposition of symmetry conditons also facilitates the alignment of the system because deviations from symmetry are easy to detect. Moreover, alignment procedures that utilize symmetry properties are much more accurate than those based on other criteria.

In order to reduce the number of aberrations from the very beginning, we require that the systems have mirror symmetry. In this case, there exists a plane midsection with respect to which the magnetic scalar potential $\psi = \psi(x, y, z)$ is antisymmetric, while the geometry and the arrangement of the polepieces about this plane are symmetric. A particle moving in the midsection will experience no force normal to this section, which we choose as the xz section of the curvilinear xyz coordinate system shown in Fig. 2.6.

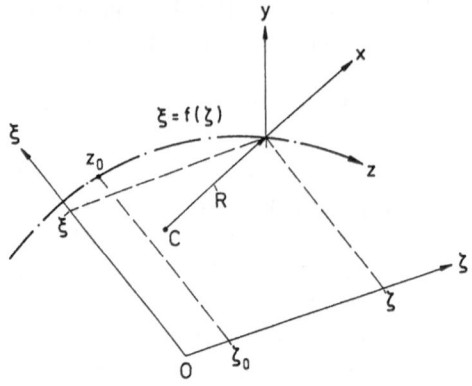

Fig. 2.6. Definition of the curvilinear (x, y, z) coordinate system when the optic axis lies in the plane $y = 0$; C denotes the local centre of curvature of the optic axis. The moving origin of the xy plane along the axis is defined by the curve $\xi = f(\zeta), y = 0$ within the frame of the spatially fixed ξ, y, ζ coordinate system.

The directions of the rectangular coordinates are chosen in accordance with the "right–hand" convention.

For the moment, we assume that the optic z–axis with curvature $\Gamma = \Gamma(z)$ need not necessarily coincide with a particle trajectory. Since the optic axis lies entirely within the plane midsection, the curvature $\Gamma = \bar{\Gamma}$ is real [2.37]. The moving origin $(x = 0, y = 0)$ of the xy plane is described by the curve $\xi = f(\zeta)$ in the spatially fixed ξ, ζ coordinate system located in the plane midsection $(y = 0)$ which contains the curved optic axis.

The metric of the curvilinear coordinate system has the form

$$ds^2 = dw\, d\bar{w} + h^2 dz^2 \tag{2.16}$$

with

$$h = 1 - \Re\{\bar{w}\Gamma\} = 1 - x/R , \tag{2.17}$$

where $R = R(z) = 1/\Gamma(z)$ represents the local radius of curvature of the optic axis; $\Re$ denotes the real part.

2.4.1 Magnetic Scalar Potential

The polepieces of most magnets are made out of iron with very high permeability, which can be assumed infinite with a sufficient degree of accuracy. In this case the pole-faces represent surfaces of constant magnetic scalar potential ψ . In the region between the pole-surfaces, ψ satisfies the Laplace equation

$$\nabla^2 \psi = 0 . \tag{2.18}$$

To obtain insight into the properties of a magnetic deflection system, we must determine the course of the trajectories in the vicinity of the optic axis. For this purpose we expand the magnetic scalar potential in a series of multipole potentials about the optic axis:

$$\psi = \Re \left\{ \sum_{\ell=0}^{\infty} \sum_{m=0}^{\infty} a_{m\ell}(z)(w\bar{w})^{\ell} \bar{w}^m \right\} . \tag{2.19}$$

This expression must satisfy the Laplace equation

$$\nabla^2 \psi = \Re \left\{ \frac{\partial}{\partial w} \left(h \frac{\partial \psi}{\partial \bar{w}} \right) \right\} + \frac{\partial}{\partial z} \left(\frac{1}{h} \frac{\partial \psi}{\partial z} \right) = 0 . \tag{2.20}$$

The complex expansion coefficient

$$a_{mo}(z) = \Psi_{m}(z) = \Psi_{mc}(z) + i\Psi_{ms}(z) \tag{2.21}$$

denotes the strength of the multipole component with multiplicity m. Once these coefficients are known, the remaining coefficients $a_{m\ell}$ with $\ell \neq 0$ may be derived from a recursion formula [2.39]. This formula can be obtained by inserting the power series (2.19) into the Laplace equation (2.20). The resulting expressions for $a_{m\ell}$ with $\ell \neq 0$ are products of the multipole strengths (2.21), the curvature Γ, and their derivatives.

For systems with midsection symmetry the power series expansion (2.19) of the magnetic scalar potential

$$\psi(x, -y, z) = -\psi(x, y, z) \tag{2.22}$$

contains only odd powers of the y coordinate. The structure of the expansion (2.19) reveals that this condition is fulfilled only if the real part of the complex coefficients $a_{m\ell}(z)$ is zero. In this case the strength of the multipole components must be imaginary:

$$\Psi_{m} = -\bar{\Psi}_{m} = i\Psi_{ms} , \quad \Psi_{mc} = 0 . \tag{2.23}$$

The corresponding multipole expansion of the magnetic scalar potential has the form

$$\psi = \Psi_{1s}y + 2\Psi_{2s}xy + \Psi_{3s}(3x^2y - y^3) + \left(2\Gamma\Psi_{2s} - \Psi_{1s}''\right)y(x^2 + y^2)/8 + \dots \tag{2.24}$$

where the dashes indicate differentiation with respect to z.

By inserting this expression into the relation

$$\boldsymbol{B} = -\text{grad}\,\psi \tag{2.25}$$

it follows that the dipole strength

$$\Psi_{1s} = \frac{\partial \psi}{\partial y}\bigg|_{\substack{y=0 \\ x=0}} = -B_y(0, 0, z) = -B(z) \tag{2.26}$$

is identical with the negative magnetic induction along the optic axis. The quadrupole component

$$\psi_2 = 2\Psi_{2s}xy \tag{2.27}$$

of the magnetic scalar potential may result from external quadrupole elements, from the fringing fields of the deflection magnets, or from magnets
with inclined inner pole-faces. Such inhomogeneous deflection magnets are
usually rotationally symmetric about an axis $x_a = -R$, where R is the constant radius of the optic axis within the magnet. As shown in Fig. 2.7 this
axis of rotation need not pass through the point x_c defined by the intersection
of the asymptotes of the conical inner pole-faces. The position of this point,

$$x_c = -R/n_1 = -(D/2)\cot\vartheta, \tag{2.28}$$

is determined by the inclination angle ϑ between the tapered inner poles and
the midsection and by the vertical distance D of the inner pole-surfaces taken
at the optic axis. The so-called field index

$$n_1 = 2(R/D)\tan\vartheta = -2\Psi_{2s}/\eta\Psi_{1s}^2 \tag{2.29}$$

determines the strength Ψ_{2s} of the quadrupole component. This index is constant within the conical magnet. The sector angle Φ of the magnet coincides
with the deflection angle only if we neglect the influence of the finite extension
of the fringing field.

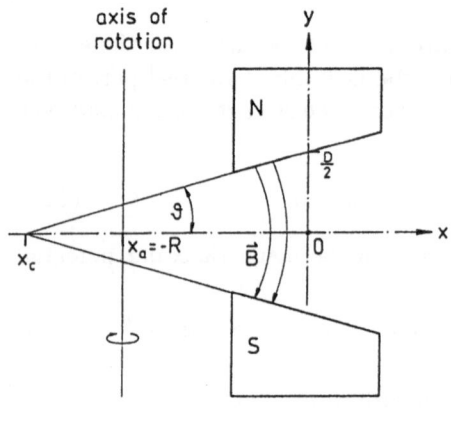

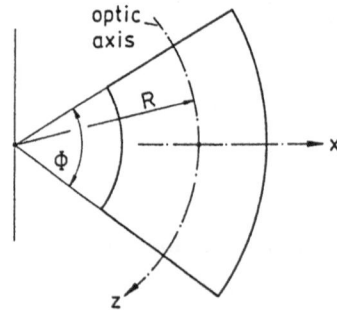

Fig. 2.7. Geometry of a conical sector magnet with tapered poles producing an inhomogeneous magnetic
field.

Inhomogeneous magnets are widely employed in strong focusing particle accelerators; they are less often used in spectrometers where magnets with plane-parallel pole-faces are preferred. All existing imaging energy filters utilize exclusively such homogeneous magnets because they can be more easily machined and more precisely aligned than conical magnets. Owing to this fact, we consider primarily magnets with plane-parallel inner pole-faces, although conical magnets will certainly be used in future ultra-dispersive energy filters [2.41].

Magnets with plane-parallel inner surfaces exert a quadrupole action in the region of the inhomogeneous fringing field if the optic axis intersects the magnetic isoinduction lines $B_y(x, z, y = 0) = const$ at an angle $\pi/2 - \theta$ that differs from 90°. As has been demonstrated previously [2.37], the strength of the fringe quadrupole is related to the derivative of the dipole strength and the local angle $\theta = \theta(z)$ enclosed by the direction of the optic axis and the normal to the magnetic isoinduction lines at a given point $x = 0, y = 0, z$ as follows:

$$\Psi_{2s} = -\frac{1}{2}\Psi_{1s}'(z)\tan\theta(z). \tag{2.30}$$

Accordingly, the quadrupole component of the fringing field about the optic axis vanishes only if this axis is perpendicular ($\theta = 0$) to the isoinduction lines within the entire region of the fringing field.

To determine the hexapole component

$$\psi_3 = \Psi_{3s}(3yx^2 - y^3) \tag{2.31}$$

of the magnetic scalar potential caused by the inhomogeneous fringing field, the local radius of curvature $\varrho = \varrho(z)$ of the isoinduction lines, the angle θ and the curvature $\Gamma = 1/R$ of the optic axis must be known everywhere along this axis. The definition of these quantities is illustrated in Fig. 2.8. The local centres of the radii of curvature ϱ and R are indicated by O_r and C_r, respectively. As outlined in detail in [2.37] the hexapole strength of a deflection magnet with plane-parallel inner pole-faces is given by the relation

$$\Psi_{3s} = \frac{1 + 3\sin^2\theta}{24\cos^3\theta}(\Psi_{1s}''\cos\theta + \Gamma\Psi_{1s}'\sin\theta) - \frac{\Psi_{1s}'\cos 2\theta}{6\varrho\cos^3\theta}. \tag{2.32}$$

Unlike the quadrupole strength (2.30), the hexapole strength Ψ_{3s} does not vanish if the isoinduction lines are straight ($\varrho = \infty$) and $\theta = 0$. The remaining hexapole term $\Psi_{3s} = \Psi_{1s}''/24$ guarantees that the third-order terms of the magnetic scalar potential will be independent of the x coordinate. This condition must be fulfilled because, in the case of infinitely extended straight boundaries, the magnetic scalar potential $\psi = \psi(y, z)$ must be two-dimensional. The last term on the right-hand side of (2.32) stems from the curvature of the isoinduction lines. Their radius of curvature ϱ is negative if the centre of curvature O is located in front of the reference plane z_r. In this case, the isoinduction lines are concave with respect to the incident

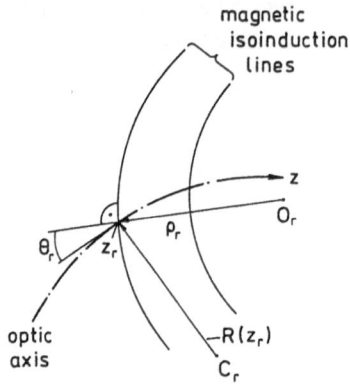

Fig. 2.8. Definition of the angle $\theta_r = \theta(z_r)$ and illustration of the radius of curvature $\rho_r = \rho(z_r)$ of the isoinduction line at an arbitrary reference point $z = z_r$ along the optic axis.

electrons. The curvature of the isoinduction lines can be adjusted by curving the entrance and/or exit faces of the deflection magnets. Since the hexapole strength depends on the derivatives of the magnetic dipole strength, the hexapole field is confined to the entrance and exit regions of the magnets.

Hexapoles are widely used for correcting the parasitic second-rank aberrations since they do not affect the paraxial path of rays. Although curving the boundaries is a well-established method for correcting the axial second-order aberrations of a single sector magnet of a spectrometer, this procedure has been proven impracticable for systems composed of many sector magnets [2.16]. Moreover, the location and the strength of the edge hexapoles are fixed. Owing to these deficiencies independent sextupole elements are increasingly being incorporated into high-performance imaging energy filters. These elements can be centred about the optic axis at arbitrary positions between, in front of, or behind the deflection magnets. An example of such an element is shown in Fig. 2.9.

2.4.2 Series Expansion of the Magnetic Vector Potential

The variational function (2.14) of a purely magnetic deflection system has the form

$$\mu = \sqrt{1 + \kappa^* + \kappa^{*2}(1 - \nu_0^2)/4} \; \sqrt{h^2 + w'\bar{w}'} - \eta \left[hA_z - \Re\{w'\bar{A}\} \right],$$

$$\text{with} \quad \nu_0 = (1 + E_0/m_0c^2)^{-1}. \quad (2.33)$$

Here

$$A = A_x + iA_y \qquad (2.34)$$

denotes the complex off-axial component of the vector potential, and

$$\kappa^* = \frac{1 + E_0/m_0c^2}{1 + E_0/2m_0c^2} \; \kappa \qquad (2.35)$$

Fig. 2.9. Sextupole element for producing an adjustable hexapole field.

is the relativistically modified relative energy deviation of an electron whose energy $E_0 - E$ differs from the nominal energy E_0. Since the variational function forms the basis of the eikonal approach, a suitable form of the vector potential is required. By choosing an appropriate gauge [2.40], one obtains power series for A and A_z, where the expansion coefficients can be expressed in terms of the curvature Γ of the optic axis, the multipole strength $\Psi_m(z)$ and their derivatives. Since these coefficients are experimentally accessible, one obtains a meaningful connection between the expansion coefficients of the vector potential and the geometry and the excitation of the magnets. For determining the optical properties of magnetic energy filters up to the second-rank aberrations, it is sufficient to consider terms up to the second degree inclusively in x and y or w and $\bar{w}$ for the off-axial component $\bar{A}$. Again assuming mirror symmetry, we eventually obtain

$$\bar{A} = \mathrm{i}\,\Psi_{1s}'\bar{w}^2/4 + \dots\ , \tag{2.36}$$

$$\begin{aligned}
h\,A_z &= \Psi_{1s}x - \Gamma\Psi_{1s}x^2/2 + \Psi_{2s}(x^2 - y^2) \\
&+ (\Psi_{3s} - \Gamma\Psi_{2s}/3)(x^3 - 3xy^2) \\
&- (2\Psi_{2s}\Gamma + \Psi_{1s}'')(x^3 + xy^2)/8 + \dots\ .
\end{aligned} \tag{2.37}$$

It should be noted that this representation of the vector potential does not satisfy the standard Coulomb gauge, except in the case of a straight optic axis $(\Gamma = 0)$.

2.4.3 Fringing Fields
of Homogeneous Magnetic Deflection Systems

Sector magnets with plane-parallel inner pole-faces and straight entrance and exit boundaries are extensively used as the deflection elements of magnetic imaging energy filters. The extent of the fringing field of unshielded sector magnets is very large and comparable to the size of the magnet. As a result, the distribution of the fringing field of such magnets is strongly affected by any iron located in this region. To obtain a well-defined fringing field, high-performance magnets are equipped with shielding plates, which cut off the long tail of the fringing field. If the shielding plates are placed at the same distance from the midsection as the inner faces of the magnet, the shielding plates are called mirror plates because they are arranged symmetrically with respect to the pole-plates. Each mirror plate is separated from the boundary of the magnet by a slit which contains the coil for exciting the magnet. The width S of the slit should not be significantly smaller than the distance D between the pole-plates, in order to avoid saturation of the iron at the inner corners of these plates. To obtain a well defined magnetic scalar potential the mirror plates should be connected with the pole-plates of the magnet at some distance from the inner pole-surfaces.

The fringing-field effects are most conveniently described by choosing a ξ, y, ζ coordinate system, which is fixed relative to the straight boundaries of the magnet. The origin of this coordinate system is placed at the midsection midway between the mirror plates and the plates of the magnet. The ξ axis is parallel to the pole-face, and the ζ axis is perpendicular to it.

As shown in Fig. 2.10, the distribution of the fringing field depends significantly on the position of the coil within the groove. In the experiment, a slit with a width $S = 5$ mm and a height of 16 mm was used. The cross-section of the coil was 5 mm × 3 mm and the distance between the inner pole-plates $D = 5$ mm. The magnetic induction

$$B(\zeta) = B_y(\zeta, y = 0) = -\left.\frac{\partial \psi}{\partial y}\right|_{y=0} \qquad (2.38)$$

was measured in the midsection along the ζ axis perpendicular to the straight exit face of the magnet for two positions of the coil. In the first case (B) the coil was placed at the bottom of the groove at a distance $D/2 = 2.5$ mm from the midsection, while in the other experiment (T) the coil was shifted up to the top of the slit. The measured normalized distributions of the fringing field demonstrate convincingly that the position of the coils within the groove must be considered in any realistic fringing-field model.

Schwartz–Christoffel Magnet. As a special case of a homogeneous magnet with mirror plates, we consider an arrangement where the depth of the slits is very large compared with the gap width S and the coils are positioned at the top of the slits. Moreover, we assume that the slits are infinitely extended in the ξ direction so that the magnetic field can be considered as planar depending only on the coordinates y and ζ. In addition, we assume that the pole and mirror plates consist of magnetic material with very high permeability and negligibly small hysteresis effects. Accordingly, their plane

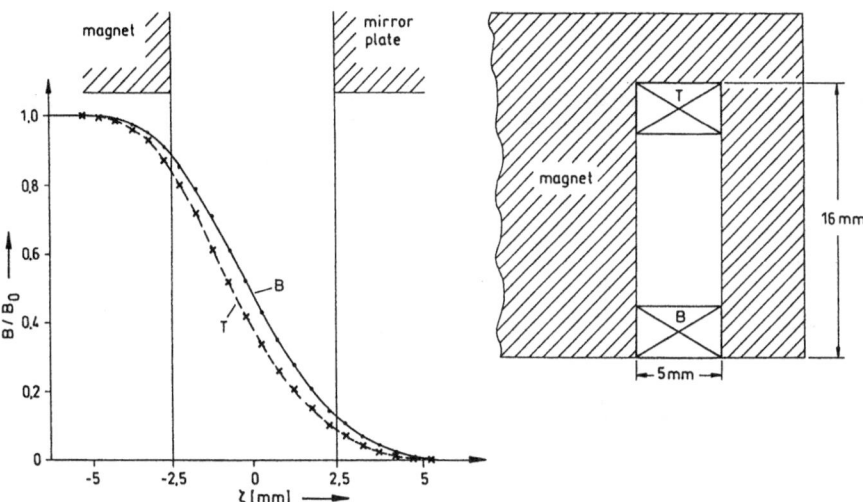

Fig. 2.10. Dependence of the fringing field of a shielded homogeneous magnet on the position of the activating coil within the slit between the pole and mirror plates. The measured curves T and B correspond to the top and bottom positions of the coils, respectively.

boundaries can be regarded with a sufficient degree of accuracy as surfaces of constant magnetic scalar potential. Owing to the required antisymmetry of the magnetic scalar potential, the mirror plates are at potential $\psi = 0$, while the plane-parallel pole-plates of the magnet are at potential $\psi = -\psi_0$ and ψ_0, respectively. For this boundary condition the two-dimensional Laplace equation can be solved by employing the Schwartz–Christoffel formula of conformal mapping. Unfortunately, the solution for the magnetic induction (2.38) along the ζ axis is obtained in an implicit form. Moreover, the different analytic representations of the solution given by *Herzog* [2.42] and *Plies* [2.43] are rather complicated. However, by performing some appropriate algebraic manipulations, both representations can be transformed into the much simpler form

$$B(\zeta) = \frac{B_0}{2} \left(\frac{\tau}{r} + \sqrt{\frac{1 + \tau^2}{1 + r^2}} \right) \ , \ r = S/D \ , \tag{2.39}$$

where the auxiliary variable τ is connected with the ζ coordinate via the relation

$$\pi\zeta/D = r \arctan\tau + \operatorname{artanh}(\tau/r) \ . \tag{2.40}$$

The variable τ can only vary in the range $-r \leq \tau \leq r$, because the relation (2.40) maps the entire ζ axis onto the section $-r \leq \tau \leq r$ of the τ axis.

Within the magnet the induction $B(\zeta)$ approaches the constant value

$$B(\infty) = B_0 = -2\psi_0/D \ , \tag{2.41}$$

at the limit $\zeta \gg S + D$, as depicted schematically in Fig. 2.11. The magnetic induction

$$B(0) = \frac{B_0}{2} \frac{1}{\sqrt{1+r^2}} = -\frac{\psi_0}{\sqrt{S^2+D^2}} \qquad (2.42)$$

at the centre of the gap ($\zeta = \tau = 0$) is always smaller than half the maximum induction B_0.

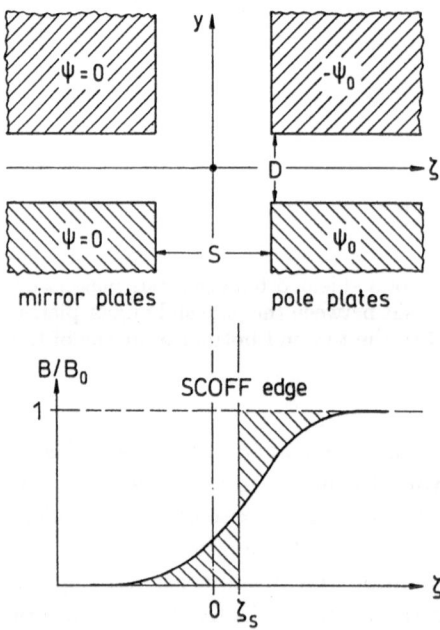

Fig. 2.11. Cross-section through the entrance region of a homogeneous magnet with mirror plates. The SCOFF edge ζ_s equalizes the hatched areas.

It is known from the theory of functions of a complex variable that both the real part and the imaginary part,

$$\chi = \chi(\zeta, y) = \Re\{\Pi\} \;, \quad \psi = \psi(\zeta, y) = \Im\{\Pi\}, \qquad (2.43)$$

of a function $\Pi = \Pi(\zeta + iy)$ are solutions of the two-dimensional Laplace equation. Moreover, χ and ψ are interrelated by the Cauchy–Riemann relations

$$\frac{\partial \chi}{\partial \zeta} = \frac{\partial \psi}{\partial y} \;, \quad \frac{\partial \psi}{\partial \zeta} = -\frac{\partial \chi}{\partial y} \;. \qquad (2.44)$$

Using the definitions (2.38) and (2.43), the first Cauchy–Riemann equation can be written at the midsection $y = 0$ as

$$\frac{\partial \chi(\zeta)}{\partial \zeta} = \frac{\partial \psi}{\partial y}\bigg|_{y=0} = -B(\zeta) = \frac{\partial \Pi(\zeta)}{\partial \zeta} \;, \qquad (2.45)$$

where $\chi(\zeta) = \chi(\zeta, y = 0)$. The last relation can directly be integrated to yield

$$\chi(\zeta) = \Pi(\zeta) = -\int_{-\infty}^{\zeta} B(\zeta)d\zeta = -\int_{-r}^{\tau(\zeta)} B(\tau)\frac{d\zeta}{d\tau}d\tau. \qquad (2.46)$$

The choice of the lower integration limit is arbitrary and represents a particular choice of gauge of the real part of the complex potential. The last integral can be evaluated analytically by inserting the expressions (2.39) and (2.40) into the integrand. After a rather lengthy calculation we eventually arrive at

$$\Pi(\zeta) = \frac{\psi_0}{2\pi}\left[\ln(1+\tau^2) - 2\ln\frac{r-\tau}{2r} + \ln\frac{1+r\tau+\sqrt{1+r^2}\sqrt{1+\tau^2}}{1-r\tau+\sqrt{1+r^2}\sqrt{1+\tau^2}}\right], \qquad (2.47)$$

in which we have made use of the relation (2.41). If we replace the variable τ in (2.40) by the complex variable $\tau + i\sigma$, the resulting relation

$$(\zeta + iy)\pi/D = \frac{r}{2}\left(\arctan\frac{\tau}{1-\sigma} + \arctan\frac{\tau}{1+\sigma}\right) + \frac{1}{4}\ln\frac{(r+\tau)^2+\sigma^2}{(r-\tau)^2+\sigma^2}$$
$$+\frac{i}{2}\left[\arctan\frac{\sigma}{r+\tau} + \arctan\frac{\sigma}{r-\tau} + \frac{r}{2}\ln\frac{(1+\sigma)^2+\tau^2}{(1-\sigma)^2+\tau^2}\right] \qquad (2.48)$$

represents a conformal mapping of the τ–σ plane onto the ζ–y plane. Accordingly, if we substitute in (2.47) the complex variable $\tau + i\sigma$ for τ we obtain together with (2.48) an implicit analytical representation of the two-dimensional complex potential $\Pi(\zeta + iy) = \chi(\zeta, y) + i\psi(\zeta, y)$. Although the separation of the right-hand side of the modified complex equation (2.47) into a real and an imaginary part is very involved, it can be done analytically by utilizing the definitions of arctan x and artanh x.

The sharp cut-off fringing field (SCOFF) approximation replaces the smooth distribution of the magnetic induction along the ζ axis by a step function of equal area. The location ζ_s of the SCOFF edge is defined by the condition

$$\lim_{\zeta\to\infty}\left[B_0(\zeta - \zeta_s) - \int_{-\infty}^{\zeta} B(\zeta)d\zeta\right] = 0 . \qquad (2.49)$$

This condition states that the two hatched areas in Fig. 2.11 are equal. By inserting the relations (2.40) and (2.47) into the expression (2.49) we obtain for the location of the SCOFF edge of a shielded homogeneous magnet the expression

$$\zeta_s = (D/2\pi)\left[2r\arctan r - \ln(1+r^2)\right] . \qquad (2.50)$$

It should be remembered that this formula is based on the assumption that the coils are placed at a large distance from the inner pole surfaces.

To elucidate the meaning of the real part $\chi = \chi(\zeta, y)$ of the complex magnetic potential, we remember that the two-dimensional magnetic field is formed by currents which flow in the positive or negative ξ–direction. Accordingly, the magnetic vector potential can be chosen in such a way that it has only a single component in this direction:

$$\boldsymbol{A} = A_\xi(\zeta, y)\boldsymbol{e}_\xi \ . \tag{2.51}$$

By inserting this expression into the relation

$$\text{curl} \ \boldsymbol{A} = - \ \text{grad} \ \psi \tag{2.52}$$

we obtain the two equations

$$\frac{\partial \psi}{\partial y} = -\frac{\partial A_\xi}{\partial \zeta} \ , \quad \frac{\partial \psi}{\partial \zeta} = \frac{\partial A_\xi}{\partial y} \ . \tag{2.53}$$

These Cauchy–Riemann differential equations become identical with those defined in (2.44) if we replace in (2.53) A_ξ by $-\chi$. Hence the negative value of the real part of the complex potential Π represents the component of the magnetic vector potential:

$$A_\xi(\zeta, y) = -\chi(\zeta, y) \ . \tag{2.54}$$

This important result is valid for arbitrary two-dimensional magnetic fields.

Magnet with Flat Surface Coils. In the preceding section we have treated the limiting case where the coils are placed far from the inner surfaces of the magnet. The other extreme case is realized by placing infinitely thin coils at the bottom of the slits separating the mirror plates from the pole-plates (situation B in Fig. 2.10). In this case, the sheet current and the inner surfaces of the mirror plate and of the magnet are located in the common plane $y = D/2$ and $y = -D/2$, respectively. For a homogeneous sheet current, the magnetic scalar potential increases linearly across the sheet. Accordingly, we call this magnet the "linear potential" (LP) magnet. The boundary conditions for ψ are as follows:

$$\psi(\zeta, y = D/2) = \left\{ \begin{array}{ll} 0 & \text{for} \quad -\infty \leq \zeta \leq -S/2 \\ -\psi_0(2\zeta/S + 1)/2 & \text{for} \quad -S/2 \leq \zeta \leq S/2 \\ -\psi_0 & \text{for} \quad S/2 \leq \infty \end{array} \right\} . \tag{2.55}$$

The boundary condition at the plane $y = -D/2$ is obtained from this expression by replacing ψ_0 by $-\psi_0$. For these boundary conditions, the two-dimensional Laplace equation can be solved most conveniently by representing the solution as a Fourier integral. Considering the antisymmetry of ψ with respect to the midsection $y = 0$, the solution can be written as

$$\psi(\zeta, y) = \frac{1}{2\pi} \int_{-\infty}^{\infty} C(k) \sinh(ky) \exp(ik\zeta) \ dk \ . \tag{2.56}$$

The function $C(k)$ is obtained by taking the Fourier transform of (2.56) with respect to ζ at the boundary $y = D/2$. Using the expression (2.55) for the left-hand side of (2.56), all integrations can be performed analytically. If we insert the resulting expression for $C(k)$ into (2.56), we obtain for the solution the integral representation

$$\psi(\zeta, y) = \frac{i\psi_0}{\pi S} \int_{-\infty}^{\infty} \frac{\exp(ik\zeta)}{k^2} \frac{\sinh(ky) \sin(kS/2)}{\sinh(kD/2)} dk .$$ (2.57)

This integral can be evaluated by evaluating the residues at the zeros of the denominator of the integrand. As a result of the rather lengthy calculation, we find

a) $\zeta \leq -S/2$:

$$\psi = \frac{\psi_0}{r\pi^2} \sum_{n=1}^{\infty} (-1)^n \frac{\exp(2n\pi\zeta/D)}{n^2} \sinh(n\pi r) \sin(2n\pi y/D);$$ (2.58)

b) $-S/2 \leq \zeta \leq S/2$:

$$\psi = -\psi_0 \frac{2\zeta + S}{S} \frac{y}{D} - \frac{\psi_0}{r\pi^2} \sum_{n=1}^{\infty} (-1)^n \frac{\exp(-n\pi r)}{n^2} \sinh(2n\pi\zeta/D) \sin(2n\pi y/D);$$ (2.59)

c) $\zeta \geq S/2$:

$$\psi = -\psi_0 \left[\frac{2y}{D} + \frac{1}{r\pi^2} \sum_{n=1}^{\infty} (-1)^n \frac{\exp(-2n\pi\zeta/D)}{n^2} \sinh(n\pi r) \sin(2n\pi y/D) \right].$$ (2.60)

The series obtained from these expressions for the magnetic induction $B(\zeta) = -\partial\psi/\partial y|_{y=0}$ at the midsection $y = 0$ can be summed. Since each of the three sums yields the same expression

$$B(\zeta) = \frac{B_0}{2\pi r} \ln \left\{ \frac{1 + \exp[\pi(2\zeta + S)/D]}{1 + \exp[\pi(2\zeta - S)/D]} \right\},$$ (2.61)

this formula is valid for the entire range $-\infty \leq \zeta \leq \infty$. It follows directly from this relation that the difference $B(\zeta) - B_0/2$ is antisymmetric with respect to the centre of the slit $\zeta = 0$. Hence, the SCOFF edge for the LP magnet is always located at this plane regardless of the ratio $r = S/D$. This result differs considerably from that of (2.50) obtained for the "Schwartz–Christoffel" (SC) magnet. The difference reflects the influence of the location of the coil within the vertical slit.

The SC magnet and the LP magnet form two limiting cases which are never exactly realized in practice. For a real magnet the curve $B(\zeta)$ will be located somewhere between the corresponding curves for the SC and the LP magnet. The real distribution depends on the location and the cross-section of the coil and on the permeability of the iron. To survey the deviations that can occur, we have plotted in Fig. 2.12 the normalized induction $B(\zeta)/B_0$

for both the SC and the LP magnet in the case $S = D = 5$ mm. The comparison with the corresponding experimental curves depicted in Fig. 2.10 demonstrates that the experimental curves are indeed embedded in the region bordered by the two theoretical extremes. Moreover, the experimental curve B (coil at the bottom of the slit) is fairly well approximated by the corresponding theoretical curve LP despite the fact that the coil used in the experiment was rather thick. The good agreement between the results of the theoretical model and the measurement also holds true for the other limiting case. The theoretical curve SC for the magnetic induction $B(\zeta)$ derived by employing the Schwartz–Christoffel formula approximates sufficiently well the experimental curve T (coil at the top of the slit) for the magnet shown in Fig. 2.10.

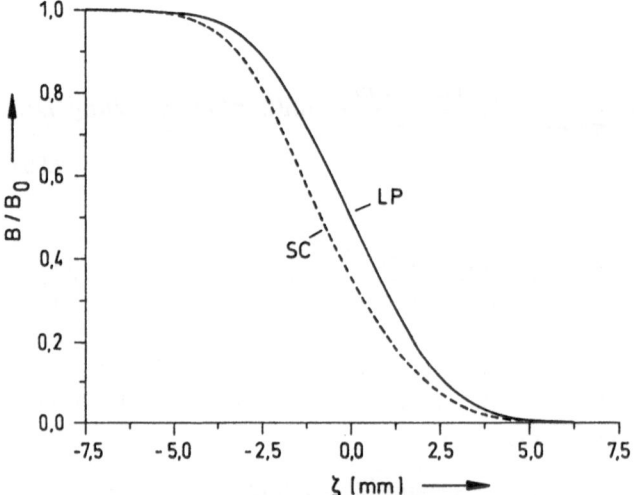

Fig. 2.12. Theoretical fringing field distributions of a shielded homogeneous magnet in the case S = D for (**a**) the Schwartz–Christoffel model (SC) and (**b**) the linear-potential model (LP). The distribution of the fringing field of a real magnet is embedded in the region between these two extremes.

For the actual design of an imaging energy filter it is advantageous to choose magnets which closely resemble one of the two theoretical models because these models possess analytical solutions for the fringing field. This behaviour facilitates considerably the exact calculation of the geometry and the arrangement of the deflection magnets as well as the determination of their optical properties. In practice, one should always measure the distribution $B(\zeta)$ of the magnetic induction at the midsection of each magnet. If the measured field distribution differs appreciably from the distributions of the two model fields, it is advantageous to calculate the effect of the fringing field numerically. The same holds true if other configurations, for example inhomogeneous conical deflection magnets, are employed. Owing to the recent advancement in numerical field calculation the optics of those magnets can be calculated nowadays without an unduly large expenditure of effort.

2.5 Gaussian Optics

The path equations (2.15) form a pair of coupled nonlinear differential equations for the off-axial components x, y of the particle trajectories. These equations can be solved analytically only in a few trivial cases. However, for an efficient analysis of the optical properties of systems comprising many elements and for optimizing such systems, analytical procedures are very desirable. Fortunately, in most electron optical systems the electrons are confined to the vicinity of a central trajectory which may be chosen as the appropriate optic axis. The behaviour of the optical elements are then primarily characterized by the paraxial trajectories. These trajectories are obtained from the linearized path equations which form a pair of second–order linear equations. Owing to this linearity any linear combination of two arbitrary solutions form a possible paraxial trajectory. This remarkable property facilitates significantly the investigation of lens systems such as imaging energy filters.

2.5.1 Paraxial Rays and Dispersion

We restrict our investigations to imaging energy filters with midsection symmetry. To derive the paraxial or Gaussian path equations we expand the variational function (2.33) in a power series with respect to the off-axial position and slope components x, y, x', y' and the chromatic parameter κ^*:

$$\mu = \mu^{(0)} + \mu^{(1)} + \mu^{(2)} + \mu^{(3)} + \cdots , \tag{2.62}$$

where $\mu^{(k)}$ denotes a homogeneous polynomial of degree k in the five expansion quantities. Each polynomial consists of a geometrical part ($\kappa^* = 0$) and a chromatic part:

$$\mu^{(k)} = \mu_g^{(k)} + \mu_c^{(k)} . \tag{2.63}$$

The structure of these polynomials is readily obtained by inserting the expansions (2.36) and (2.37) for the components of the vector potential into the variational function (2.33) and by expanding the remaining part of this function in a power series. We can ignore the polynomial $\mu^{(0)}$ because it does not contain any off-axial components and, hence, does not contribute to the Euler–Lagrange equations (2.15). The chromatic term of the polynomial

$$\mu^{(1)} = \kappa^*/2 - x(\Gamma + \eta\Psi_{1s}) \tag{2.64}$$

likewise does not give a contribution to these equations. If we require that the z axis coincides with a possible ray, the geometrical term must vanish [2.29], resulting in the relation

$$\Gamma = -\eta\Psi_{1s} = -1/R \tag{2.65}$$

for the curvature Γ of the optic axis. On using this relation, the polynomial of second-degree adopts the form

$$\mu^{(2)} = \mu_g^{(2)} + \mu_c^{(2)}, \tag{2.66}$$

$$\mu_g^{(2)} = (x'^2 + y'^2 - \eta^2\Psi_{1s}^2 x^2)/2 - \eta\Psi_{2s}(x^2 - y^2), \tag{2.67}$$

$$\mu_c^{(2)} = -\nu_0^2\kappa^{*2}/8 + \kappa^* x\eta\Psi_{1s}/2 . \tag{2.68}$$

The polynomial (2.66) determines fully the Gaussian optics of the system. By substituting this polynomial for the variational function in the Euler–Lagrange equations (2.15), we obtain the paraxial path equations

$$x'' + (\eta^2\Psi_{1s}^2 + 2\eta\Psi_{2s})x = \kappa^*\eta\Psi_{1s}/2 , \tag{2.69}$$

$$y'' - 2\eta\Psi_{2s} y = 0 . \tag{2.70}$$

The inhomogeneous term of the first equation (2.69) reflects the fact that electrons that initially move along the optic axis are deflected from this axis if their energy deviates from the nominal energy E_0. This behaviour results in a dispersion of the beam. The paraxial ray equations become homogeneous for monochromatic electrons with the nominal energy ($\kappa^* = 0$).

Each of these equations has two linearly independent solutions. Since we restrict our investigations solely to the imaging energy filter and do not consider in detail the optics of the other lenses of the electron optical system, it is advantageous to define the fundamental rays at two distinct planes in front of the filter. In accordance with Fig. 2.13 we require that these planes be the intermediate images z_i and z_d of the object and the diffraction plane, respectively.

As the fundamental rays x_α and y_β, we choose the solutions that intersect the centre of the intermediate image z_i with unit slope:

$$x_\alpha(z_i) = x_{\alpha i} = 0 , \quad y_{\beta i} = 0 ,$$

$$x'_{\alpha i} = 1 , \quad y'_{\beta i} = 1 \tag{2.71}$$

Since these rays originate from the centre of the object plane z_o with slope

$$x'_{\alpha o} = y'_{\beta o} = 1/M_i, \tag{2.72}$$

where M_i is the magnification at the intermediate image z_i in front of the filter, these rays represent the axial fundamental rays up to a factor $1/M_i$. If the image is located within the filter, the conditions (2.71) define the asymptotes of the axial fundamental rays at the virtual image plane.

The two other fundamental rays x_γ and y_δ pass through the centre of the intermediate image z_d of the diffraction plane, which coincides with the back-focal plane of the objective lens in the case of parallel illumination. Accordingly, these rays can be considered as field rays, because they start parallel to the optic axis from an off-axial point of the object. Here we fix these rays at the plane z_d by imposing the conditions

$$x_\gamma(z_d) = x_{\gamma d} = 0 \quad , \quad y_{\delta d} = 0 \ ,$$
$$x'_{\gamma d} = 1 \quad , \quad y'_{\delta d} = 1 \ , \tag{2.73}$$

which differ from the usual initial conditions imposed on the field rays. Owing to this choice of constraint, all the fundamental rays have the dimension of a length.

Accordingly, the constant

$$L = -x_{\alpha d} = x_{\gamma i} = -y_{\beta d} = y_{\delta i} \tag{2.74}$$

of the two Helmholtz formulas (Wronskians)

$$x_\gamma x'_\alpha - x_\alpha x'_\gamma = L \quad , \quad y_\delta y'_\beta - y_\beta y'_\delta = L \tag{2.75}$$

represents a particular length. If the intermediate images are located in the field-free region as depicted in Fig. 2.13, the length L is the distance between the planes z_i and z_d.

Owing to the chosen properties (2.71) and (2.73) of the fundamental rays we define a ray by its slope components

$$\alpha = x'_i \ , \ \beta = y'_i \ , \ \gamma = x'_d \ , \ \delta = y'_d \tag{2.76}$$

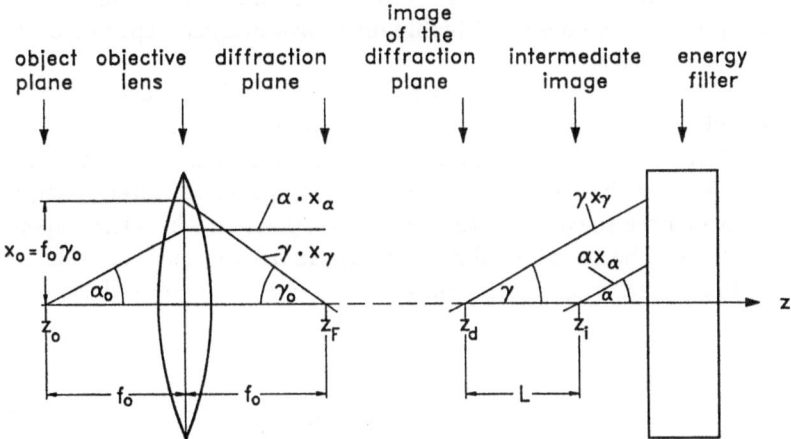

Fig. 2.13. Definition of the paraxial fundamental rays x_α and x_γ and the corresponding ray parameters α and γ. Their connection with the conventional beam defining parameters α_o and x_o at the object plane is shown on the left-hand side.

at the planes z_i and z_d. In this case the off-axial components of the trajectory are described by the linear combinations

$$x = x^{(1)} = \alpha x_\alpha + \gamma x_\gamma + \kappa^* x_\kappa,$$
$$y = y^{(1)} = \beta y_\beta + \delta y_\delta \quad , \qquad (2.77)$$

where the superscript (1) denotes the first-order approximation.

In front of the filter the beam has no dispersion, and the dispersion ray x_κ and its derivative x'_κ vanish. This ray represents up to the factor κ^* the inhomogeneous solution of the equation (2.69) for the x component of the paraxial trajectory. Assuming that the solutions x_α and x_γ are known, the dispersion ray can be calculated readily by employing the method of variation of constants. The result has the form

$$x_\kappa = \left[x_\alpha \int_{-\infty}^{z} \eta \Psi_{1s} x_\gamma dz - x_\gamma \int_{-\infty}^{z} \eta \Psi_{1s} x_\alpha dz \right] / 2L . \qquad (2.78)$$

In the case of stigmatic imaging, the axial fundamental rays x_α and y_β are zero at the final image plane z_I. To guarantee that this image will be free of dispersion, the second integral on the right-hand side of (2.78) must vanish for $z = z_I$. This occurs, for example, if the integrand $\eta \Psi_{1s} x_\alpha$ is antisymmetric with respect to the midplane of the magnetic imaging filter.

Filtering is performed at an image of the diffraction plane located within or outside of the filter because, at this plane, the selection slit does not cause vignetting at the image. The dispersion at the energy-selection plane z_E, where x_γ is zero, is proportional to the area enclosed by the curve $\Psi_{1s} x_\gamma$ and the optic axis. Since Ψ_{1s} and x_γ may change sign in a deflection system composed of many elements, the system must be designed in such a way that a change of the sign of x_γ is accompanied by a reversal of the curvature of the optic axis in order to maximize the dispersion. This important result forms the basic criterion for the design of highly dispersive magnetic spectometers or imaging energy filters.

Achromats are deflection systems which do not introduce a dispersion at any plane behind the filter. To achieve such a complete achromatism both integrals in (2.78) must vanish simultaneously in the region outside the deflection magnets. If such systems are utilized as imaging energy filters, energy selection must be performed inside the system at a position where the dispersion is large [2.38]. The geometrical ray parameters $\alpha, \beta, \gamma,$ and δ are related to the initial parameters $\alpha_o, \beta_o, x_o,$ and y_o at the object plane as illustrated in Fig. 2.13:

$$\alpha_o = M_i \alpha \quad , \quad \beta_o = M_i \beta,$$
$$x_o = f_o M_d \gamma \quad , \quad y_o = f_o M_d \delta. \qquad (2.79)$$

Here f_o is the focal length of the objective lens, and

$$M_d = L/f_o M_i \qquad (2.80)$$

is the (de)magnification of the diffraction plane z_F at the conjugate plane z_d. The expression (2.80) follows from the Helmholtz relation.

2.5.2 SCOFF Approximation

The SCOFF approximation replaces the magnetic dipole strength $\Psi_{1s}(z)$ of a sector magnet by a box-shaped distribution

$$\Psi_{1s} = \left\{ \begin{array}{ll} -B_0 & \text{for} \quad z_1 \leq z \leq z_2 \\ 0 & \text{elsewhere} , \end{array} \right. \tag{2.81}$$

where B_0 is the constant induction in the interior of the magnet. For an inhomogeneous conical magnet the equivalent approximation is also made for the quadrupole strength. With the assumption (2.81) the course of the optic axis is approximated by straight lines in the region outside the effective field boundaries and by a circular arc with curvature $\Gamma_0 = 1/R_0 = -\eta B_0$ inside these boundaries. In a real dipole magnet the curvature Γ of the optic axis changes steadily from zero to its constant value Γ_0 when traversing the extended fringing field, as depicted schematically in Fig. 2.14. As a consequence, the angles $\theta_1 = \theta(z_1)$ and $\pi - \theta_2 = \pi - \theta(z_2)$ taken at the effective field boundaries z_1 and z_2 differ from the asymptotic tilt angles ϵ_1 and ϵ_2 which are enclosed by the initial and final direction of the optic axis and the normal to the entrance and exit pole-face, respectively. These angles are positive if the optic axis outside the magnetic field lies on the same side of the normal as does the centre of curvature C of the optic axis. Neglecting the extent of the fringing field, we have

$$\theta_{\text{in}} = \epsilon_1 = \theta_1 , \quad \theta_{\text{fin}} = \theta_2 = \pi - \epsilon_2 . \tag{2.82}$$

The derivative

$$\Psi'_{1s} = [\delta(z - z_1) - \delta(z - z_2)]\, \Gamma_0/\eta \tag{2.83}$$

of the effective dipole strength (2.81) is composed of two delta functions; one of them is located at the entrance plane z_1 , the other at the exit plane z_2. Accordingly, each edge quadrupole can be considered as a thin lens which deflects the electrons toward the axis in one of its two principal sections and defocuses them in the other section. The focal lengths f_x and f_y of the short edge quadrupole are obtained most conveniently by means of the paraxial path equations. In front of the discontinuity $z < z_1$ the focal ray with off-axial components $x = y = 1$ runs parallel to the optic axis $[x'(-\infty) = y'(-\infty) = 0]$ within the field-free region ($\Psi_{1s} = 0, \Psi_{2s} = 0$). Inserting the expression (2.30) for the quadrupole strength Ψ_{2s} together with the SCOFF relation (2.83) into the paraxial path equations (2.69) and (2.70) and integrating over the discontinuity at z_1, we obtain

$$x'_1 = -y'_1 = -f_{x1} = f_{y1} = \Gamma_0 \tan \epsilon_1 . \tag{2.84}$$

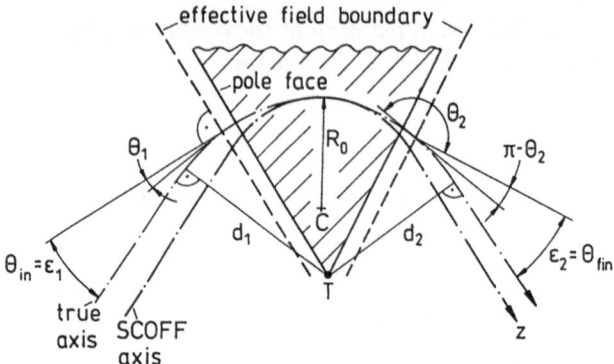

Fig. 2.14. Defining parameters for the true axis and its SCOFF approximation of a homogeneous bending magnet with extended fringing fields.

Within the region $z_1 < z < z_2$ the magnetic field is constant. In this case the paraxial ray equations form a pair of linear second-order differential equations with constant coefficients:

$$R_0^2 x'' + (1 - n_1)x = \kappa^* R_0/2 \quad,$$
$$R_0^2 y'' + n_1 y = 0 \quad. \tag{2.85}$$

The field index n_1 (2.29) is zero for sector magnets with plane-parallel inner surfaces. Hence these magnets focus in the y direction only via their edge quadrupoles. The solution of (2.83) adopts the simple form

$$x = x_1 \cos \Phi + R_0 x_1' \sin \Phi + \kappa^* R_0 (1 - \cos \Phi)/2 \quad,$$
$$y = y_1 + y_1' R_0 \Phi \quad, \tag{2.86}$$

in the case $n_1 = 0$ where

$$R_0 \Phi = z - z_1 \tag{2.87}$$

is the length of the optic axis measured from the entrance plane z_1, and Φ is the deflection angle at the plane z. The slopes x_1' and y_1' taken behind the effective field boundary at the entrance differ from the corresponding slopes x_0' and y_0', respectively, in front of the entrance plane due to the focusing effect of the fringe quadrupole. When an electron traverses the effective field boundary the slope components of its trajectory change abruptly by the amounts

$$x_1' - x_0' = x_1 \Gamma_0 \tan \epsilon_1 \quad,$$
$$y_1 - y_0' = -y_1 \Gamma_0 \tan \epsilon_1 \quad. \tag{2.88}$$

A similar second refraction occurs when the ray passes through the effective field boundary at the exit plane z_2.

The four ray parameters x_1, y_1, x_0' and y_0' in front of the effective field boundary z_1 define the trajectory. Since we have fixed the ray by its slope components α, β and γ, δ at the planes z_i and z_d, respectively, the ray parameters at $z_1 - \epsilon R_0$, $\epsilon \ll 1$, are linear combinations of the initial slope parameters. The effect of the edge quadrupoles and the central homogeneous region of a sector magnet is represented most concisely in terms of transfer matrices. For example, the relation between (x_2, x_2') immediately behind the exit plane z_2 and (x_1, x_0') in front of the effective field boundary z_1 is described by the product of three transfer lenses; two consider the effect of the edge quadrupoles and the third the propagation between the boundary planes. Since the corresponding transfer matrices are listed in most recent textbooks on charged particle optics [2.44,45], there is no need to restate these matrices here.

2.5.3 Extended Fringing Fields

The results obtained from the SCOFF calculations form only a crude approximation and can, therefore, only serve as a rough guide for the design of magnetic imaging energy filters. Experiments have shown that filters designed using the SCOFF data neither produce the required deflection angles Φ_i for a given radius R_{0i} of the i-th magnet nor yield stigmatic first-order imaging. The reason for the substantial differences between the experimental results and the predictions of the SCOFF calculations stems from the fact that for the filter magnets the ratio

$$\delta = D/R_0 \tag{2.89}$$

is of the order of 0.1. For the spectrometers used in nuclear physics, this ratio is one order of magnitude smaller. Owing to the relatively large value of δ, the path length through the fringing field of a filter magnet amounts to more than 10% of the total path through the entire field of the magnet. Hence, to obtain more reliable theoretical data, it is necessary to take into account the extent of the fringing fields accurately. When realistic extended fringing fields are used, we speak of extended fringing field (EFF) calculations. The entrance and exit field regions can be treated separately because the magnetic induction approaches its constant interior value B_0 very rapidly inside the magnet.

The straight boundaries of the magnet enclose the prism angle Φ_p which is connected with the deflection angle Φ of the optic axis and the asymptotic tilt angles ϵ_1 and ϵ_2 via the relation

$$\Phi_p = \Phi - \epsilon_1 - \epsilon_2 \quad . \tag{2.90}$$

For reasons of simplicity we assume that the mirror plates are separated from the entrance and exit pole-faces by the same distance S.

Course of the True Optic Axis. The course of the optic axis in the region of the entrance fringing field is described most suitably by means of the ξ, y, ζ coordinate system introduced in Sect. 2.4.3. For the exit region we choose an equivalent $\tilde{\xi}, y, \tilde{\zeta}$ coordinate system which is fixed relative to the straight exit boundary of the magnet as depicted in Fig. 2.15.

It should be noted that our EFF calculation procedure for determining the precise course of the optic axis and the geometry of the magnetic poles differs from the method developed by *Lanio* [2.12]. The iteration procedure which we propose here is expected to provide significantly faster convergence than the Lanio method because it requires fewer iteration steps. Both methods start from the SCOFF data, which serve as the zeroth-order approximation for the successive iterations.

The steadily increasing curvature of the optic axis within the realistic fringing field in front of the sharp SCOFF edge ζ_S reduces the effective tilt angle $\theta_1 = \theta(\zeta_S)$ with respect to the corresponding asymptotic tilt angle ϵ_1, as illustrated in Fig. 2.14. To obtain $\epsilon_1 = \theta(\zeta = -\infty)$ we need to know the slope angle $\theta = \theta(\zeta)$ of the optic axis as a function of the ζ coordinate. Assuming that $R_0 S$ and $R_0 \gg D$, the entrance and the exit fringing field

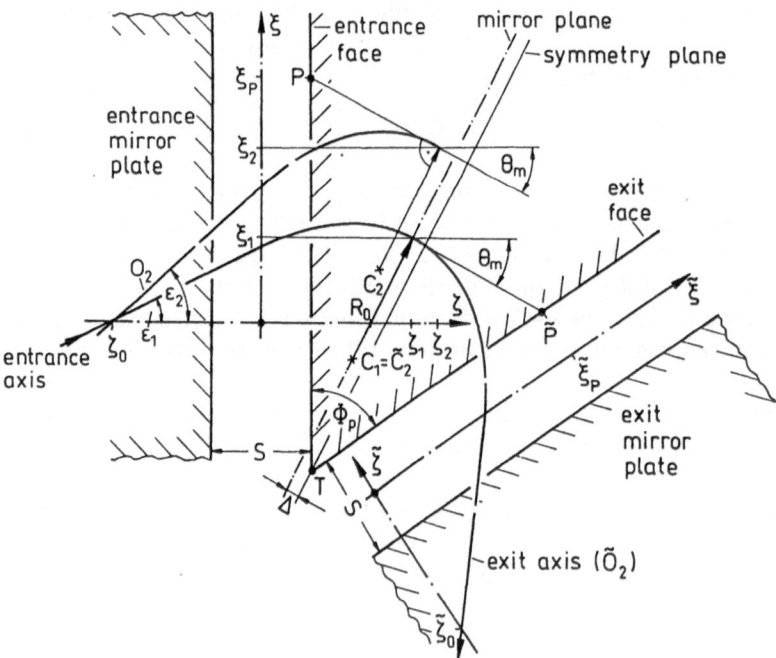

Fig. 2.15. Illustration of the procedure employed for determining the exact location and the true course of the curved optic axis within the midsection of a homogeneous magnet with symmetric shielding of the entrance and exit fringing fields. The quantities defining the optic axis in the exit region are marked by a tilde.

can be considered as planar $(\partial/\partial\xi = 0)$. It follows from Hamilton's canonical equations that in this case the ξ–component of the canonical momentum $p = mv - eA$ is conserved:

$$p_\xi = mv_\xi - eA_\xi = p_{\xi 0} = const. \tag{2.91}$$

In addition, the absolute value of the particle velocity also remains unchanged during the motion of the electron in the static field of the magnet:

$$v = v_0 \quad . \tag{2.92}$$

Accordingly, the relativistic mass, and the kinetic momentum mv of the electron are constant within the magnetic field.

For calculating the course of the optic axis, we assume that this axis intersects the point $\zeta = \zeta_0, y = 0, \xi = 0$ which is located well outside the fringing field as depicted in Fig. 2.15. The choice $\xi = 0$ is arbitrary because the magnetic field does not depend on this coordinate.

Recalling the geometric relation

$$\frac{v_\xi}{v} = \frac{d\xi}{dz} = \sin\theta \quad , \tag{2.93}$$

the expression (2.91) can be written at the mid-section $y = 0$ in the form

$$\sin\theta - \frac{e}{mv_0}A_\xi(\zeta, y = 0) = \sin\theta + w(\zeta)\delta = \sin\theta_0 \quad , \tag{2.94}$$

where $\theta_0 = \theta_{in} = \epsilon_1$ is the initial (asymptotic) slope angle of the optic axis with repect to the normal to the entrance pole-face. The function

$$w(\zeta) = \chi(\zeta)/2\psi_0 = \Pi(\zeta)/2\psi_0 \tag{2.95}$$

represents the normalized real part of the complex magnetic scalar potential (2.46). At the SCOFF boundary $\zeta = \zeta_S$, the effective tilt angle $\theta(\zeta_S) = \theta_1$ satisfies the condition

$$\sin\theta_1 = \sin\epsilon_1 - w(\zeta_S)\delta \quad . \tag{2.96}$$

For the LP field model discussed in Sect. 2.4.3, the SCOFF edge is located at the centre of the slit $(\zeta_S = 0)$. The value $w(0)$ for this field model has been derived by inserting the expression (2.61) into the integrand of (2.46) and perfoming the integration analytically:

$$w(0) = \frac{1}{8\pi^2\rho}\left\{ f\left(\frac{1}{1+e^{2\pi\rho}}\right) - \frac{\pi^2}{12} + \pi^2\rho^2 + \frac{1}{2}\left[\ln\left(1+e^{-2\pi\rho}\right)\right]^2 \right.$$

$$\left. +\pi\rho\ln\left(1+e^{-2\pi\rho}\right)\right\} \quad . \tag{2.97}$$

Here

$$f(x) = -\int_1^x \frac{\ln t}{t-1} dt \qquad (2.98)$$

denotes the dilogarithm function, which is tabulated in [2.46]. The value $\omega(\zeta_S)$ for the SC-model outlined in Sect. 2.4.3 can readily be obtained from formulae (2.47), (2.40) and (2.50). Accordingly, the asymptotic tilt angle ϵ_1 can be determined directly for the two field models for any value ρ, δ and θ_1. The same behaviour holds for the exit fringing field. Since the $\tilde\zeta$ direction points toward the magnet, the formula (2.96) can also be used to determine the asymptotic exit tilt angle ϵ_2 for a given SCOFF tilt angle $\tilde\theta = \pi - \theta_2$, where θ_2 is measured with respect to the z direction, as shown in Fig. 2.14. Equation (2.96) also demonstrates that the difference $\sin\epsilon_1 - \sin\theta_1$ is proportional to the ratio $\delta = D/R_0$, as has been argued previously on physical grounds.

To determine the realistic course of the optic axis in the entrance region of a dipole magnet, we start from the slope

$$\frac{d\xi}{d\zeta} = \tan\theta(\zeta) = \frac{\sin\theta}{\sqrt{1 - \sin^2\theta}} \qquad (2.99)$$

of the axis measured in the ξ–ζ coordinate system. Substituting the expression (5.33) for $\sin\theta$, we obtain

$$\xi = \int_{\zeta_0}^{\zeta} \frac{\sin\theta_0 - \omega(\zeta)\delta}{\sqrt{1 - [\sin\theta_0 - \omega(\zeta)\delta]^2}} \, d\zeta \quad , \qquad (2.100)$$

where we have put $\xi(\zeta_0) = \zeta_0 = 0$. Unfortunately, this integral cannot be evaluated analytically for the fringing field models discussed in Sect. 2.4. However, it can easily be calculated numerically by means of standard integration routines. The starting point $\zeta_0, \xi_0 = 0$ is placed outside of the fringing field, where the starting angle θ_0 coincides with the asymptotic tilt angle ϵ_1. The precise course of the optic axis through a sector magnet is defined by the asymptotic tilt angles ϵ_1 and ϵ_2 , the deflection angle Φ, the radius of curvature R_0 inside the magnet, and the ratios $\rho = S/D$ and $\delta = D/R_0$. In order to adjust the individual magnets of the filter correctly, one must know the exact position of each magnet with respect to the entrance asymptote of the optic axis as well as the position of the exit asymptote. As illustrated in Fig. 2.14, the alignment of the magnets can be performed very conveniently if the vertical distances d_1 and d_2 between the tip T of the magnetic prism and the entrance and the exit asymptote of the optic axis, respectively, are known. Unfortunately these distances can only be determined after the exact course of the beam axis through the entire magnet has been calculated. To obtain this course, we first calculate the entrance branch $\xi = \xi(\zeta)$ of the curved axis from an arbitrary starting point $\xi_0 = 0, \zeta_0$ outside the magnetic fringing field up to the endpoint $\zeta_1, \xi_1 = \xi(\zeta_1)$, where the tilt angle

$$\theta(\zeta_1) = \theta_m = -\Phi_p/2 = (\epsilon_1 + \epsilon_2 - \Phi)/2 \qquad (2.101)$$

between the tangent on the optic axis and the ζ direction is negative and its absolute value amounts to half the prism angle. Inserting (2.101) into (2.94) we obtain the relation

$$\omega(\zeta_1)\delta = \sin\epsilon_1 + \sin\Phi_p/2 \qquad (2.102)$$

which defines the coordinate ζ_1 at the endpoint. Once this coordinate is known, the remaining coordinate ξ_1 can then be derived by evaluating the integral

$$\xi_1 = \int_{\zeta_0}^{\zeta_1} \frac{\sin\epsilon_1 - \omega(\zeta)\delta}{\sqrt{1 - [\sin\epsilon_1 - \omega(\zeta)\delta]^2}} \; d\zeta \quad . \qquad (2.103)$$

In the next step of the calculation we determine the exit branch of the optic axis. For this purpose it is advantageous to determine the course of another optic axis in the entrance fringing field for an asymptotic tilt angle $\theta_{in} = \epsilon_2$, as illustrated in Fig. 2.15. This procedure is permissible because we have assumed that the exit fringing field has the same distribution as the entrance fringing field. For simplicity we start from the same initial point $\zeta = \zeta_0, \xi_0 = 0$. The endpoint $\xi_2 = \xi(\zeta_2), \zeta_2$ is determined by the condition that the slopes of the entrance branch and the exit branch must coincide at the junction ξ_1, ζ_1 :

$$\theta(\zeta_2) = \theta_m = -\Phi_p/2 \quad . \qquad (2.104)$$

Since the asymptotic tilt angle ϵ_2 may differ from ϵ_1, the endpoint ζ_2, ξ_2 will not be located at the same position as the endpoint ξ_1, ζ_1 of the entrance branch. The coordinates ζ_2, ξ_2 are obtained by replacing the index "1" by the index "2" in (2.101) and (2.102).

In order to obtain the entire course of the optic axis and the location of the pole-faces, we shift the branch for the tilt angle ϵ_2 together with the accompanying coordinate system and the entrance boundaries in the ζ direction by the distance $\zeta_1 - \zeta_2$ and in the ξ direction by the distance $\xi_1 - \xi_2$ in such a way that the shifted centre of curvature $\tilde{C}_2$ coincides with the centre of curvature C_1 of the entrance branch at the endpoint ξ_1, ζ_1 . It should be noted that this coincidence occurs only if the magnetic field strengths at the two endpoints are identical. Since both endpoints are located well within the magnet, this requirement is fulfilled with a sufficient degree of accuracy.

The plane that contains the straight line through the centre of curvature $C_1 = \tilde{C}_2$ and the endpoint of the entrance branch forms the mirror plane for the shifted branch. By taking the mirror image of this branch together with that of the shifted boundaries, we obtain the exit branch of the optic axis and the location of the exit boundaries of the magnet and the shielding plate. The mirror image of the shifted $\xi - \zeta$ coordinate system forms the appropriate $\tilde{\xi} - \tilde{\eta}$ coordinate system for the exit fringing field. Accordingly, the directions of both the ζ axis and the $\tilde{\zeta}$ axis point toward the boundary planes of the magnet.

In the case $\epsilon_2 \neq \epsilon_1$, the mirror plane does not coincide with the bisecting line of the prism. The symmetry plane of the prism is displaced from the mirror plane by a distance

$$\Delta = (\zeta_1 - \zeta_2) \tan(\Phi_p/2) \quad . \tag{2.105}$$

The two planes coincide ($\Delta = 0$) only for the symmetric arrangement ($\epsilon_1 = \epsilon_2$) . Since the procedure outlined above completely defines the course of the optic axis with respect to the geometric pole boundaries of the magnet, the vertical distances d_1 and d_2 of the entrance and exit asymptote, respectively, from the tip T of the prism can be determined by means of entirely geometrical considerations. The distance d_1 is given by the expression

$$
\begin{aligned}
d_1 \;=\;& [a_1 + (\zeta_2 - \zeta_1) \tan(\Phi_p/2)] \cos \epsilon_1 \\
+\;& (\zeta_1 - S/2)\cos \left[\frac{(\Phi + \epsilon_1 - \epsilon_2)/2}{\cos(\Phi_p/2)} \right],
\end{aligned} \tag{2.106}
$$

where

$$
\begin{aligned}
a_1 \;=\;& \lim_{\zeta_0 \to -\infty} [(\zeta_1 - \zeta_0) \tan \epsilon_1 - \xi_1] \\
\;=\;& \int_{-\infty}^{\zeta_1} \left\{ \tan \epsilon_1 - \frac{\sin \epsilon_1 - \omega(\zeta)\delta}{\sqrt{1 - [\sin \epsilon_1 - \omega(\zeta)\delta]^2}} \right\} d\zeta . \tag{2.107}
\end{aligned}
$$

The relation for the distance d_2 of the exit asymptote of the optic axis from the tip T of the prism is derived from these expressions by exchanging the indices 1 and 2.

Paraxial Trajectories. The calculation procedure outlined in the preceding section has the advantage of directly yielding the positions of the entrance and exit asymptotes of the optic axis relative to the boundaries of the magnet. However, the finite extent of the fringing field also influences the path of the paraxial rays which determine the imaging properties of the magnet. To take this effect into consideration, the paraxial trajectories must be recalculated along the realistic axis determined in the first step of the iteration procedure. Using the appropriate realistic field model for the extended fringing field, the paraxial rays are obtained by numerical integration of the paraxial path equations (2.69) and (2.70). Since the fringing field is defined in the ζ–y coordinate system, while the trajectories are referred to the curved x, y, z coordinate system, we must determine the position z along the curved optic axis as a function of the ζ coordinate and the $\tilde{\zeta}$ coordinate.

Using the relations

$$\frac{d\zeta}{dz} = \cos \theta \quad , \quad \frac{d\tilde{\zeta}}{dz} = -\cos \tilde{\theta} \quad , \tag{2.108}$$

together with (2.93) and placing the origin $z = 0$ at the endpoint ξ_1, ζ_1 of the entrance branch of the optic axis we obtain the representation

$$z = -\int_{\zeta}^{\zeta_1} \frac{d\zeta}{\sqrt{1 - [\sin\epsilon_1 - \omega(\zeta)\delta]^2}} \qquad \text{for} \quad z < 0 \quad ,$$

$$z = \int_{\tilde{\zeta}}^{\zeta_2} \frac{d\zeta}{\sqrt{1 - [\sin\epsilon_2 - \omega(\zeta)\delta]^2}} \qquad \text{for} \quad z > 0 \quad . \qquad (2.109)$$

In the last integral we have substituted ζ for the variable $\tilde{\zeta}$ of the exit branch. This replacement is permissible because the fringing field of the exit region is the same as that of the entrance region. Unfortunately, the integrals cannot be evaluated analytically for the field models proposed in Sect. 2.4. For efficient numerical calculation of the paraxial trajectories through each dipole magnet, it is advantageous to replace the z coordinate by the ζ coordinate employing the relation [2.47], because the tilt angle θ and the magnetic dipole strengths Ψ_{1s} are given as functions of ζ . The resulting paraxial path equations

$$\ddot{x}\cos^2\theta - \eta(\Psi_{1s}x)^{\cdot}\sin\theta + \eta^2\Psi_{1s}^2 x = \kappa^*\eta\Psi/2 \quad ,$$

$$\ddot{y}\cos^2\theta - \eta\Psi_{1s}\dot{x}\sin\theta + \eta\dot{\Psi}_{1s}x\sin\theta = 0 \qquad (2.110)$$

for the entrance branch ($\zeta \le \zeta_1$) are also valid for the exit branch $\tilde{\zeta} \le \zeta_2$ because the replacement of θ by $\pi - \tilde{\theta}$ does not change the form of the differential equations. Dots indicate differentiation with respect to ζ or $\tilde{\zeta}$, as appropiate. The solutions of (2.110) and (2.109) yield the parametric representation

$$x = x(\zeta) \quad , \quad y = y(\zeta) \quad , \quad z = z(\zeta) \qquad (2.111)$$

for the paraxial trajectories.

These trajectories differ from those obtained initially from the SCOFF calculations. As a result the realistic ray-paths will be astigmatic in the region behind the filter. This first-order misalignment can be eliminated, in principle, by a proper change of the asymptotic tilt angles ϵ_1 and ϵ_2 . Unfortunately, the correct adjustment is not known, because a change of the tilt angles also affects the course of the optic axis. Therefore, a trial-and-error method must be used. Fortunately, the required changes of the tilt angles are small and do not influence appreciably the course of the optic axis. It can be expected that a few iteration steps will suffice to adjust correctly the path of the fundamental rays and the course of the optic axis through a magnet. For systems consisting of several magnets, more iteration steps are necessary because the path of the rays through a given magnet depends on the alignment of these rays within the magnets located in front of the magnet under consideration. In systems with midplane symmetry the expenditure of effort is reduced considerably because only half of the constituent magnets must be "aligned" by the numerical iteration procedure. In the case of several magnets the distances between adjacent magnets are additional free parameters for the alignment. Accordingly, the number of free parameters is generally larger than the number of conditions imposed on the system. Hence, many different arrangements exist which fulfil the paraxial imaging conditions. Unfortunately, one does not generally know in advance which parameters should be varied in order to obtain the fastest convergence with the smallest changes of the parameters.

Although the procedure proposed in this section has not yet been tested, it should converge much more rapidly than the Lanio method, which requires up to 30 iteration steps to determine accurately the geometry and the path of rays for a symmetric system consisting of four magnets when realistic fringing fields are considered [2.12]. Finally, it should be emphasized that the procedure proposed in this section is valid for arbitrary two-dimensional fringing fields. Consequently, planar field distributions obtained by the charge-simulation method or the finite-element method can also be used as well as measured distributions such as those depicted in Fig. 2.10.

2.6 Aberrations

The paraxial approximation describes the imaging properties of electron optical systems for narrow bundles, confined to the vicinity of the optic axis. Owing to the linearity of the paraxial path equations (2.69) and (2.70) any linear combination of two rays forms another possible trajectory.

In high-performance imaging energy filters, the size of the transferred field of view and/or the angular width of the beam exceed the limits of validity of the paraxial approximation. As a result, the nonlinear terms in the general path equations (2.15) can no longer be neglected.

The theory of aberrations is well established. Two different methods of calculating the aberrations are primarily employed, the trajectory method and the eikonal method. The trajectory method, which is best suited for numerical and computer algebraic calculations, starts from the Euler–Lagrange equations (2.15) and considers the nonlinear terms in these equations as perturbations. The eikonal approach is based on Hamilton's characteristic functions and derives the aberrations by means of appropriate perturbation procedures [2.29,35]. The eikonal approach has a distinct advantage because it reveals automatically any interrelations between the various aberration coefficients. These relations are generally not obvious when the trajectory method is adopted. In the following we differentiate between *deviations* and *aberrations*. We define as the path deviation of rank r the polynomial

$$w^{(r)} = x^{(r)}(z) + iy^{(r)}(z) \qquad (2.112)$$

of the power series expansion

$$w - w^{(1)} = \sum_{r=2}^{\infty} w^{(r)}(z) \qquad (2.113)$$

of the true complex trajectory $w = w(\alpha, \beta, \gamma, \delta, \kappa^*; z)$. The polynomial $w^{(r)}$ comprises all terms of rank r in the expansion parameters $\alpha, \beta, \gamma, \delta,$ and κ^* .

The aberration of rank r is the value of $w^{(r)}$ at a distinct plane z , typically the final image plane z_I or the energy-selection plane z_E . The coefficient of a constituent monomial is the aberration coefficient up to a factor that represents the magnification.

A constant eikonal characterizes a fixed wave surface in space. In the field-free region the rays are the orthogonal trajectories of the continuous set of surfaces. In order to achieve ideal stigmatic imaging, an outgoing spherical wave emanating from a specific point of the object must be transformed by the optical system into an incoming spherical wave, whose centre determines the conjugate image point. Unfortunately, no such ideal imaging system actually exists.

In practice, the wave surfaces in the image space depart from the ideal spherical shape, causing the rays to miss their ideal image point, thus creating aberrations. Expressions for the deviations of rank r have been obtained by means of a rather involved iteration procedure [2.29]. Without going into the details of the calculations, we simply state the final result. For magnetic systems with midsection symmetry, the components of the r-th rank deviation can be written as

$$x^{(r)} = \frac{1}{L}\left\{x_\alpha G_\gamma^{(r)} - x_\gamma G_\alpha^{(r)}\right\} ,$$

$$y^{(r)} = \frac{1}{L}\left\{y_\beta G_\delta^{(r)} - y_\delta G_\beta^{(r)}\right\} . \tag{2.114}$$

The components of the r-th rank deviation of the normalized off–axial canonical momentum have the form

$$p_x^{(r)} = \frac{1}{L}\left\{x_\alpha' G_\gamma^{(r)} - x_\gamma' G_\alpha^{(r)}\right\} ,$$

$$p_y^{(r)} = \frac{1}{L}\left\{y_\beta' G_\delta^{(r)} - y_\delta' G_\beta^{(r)}\right\} . \tag{2.115}$$

It should be noted that the r-th rank momentum deviation is not identical with the derivative of the corresponding position deviation (2.114) because the quantities

$$G_\nu^{(r)} = \frac{\partial E^{(r+1)}}{\partial \nu} - \sum_{\ell=2}^{r-1}\left\{p_x^{(\ell)}\frac{\partial x^{(r+1-\ell)}}{\partial \nu} + p_y^{(\ell)}\frac{\partial y^{(r+1-\ell)}}{\partial \nu}\right\} , \tag{2.116}$$

with $\nu = \alpha, \beta, \gamma, \delta$, are functions of z . The expansion terms

$$E^{(r+1)} = \int_{z_0}^{z} m^{(r+1)} \, dz \tag{2.117}$$

of the perturbation eikonal

$$E_p = \sum_{r=2}^{\infty} E^{(r+1)} \tag{2.118}$$

are polynomials of degree $r+1$ in the five expansion parameters $\alpha, \beta, \gamma, \delta$, and κ^*. The integrand $m^{(r+1)}$ can be expressed in terms of the expansion polynomials $\mu^{(k)}$ of the variational function (2.62) and the path deviations $w^{(\ell)} = x^{(\ell)} + iy^{(\ell)}$ up to the rank $r-1$. The "deviation" $\ell = 1$ represents the paraxial trajectory

$$w^{(1)} = x^{(1)} + iy^{(1)} \qquad . \tag{2.119}$$

Using operator notation, we can express the polynomials

$$\tilde{m}^{(r+1)} = m^{(r+1)} + D^{(r)}\mu_1^{(2)} \tag{2.120}$$

in terms of the variational polynomials

$$\mu_1^{(k)} = \mu^{(k)} \left(x^{(1)}, y^{(1)}, x^{(1)'}, y^{(1)'}; z \right) \tag{2.121}$$

and the operators

$$D^{(h)} = x^{(h)} \frac{\partial}{\partial x_1^{(1)}} + y^{(h)} \frac{\partial}{\partial y^{(1)}} + x^{(h)'} \frac{\partial}{\partial x^{(1)'}} + y^{(h)'} \frac{\partial}{\partial y^{(1)'}} \tag{2.122}$$

as

$$\tilde{m}^{(r+1)} = \frac{1}{(r+1)!} \left\{ \frac{\partial^{r+1}}{\partial \beta^{r+1}} \sum_{k=2}^{r+1} \beta^k \exp\left[\sum_{h=2}^{\infty} \beta^{h-1} D^{(h)} \right] \mu_1^{(k)} \right\}_{\beta=0} \qquad . \tag{2.123}$$

The operator $D^{(h)}$ replaces one of each of the paraxial ray components in (2.121) by the corresponding components of the h-th rank deviation. The polynomials $\tilde{m}^{(\ell+1)}$ satisfy the integral relation

$$\int_{z_o}^{z} D^{(h)} \tilde{m}^{(\ell+1)} \mathrm{d}z = p_x^{(\ell)} x^{(h)} + p_y^{(\ell)} y^{(h)} \qquad . \tag{2.124}$$

Owing to this behaviour, different representations exist for the higher-rank aberrations. Employing (2.122) together with (2.120), we obtain for the first three variational polynomials $m^{(r+1)}$ of the perturbation eikonal (2.118) the expressions:

$$\begin{aligned}
m^{(2)} &= 0 \quad , \\
m^{(3)} &= \mu_1^{(3)} \quad , \\
m^{(4)} &= \mu_1^{(4)} + D^{(2)}\mu_1^{(3)}/2 + D^{(2)}\tilde{m}^{(3)}/2 \\
&= \mu_1^{(4)} - \mu_{g2}^{(2)} + D^{(2)}\tilde{m}^{(3)} \quad .
\end{aligned} \tag{2.125}$$

In the last equation we have made use of (2.120) and the identity

$$\mu_{g2}^{(2)} = \mu_g^{(2)}\left(w^{(2)}, w^{(2)'}, z\right) \quad = \quad D^{(2)^2}\mu_1^{(2)}/2 \ . \qquad (2.126)$$

Here we have used the fact that the chromatic term $\mu_c^{(2)}$ of $\mu_1^{(2)}$ depends only linearly on the position coordinates. The first term of each expression on the right-hand side of (2.125) describes the contribution that is obtained in the first step of the iteration procedure.

2.6.1 Primary Deviations

The primary deviations are obtained in the first step of the iteration procedure. Accordingly, they are completely defined by the first non-vanishing expansion polynomial $m^{(3)} = \mu_1^{(3)}$ of the perturbation eikonal. For magnetic deflection systems with midsection symmetry the cubic term of the variational function (2.62) has the form

$$\begin{aligned}
\mu^{(3)} \quad = \quad & \Gamma x(x'^2 + y'^2)/2 + \eta\Psi_{1s}'(x^2 x' + 3x'y^2 - 2xyy')/8 \\
& - (\eta\Psi_{3s} - \eta\Psi_{2s}\Gamma/3)(x^3 - 3xy^2) + \eta\Psi_{2s}\Gamma x(x^2 + y^2)/4 \\
& + \kappa^*(x'^2 + y'^2)/4 - \nu_0^2\kappa^{*2}\Gamma x/8 \ . \qquad (2.127)
\end{aligned}$$

This cubic term produces a quadratic nonlinearity in the Euler–Lagrange equations. For $r = 2$ the perturbation function $G_\nu^{(r)}$ (2.116) adopts the simple form

$$G_\nu^{(2)} = \frac{\partial E^{(3)}}{\partial\nu} \ , \qquad (2.128)$$

where

$$E^{(3)} = \int_{z_o}^z \mu^{(3)}\left(x^{(1)}, y^{(1)}, x^{(1)'}, y^{(1)'}; z\right) dz \qquad (2.129)$$

represents the well known primary aberration eikonal, which is a polynomial of rank three in the expansion parameters. Ordering with respect to these parameters gives

$$\begin{aligned}
E^{(3)} \quad = \quad & \left(A_{\alpha\alpha\alpha}\alpha^3 + 3A_{\alpha\alpha\gamma}\alpha^2\gamma + 3A_{\alpha\gamma\gamma}\alpha\gamma^2 + A_{\gamma\gamma\gamma}\gamma^3\right)/3 \\
& + \left(B_{\alpha\beta\beta}\alpha\beta^2 + 2B_{\alpha\beta\delta}\alpha\beta\delta + B_{\alpha\delta\delta}\alpha\delta^2\right)/2 \\
& + \left(B_{\gamma\beta\beta}\gamma\beta^2 + 2B_{\gamma\beta\delta}\beta\gamma\delta + B_{\gamma\delta\delta}\gamma\delta^2\right)/2 \\
& + \left(C_{\alpha\alpha\kappa}\alpha^2\kappa^* + 2C_{\alpha\gamma\kappa}\alpha\gamma\kappa^* + C_{\gamma\gamma\kappa}\gamma^2\kappa^*\right)/2 \\
& + \left(C_{\beta\beta\kappa}\beta^2\kappa^* + 2C_{\beta\delta\kappa}\beta\delta\kappa^* + C_{\delta\delta\kappa}\delta^2\kappa^*\right)/2 \\
& + C_{\alpha\kappa\kappa}\alpha\kappa^{*2} + C_{\gamma\kappa\kappa}\gamma\kappa^{*2} \ . \qquad (2.130)
\end{aligned}$$

The eikonal coefficients $A_{\mu\nu\sigma}$, $B_{\mu\nu\sigma}$ and $C_{\mu\nu\sigma}$ are functions of the z coordinate. They determine the variation of the cubic deformation of the wave surface along the optic axis. It follows from (2.127) and (2.129) that the eikonal coefficients vary only in the region of the magnetic dipole and/or hexapole fields. In the other regions the coefficients are constant. Hence, if

both the energy selection plane and the final image plane are located outside
the imaging energy filter, the aberrations at these planes are governed by the
same eikonal. If we insert (2.130) and (2.128) into the expressions (2.114)
for the path deviations and represent the result as a sum of second-degree
monomials in the expansion parameters, the second-rank deviations adopt
the perspicuous form

$$
x^{(2)} = \alpha^2 x_{\alpha\alpha} + \beta^2 x_{\beta\beta} + \alpha\gamma x_{\alpha\gamma} + \beta\delta x_{\beta\delta} + \gamma^2 x_{\gamma\gamma}
$$
$$
+ \delta^2 x_{\delta\delta} + \alpha\kappa^* x_{\alpha\kappa} + \gamma\kappa^* x_{\gamma\kappa} + \kappa^{*2} x_{\kappa\kappa} \quad , \tag{2.131}
$$

$$
y^{(2)} = \alpha\beta y_{\alpha\beta} + \alpha\delta y_{\alpha\delta} + \beta\gamma y_{\beta\gamma} + \gamma\delta y_{\gamma\delta}
$$
$$
+ \beta\kappa^* y_{\beta\kappa} + \delta\kappa^* y_{\delta\kappa} \quad . \tag{2.132}
$$

The "coefficients" $x_{\mu\nu} = x_{\mu\nu}(z)$ and $y_{\mu\nu} = y_{\mu\nu}(z)$ can be considered as
the second-rank fundamental rays in analogy with the representation (2.77)
of the paraxial approximation in terms of the first-rank fundamental rays
$x_\alpha, y_\beta, x_\gamma, y_\delta$, and x_κ . Generally fifteeen complex second-rank fundamental
rays

$$
w_{\mu\nu} = x_{\mu\nu} + i\, y_{\mu\nu} \tag{2.133}
$$

exist, resulting in 30 components. Owing to the required midsection symmetry 15 components are zero. Moreover, the x component (2.131) does not
possess mixed terms containing ray parameters belonging to both the x and
the y components of the paraxial ray (2.77). The opposite behaviour holds
for the y component (2.132), which contains exclusively mixed terms. Each
second-rank fundamental ray fulfils the initial conditions

$$
x_{\mu\nu}(z_o) = 0 \quad , \quad x'_{\mu\nu}(z_o) = 0 \quad ,
$$
$$
y_{\mu\nu}(z_o) = 0 \quad , \quad y'_{\mu\nu}(z_o) = 0 \quad . \tag{2.134}
$$

Hence, the second-rank deviation $w^{(2)} = x^{(2)} + iy^{(2)}$ originates within the
deflection system with zero slope in the same way as the dispersion ray x_κ.

The second-rank fundamental rays have the peculiar structure

$$
x_{\mu\nu} = \frac{2 - \delta_{\mu\nu}}{2L} \frac{\partial^2}{\partial\mu\partial\nu} \left[\frac{\partial x^{(1)}}{\partial\alpha} \frac{\partial E^{(3)}}{\partial\gamma} - \frac{\partial x^{(1)}}{\partial\gamma} \frac{\partial E^{(3)}}{\partial\alpha} \right] \quad ,
$$
$$
y_{\mu\nu} = \frac{1}{L} \frac{\partial^2}{\partial\mu\partial\nu} \left[\frac{\partial y^{(1)}}{\partial\beta} \frac{\partial E^{(3)}}{\partial\delta} - \frac{\partial y^{(1)}}{\partial\delta} \frac{\partial E^{(3)}}{\partial\beta} \right] \quad , \tag{2.135}
$$

which reveals the symplectic nature of the trajectories [2.47]; $\delta_{\mu\nu}$ is the well
known Kronecker symbol ($\delta_{\mu\nu}$=1 for $\mu=\nu$, $\delta_{\mu\nu} = 0$ for $\mu \neq \nu$). Each second-
rank fundamental ray $x_{\mu\nu}$ represents up to the factor $(2-\delta_{\mu\nu})/2L$ the second
derivative of the Poisson bracket

$$
\left[x^{(1)}, E^{(3)} \right] = \frac{\partial x^{(1)}}{\partial\alpha} \frac{\partial E^{(3)}}{\partial\gamma} - \frac{\partial x^{(1)}}{\partial\gamma} \frac{\partial E^{(3)}}{\partial\alpha} \tag{2.136}
$$

with respect to two distinct ray parameters μ and ν , where $\mu, \nu = \alpha, \beta, \gamma, \delta,$ and κ^* . The same behaviour holds for $y_{\mu\nu}$ except at the different factor $1/2$. Inserting the expression (2.130) for $E^{(3)}$ into (2.135), expressions for the second-rank fundamental rays in the form of linear combinations of the eikonal coefficients are found:

$$
\begin{aligned}
Lx_{\mu\nu} &= (2 - \delta_{\mu\nu})(x_\alpha A_{\mu\nu\gamma} - x_\gamma A_{\alpha\mu\nu}) , & \mu, \nu &= \alpha, \gamma; \\
Lx_{\sigma\tau} &= (x_\alpha B_{\gamma\sigma\tau} - x_\gamma B_{\alpha\sigma\tau})/(1 + \delta_{\sigma\tau}), & \sigma, \tau &= \beta, \delta; \\
Lx_{\mu\kappa} &= x_\alpha C_{\gamma\mu\kappa} - x_\gamma C_{\alpha\mu\kappa} & \mu &= \alpha, \gamma, \kappa; \quad (2.137)
\end{aligned}
$$

$$
\begin{aligned}
Ly_{\mu\sigma} &= y_\beta B_{\mu\sigma\delta} - y_\delta B_{\mu\beta\sigma} , & \mu = \alpha, \gamma, \sigma = \beta, \delta; \\
Ly_{\sigma\kappa} &= y_\beta C_{\sigma\delta\kappa} - y_\delta C_{\beta\sigma\kappa} , & \sigma = \beta, \delta . \quad (2.138)
\end{aligned}
$$

These eikonal coefficients are obtained most readily by substituting (2.127) into the integrand of the integral (2.129) and expressing the curvature Γ by means of (2.65). As a result we obtain

$$
\begin{aligned}
A_{\lambda\mu\nu} = -\frac{\eta}{2} \int_{z_0}^{z} \Big[& (6\Psi_{3s} + 7\eta\Psi_{1s}\Psi_{2s}/2) \, x_\lambda x_\mu x_\nu \\
& + \Psi_{1s} \left(x'_\lambda x'_\mu x_\nu + x'_\lambda x_\mu x'_\nu + x_\lambda x'_\mu x'_\nu \right) \\
& + \Psi'_{1s} \left(x_\lambda x_\mu x'_\nu + x_\lambda x_\nu x'_\mu + x_\mu x_\nu x'_\lambda \right)/4 \Big] \mathrm{d}z, \quad (2.139)
\end{aligned}
$$

$$
\begin{aligned}
B_{\mu\sigma\tau} = -\eta \int_{z_0}^{z} \Big[& \Psi_{1s} x_\mu y'_\sigma y'_\tau + \Psi'_{1s} \left(3x'_\mu y_\sigma y_\tau - x_\mu y_\sigma y'_\tau - x_\mu y'_\sigma y_\tau \right)/4 \\
& - (6\Psi_{3s} + 3\eta\Psi_{1s}\Psi_{2s}/2) \, x_\mu y_\sigma y_\tau \Big] \mathrm{d}z , \quad (2.140)
\end{aligned}
$$

$$
C_{\nu\varrho\kappa} = 2A_{\kappa\varrho\nu} + \frac{1}{2} \int_{z_0}^{z} x'_\nu x'_\varrho \, \mathrm{d}z , \quad (2.141)
$$

$$
C_{\sigma\tau\kappa} = B_{\kappa\sigma\tau} + \frac{1}{2} \int_{z_0}^{z} y'_\sigma y'_\tau \, \mathrm{d}z , \quad (2.142)
$$

$$
C_{\nu\kappa\kappa} = A_{\nu\kappa\kappa} + \frac{1}{2} \int_{z_0}^{z} x'_\nu x'_\kappa \, \mathrm{d}z - \frac{3\nu_0^2}{8} \int_{z_0}^{z} \eta\Psi_{1s} x_\nu \, \mathrm{d}z \quad (2.143)
$$

with $\lambda, \mu = \alpha, \gamma, \kappa$; $\nu, \varrho = \alpha, \gamma$; $\sigma, \tau = \beta, \delta$.

For homogeneous deflection magnets, Ψ_{2s} describes the strength (2.30) of the fringing field quadrupole, while Ψ_{3s} represents the sum of the strengths of the edge hexapole (2.32) and of the actual sextupole elements centred along the axis. The resulting formulae listed in [2.37] are especially suited for the SCOFF approximation because they can be evaluated analytically in this case [2.14].

2.6.2 Second-Rank Aberrations

The lateral aberrations describe the off-axial displacement of the trajectories from their ideal Gaussian points at the observation plane. In an energy-filtering electron microscope this plane is either the final image plane z_I or the energy-selection plane z_E, which coincides with an image of the diffraction plane. Since at each of these planes two of the four paraxial fundamental rays $x_\alpha, y_\beta, x_\gamma$, and y_δ vanish, the second-rank (primary) aberrations adopt a rather simple form. At the final image plane $z = z_I$ we obtain from (2.114) and (2.128) the expressions

$$x^{(2)}(z_I) = x_I^{(2)} = -\frac{x_{\gamma I}}{L}\frac{\partial E_I^{(3)}}{\partial\alpha} \quad ,$$

$$y_I^{(2)} = -\frac{y_{\delta I}}{L}\frac{\partial E_I^{(3)}}{\partial\beta} \quad , \tag{2.144}$$

where $E_I^{(3)} = E^{(3)}(z_I)$ is the third-rank perturbation eikonal at the image plane z_I.

The ratio

$$M_I = |x_{\gamma I}/L| = |y_{\delta I}/L| \tag{2.145}$$

represents the magnification at the final image plane of the intermediate image z_i located in front of the filter. Hence M_I is the magnification of the projection system behind the filter. If we refer the image aberrations back to the object plane, the relations (2.144) must be divided by the total magnification

$$M = M_i \, M_I \tag{2.146}$$

of the object at the final image.

At the energy selection plane z_E the fundamental field rays x_γ and y_δ are zero, and we obtain the second-rank aberrations as

$$x_E^{(2)} = \frac{x_{\alpha E}}{L}\frac{\partial E_E^{(3)}}{\partial\gamma} \quad , \quad y_E^{(2)} = \frac{y_{\beta E}}{L}\frac{\partial E_E^{(3)}}{\partial\delta} \quad , \tag{2.147}$$

where

$$|x_{\alpha E}/L| = |y_{\beta E}/L| = M_E \tag{2.148}$$

is the magnification of the diffraction plane z_d at the energy selection plane z_E. For symmetric filters we have $M_E = 1$ if the energy selection is performed at the first conjugate plane in front of the lenses of the projection system. The primary perturbation eikonal $E_{\mathrm{E}}^{(3)} = E^{(3)}(z_{\mathrm{E}})$ at the energy selection plane coincides with $E^{(3)}(z_{\mathrm{I}})$ at the image plane z_{I} if both planes are located behind the filter. Consequently, the aberrations at these planes are interrelated.

Since there exist 15 second-rank fundamental rays, which are functions of the z coordinate, 15 second-rank aberrations are produced at each of the two planes, resulting in 30 aberration coefficients. Owing to the constant value of the eikonal $E^{(3)}$ behind the filter, 12 linear relations exist between the 30 coefficients because $E^{(3)}$ is composed of only 18 monomials (2.130). The different contribution of the 18 eikonal terms to the aberrations at the planes z_{E} and z_{I} follows directly from the relations (2.137) and (2.138) for the second-rank fundamental rays. The ten coefficients $A_{\lambda\mu\nu}, B_{\lambda\mu\nu}$ of the eikonal (6.19) are of entirely geometric nature and determine the second-order (geometrical) aberrations. The remaining 8 coefficients $C_{\lambda\mu\nu}$ describe the second-rank chromatic aberrations composed of the familiar first-order and first-degree chromatic aberrations and the second-degree (quadratic) dispersion. For the discussion of the aberrations at the different planes it is advantageous to start from the second-rank deviations (2.121) and (2.122). These expressions demonstrate that both the geometric aberrations and the chromatic aberrations can be subdivided into three characteristic groups according to their dependence on the ray parameters. It follows from (2.137) and (2.138) that the second-rank fundamental rays are directly proportional to a specific eikonal coefficient at both the image plane ($x_{\alpha\mathrm{I}} = y_{\beta\mathrm{I}} = 0$) and the energy-selection plane ($x_{\gamma\mathrm{E}} = y_{\delta\mathrm{E}} = 0$). As a result, a distinct eikonal coefficient can be the coefficient of two different types of aberrations, one observed at the image plane and the other at the energy-selection plane. This behaviour is illustrated in Table 1. At the image plane, the geometric aberrations that depend exclusively on the aperture parameters α and β represent the second-order aperture aberrations. The mixed geometric aberrations are bilinear in α and γ and in β and δ, respectively. These aberrations can be classified as "inclination of the image field" and as "second-order field astigmatism". The terms that contain only the field parameters γ and δ produce the second-order distortions.

On going from the image plane z_{I} to the energy–selection plane z_{E}, the meaning of the geometric ray parameters must be exchanged, because an image of the diffraction plane is located at z_{E}. Hence γ and δ form in this case the initial axial slope components, while α and β determine the field of view of the diffraction pattern. The same behaviour holds for the chromatic aberrations, as demonstrated in Table 2.1. The chromatic aberrations are subdivided into the familiar axial chromatic aberration, the chromatic dependence of magnification, and the second-degree dispersion. The integral representations (2.141–143) of the chromatic aberration coefficents $C_{\lambda\mu\nu}$ demonstrate that all imaging elements located between the object plane z_{o} and the observation plane contribute to the second-rank chromatic aberrations. This fact becomes obvious if we reshape by partial integrations the integrals whose

Table 2.1. Influence of the primary eikonal coefficients on the second-rank aberrations at the image and energy-selection plane, respectively.

IMAGE PLANE						
	geometric aberrations			chromatic aberrations		
type	aperture aberration	distortion	mixed aberration	axial chromatic aberration	energy-dependent magnification	second-degree dispersion
coefficient	$A_{\alpha\alpha\alpha}$ $B_{\alpha\beta\beta}$	$A_{\alpha\gamma\gamma}$ $B_{\beta\delta\gamma}$ $B_{\alpha\delta\delta}$	$A_{\alpha\alpha\gamma}$ $B_{\alpha\beta\delta}$ $B_{\beta\beta\gamma}$	$C_{\alpha\alpha\kappa}$ $C_{\beta\beta\kappa}$	$C_{\alpha\gamma\kappa}$ $C_{\beta\delta\kappa}$	$C_{\alpha\kappa\kappa}$
ENERGY-SELECTION PLANE						
	geometric aberrations			chromatic aberrations		
type	aperture aberration	distortion	mixed aberration	axial chromatic aberration	energy-dependent magnification	second-degree dispersion
coefficient	$A_{\gamma\gamma\gamma}$ $B_{\gamma\delta\delta}$	$A_{\alpha\alpha\gamma}$ $B_{\alpha\beta\delta}$ $B_{\beta\beta\gamma}$	$A_{\alpha\gamma\gamma}$ $B_{\beta\delta\gamma}$ $B_{\alpha\delta\delta}$	$C_{\gamma\gamma\kappa}$ $C_{\delta\delta\kappa}$	$C_{\alpha\gamma\kappa}$ $C_{\beta\delta\kappa}$	$C_{\gamma\kappa\kappa}$

integrands are bilinear in the derivatives of the paraxial fundamental rays and recall that the paraxial path equations in the region outside the filter are those of round lenses or quadrupoles. As a result, the second-degree dispersion consists of two contributions, one originating within the filter, the other resulting from the combination of the first-degree dispersion with the chromatic aberration of the round lenses located between the filter and the image plane. This combination aberration is not present at the energy selection plane z_E, since this is situated in front of the projection system. However, strong chromatic combination aberrations may occur if the energy spectrum is transferred to another plane. This occurs, for example, if a magnified image of the energy spectrum is matched to the CCD array of a parallel detector for spectrum-imaging.

The eikonal terms that depend solely on the field parameters γ and δ determine the second-order axial aberration at the energy-selection plane. To elucidate the structure of this aberration, we consider an annular grid centred on the optic axis at the object plane and assume parallel illumination. In the absence of the aberration, the annular beamlets intersect the centre of the energy dispersion plane. If the second-order axial aberration does not vanish, each annular beamlet forms an ellipse at the energy selection plane as shown

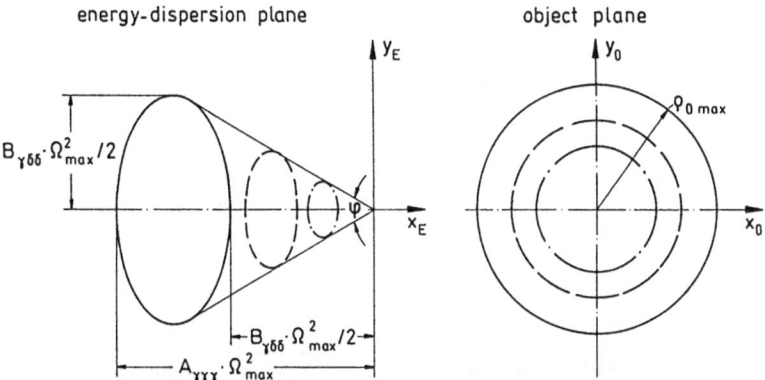

Fig. 2.16. Elliptic axial coma produced by an uncorrected imaging energy filter at the energy-selection plane z_E. In the case of parallel illumination, each circle about the centre of the object plane forms an ellipse at z_E.

in Fig. 2.16. The centre of this ellipse is displaced in the x direction by the distance

$$\Delta x = (A_{\gamma\gamma\gamma}/2 + B_{\gamma\delta\delta}/4)\,\Omega^2 \qquad (2.149)$$

from the optic axis. The angle

$$\Omega = \left(\delta^2 + \gamma^2\right)^{1/2} = M_i \varrho_o / L \qquad (2.150)$$

is proportional to the radius ϱ_o of the annulus at the object plane and to the magnification M_i of the intermediate image in front of the energy filter. For different values of ϱ_o we obtain a family of shifted ellipses. The size and the displacement of each ellipse increases quadratically with ϱ_o or Ω, respectively. As illustrated in Fig. 2.16, the aberration figure resulting from many beamlets resembles the well-known coma of round lenses with the difference that the spots are ellipses instead of circles. Owing to this behaviour, we refer to the second-order axial aberration as axial elliptic coma. The tangents to the ellipses intersect each other at the optic axis at an angle

$$\varphi = 2\arctan\sqrt{\frac{B_{\gamma\delta\delta}}{2A_{\gamma\gamma\gamma}}}\;. \qquad (2.151)$$

This relation indicates that the aberration has a coma-like shape only if the eikonal coefficients $B_{\gamma\delta\delta}$ and $A_{\gamma\gamma\gamma}$ have the same sign. This is the case for all uncorrected imaging energy filters investigated so far. For $A_{\gamma\gamma\gamma} = 3B_{\gamma\delta\delta}/2$ the ellipses degenerate to circles, and the elliptic coma adopts the shape of the familiar round-lens coma. It should be noted that it is always possible to express the axial elliptic coma as a conventional axial coma together with a threefold axial astigmatism [2.36].

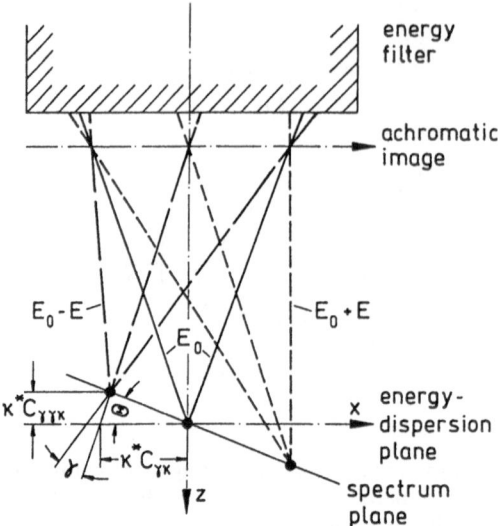

energy
filter

achromatic
image

$E_0 - E$ $E_0 + E$

E_0

$\kappa^* C_{\gamma\gamma\kappa}$

$-\kappa^* C_{\gamma\kappa}$

x energy-
dispersion
plane

spectrum
plane

Fig. 2.17. Schematic illustration of the combined action of dispersion and axial chromatic aberration, resulting in a tilt of the spectrum plane.

For parallel recording of the energy loss spectrum with high-energy resolution, the spectrum plane must be perpendicular to the optic axis. As illustrated in Fig. 2.17 the combined effect of the axial chromatic aberration and the dispersion at the energy-selection plane z_E causes a tilt of the spectrum plane with respect to the plane $z_E = \text{const}$. The dispersion $\kappa^* C_{\gamma\kappa}$ at the energy-selection plane causes a lateral shift of the rays, which is proportional to κ^* and to the dispersion coefficient

$$C_{\gamma\kappa} = \int_{z_o}^{z_E} \eta \Psi_{1s} x_\gamma dz / 2 \quad . \tag{2.152}$$

Owing to the axial aberration, the trajectories of a homocentric bundle of rays with a relative energy loss κ do not intersect each other in a single point at the energy-selection plane but form instead two line foci located above or beneath this plane at distances $\Delta z_x = \kappa^* C_{\gamma\gamma\kappa}$ and $\Delta z_y = \kappa^* C_{\delta\delta\kappa}$, respectively. Since the energy resolution does not depend on the focusing properties perpendicular to the direction of the dispersion, the spectrum is focused in a plane that is rotated through the angle

$$\Theta = \arctan(C_{\gamma\gamma\kappa}/C_{\gamma\kappa}) \tag{2.153}$$

about the y_E axis of the energy-selection plane. If the spectrum is transferred with magnification M_s to another plane, the tilt angle increases by the factor M_s. Unfortunately, spectrum-imaging with a CCD array always requires magnification of the spectrum. Hence, to obtain high energy resolution, the correction of $C_{\gamma\gamma\kappa}$ is mandatory in this case.

The axial aberration at the energy-selection plane prevents isochromatic energy filtering. As a consequence, the selected energy depends on the position of the individual object elements. To understand this unfortunate be-

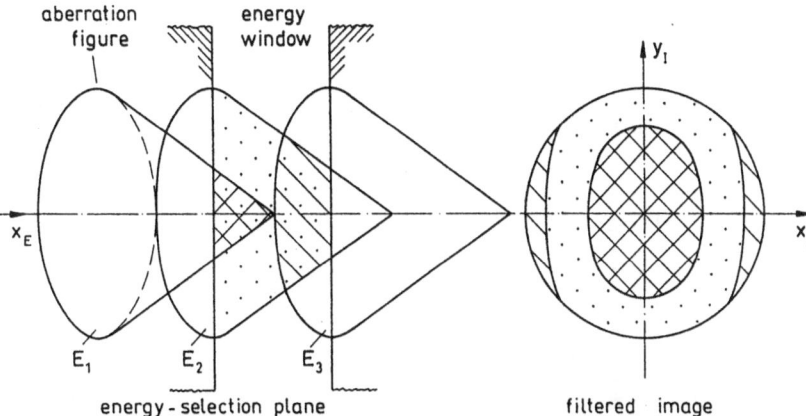

Fig. 2.18. Illustration of the effect of non-isochromatic energy selection caused by the axial second-order aberrations at the energy-selection plane.

haviour, we consider a simple energy loss spectrum consisting of three characteristic energy losses $E_1, E_2 = 2E_1$ and $E_3 = 3E_1$; the dispersion constant $C_{\gamma\kappa}$ is $(E_0/2E_1)A_{\gamma\gamma\gamma}\Omega_{max}^2$, and $A_{\gamma\gamma\gamma} = B_{\gamma\delta\delta}$. The resulting aberration figures at the energy-selection plane are depicted on the left-hand side of Fig. 2.18 together with the energy-selection slit. The width of this energy window has been chosen to be $A_{\gamma\gamma\gamma}\Omega_{max}^2/2$.

Electrons that intersect an edge of the slit originate at the object from an elliptic ring centred on the optic axis. The corresponding filtered image shown on the right-hand side of Fig. 2.18 consists of three regions. The central region is formed by the electrons with energy loss $E = E_1$, the intermediate region bordered by two ellipses and the circular edge of the field aperture is produced by the electrons with $E = E_2$, while the outer two sickle-shaped areas are exclusively formed by the electrons with energy loss E_3. As a consequence of this behaviour, material with the characteristic energy loss E_1 will only become visible in the inner region, although it may be present in the outer region of the object. This example demonstrates rather convincingly that the axial aberration at the energy selection plane must either be eliminated or sufficiently suppressed by reducing the field of view in order to obtain reliable elemental information about the object area being imaged.

We now discuss the effect of the second-rank aberrations at the final image plane. Imaging energy filters are usually placed in front of the projector lens system, where the field rays are far from the axis due to the high magnification M_i, while the axial rays are strongly confined to the central trajectory of each beamlet originating from a distinct point of the object. The axial aberrations of the filter can hence be neglected, whereas the distortions are of major concern, as is the case for the projector lens of a conventional TEM.

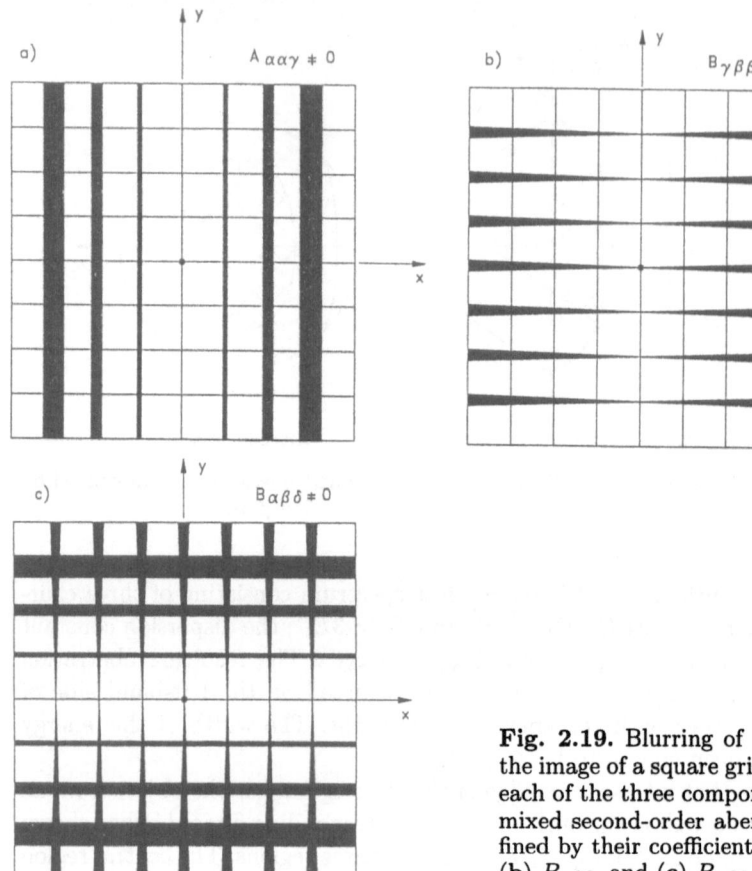

Fig. 2.19. Blurring of the lines in the image of a square grid caused by each of the three components of the mixed second-order aberrations defined by their coefficients (a) $A_{\alpha\alpha\gamma}$, (b) $B_{\gamma\beta\beta}$, and (c) $B_{\alpha\beta\delta}$.

The second-order distortions referred back to the object plane

$$x_D^{(2)} = \left(\gamma^2 A_{\alpha\gamma\gamma} + \delta^2 B_{\alpha\delta\delta}/2\right)/M_i \quad ,$$
$$y_D^{(2)} = \gamma\delta B_{\gamma\beta\delta}/M_i \tag{2.154}$$

have three components with coefficients $A_{\alpha\gamma\gamma}$, $B_{\alpha\delta\delta}$ and $B_{\gamma\beta\delta}$. These distortions do not affect the resolution but merely destroy the linearity between the object and the image.

To survey the influence of the individual distortion components, we consider the image of a square grid which is oriented in the object plane parallel to the coordinate axes. The component $\gamma^2 A_{\alpha\gamma\gamma}/M_i$ deforms each square mesh into a rectangle with quadratically increasing length in the x direction. The aberration $\delta^2 B_{\alpha\delta\delta}/2M_i$ results in a parabolic curvature of the grid lines x_o = const. , while the trapezium distortion $\gamma\delta B_{\gamma\beta\delta}/M_i$ deforms the square grid into a trapezoid [2.26].

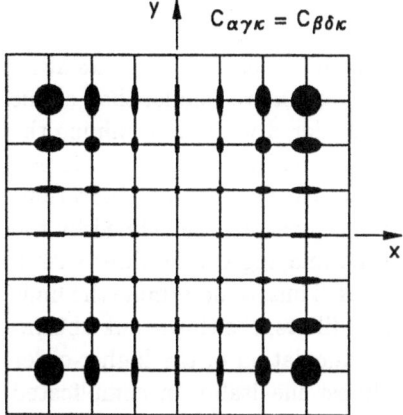

Fig. 2.20. Effect of the chromatic aberration of magnification on the image of a perforated opaque foil with a square lattice of holes.

The mixed second-order field aberrations referred back to the object plane

$$x_F^{(2)} = (2\alpha\gamma A_{\alpha\alpha\gamma} + \beta\delta B_{\alpha\beta\delta})/M_i \quad ,$$

$$y_F^{(2)} = (\alpha\delta B_{\alpha\beta\delta} + \beta\gamma B_{\gamma\beta\beta})/M_i \qquad (2.155)$$

blur the image. The size of the blurred image spot increases linearly with the distance from the axis. These aberrations reduce primarily the resolution of the outer image points and, hence, limit the field of view. The effect of the three aberration components is convincingly depicted in Fig. 2.19, which shows the blurring of the meshes in the image of a square grid. For $B_{\alpha\beta\delta} = B_{\gamma\beta\beta} = 0$ only the lines of the grid running in the y direction are blurred (Fig. 2.19a), while in the case $A_{\alpha\alpha\gamma} = B_{\alpha\beta\delta} = 0$ only the x lines are broadened (Fig. 2.19b). A blurring in both directions occurs for $A_{\alpha\alpha\gamma} = B_{\gamma\beta\beta} = 0$ since the eikonal coefficient $B_{\alpha\beta\delta}$ affects both directions. By employing the complex representation, the mixed field aberrations can be subdivided into the "inclination of the image field" and the second-order field astigmatism [2.36].

The chromatic aberrations also affect the imaging properties of the imaging filter. Although the first-degree dispersion vanishes at the image plane this is generally not the case for the dispersion of second degree. Fortunately, this aberration is negligibly small for the energy windows used for zero-loss imaging. The axial chromatic aberration of the filter can also be neglected because its contribution to this aberration is inversely proportional to the square of the intermediate magnification M_i. This behaviour differs from that of the chromatic aberration of magnification. There, the contributions of the constituent elements of the microscope add up independently of the respective magnifications. The effect of this aberration is illustrated in Fig. 2.20, which shows the image of a perforated opaque foil with a square lattice of holes for the special case $C_{\alpha\gamma\kappa} = C_{\beta\delta\kappa}$. Since these coefficients are usually significantly larger than those of the round lenses, the chromatic aberration of magnification of the filter must be eliminated.

2.6.3 Secondary (Third-Rank) Aberrations

The performance of an electron optical instrument is determined by its aberrations. Because the beam parameters $\alpha, \beta, \gamma, \delta$ and κ are smaller than about 0.01, only the primary aberrations need be considered when determining the quality of an uncorrected system. To improve the performance of the system, one must correct the limiting primary aberrations. Once these aberrations have been eliminated, the efficiency of the instrument will be defined by the next-higher-rank aberrations. In the case of an imaging energy filter with a curved axis, these are the third-rank aberrations. Thus, to determine the limit of the performance of corrected imaging energy filters, the third-rank aberrations must be calculated. Unfortunately, the calculation of the higher-order aberrations is rather lengthy and results almost inevitably in complicated and involved formulae. Nevertheless, a very compact representation has been obtained by adopting a Lie-algebraic approach. For this, we must choose the form

$$m^{(4)} = m^{*(4)} + D^{(2)} \tilde{m}^{(3)}/2 \quad , \tag{2.156}$$

for the fourth-rank variational function (2.125) of the perturbation eikonal $E^{(4)}$. The function

$$m^{*(4)} = \mu_1^{(4)} + D^{(2)} \mu_1^{(3)}/2 \tag{2.157}$$

is the integrand of the modified fourth-rank perturbation eikonal

$$E^{*(4)} = \int_{z_o}^{z} m^{*(4)} \, dz \quad . \tag{2.158}$$

Using this representation and the relation (2.124), the functions $G_\nu^{(3)}$, with $\nu = \alpha, \beta, \gamma, \delta$ can be written as

$$G_\nu^{(3)} = \frac{\partial E^{*(4)}}{\partial \nu} - \left[E^{(3)}, \frac{\partial E^{(3)}}{\partial \nu} \right] / 2L \quad . \tag{2.159}$$

Here the bracket

$$[f, g] = \frac{\partial f}{\partial \alpha} \frac{\partial g}{\partial \gamma} - \frac{\partial f}{\partial \gamma} \frac{\partial g}{\partial \alpha} + \frac{\partial f}{\partial \beta} \frac{\partial g}{\partial \delta} - \frac{\partial f}{\partial \delta} \frac{\partial g}{\partial \beta} \tag{2.160}$$

denotes the familiar Poisson bracket operation of classical mechanics. Inserting (2.159) into (2.114) and performing some algebraic manipulations, the third-rank deviations eventually reduce to the concise Lie-algebraic form

$$
\begin{aligned}
x^{(3)} &= \left[x^{(1)}, E^{*(4)} \right] /L + \left[\left[x^{(1)}, E^{(3)} \right], E^{(3)} \right] /2L^2 \quad , \\
y^{(3)} &= \left[y^{(1)}, E^{*(4)} \right] /L + \left[\left[y^{(1)}, E^{(3)} \right], E^{(3)} \right] /2L^2 \quad .
\end{aligned} \tag{2.161}
$$

These path deviations are valid for an arbitrary reference plane z. Accordingly, by choosing $z = z_E$ and $z = z_I$, we obtain immediately the third-rank aberrations at the energy-selection plane and the final image plane, respectively. The second term on the right-hand side of each of the two equations (2.161) represents the "concatenation rule" for the combination of the second-rank deviations. These terms produce the dominant part of the third-rank combination aberrations. As a result, each third-rank aberration coefficient is composed of a fourth-rank eikonal coefficient and products of any two third-rank eikonal coefficients. Therefore the relations existing between the higher-order aberration constants are generally nonlinear. They become linear only if the lower-rank aberrations have been corrected. If the energy-selection plane is located outside the magnetic field of the filter, the contributions of the filter to the aberrations at the image plane and at the energy-selection plane are governed by the same eikonals, $E_F^{*(4)}$ and $E^{(3)}$, because both are constant behind the filter. However, we must bear in mind that $E_F^{*(4)}$ represents only the part of the fourth-rank perturbation eikonal

$$E^{*(4)} = E_F^{*(4)} + E_R^{*(4)} \tag{2.162}$$

that is produced within the filter. The round lenses of the microscope add the part $E_R^{*(4)}$, which represents the third-order geometrical aberrations and the chromatic aberrations produced by the first-degree dispersion and the aberrations of the round lenses located behind the filter.

Canonical representations of the kind (2.161) also exist for the fourth- and higher-rank path deviations. Finally, it should be noted that the elegance and simplicity of the relations (2.161) clearly refute the many objections which have been raised concerning the usefulness and power of the eikonal method.

2.7 Correction of Aberrations

Unlike the ideal electron optical system, real systems are always affected by first-order aberrations caused by misalignment, mechanical inaccuracies, and inhomogeneities of materials. Since these parasitic first-order aberrations limit most severely the performance of the instrument, they must be eliminated first. To correct the first-order distortions, the imaging energy filter must be aligned very precisely with respect to the optic axis of the electron microscope. In addition, at least one quadrupole stigmator must be incorporated to compensate for the axial astigmatism. A detailed description of the first-order alignment will be given in the following sections.

Assuming perfect first-order alignment, the performance of the system is limited by the second-rank aberrations. An imaging energy filter that is suitable for a routine instrument must be designed in such a way that the number of aberrations is reduced as much as possible from the very beginning without any correctors. The number of correctors required should be kept as small as possible. The correctors must be placed at appropriate positions along the axis in order to prevent the correction of the primary aberrations from introducing large secondary (third-rank) aberrations, which would hinder any appreciable improvement of the instrumental performance. An effective correction can be achieved only if the fundamental paraxial rays differ substantially at the planes of the correctors. To meet this constraint, a strongly astigmatic path of the fundamental paraxial rays is necessary in the regions between the deflection elements. Therefore one must design the filter in such a way that it possesses appropriate paraxial ray-paths. In the ideal case, all the correctors are decoupled, so that each corrector affects only a single, distinct eikonal term. Unfortunately, this is not always possible. One must then seek an arrangement that suppresses as far as possible "cross talk" between the individual correctors and hence results in small strengths of the multipole correctors. The slopes of the secondary fundamental rays will then be changed only moderately by the correcting elements. Accordingly, the second-rank deviations can be eliminated behind the filter without unduly enlarging the off-axial distances of these rays within the filter. As a result the correction of the primary aberrations does not increase the higher-rank aberrations beyond a tolerable limit.

2.7.1 Aberration Correction by Imposing Symmetry Conditions

The number of aberrations that appear at the final image plane and the energy-selection plane behind the deflection system can be significantly reduced from the very beginning by introducing symmetry planes in addition to the required midsection symmetry. These planes must be symmetry planes for the fundamental paraxial rays and the multipole fields. Only in this case do some of the integrands of the aberration integrals (2.139–143) become antisymmetric functions, whereupon the corresponding aberration coefficients vanish.

Usually one imposes a single midplane symmetry. In this case two fundamental rays are symmetric, while the other two are antisymmetric about the central plane, as shown in Fig. 2.21 for the Berlin Omega filter. The strengths Ψ_{1s}, Ψ_{2s}, and Ψ_{3s} of the dipole, quadrupole, and hexapole components of the magnetic field are symmetric with respect to the midplane z_m. Accordingly, the integrands of the geometrical aberration coefficients $A_{\alpha\alpha\alpha}$, $A_{\alpha\gamma\gamma}$, $B_{\alpha\beta\beta}$, $B_{\alpha\delta\delta}$, and $B_{\gamma\delta\delta}$ are antisymmetric, whereas the integrands of $A_{\alpha\alpha\gamma}$, $A_{\gamma\gamma\gamma}$, $B_{\alpha\beta\beta}$, $B_{\gamma\beta\beta}$, and $B_{\gamma\delta\delta}$ are symmetric. This behaviour results from the fact that the fundamental rays are linearly independent. Thus only two rays, one in the x-z the other in the y-z section, can exhibit the same symmetry about a given plane. The other two rays show the opposite symmetry behaviour.

Introduction of a single symmetry can, therefore, only eliminate half of the second-order (geometrical) aberrations, for example the aperture aberrations and the distortions. By introducing an additional symmetry plane for each half of the bending system, the integrands that are symmetric with respect

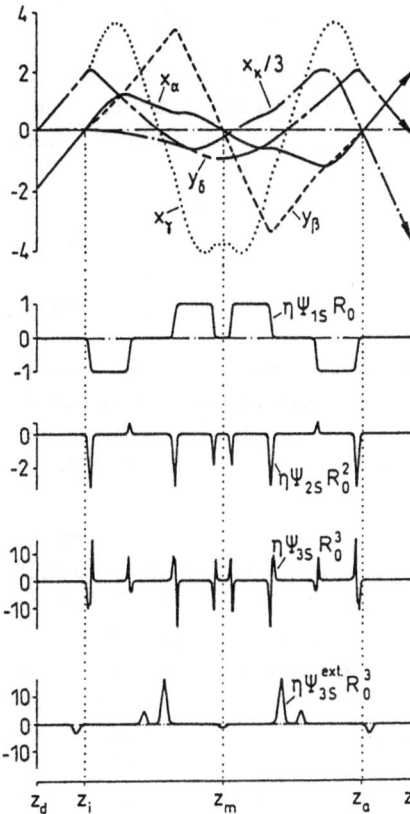

Fig. 2.21. Course of the fundamental paraxial rays and distribution of the normalized dipole, quadrupole, and sextupole strengths along the straightened optic axis of the corrected Ω filter. The curve Ψ_{3s} describes the strength of the internal hexapole field produced by the fringing fields, while the strength of the external adjustable sextupoles is represented by Ψ_{3s}^{ext}.

to z_m can be made antisymmetric with respect to the central plane of each half of the system. For these two-fold-symmetric systems, all second-order aberrations cancel out as well as the chromatic aberration of magnification. Unfortunately, the dispersion ray x_κ also vanishes outside this achromatic system. Such imaging filters therefore require internal energy selection. Because achromatic systems do not produce a spectrum outside the filter, spectrum imaging is not possible without changing the magnetic excitation of the filter elements. Owing to this shortcoming, such systems have not been used as imaging energy filters so far. Since all second-order aberrations are absent from these twofold symmetric systems, the geometrical second-rank deviations must vanish outside these filters. Owing to the required double symmetry of the magnetic field and the paraxial rays, the geometrical second-order fundamental rays are either symmetric or antisymmetric with respect to the midplane z_m. The same holds true for the dispersion.

2.7.2 Correction of Aberrations by Means of Sextupoles

In the preceding section we have shown that it is not possible to correct all aberrations in the energy-selection plane by means of symmetry conditions without cancelling the dispersion at this plane as well. In order to eliminate the disturbing axial aberrations at the energy-selection plane, it therefore is essential to introduce additional hexapole fields. Hexapole fields generated by curving the entrance and/or exit pole-faces of the magnets are not suitable in systems composed of several deflection magnets because they create chaotic behaviour during the alignment of the paraxial ray-paths. For such systems, adjustable sextupole elements are mandatory.

However, one cannot always correct arbitrary aberration components by means of the sextupoles. For example, it is not possible to correct the axial coma by means of a sextupole placed in the region where the paraxial ray-paths are rotationally symmetric. This situation arises, for example, in systems composed of $n_1 = 1/2$ bending magnets.

Correction of the Geometrical Second-Order Aberrations. Simultaneous correction of all remaining second-order aberrations requires a strongly astigmatic path of the paraxial rays in the drift spaces between the deflection magnets. Unfortunately, the individual aberration components cannot in general be corrected independently. Since this coupling hampers the correction procedure, one aims for arrangements in which the correction of the aberration components is as decoupled as possible. Most aberration components, for example the coefficients $A_{\alpha\alpha\gamma}$, $B_{\alpha\beta\delta}$, and $B_{\gamma\beta\beta}$ of the image tilt and the field astigmatism, can be eliminated largely independently if astigmatic images of both the object plane and the diffraction plane are situated in the drift spaces between the constituent deflection magnets of the filter. Since half of the geometrical second-rank aberrations have been cancelled out by symmetry, it is necessary to incorporate the sextupoles in pairs in such a way that the two sextupoles of each pair are placed symmetrically about the central symmetry plane. In this case the sextupoles will not introduce distortion and aperture aberrations regardless of their position. A sextupole centred at the midplane need not be split, since it automatically fulfils the symmetry condition. As an example we assume that the lines of the astigmatic images of the object and the diffraction plane are perpendicular to the plane-parallel inner pole-faces. The sextupoles of the first pair are placed at a position where the fundamental rays x_α and x_γ are both much larger than the other two rays x_β and y_δ. This pair compensates for the coefficient $A_{\alpha\alpha\gamma}$, thereby slightly altering the coefficients $B_{\alpha\beta\delta}$ and $B_{\gamma\beta\beta}$. In the next step the coefficient $B_{\gamma\beta\beta}$ is eliminated by means of the second pair, whose elements are situated at conjugate line images ($x_\alpha = 0$) of the object plane, without affecting the first correction. The sextupoles of the third pair are placed at conjugate vertical astigmatic images of the diffraction plane ($x_\gamma = 0$). They eliminate the remaining coefficient $B_{\alpha\beta\delta}$ without disturbing the preceding corrections. The first and second corrections introduce axial aberrations at the energy-selection plane. These aberrations must hence be compensated in the last step of the correction procedure in such a way that no other aberrations are introduced. This requirement can be met if a strongly first-order-distorted stigmatic image of the object plane is formed at the central plane of the filter. In this case we need three sextupoles, one pair and a single one which is placed at the midplane, to eliminate the remaining geometrical coefficents $A_{\gamma\gamma\gamma}$ and $B_{\gamma\delta\delta}$. The two other sextupoles are put in front of and behind the filter at conjugate distortion-free stigmatic images of the object. At these planes $x_\gamma = y_\delta$ while at the central plane $x_\gamma \neq y_\delta$. The smaller the ratio x_γ/y_δ at this plane, the less the correction of the coefficient $B_{\gamma\delta\delta}$ will affect the coefficient $A_{\gamma\gamma\gamma}$. Although the correction of these coefficients is slightly coupled, it has the great advantage that the elimination does not introduce any other second-order aberration.

For the example shown in Fig. 2.21, this correction scheme has been employed except that the coefficients of the mixed aberrations have not been compensated independently. Instead, the external sextupoles have been placed optimally such that the three coefficients $A_{\alpha\alpha\gamma}$, $B_{\alpha\beta\delta}$, and $B_{\gamma\beta\beta}$ are eliminated by only two sextupole pairs. The precise location of theses sextupoles is indicated in Fig. 2.21 by the course of the corresponding normalized sextupole strength $\eta\Psi_{3s}^{ext}R_0^3$ along the straightened filter axis. The distribution of the strength of the edge sextupoles is given by the curve $\eta\Psi_{3s}R_0^3$.

The correction of the mixed second-order aberrations alters the axial aberrations at the energy-selection plane. Therefore, one should first cancel the second-order aberrations in the image plane and subsequently the remaining geometrical aberrations at the energy-selection plane. To survey the effect of these corrections we have plotted in Fig. 2.22 the secondary geometrical fundamental rays $x_{\mu\nu}$ and $y_{\mu\nu}$ of the filter proposed in Fig. 2.21 before and after the correction by the external sextupoles. Without this correction the course of all secondary fundamental rays is entirely asymmetric with respect to the midplane z_m of the filter. Owing to the imposed symmetry, the axial secondary fundamental rays $x_{\alpha\alpha}, x_{\beta\beta}$, and $y_{\alpha\beta}$ intersect at the centre of the achromatic image plane z_a and, therefore, do not cause second-order axial aberrations at the final image. The same is true of the field rays $x_{\gamma\gamma}$, $x_{\delta\delta}$, and $y_{\gamma\delta}$ as illustrated on the left-hand side of Fig. 2.22c. The mixed secondary fundamental rays $x_{\alpha\gamma}, x_{\beta\delta}, y_{\alpha\delta}$, and $y_{\beta\gamma}$ shown in Fig. 2.22b do not intersect at the centre of the achromatic image plane and hence do produce aberrations (image tilt and field astigmatism) at the final image. However, since these rays intersect at the centre of the energy-selection plane, they do not introduce aberrations at this plane.

Summarizing, we can state that, owing to the imposed symmetry, the geometrical secondary fundamental rays pass through either the centre of the achromatic image plane or the centre of the energy-selection plane. If we require that the secondary fundamental rays vanish at two non-conjugate planes behind the filter, they must be zero in the entire region outside the fil-

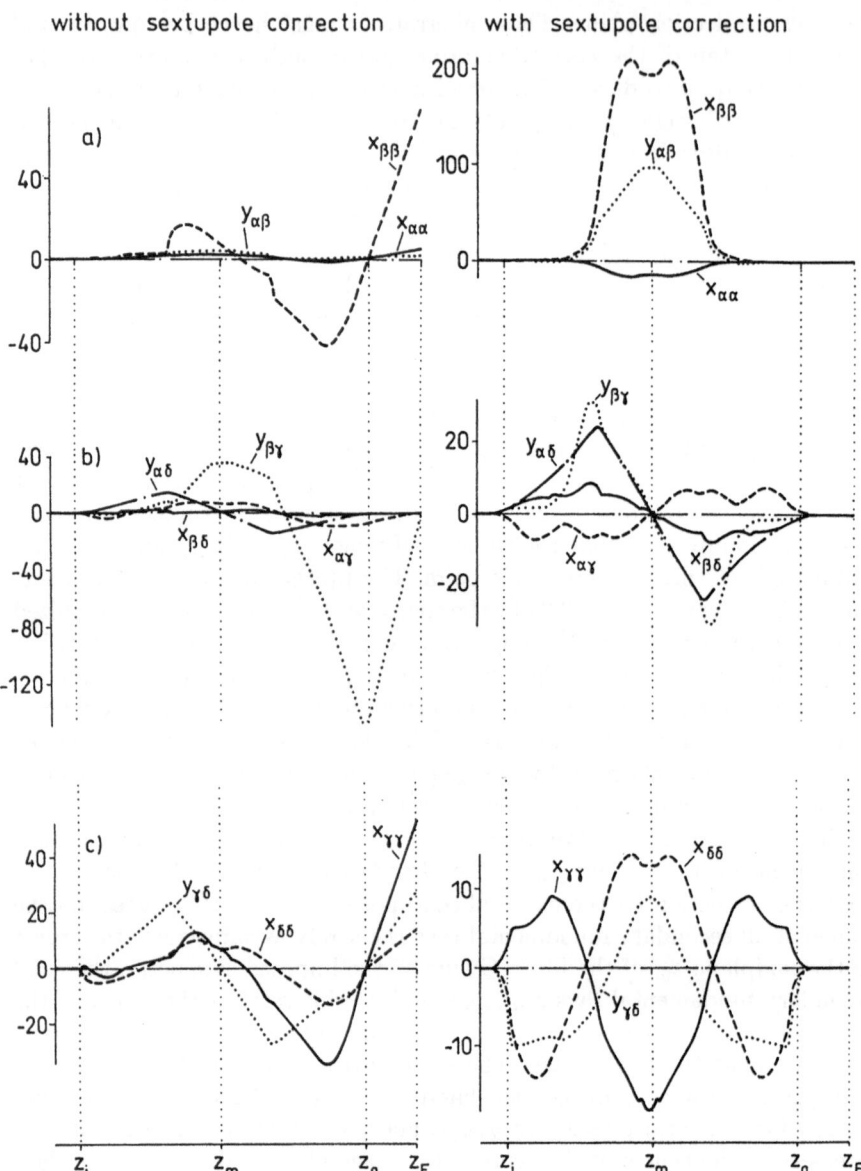

Fig. 2.22. Course of the fundamental secondary axial (**a**), mixed (**b**), and field (**c**) rays before (left-hand side) and after (right-hand side) correction of the second-order aberrations for the system outlined in Fig. 2.21. All rays are normalized with respect to the radius of curvature R_0 of the deflection magnets.

ter. This behaviour is convincingly demonstrated by the path of the secondary fundamental rays in the case of complete sextupole correction, as depicted on the right-hand side of Fig. 2.22. Owing to the symmetric arrangement of the magnets and the sextupoles, the second-order geometrical deviations must vanish at the exit of the deflection system with zero slope in exactly the same way as they originate at the entrance to the system. Accordingly, the geometrical secondary fundamental rays must be either symmetric or antisymmetric with respect to the midplane z_m. As can be seen from Fig. 2.22, the course of the mixed rays $x_{\alpha\gamma}$, $x_{\beta\delta}$, $y_{\alpha\delta}$, and $y_{\beta\gamma}$ is antisymmetric, while the course of the axial rays and the field rays $x_{\gamma\gamma}$, $x_{\delta\delta}$, and $y_{\gamma\delta}$ is symmetric with respect to the midplane z_m.

An indication for the proper position of the external sextupoles is to be found in the fact that the average off-axial distances of the secondary fundamental rays within the filter are reduced by the sextupoles with the exception of the axial rays $x_{\alpha\alpha}$, $x_{\beta\beta}$ and $y_{\alpha\beta}$, which are significantly enlarged by the correction.

Compensation of the Chromatic Aberrations. The correction of the geometrical aberrations also influences the chromatic second-rank aberrations. Replacing the ray parameters α, β, γ, and δ by those (2.79) at the object plane, it becomes obvious that the axial chromatic aberrations at the achromatic image plane are negligibly small because these aberrations referred back to the object plane are inversely proportional to M_i^2. On the other hand, the chromatic aberration of magnification is independent of M_i and must, therefore, be eliminated or sufficently suppressed to guarantee a large field of view. Hence one would expect that two additional pairs of sextupoles are required to eliminate the corresponding coefficents $C_{\alpha\gamma\kappa}$ and $C_{\beta\delta\kappa}$. Fortunately, this is not the case because in symmetric systems which are free of (geometrical) second-order aberrations the chromatic aberration of magnification vanishes as well. To prove this, we consider that, for symmetric systems, according to (2.141) and (2.142), the coefficients of the chromatic aberration of magnification fulfil the relations

$$C_{\alpha\gamma\kappa} = 2A_{\kappa\alpha\gamma} \quad , \quad C_{\beta\delta\kappa} = B_{\kappa\beta\delta} \tag{2.163}$$

where the coefficients $A_{\kappa\alpha\gamma}$ and $B_{\kappa\beta\delta}$ are given by the integral expressions (2.139) and (2.140), respectively. The dispersion ray (2.78) can be split up into a symmetric part $x_{\kappa s}$ and an antisymmetric part $x_{\kappa a}$ with respect to the central plane z_m. Using the abbreviation

$$s = \eta \Psi_{1s}/2L \tag{2.164}$$

the dispersion ray can be written as

$$x_\kappa = x_{\kappa s} + x_{\kappa a} \tag{2.165}$$

where

$$x_{\kappa s} = x_\alpha \int_{z_m}^{z} s x_\gamma \mathrm{d}z - x_\gamma \int_{-\infty}^{z} s x_\alpha \mathrm{d}z \quad , \tag{2.166}$$

$$x_{\kappa a} = x_\alpha \int_{-\infty}^{z_m} s x_\gamma \mathrm{d}z = x_\alpha C_{\gamma\kappa}/2L \quad . \tag{2.167}$$

Since the fundamental rays x_γ and y_δ are symmetric and the rays x_α and y_β are antisymmetric with respect to z_m, each integrand of the coefficients (2.163) can be written as a sum of an antisymmetric term containing the part $x_{\kappa s}$ of the dispersion ray and a symmetric term which contains the part $x_{\kappa a}$. The integrals over the antisymmetric integrands vanish and hence do not contribute to the aberration coefficients. Inserting the antisymmetric part $x_{\kappa a}$ into the integral expressions (2.139) and (2.140), we find

$$
\begin{aligned}
C_{\alpha\gamma\kappa} &= A_{\kappa\alpha\gamma} &= A_{\alpha\alpha\gamma} C_{\gamma\kappa}/L \quad , \\
C_{\beta\delta\kappa} &= B_{\kappa\beta\delta} &= B_{\alpha\beta\delta} C_{\gamma\kappa}/2L \quad .
\end{aligned}
\tag{2.168}
$$

Accordingly, the coefficients of the chromatic aberration of magnification vanish simultaneously with the mixed geometrical aberrations, which have been eliminated by the sextupoles. This behaviour is demonstrated in Fig. 2.23 where the chromatic second-rank fundamental rays are plotted before and after the correction of the geometrical aberrations. As can be seen from Fig. 2.23b, the asymptotes of the secondary chromatic rays $x_{\gamma\kappa}$ and $y_{\delta\kappa}$ originate in the centre of the achromatic image z_a. Hence these rays vanish at the conjugate final image z_I if we disregard the contributions of the round lenses. As illustrated in Fig. 2.23a, the sextupole correction of the geometrical aberration also influences the path of the axial secondary chromatic rays $x_{\alpha\kappa}$ and $y_{\beta\kappa}$. After this correction has been made, $x_{\alpha\kappa}$ and $y_{\beta\kappa}$ vanish at the energy-selection plane z_E. Moreover, the second-degree dispersion $x_{\kappa\kappa}$ becomes negligibly small at this plane so that the energy spectrum is almost linearly displayed.

Unlike the geometrical secondary fundamental rays, the chromatic secondary rays do not vanish outside the filter after the elimination of the second-order fundamental rays by the sextupoles. For complete correction of the latter deviations, additional sextupoles are mandatory.

2.7.3 Correction of Third-Order Aberrations

The performance of an imaging energy filter corrected to second order is determined by the third-order aberrations. Since the filter is placed in front of the projector lens, the third-order distortions are expected to impair most strongly the image quality. Fortunately, this is not the case for a symmetric energy filter whose second-order (geometrical) aberrations have been eliminated. The correction of these aberrations simultaneously eliminates the third-order distortions and the third-order comas. This behaviour can be proven directly by means of symmetry considerations. For a second-order-corrected filter, all geometrical coefficients of the third-rank eikonal $E^{(3)}$ vanish. Hence this eikonal term does not contribute to the third-order combination aberrations. The third-order aberrations (2.161) are then determined

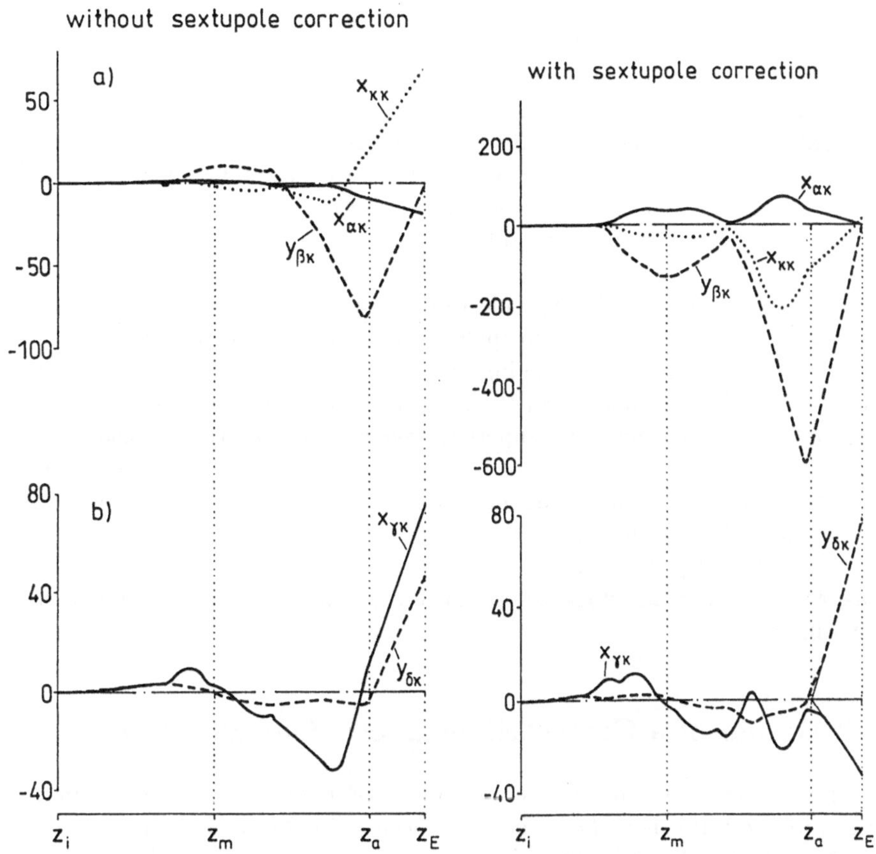

Fig. 2.23. Course of the fundamental second-rank chromatic axial (**a**) and field (**b**) rays before (left-hand side) and after (right-hand side) correction of the second-order (geometrical) aberrations for the Berlin filter. All rays are normalized with respect to the radius of curvature R_0.

entirely by the fourth rank-perturbation eikonal (2.158). Its integrand (2.157) is composed of a fourth-degree polynomial of the paraxial rays and a third-degree polynomial, which is linear in the secondary fundamental rays and bilinear in the paraxial rays. Accordingly, the third-order distortions are determined (a) by a polynomial that is cubic in the field rays x_γ and y_δ and linear in x_α or y_β and (b) by a polynomial that contains the ray combinations $x_{\gamma\gamma}x_\alpha x_\gamma$, $x_{\gamma\gamma}y_\beta y_\delta$, $x_{\delta\delta}x_\alpha x_\gamma$, $x_{\delta\delta}y_\beta y_\delta$, $y_{\gamma\delta}y_\beta x_\gamma$, and $y_{\beta\gamma}x_\gamma y_\delta$ multiplied with powers of the multipole strengths. For the corrected system, these ray products as well as the products of the paraxial rays belonging to the first polynomial are antisymmetric functions with respect to the central plane z_m. Since the multipole strengths are even functions with respect to z_m, the part of the integrand $m^{*(4)}$ (2.158) that determines the third-order distortions is antisymmetric. Hence, the eikonal terms that cause the third-order distortions are not present. Therefore, any third-order distortions at the final image plane are produced by the round lenses and not by the filter. The high degree of correction of the symmetric filters strongly favours their application for high-performance spectroscopic imaging.

Unfortunately, the third-order axial aberrations are not eliminated. Although they hardly affect the image resolution, they will be dominant at the energy-selection plane. Fortunately, the correction of the second-order aberrations does not enlarge these aberrations. As can be seen from Fig. 2.22c, the secondary fundamental rays $x_{\gamma\gamma}, x_{\delta\delta}$, and $y_{\gamma\delta}$ which contribute to the axial aberrations at the energy selection plane, are on average reduced by at least a factor of two by the sextupoles. If these aberrations are so large that they prevent isochromatic imaging of the required field of view, the deleterious aberration component in the x-direction can be compensated by two octopole fields; one must be superimposed onto the sextupole field at the central plane z_m and the other at the intermediate image plane z_i or z_a. For activation these fields independently, it is advantageous to place twelve-pole elements at these planes. As the octopole fields are centred about stigmatic images of the object, the proposed correction of the axial aberration at the energy-selection plane does not produce any third-order distortions at the final image.

2.8 Design of a Corrected Magnetic Omega Filter

Imaging energy filters have been the subject of many designs reported in the literature [2.9–16]. Few, however, have been built. Most of the proposed or existing filters only partially fulfil the conditions necessary for achieving optimum performance. The reasons stem from inadequate calculation procedures or from large second-rank aberrations.

In the following sections we shall outline the performance of a corrected magnetic imaging energy filter, which has been built and tested at the Fritz-Haber-Institut of the Max-Planck-Gesellschaft according to calculations performed at the Technische Hochschule Darmstadt [2.13,14,48]. We shall discuss the experimental aspects of aligning and correcting this system which we call the FHI filter. The procedure can serve as a guide for aligning arbitrary deflection systems with midplane symmetry.

The calculation of the filter parameters realistically takes into account the influence of the fringing field extension on the imaging properties of the filter. For properly alignment of the energy filter relative to the axis of the electron microscope, the first-order imaging properties of the filter itself can be utilized. Moreover, the possibility of observing the blurred images of a grid in the final (achromatic) image plane and in the energy-selection plane enables us to measure and correct the second-rank aberrations successfully. The first prototype of the FHI filter had curved edges and sextupole correctors to correct the second-order aberrations. The experiments revealed significant differences between the calculated and measured imaging properties of the filter [2.16]. The disparities are caused by deviations between the calculated and measured quadrupole fields at the pole boundaries and result in a displacement of the optic axis. Owing to the use of curved pole-faces, any displacement of the optic axis changes the edge angles. As a result, the quadrupole fields are changed and the first-order imaging properties are altered. Hence curved edges prevent correct first-order alignment of the electron beam within the filter. The experiments clearly revealed that the standard SCOFF calculation is not sufficiently accurate for determination of the geometry of an optimum filter. In the light of this experience, the calculation procedure has been refined to take into account the actual fringing field extension (EFF) and only deflection magnets with straight edges are employed.

When determining the precise geometry of the optimum filter, it is advantageous to start with the SCOFF calculation. From a larger number of possible arrangements, we have chosen a system that is rather insensitive to small variations of the geometrical parameters. The definition of the geometrical parameters of the symmetric filter is shown in Fig. 2.24, and the corresponding design parameters are listed in Table 2.2 for both the SCOFF and the EFF calculations. The latter procedure considers the extension of the fringing fields and allows the geometry of the magnetic polepieces to be determined accurately.

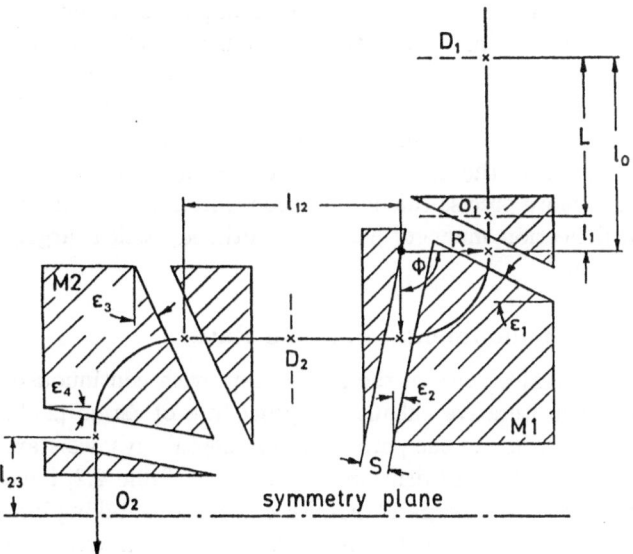

Fig. 2.24. Parameters defining the geometry of an imaging energy filter. The elements of the second half of the filter are arranged symmetrically about the midplane of the filter.

2.8.1 Calculation of the Filter Axis

In the first step of the calculation procedure, the approximate course of the optic axis is determined by employing the SCOFF approximation. The resulting curved axis shown in Fig. 2.25 is composed of straight lines outside the magnetic fields and circular arcs within the filter. In reality, however, the curvature of the optic axis does not jump at the "effective" field boundaries but changes steadily from its initial to its final value when traversing the fringing field regions. In the next step, the course of the optic axis is calculated anew, taking into account the real fringing field extensions, by solving numerically the equation of motion. The extension of the fringing fields has been considered by the Schwartz–Christoffel method of conformal mapping [2.42,43] (Sect. 2.4.3). The validity of the calculated fringing fields has been confirmed by measurements. The deviations were less than 2%. The result of the calculations shows clearly that the optic axis deviates from the predicted axis obtained for the SCOFF geometry (Fig. 2.25). The total deflection angle Φ differs somewhat from the required angle $\Phi = 90°$. As shown in the figure, the correct angle can be obtained by shifting the exit pole-face in the direction of the required deflection by a distance δ. This shift is associated with a lateral displacement λ of the true optic axis from the SCOFF axis. In the case of straight pole-faces, a lateral shift does not alter the focusing properties of the edge quadrupoles. However, this shift must be compensated for by an equal lateral shift of the next magnet. For example, to achieve the deflection of $\Phi = 90°$ in the first magnet of the energy filter presented here with a slit width $S = D = 0.1R_0$ between the mirror plates and pole-faces, the exit pole-face of the first magnet was shifted by a distance $\delta_1 = 0.0158R_0$ with respect to that of the SCOFF model. The longitudinal shift of the edge caused a lateral displacement of the filter axis $\lambda_1 = 0.0187R_0$. In Sect. 2.4.3 a somewhat different but more precise calculation procedure for determining the true optic axis of deflection magnets has been outlined, which largely avoids such a readjustment.

2.8.2 Calculation of the Fundamental Paraxial Rays

To determine the electron trajectories in the real system with continuously varying magnetic fields, the iterative calculation procedure of *Lanio* [2.48] has been used. For the calculation of the paraxial fundamental rays, we start from a system whose edge angles and drift spaces, listed in Table 2.2, have been determined by standard (approximate) SCOFF calculations, taking into account the shifts δ_n and the lateral displacements λ_n of the beam axis. Using these values, the calculated paraxial rays of the realistic system deviate significantly from the required ray-paths resulting in deleterious first-order aberrations. Since these deviations do not result from incorrect locations of the effective field boundaries, it is not possible to adjust the required ray-paths within the filter merely by varying the distances between the magnets.

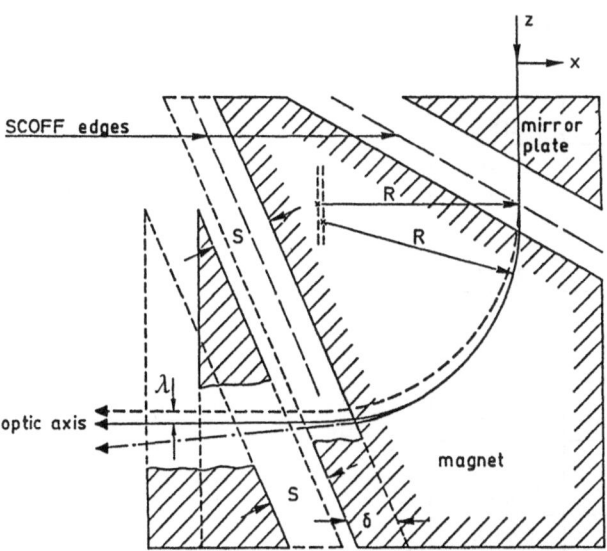

Fig. 2.25. Course of the optic axis within a parallel-plate magnet obtained by SCOFF approximation (- - -), extended fringing field (EFF) calculation (- · - ·), and extended fringing field calculation for a magnet adjusted by an edge shift δ accompanied by a lateral displacement λ (—).

Such a procedure is only capable of adjusting the rays in one principal section at the expense of an increase of the path deviations in the other section. To obtain the required path of the fundamental rays in both sections, it is, therefore, necessary to change the entrance and exit angles and the distances between the magnets. Fortunately, there are more parameters available than conditions that must be met. For example, the exit angle ϵ_2 of the first magnet of the filter proposed in Table 2.2 has been kept fixed because its variation hardly affects the paraxial ray-paths. After changing the geometric parameters appropriately, the rays are computed anew. The entire procedure is repeated until the fundamental rays fulfil the required symmetry conditions. Table 2.2 contains the adjusted design parameters of the final Ω–filter together with the parameters obtained by the SCOFF calculations, which served as the starting values of the iteration procedure.

The flattened paths of the fundamental rays $x_\alpha, y_\beta, x_\gamma, y_\delta$ and the fundamental dispersion ray x_κ of the adjusted FHI energy filter are shown in Fig. 2.21. The axial fundamental rays x_α and y_β are antisymmetric with respect to the central plane, while the fundamental rays x_γ and y_δ are symmetric. The flattened course of the fundamental rays demonstrates that the intermediate image plane z_i and the diffraction plane z_d are imaged stigmatically with unit magnification into the planes z_a and z_E, respectively. Moreover, the image plane z_i is transferred by the filter to the plane z_a without inversion. An inverted and strongly distorted image with $M_x/M_y \approx 4$

Table 2.2. Geometrical design parameters of the FHI filter. All lengths have been normalized with respect to the deflection radius R_0.

	SCOFF	EFF
ε_1	40 °	43.24 °
ε_2	−7 °	−7.00 °
ε_3	40 °	41.55 °
ε_4	24 °	21.84 °
l_0	2.089	2.070
l_1	0.190	0.171
l_{12}	1.845	1.801
l_{23}	0.512	0.676
D	−	0.1
S_1, S_2, S_3, S_4	−	0.1
δ_1	−	0.0518
λ_1	−	0.0187
δ_2	−	0.0120
λ_2	−	−0.0184

is formed at the midplane z_m of the filter. The intermediate images of the diffraction plane, which are located between the first and second and the third and fourth magnets, are perturbed by a slight axial astigmatism. The strongly astigmatic path of the paraxial rays within the filter is a prerequisite for complete correction of the second-order aberrations. Configurations whose aberrations can be eliminated largely independently of each other by means of sextupole correctors are especially desirable. This condition is partly fulfilled for the FHI filter.

Owing to the midplane symmetry of the magnetic fields and the paraxial rays, the dispersion vanishes at the achromatic image plane z_a. This behaviour follows directly from the integral representation (2.78) for the dispersion ray x_κ. As a result, no chromatic displacement occurs at the final image plane. The value of the dispersion ray at the energy-selection plane z_E defines the constant $C_{\gamma\kappa}$ (2.152) of the first-degree dispersion $C_{\gamma\kappa}/E_0$. In order to maximize the dispersion, the integrand of the relation (2.152) has been made positive throughout the filter up to the energy-selection plane. This behaviour explains the large dispersion of 6 μm/eV for the FHI filter at an accelerating voltage of 100 kV and a radius of curvature $R_0 = 5$ cm.

2.8.3 Actual Filter Design

The actual design of the FHI filter together with the intermediate lens and the additional transfer lens is shown in Fig. 2.26. The filter is composed of four sector magnets arranged symmetrically about the midplane. The profiles of the magnets have been manufactured with high accuracy by computer-controlled spark erosion techniques. To achieve largely homogeneous magnetic fields within the magnets, the deflection plates are surrounded by the exciting coils. With such an arrangement, the deviations of the magnetic field strength from its constant nominal value can be held below 1%. The incorporated mirror plates define rather accurately the extent of the fringing fields. The magnetic field of each magnet deflects the beam axis by an angle $\Phi = 90°$. Before the magnets were fixed on a polished ground plate they were adjusted with respect to each other by means of length standards.

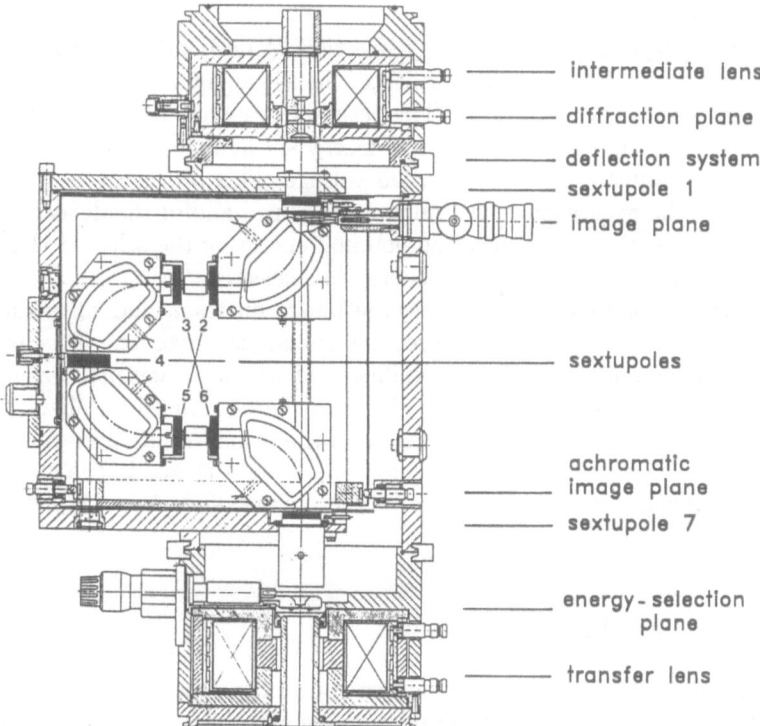

intermediate lens

diffraction plane

deflection system

sextupole 1

image plane

sextupoles

achromatic image plane

sextupole 7

energy-selection plane

transfer lens

Fig. 2.26. Cross-section through the sextupole-corrected FHI filter and the adjacent lenses located in front of and behind the filter.

For eliminating the remaining second-order aberrations at the final image plane and the energy-selection plane, three sextupole pairs are arranged symmetrically about the central plane z_m of the filter. Two of the sextupole pairs, S2/S6 and S3/S5, are located in the region between the magnets, while the sextupoles of the third pair S1/S7 are placed in front of and behind the filter. A twelve-pole arrangement has been chosen for the latter element to allow the excitation of an additional quadrupole field without producing higher-order multipole components [2.25]. This quadrupole field is used to form a band-like spectrum at the energy-selection plane. The sextupole field S4 produced by a twelve-pole element placed at the central plane of the filter fulfils the symmetry condition automatically. An additional quadrupole field created by the central twelve-pole corrector acts as a stigmator compensating for any axial astigmatism at the energy-selection plane resulting from machining tolerances, misalignment, and parasitic fields.

A deflection system located between the diffraction plane and the image plane is used to "wobble" the illumination aperture in the x or the y direction. This procedure is needed to measure the second-rank aberration coefficients and allows the system to be adjusted and the state of correction of the second-order aberrations to be determined.

The energy dispersion at the energy-selection plane is utilized to select electrons with distinct energies. The width of the slit aperture placed in this plane determines the energy window. The desired energy loss can be selected without displacing the slit by shifting the energy spectrum across the opening by properly adjusting the acceleration voltage. This procedure has the advantage that the excitation of the microscope lenses and of the deflection magnets can be kept fixed because the nominal energy of the selected electrons stays the same regardless of the energy loss selected.

Unlike other proposed energy filters, the image planes z_i and z_a of the FHI filter are located in the field-free regions outside the magnets. This arrangement enables one to insert a metalline grid at the image plane z_i in front of the filter. The shadow image of this grid can be utilized for aligning the filter and to check the state of correction of the second-rank aberrations. The transfer lens beneath the filter transfers the energy-filtered image or alternatively the energy spectrum into the "object" plane of the final projector lens.

2.9 First-Order Alignment

At first sight the task of adjusting the curved optic axis of the energy filter to the straight optic axis of the electron microscope with high precision seems to be extremely difficult. However, the entire filter is a straight-vision system, whose entrance and exit axes form a common straight line. Thus, if the filter is internally adjusted, only the asymptotic filter axes need be matched to the optic axis of the microscope. The coincidence is achieved by the first alignment step. In the next step the symmetric path of the beam within the energy filter is adjusted as well as the correct transfer of the achromatic image plane by the transfer lens. Finally, the second-order aberrations at the achromatic image and the energy-selection plane are precisely corrected.

Fortunately, the first- and second-order imaging properties of the filter allow one to check the alignment status. This possibility is utilized for aligning the filter with a very high accuracy. The adjustment is performed after the microscope has been aligned properly. In order to achieve exact adjustment of the microscope, the deflection magnets of the filter must be initially demagnetized.

The point of intersection of the microscope axis with the plane of observation is marked by the centre C of the grid placed at the image plane z_i in front of the filter, as shown in Fig. 2.27. This centre serves as a "reference point". The bars of the square grid must be oriented excactly parallel to the plane inner pole-faces of the filter. The bars intersecting the centre define the x,y Cartesian coordinate system, where the x coordinate is parallel to the inner pole-faces while the y coordinate is chosen perpendicular to these faces. The form of the shadow image of the grid and the position of its centre C^* serve as criteria for the alignment procedure. The directions of the bars intersecting the centre of the shadow image define the coordinate system transferred to the observation plane.

2.9.1 Adjustment of the Filter Axis

After the energy filter has been incorporated into the microscope, the optic axis of the filter will generally be skewed with respect to the microscope axis. At most six adjustment steps are necessary to align the filter with respect to the microscope because the location of a rigid body in space is defined by six coordinates: three position coordinates and three Eulerian angles. Hence, three of the alignment steps are rotations about the normal n, the binormal b, and the asymptotic axis of the filter. The normal n is perpendicular to the midsection of the filter. The two rotations about the normal and the binormal align the filter axis parallel to the axis of the microcope. A rotation about the filter axis is unnecessary because it does not affect the imaging properties. In a subsequent step the filter is shifted in two directions perpendicular to the filter axis in order to bring both axes into coincidence. The last degree of freedom concerns a shift of the filter along the optic axis but such a misalignment can be eliminated more efficiently by changing the focus of the round lenses in front of the filter.

The successive steps for correcting the tilt of the filter axis are shown schematically in Fig. 2.27. A skew position of the filter axis with respect to the microcope axis has been assumed. The projections of the misaligned filter axis onto the x-z and y-z sections are depicted in Figs. 2.27a,c. Since the axis of the imperfectly adjusted filter differs from the microscope axis, the centre C placed on the optic axis is an off-axial object point for the filter. With respect to the coordinate system of the filter, this centre lies at an off-axial position with coordinates x_i and y_i. The filter forms an image of the grid located at the plane z_i with magnification $M_a = +1$ at the achromatic image plane z_a. Owing to the tilt of the filter axis, the image C^* of the centre C is displaced with respect to the axis of the microscope by a distance with components a_x and a_y, respectively.

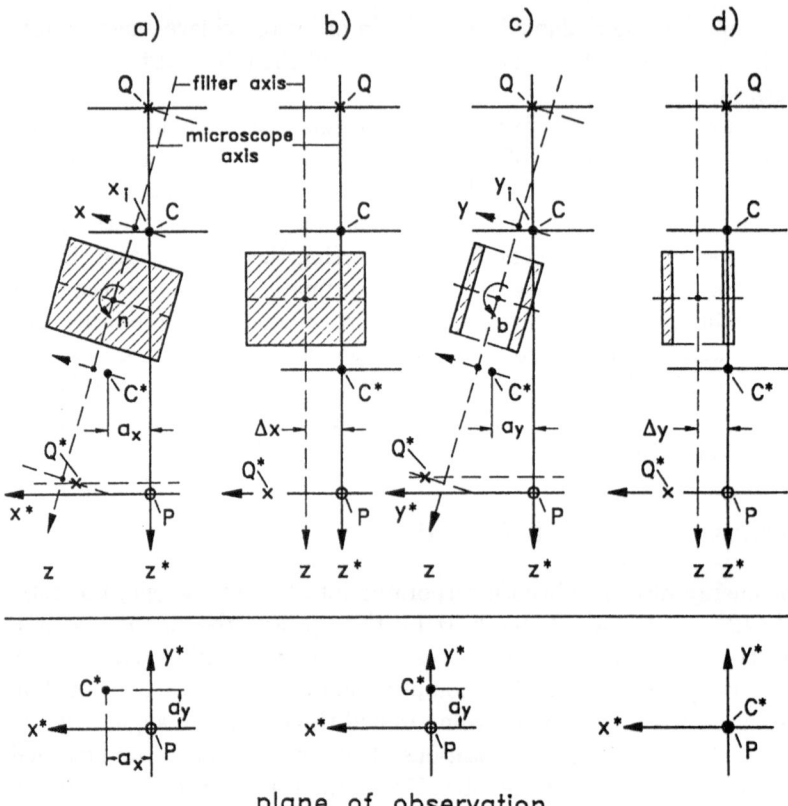

plane of observation

Fig. 2.27. Alignment of the filter axis with respect to the axis of the microscope; (a) and (c) show the projections of the initial course of the filter axis onto the x-z section and onto the y-z section, respectively. The centre C of the grid placed at the intermediate image plane of the electron microscope is imaged by the misaligned filter into the point C^*, which is displaced from the microscope axis by the distances a_x and a_y. The rotations about the normal n (a) and the binormal b (c) parallel the asymptotes of the filter at distances Δx and Δy with respect to the microscope axis, as shown in (b) and (d).

In the first alignment step (Fig. 2.27a), the filter is rotated about the normal n, which has been placed in the centre of gravity, until the deviation a_x is cancelled (Fig. 2.27b). After this rotation the projection of the filter axis onto the x-z section is parallel to the microscope axis. This alignment step presupposes that the path of rays within the filter is adjusted perfectly. However, this assumption is generally not true in practice. Any deviation from the ideal symmetric course of the rays introduces an additional shift of the image C^* as the true beam axis differs from its nominal course. Fortunately, this additional misalignment does not affect the achromatic image because an internal non-symetrically aligned filter acts roughly like an energy-selection system trimmed to a somewhat different nominal energy. As the filter transfers the shadow image achromatically to the corresponding image plane behind the filter (i.e., the image position will be independent of the energy of the electrons), no additional image shift occurs in this plane if the ray-paths are slightly asymmetric within the filter. In order to ensure that the imperfect internal alignment of the filter does not affect the mechanical alignment of the filter, the transfer lens must be properly focused onto the achromatic image plane. The appropriate focusing of the transfer lens is obtained by slightly changing the accelerating voltage and eliminating the resulting image shift caused by the first-degree dispersion by fine tuning of the lens.

In the second step of the alignment (Fig. 2.27c) the filter is rotated about the binormal **b** until the image point C^* is positioned on the axis of the microscope. The filter axis then runs parallel to the microscope axis, displaced in the x and y directions by the distances Δx and Δy (Figs. 2.27b,d).

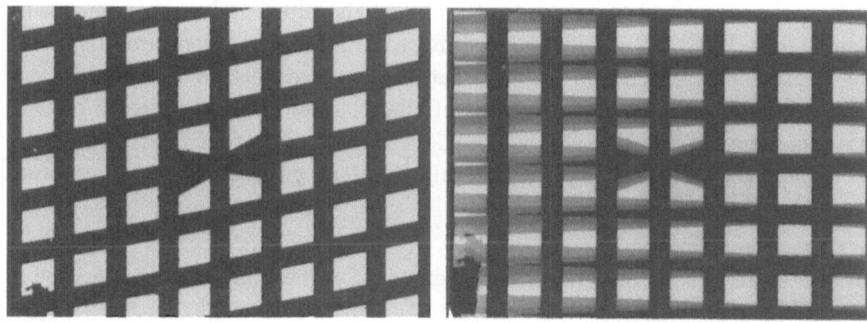

Fig. 2.28. (left) Angular distortion caused by a lateral displacement of the filter axis in the y direction.

Fig. 2.29. (right) Displacement of the centre of symmetry of the aberration with coefficient $B_{\beta\beta\gamma}$ caused by a lateral shift of the filter axis in the x direction.

The lateral displacement of the filter axis is eliminated by shifting the prealigned filter towards the optic axis of the microscope. In order to match the two axes exactly, we need a sensitive measurement of the actual state of the alignment. Fortunately, the shadow image of the grid is very sensitive to parallel shifts of the filter axis. For example, a deviation in the y direction shows up as the angular distortion of the shadow image illustrated in Fig. 2.28. This effect serves as an accurate correction criterion for the elimination of the y shift. The distortion arises from a parasitic quadrupole field about the displaced beam axis. The magnetic scalar potential of the ideal filter contains a sextupole term created by the fringing fields at the edges of the deflection magnets. With respect to the midsection of the filter, the sextupole term has the form (2.31)

$$\psi_3 = \Psi_{3s}(3x^2y - y^3) \quad . \tag{2.169}$$

The optic axis of the microscope and hence the true beam axis are shifted by the distance Δy from the midsection of the laterally displaced filter. Accordingly, the multipole expansion about the beam axis is

$$\psi_3 \approx \Psi_{3s}(3x^2y - y^3) + 3\Delta y\Psi_{3s}(x^2 - y^2) \quad , \tag{2.170}$$

where we have neglected higher-order terms in Δy. The first term on the right-hand side is the sextupole field, while the second term describes a parasitic quadrupole which is rotated through an angle of 45° with respect to the edge quadrupole of the magnets. This quadrupole field is responsible for the angular distortion of the shadow image.

Unfortunately, the deviation Δx in the x direction does not result in a quadrupole term, because the two-dimensional fringing field of each magnet does not depend on the x coordinate. To compensate for the misalignment Δx we consider that the second-order aberrations result from all elements of the filter and are centred about the optic axis of the filter, as has been shown in Sect. 2.6.2. (Fig. 2.19). A lateral x shift of the filter axis, therefore, displaces the centre of symmetry of the aberrations with respect to the microscope axis, which is defined by the centre of the shadow image. In Fig. 2.29 this effect is demonstrated by activating the mixed second-order aberration with coefficient $B_{\beta\beta\gamma}$, which displaces the image in the y direction by the amount

$$\Delta y = \beta\gamma \cdot B_{\beta\beta\gamma} \quad . \tag{2.171}$$

To obtain an observable image displacement Δy, the illumination aperture β has been wobbled by sending square wave currents through the coils of the deflection system located between the image and the diffraction plane. The β–wobble also causes an additional displacement in the x direction introduced by the aberration coefficient $B_{\alpha\beta\delta}$. However, this shift is negligibly small, as has been confirmed by computer calculations of the total image displacement taking into account the two second-order aberrations which are activated by the β–wobble. Since this wobble does not affect the centre of the aberration figure, this centre can easily be observed. By shifting the pole-plates of the magnets the centre is matched to the reference point. At this position the filter axis coincides with the axis of the microscope.

It should be noticed that this alignment technique is only useful if the intermediate image plane z_i is located in the field-free region in front of the filter. The procedure is not practicable for Ω filters with virtual images located inside the filter.

2.9.2 Internal Paraxial Alignment

The alignment of the filter axis has been achieved regardless of any internal paraxial misalignment. To adjust the required symmetric course of the paraxial rays precisely, two further adjustments are needed. If the symmetry condition is not fulfilled the beam axis behind the filter is inclined in the x-z section with respect to the axis of the microscope. This mismatch causes a deviation of the centre of the diffraction image at the energy-selection plane relative to the reference point. After the transfer lens has been focused onto this plane, the symmetry of the rays is adjusted by changing the currents through the windings of the magnets until the centre of the aberration figure meets the reference point.

The refocusing of the transfer lens onto the achromatic image plane can be checked by slightly changing the high voltage. If the focus is not perfect, the image is shifted by the non-vanishing first-degree dispersion. This displacement is subsequently eliminated by fine focusing of the transfer lens.

It should be noticed that the procedure outlined aligns the filter axis with a very high precision. Experiments have shown that the displacements oberserved at the final plane of the microscope can be eliminated with an accuracy of about ± 0.5 mm with respect to the reference point. Assuming a post-magnification $M_{\mathrm{I}} = 100$ of the achromatic image by the round lenses behind the filter, the accuracy referred to the achromatic image plane will thus be $\pm 5 \mu$m. Taking into account a distance of 200 mm between the image plane and the achromatic image plane, the residual tilt angle of the filter axis with respect to the microcope axis will be smaller than $\pm 1.4 \times 10^{-3}$ degrees. The axis shifts Δx and Δy can be corrected with an accuracy of $\pm 5 \mu$m. After the energy filter has been aligned to first order the disturbing aberrations must be eliminted at the achromatic image plane and the energy-selection plane.

2.10 Experimental Correction of Aberrations

The theoretical considerations for eliminating the second-rank aberrations have been outlined in Sect. 2.7. In this section, we discuss methods of measuring and compensating for the aberrations in practice. The appropriate experimental procedures have been developed in the context of optimizing the performance of the FHI energy filter. The distances Δx and Δy of the true ray from its paraxial approximation at the Gaussian image plane are the result of all the individual aberrations. To isolate a distinct aberration from all the others, we have developed a special method for the measurement and the correction of the second-rank aberrations [2.49].

We choose the constant radius of curvature $R = R_0$ of the filter axis within the magnets as the unit of length. The ray parameters (2.76)

$$\gamma = x_{\mathrm{i}}/L \quad , \quad \delta = y_{\mathrm{i}}/L \qquad (2.172)$$

can be expressed in terms of the distance L between the image plane z_{i} and the diffraction plane z_{d} and the off-axial distances x_{i} and y_{i} of the ray at the image plane z_i (Fig. 2.30). The limiting illumination angles at the intermediate image are defined as

$$\alpha = \beta = r_d/L \tag{2.173}$$

where r_d is the radius of the aperture at the diffraction plane z_d.

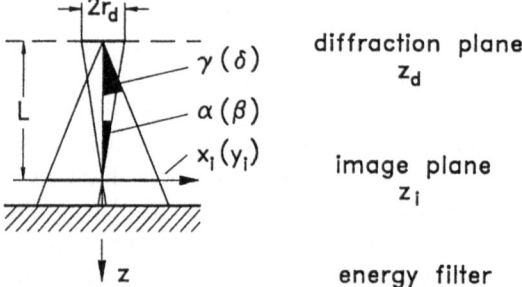

Fig. 2.30. Definition of the illumination aperture angles α, β and the lateral limiting components x_i, y_i of the beam expressed as angle variables γ, δ.

2.10.1 Aberrations and Symmetry

The symmetry conditions imposed on the FHI filter are such that the magnetic fields and the fundamental rays x_γ, y_δ are symmetric and the rays x_α, y_β are antisymmetric with respect to the mid-plane of the filter (Fig. 2.21). For this symmetry the aperture aberrations $(A_{\alpha\alpha\alpha}, B_{\alpha\beta\beta})$ and the distortions $(A_{\alpha\gamma\gamma}, B_{\beta\gamma\delta}, B_{\alpha\delta\delta})$ vanish at the achromatic image plane z_a as outlined in Sect. 2.7.1. In addition the mixed aberrations at the energy-selection plane cancel out since they are related to the distortion coefficients. Furthermore, the integrands of the remaining second-order geometrical coefficients become symmetric. Hence the coefficients $A_{\alpha\alpha\gamma}, B_{\alpha\beta\delta}, B_{\beta\beta\gamma}$ of the inclinations of the image field and the field astigmatism are non-zero. The same behaviour holds for the coefficients $A_{\gamma\gamma\gamma}, B_{\gamma\delta\delta}$ of the axial aberrations at the energy-selection plane.

Besides these coefficients we have to consider the non-vanishing second-rank chromatic aberrations. At the achromatic image plane these components are composed of the well-known axial chromatic aberration $(C_{\alpha\alpha\kappa}, C_{\beta\beta\kappa})$, the chromatic aberration of magnification $(C_{\alpha\gamma\kappa}, C_{\beta\delta\kappa})$, and the second-degree dispersion $(C_{\alpha\kappa\kappa})$, which depends quadratically on the relative energy deviation κ. At the energy-selection plane the most disturbing chromatic aberrations result from the coefficients $C_{\gamma\gamma\kappa}$ and $C_{\gamma\kappa\kappa}$. These coefficients cause an inclination of the energy spectrum and a non-linear shift of the individual energy losses. The theoretical values of the non-vanishing normalized second-rank aberration coefficients of the FHI filter are listed in Table 2.3.

Table 2.3. Calculated and measured normalized second-rank aberration coefficients of the non-corrected FHI filter.

IMAGE PLANE					
geometric aberrations			chromatic aberrations		
coefficient	calculated	measured	coefficient	calculated	measured
$A_{\alpha\alpha\gamma}$	−3.3	−3.2	$C_{\alpha\alpha\kappa}$	8.0	−
$B_{\alpha\beta\delta}$	−0.8	−	$C_{\beta\beta\kappa}$	74.8	76.2
$B_{\beta\beta\gamma}$	−141.0	−147.6	$C_{\alpha\gamma\kappa}$	−17.9	−17.9
			$C_{\beta\delta\kappa}$	−2.2	−
			$C_{\alpha\alpha\kappa}$	23.3	23.9
ENERGY−SELECTION PLANE					
geometric aberrations			chromatic aberrations		
coefficient	calculated	measured	coefficient	calculated	measured
$A_{\gamma\gamma\gamma}$	−54.0	−44.0	$C_{\gamma\gamma\kappa}$	79.0	
$B_{\gamma\delta\delta}$	−29.0	−34.4	$C_{\delta\delta\kappa}$	50.0	−
$A_{\alpha\alpha\gamma}$	−3.3	−3.2	$C_{\alpha\gamma\kappa}$	−17.9	−17.9
$B_{\alpha\beta\delta}$	−0.8	−	$C_{\beta\delta\kappa}$	−2.2	−
$B_{\beta\beta\gamma}$	−141.0	−147.6	$C_{\gamma\kappa\kappa}$	−70.0	−

2.10.2 Measurement of the Second-Rank Aberrations

To facilitate the discussion of the experimental results, we start with the monochromatic case ($\kappa = 0$), which is realized at present-day resolutions by the standard illumination. In the next step, the chromatic terms are visualized by changing the high voltage. To confirm the measurements of the aberration coefficients, the results have been compared with simulated images taking into account the geometric and the chromatic aberrations of the non-corrected energy filter.

2.10.3 Achromatic Image Plane

For a fixed point ($x_i = L\gamma$, $y_i = L\delta$) of the intermediate image, the second-order aberrations of the symmetric FHI filter at the achromatic image plane are linear functions of the illumination angles α and β. To visualize the blurring of the image, the illumination angles are varied by wobbling α or β. This procedure is realized by passing square-wave currents through the deflection system in front of the filter. The wobbling amplitude can be measured at the energy-selection plane as the distance between the two caustic figures cre-

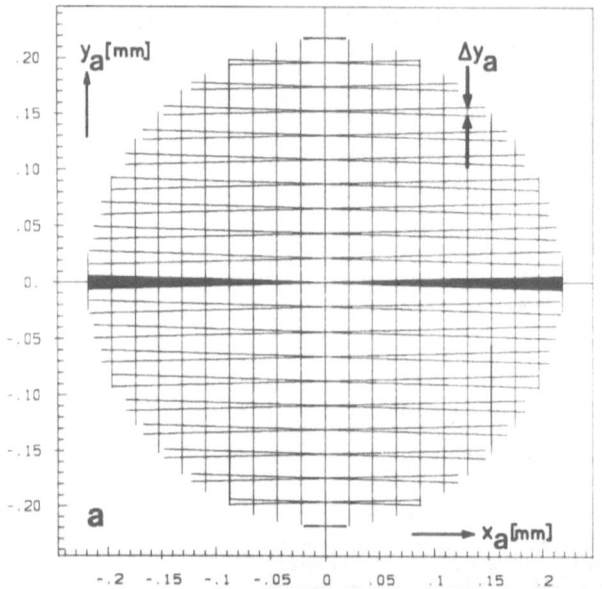

Fig. 2.31. Action of the mixed aberration coefficient $B_{\beta\beta\gamma}$ at the achromatic image plane visualized by a β–wobble with $\Delta\beta = \pm 7 \cdot 10^{-4}$ [2.49]; (**a**) computer simulated, (**b**) experimental, (**c**) after sextupole correction.

ated by the wobbling procedure. The width of the amplitude must be chosen carefully in order to avoid contributions from the higher-order aberrations.

By altering the illumination angle β, lateral displacements

$$\Delta y = \beta\gamma B_{\beta\beta\gamma} \quad , \quad \Delta x = \beta\delta B_{\alpha\beta\delta} \tag{2.174}$$

are produced, as shown in Fig. 2.31. The shift caused by the coefficient $B_{\alpha\beta\delta}$ is too minute to permit an experimental evaluation of this aberration coefficient. This is corroborated by the calculated aberration coefficients and by the computer-simulated images (Fig. 2.31a). Hence the β–wobble leads to the isolation of a single aberration whose coefficient $B_{\beta\beta\gamma}$ is accessible to measurement, provided the angle γ is known. The parameters γ and δ can be measured by choosing a ruled object. When a properly oriented square grid is inserted into the plane z_i, the bars will appear at the final image as discrete values of γ and δ. Accordingly, this image can serve as a gradation. By measuring the width of the penumbral region of the bars caused by either the β–wobble or the corresponding α–wobble we find

$$B_{\beta\beta\gamma} = -147.6R_0 \quad , \quad A_{\alpha\alpha\gamma} = -3.2R_0 \quad . \tag{2.175}$$

Because all aberration coefficients affect the image simultaneously, it is possible to measure the chromatic coefficients if the geometric coefficients are known. For example, to determine the coefficient $C_{\beta\beta\kappa}$, we select a fixed chromatic deviation κ by changing the acceleration voltage and apply the already introduced wobbling of the aperture β. Computer-simulated images show that only $B_{\beta\beta\gamma}$ and $C_{\beta\beta\kappa}$ contribute to the displacement

$$\Delta y = \beta\gamma B_{\beta\beta\gamma} + \beta\kappa C_{\beta\beta\kappa} \quad . \tag{2.176}$$

In this equation the coefficient $B_{\beta\beta\gamma}$ is known. As shown in Fig. 2.32, the displacement vanishes for a distinct value γ_0 of the parameter γ. Using this value together with $B_{\beta\beta\gamma} = -147.6R_0$ we find $C_{\beta\beta\kappa} = 76.2R_0$. The second-degree dispersion ($C_{\alpha\kappa\kappa}$) is also activated by changing the high voltage. However, it is not recognizable by the wobbling technique because it introduces a constant shift of the entire image. To surmount this obstacle, a second procedure has been devised which allows one to determine the coeffients $C_{\alpha\gamma\kappa}$ and $C_{\alpha\kappa\kappa}$ simultaneously. This method consists of a double exposure with two different energy settings. The first exposure is taken at the nominal energy ($\kappa = 0$) and a second at $\kappa \neq 0$. Though all second-rank chromatic aberration coefficients are activated, only the coeffients $C_{\alpha\gamma\kappa}$ and $C_{\alpha\kappa\kappa}$ cause a measurable shift

$$\Delta x = \gamma\kappa C_{\alpha\gamma\kappa} + \kappa^2 C_{\alpha\kappa\kappa} \tag{2.177}$$

at the chromatic image plane of the FHI filter, which is shown in Fig. 2.33 as a function of the parameter γ.

Along the y axis ($\gamma = 0$) only the second–degree dispersion shows up. The corresponding x shift

$$\Delta x(\gamma = 0) = \kappa^2 C_{\alpha\alpha\kappa} \tag{2.178}$$

yields for the coefficient of the second-degree dispersion $C_{\alpha\kappa\kappa} = 23.9R_0$. From the location γ_0 at which $\Delta x = 0$, we obtain

$$C_{\alpha\gamma\kappa} = -\kappa C_{\alpha\kappa\kappa}/\gamma_0 \tag{2.179}$$

which leads to a value $C_{\alpha\gamma\kappa} = -17.9R_0$.

The remaining coefficients ($B_{\alpha\beta\delta}, C_{\alpha\alpha\kappa}, C_{\beta\delta\kappa}$) at the achromatic image plane are very small and hence generate perceptible effects only if very large angular and energy deviations are introduced. However, such large values induce strong higher-rank aberrations, which will render the measurements false.

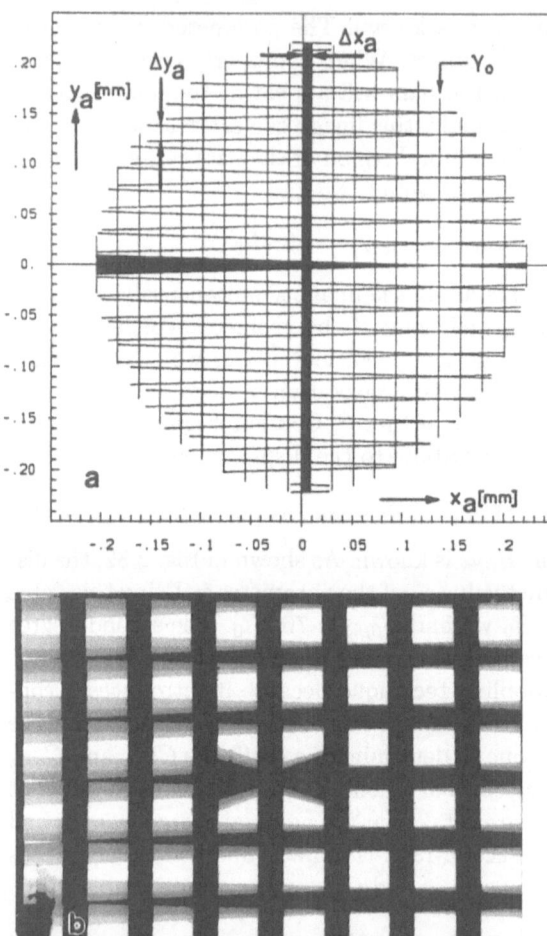

Fig. 2.32. Shift of the line (γ_o) unaffected by the β–wobble $(\Delta\beta = \pm 7 \cdot 10^{-4})$ after changing the nominal energy by $E = 200$ eV [2.49]; (**a**) computer simulated, (**b**) experimental.

2.10.4 Energy-Selection Plane

The dominant second-order aberrations at the energy-selection plane are the aperture aberrations with coefficents $A_{\gamma\gamma}$ and $B_{\gamma\delta\delta}$ and the first-order astigmatism. The resulting aberration figure can easily be calculated or measured. The only problem is posed by the unknown astigmatism coefficients $A_{\gamma\gamma}$ and $B_{\delta\delta}$. Their determination requires a comparison of computer-calculated caustic figures with those obtained experimentally. The aberration that matches the experimental aberration figure of the FHI filter (Fig. 2.34b) is obtained by choosing

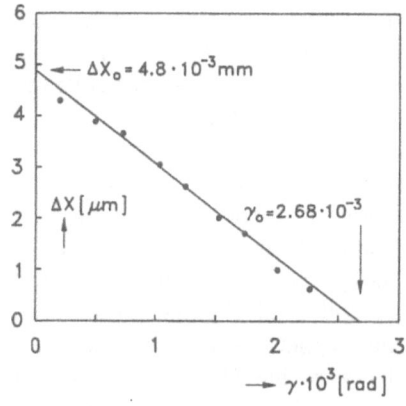

Fig. 2.33. Image shift in the x direction at the achromatic image plane z_a of the FHI filter caused by the aberration coefficients $C_{\alpha\gamma\kappa}$ and $C_{\alpha\kappa\kappa}$ after changing the nominal energy by $E = 200$ eV [2.49].

$$A_{\gamma\gamma\gamma} = -44.0R_0 \quad , \qquad B_{\gamma\delta\delta} = -34.4R_0 \quad ,$$
$$A_{\gamma\gamma} = -0.6R_0 \quad , \qquad B_{\delta\delta} = 0.05R_0 \quad . \qquad (2.180)$$

Comparison of the simulated aberration figure (Fig. 2.34a) with the experimental one shows a very good agreement.

Owing to misalignment and mechanical imperfections, the actual paraxial trajectories generally deviate from their ideal nominal course. The maximum

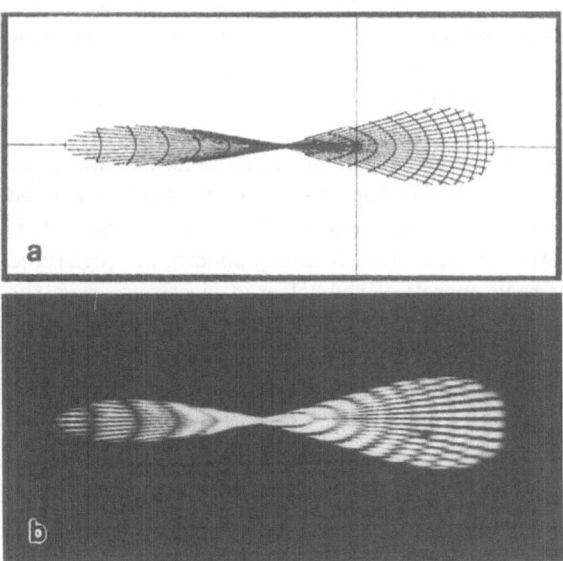

Fig. 2.34. Aberration figure at the energy-selection plane produced by the second-order aperture aberrations ($A_{\gamma\gamma\gamma}, B_{\gamma\delta\delta}$) and the first-order axial astigmatism ($A_{\gamma\gamma}, B_{\delta\delta}$) [2.49]; (**a**) computer-simulated, (**b**) experimental.

deviations at the energy-selection plane that result from inaccurate approximation of the extension of the fringing field are given by the coefficients $A_{\gamma\gamma}$ and $B_{\delta\delta}$. These astigmatic deviations occur, for example, if the filter magnets are designed on the basis of the Schwartz–Christoffel field, yet flat coils are placed at the bottom of the slits between the mirror plates and the poles of the magnets. In this case the actual field is given by the Linear Potential (LP) model, rather than by the Schwartz–Christoffel (SC) model, as discussed in detail in Sect. 2.4.3. The inaccurate consideration of the real fringing field also affects the deflection angle. Use of the values of the SC model for the LP model yields a deflection angle which differs by about 1% from the desired value $\Phi = 90°$. In practice, however, this angular misalignment can readily be eliminated by properly readjusting the magnetic field strengths of the deflection magnets. Since the fundamental ray x_γ has the largest off-axial distance of the four fundamental rays within the filter, this ray is especially sensitive to deviations from the true field. As a result, an intolerably large axial astigmatism $(A_{\gamma\gamma})$ may be introduced at the energy-selection plane, as is the case for the FHI filter. The computer simulations confirm this behaviour. On the other hand, the ray y_δ which stays close to the optic axis is rather insensitive to field inaccuracies, as demonstrated by the small value of the corresponding astigmatism coefficient $B_{\delta\delta}$. The smallness of $B_{\delta\delta}$ with respect to $A_{\gamma\gamma}$ has been confirmed for the FHI filter by measurements. In order to adjust the required symmetrical path of rays within the filter, therefore, it is necessary to incorporate a quadrupole stigmator at the midplane z_m of the filter. At this position the stigmator acts preferentially on the ray x_γ because a strongly distorted image of the object plane is located at this plane $(x_{\alpha m} = y_{\beta m} = 0, y_{\delta m} \ll x_{\gamma m})$. By means of this stigmator the required symmetric path of the fundamental rays can always be adjusted, as will be demonstrated in Sect. 2.10.6.

The chromatic aberrations at the energy-selection plane do not affect the image quality of the filtered image if the displacements are small compared with the first-order dispersion. For existing energy filters, this condition is always fulfilled. The same statement holds true if energy spectra are registered sequentially, because the average energy of the selected electrons stays the same, regardless of the selected energy loss, provided that the energy-selection is performed by varying the acceleration voltage.

The aberration coefficients of the uncorrected FHI energy filter have been obtained by evaluating the aberration integrals (2.139–143). The normalized results are listed together with the measured values in Table 2.3. The agreement between the two sets of aberration coefficients attests to the validity of the measurement procedure. This observation justifies the use of the method for the correction of the second-order aberrations by sextupoles.

2.10.5 Sextupole Alignment

Before we describe the correction of the residual second-order aberrations, we discuss the procedure necessary for properly positioning the sextupoles along the filter axis. The axis of each sextupole must coincide very precisely with the optic axis to avoid parasitic multipole fields, notably dipole, quadrupole, and Ψ_{3c} sextupole fields. The location of the sextupole correctors inside the filter can be seen from Fig. 2.26. To observe the effects of these correctors we again use the grid inserted at the plane z_i in front of the filter. The sextupoles S1 and S7 are located near the image planes of the filter and hence do not affect the transferred shadow image. For visualizing the effect of these elements, we focus the transfer lens onto a plane between the chromatic and the energy-selection plane. The symmetric caustic figure seen in this plane when the sextupole correctors are not excited is properly aligned to the reference point, which defines the point of intersection of the filter axis with the observation plane, as shown in Fig. 2.35. The sextupole magnets are first excited and aligned individually and then combined to form pairs.

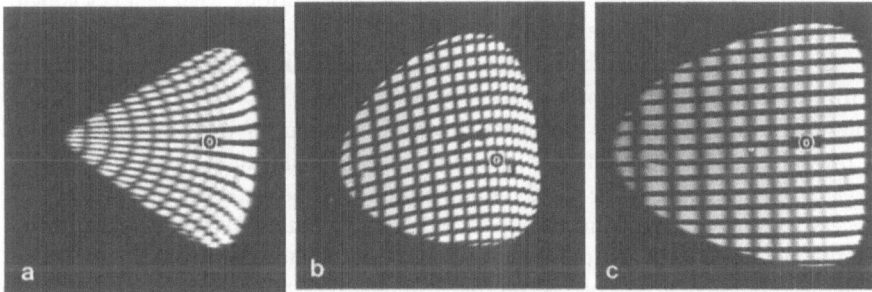

Fig. 2.35. Aberration figures used as criteria for the alignment accuracy of the sextupole correctors with respect to the filter axis. The initial aberration figure (a) (without exciting a sextupole) is symmetric. The effect of a laterally displaced sextupole (b) is indicated by a distortion due to a quadrupole Ψ_{2c} and by a shift of the entire figure caused by the dipole terms Ψ_{1c} and Ψ_{1s}. Vanishing of the distortion and of the shift indicates that the filter axis and the sextupole axis coincide (c).

As outlined in Sect. 2.6.1, Ψ_{3s} sextupole fields must be created for correcting the second-order aberrations of a deflection system with midsection symmetry. The index s indicates a sine dependence of the potential on the azimuthal coordinate. In order to obtain such sextupole fields, the corrector elements are positioned such that its polepieces are located symmetrically about the midsection $y = 0$ of the deflection magnets and two polepieces are centred on the y axis.

To study the effect of a sextupole which is displaced by shifts Δx and Δy from the optic axis of the filter, we consider the corresponding scalar magnetic potential

$$
\begin{aligned}
\psi_3 &= \Psi_{3s}\left[3(x+\Delta x)^2(y+\Delta y)-(y+\Delta y)^3\right] \\
&= \Psi_{3s}(3x^2y-y^3)+6\Psi_{3s}xy\Delta x+3\Psi_{3s}\Delta y(x^2-y^2) \quad\quad (2.181) \\
&+ 3\Psi_{3s}(\Delta x^2-\Delta y^2)y+6\Psi_{3s}x\Delta x\Delta y+\Psi_{3s}(3\Delta y\Delta x^2-\Delta y^3) \quad,
\end{aligned}
$$

where we have neglected the fringing fields.

The first term in the last relation describes the actual sextupole field, the second and third term denote parasitic quadrupoles with strengths

$$
\Psi_{2s}=3\Psi_{3s}\Delta x \quad, \quad \Psi_{2c}=3\Psi_{3s}\Delta y \quad, \quad\quad (2.182)
$$

while the fourth and fifth terms describe deleterious dipole fields whose strengths

$$
\Psi_{1s}=3\Psi_{3s}(\Delta x^2-\Delta y^2) \quad, \quad \Psi_{1c}=6\Psi_{3s}\Delta x\Delta y \quad\quad (2.183)
$$

depend quadratically and bilinearly, respectively, on the misalignment shifts.

The quadrupole strength Ψ_{2s} causes a first-order astigmatism, leaving the symmetry of the caustic figure unchanged, while the Ψ_{2c} term yields a distortion of the caustic, because the principal sections of this quadrupole are rotated by 45° with respect to those of the Ψ_{2s} quadrupole. The dipole fields displace the entire caustic in the x and y directions with respect to the reference point, as can be seen from Fig. 2.35.

In the first alignment step, the correctors are shifted mechanically in the y direction until the distortion is cancelled ($\Delta y = 0$). It follows from (2.183) that the dipole term Ψ_{1c} has then been eliminated as well. A shift in the x direction until the centre of the caustic figure meets the reference point ($\Delta x = 0$) compensates for the Ψ_{2s} quadrupole and the Ψ_{1s} dipole (Fig. 2.35). If necessary, the centre of the sextupole field can be tuned by exciting trimming dipole fields, which are created by additional windings situated on every polepiece of each sextupole corrector.

After the seven sextupoles have been aligned with the filter axis, the correction of the second-order aberrations at the achromatic and the energy-selection plane can be performed without affecting the paraxial adjustment.

2.10.6 Correction by Sextupoles

In principle, each second-rank aberration coefficient can be compensated for by a properly excited sextupole. However, the corresponding hexapole field generally produces other aberration coefficients too. This behaviour can strongly hamper the correction in practice. For example, to prevent the sextupoles from introducing contributions to aberrations compensated by symmetry, it is necessary to incorporate the correctors in pairs placed symmetrically about the central plane of the filter. The single corrector located at the midplane automatically fulfils this condition.

It is advantageous to use the normalized sextupole strengths [2.48]

$$
K_n = \epsilon\mu(2e/m_0\Phi^*)^{1/2}\cdot(NI)\ell_n R_0^2/a_n^3 \quad\quad (2.184)
$$

for describing the effect of the sextupoles. Here $(NI)_n, \ell_n, a_n$ denote the ampere-turns, the length of the sextupole field, and the bore radius of the n–th sextupole, respectively; μ is the permeability of the polepiece material; R_0 is the deflection radius of the magnets. The geometrical factor ϵ depends on the width of the gaps between the polepieces and is somewhat smaller than unity. Using for K_n the theoretical values that yield complete correction of the second-order aberrations, an experimental procedure has been developed that allows one to adjust very accurately the ratios of the sextupole strengths.

Correction at the Achromatic Image Plane. The expressions (2.139) and (2.140) for the geometric eikonal coefficients reveal that the coefficients $A_{\alpha\alpha\gamma}, B_{\alpha\beta\delta}, B_{\beta\beta\gamma}$ of the mixed second-order aberrations can be compensated by three Ψ_{3s} sextupole pairs. Their constituent elements are positioned in the drift spaces between the first and second and the third and fourth magnets, respectively. Furthermore, it is advantageous to place the sextupoles of the pair S2/S6 close to the exit pole-face of the first and to the entrance pole-face of the fourth magnet, respectively, because at these positions the sextupole fields act almost exclusively on the coefficent $B_{\beta\beta\gamma}$. This behaviour stems from the fact that at these planes the off-axial distances of the fundamental rays x_γ and y_β are considerably larger than those of the two other rays and x_α and y_δ.

Since the spaces between the magnets are limited, one can try to reduce the number of sextupole pairs by varying their position along the optic axis of the filter. Indeed, as has been shown by *Lanio* [2.48], the three linearly independent coefficients of the mixed second-order aberrations can be eliminated by means of only two sextupole pairs if they are placed at particular positions. In accordance with Lanio, we place a third sextupole pair with a normalized strength K_{add} midway between the correctors S2,S3 and S5,S6. With these assumptions, the sextupole strengths necessary for complete correction of the mixed second-order aberrations have been calculated. The resulting values $K_2 = K_6, K_3 = K_5$ and K_{add} are shown in Fig. 2.36 as functions of the normalized distance d_2/R_0 where d_2 denotes the distances of the correctors S2 and S6 from the "exit" plane of the first and the "entrance" plane of the fourth magnet, respectively. These planes are located midway between the magnetic pole-faces and the mirror plates. The curves reveal that the strength of the additional sextupole pair vanishes ($K_{add} = 0$) for a distance $d_2 = 0.594R_0$ of the sextupole pair S2/S6. Hence, for this preferential position the additional sextupole pair is unnecessary.

Because the two sextupole pairs act simultaneously on all second-order aberrations at the achromatic image plane, their theoretically determined normalized strengths

$$K_2 = K_6 = -1.26 \quad , \quad K_3 = K_5 = -4.16 \tag{2.185}$$

and their distances

$$d_2 = d_6 = 0.594R_0 \quad , \quad d_3 = d_5 = 0.391R_0 \tag{2.186}$$

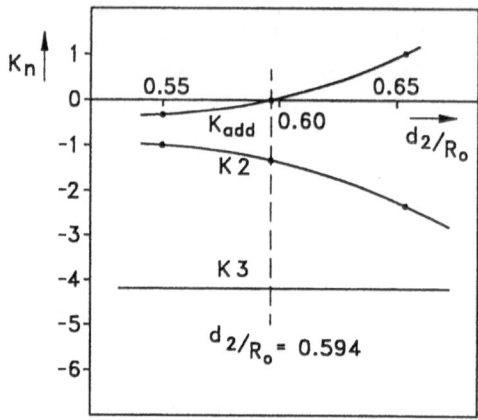

Fig. 2.36. Sextupole strengths $K_n(n = 2.3)$ and K_{add} required for correction of the geometrical second-order aberrations at the achromatic image plane as functions of the normalized distance d_2/R_0 of the sextupole S2 from the exit plane of the first deflection magnet. Notice that for the sextupole S6 the strength and the distance must be $K_6 = K_2$ and $d_6 = d_2$.

must be adjusted rather accurately at the outset to enable the second-order correction to be achieved in practice. In order to find the exact strengths experimentally, we use as a criterion the vanishing of the largest aberration coefficient, namely $B_{\beta\beta\gamma}$, activated by the wobble procedure described in Sect. 2.10.2. The behaviour of the corresponding aberration serves as an indicator of the action of the sextupole pair under consideration. For this purpose, we study how $B_{\beta\beta\gamma}$ varies when operating one sextupole pair and leaving the other one unactivated. The correction procedure must be supported by computer calculation of the resulting aberration coefficient $B_{\beta\beta\gamma}$. After having excited only a single pair with the nominal strength required for the complete correction if both sextupole pairs are excited, the results of the calculations are:

$$B_{\beta\beta\gamma}(3/5) = 102.5R_0 \quad , \quad B_{\beta\beta\gamma}(2/6) = -242.6R_0 \quad . \qquad (2.187)$$

We then compensate $B_{\beta\beta\gamma}$ experimentally by exciting successively each of the two pairs until the penumbral regions of the shadow image of the grid vanish and measure the currents flowing through the coils of the correctors. For the FHI-filter, the currents have been found to be

$$I(2/6) = 41.9\,\text{mA} \quad , \quad I(3/5) = -11.7\,\text{mA} \qquad (2.188)$$

In both cases, complete correction of $B_{\beta\beta\gamma}$ is achieved by one sextupole pair. However, if both pairs are exited, the aberrations introduced by one pair must be compensated for by the aberrations produced by the other. The correct currents I_c for achieving complete correction can be derived from the following expressions:

$$I_c(2/6) = [B_{\beta\beta\gamma}(3/5)/B_{\beta\beta\gamma}] \cdot I(2/6) \quad , \qquad (2.189)$$

$$I_c(3/5) = [B_{\beta\beta\gamma}(2/6)/B_{\beta\beta\gamma}] \cdot I(3/5) \quad . \qquad (2.190)$$

Taking into account the aberration coefficient $B_{\beta\beta\gamma} = -147R_0$ for the FHI-filter, we find for the sextupoles the currents

$$I_c(2/6) = -29.2 \text{ mA} \quad , \quad I_c(3/5) = -19.3 \text{ mA} \quad . \tag{2.191}$$

The effect of the correction by sextupoles is illustrated in Fig. 2.37 where the calculated relative values $A_{\gamma\gamma\gamma}(z)/A_{\gamma\gamma\gamma}(\infty)$ and $B_{\beta\beta\gamma}(z)/B_{\beta\beta\gamma}(\infty)$ of the eikonal coefficients $A_{\gamma\gamma\gamma}(z)$ and $B_{\beta\beta\gamma}(z)$ are shown along the optic z-axis of the filter. The experimental correction of the aberration with coefficient $B_{\gamma\beta\beta}$ is demonstrated in Fig. 2.31.

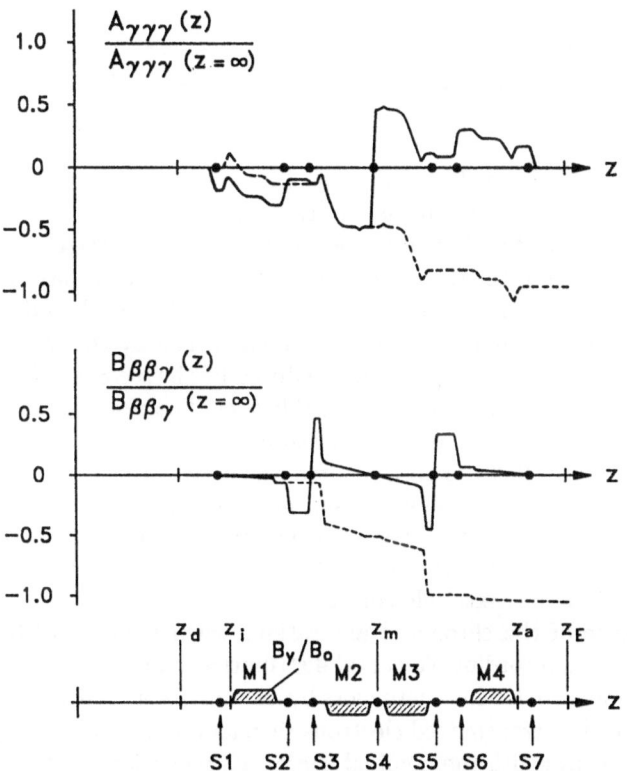

Fig. 2.37. The eikonal coefficients $A_{\gamma\gamma\gamma}(z)$ and $B_{\beta\beta\gamma}(z)$ of the FHI filter as functions of the location z without (- - -) and with (—) sextupole correction. The normalization has been performed by means of the coefficients $A_{\gamma\gamma\gamma}(\infty)$ and $B_{\beta\beta\gamma}(\infty)$ without sextupole correction.

So far we have only considered the geometrical aberrations. However, the filter also introduces second-rank chromatic aberrations. These aberrations are produced either directly or as combination aberrations resulting from the dispersion. Owing to the dispersion, the axis of energy-loss electrons is

Table 2.4. Chromatic aberration coefficients of the FHI filter remaining after the sextupole correction of the geometrical aberrations. The coefficients are normalized with respect to the deflection radius R_0.

CHROMATIC ABERRATIONS					
image plane			energy-selection plane		
coefficient	calculated	measured	coefficient	calculated	measured
$C_{\alpha\alpha\kappa}$	−40.0	−39.0	$C_{\gamma\gamma\kappa}$	−34.0	−38.2
$C_{\beta\beta\kappa}$	571.0	544.1	$C_{\delta\delta\kappa}$	79.0	−
$C_{\alpha\gamma\kappa}$	0.1	≤ −0.6	$C_{\alpha\gamma\kappa}$	≤ −0.6	≤ −0.6
$C_{\beta\delta\kappa}$	−0.2	−	$C_{\beta\delta\kappa}$	≤ −0.2	−
$C_{\alpha\kappa\kappa}$	−109.0	−101.8	$C_{\gamma\kappa\kappa}$	−18.5	−

displaced from the optic axis of the filter. This displacement induces second-rank aberrations within the deflection fields and the sextupole fields.

The correction of the geometrical aberrations strongly influences the chromatic aberrations. According to the theoretical predictions, the coefficients of the chromatic aberration of magnification ($C_{\alpha\gamma\kappa}, C_{\beta\delta\kappa}$) should vanish completely (see Sect.2.7.2). These predictions have been confirmed by the measurements. For example, the coefficient $C_{\alpha\gamma\kappa}$ was reduced by the second-order correction by almost two orders of magnitude (Table 2.4). The resulting increase of the chromatic term $C_{\beta\beta\kappa}$ has been measured using the procedure outlined in Sect. 2.10.2 (Fig. 2.38). The measured value agrees rather well with the calculated one. These convincing agreements indicate that the correction of the second-order aberrations has been performed with sufficient accuracy. In Table 2.4 we have listed the chromatic aberration coeffients calculated and measured after the sextupole correction.

To limit the influence of the chromatic aberrations when imaging with selected energy losses, when recording elemental distributions, for example, it is advantageous to compensate for the energy loss by an equivalent increase of the acceleration voltage. The transmitted electrons then travel along the true optic axis of the filter always with the nominal energy E_0 regardless of their specific energy loss E. Consequently, the chromatic aberration is determined by the width of the energy window rather than by the total energy loss.

The energy filter is placed most suitably behind one of the intermediate lenses within the column of the microscope. For example, the FHI-filter is inserted in front of the last projector lens. A strongly demagnified image of the diffraction diagram is then formed at the diffraction plane z_d. To study the influence of the dominant chromatic aberrations on the image quality, it is convenient to refer the aberrations back to the object plane, since they are then independent of the total magnification.

The components of the second-rank axial chromatic aberrations become

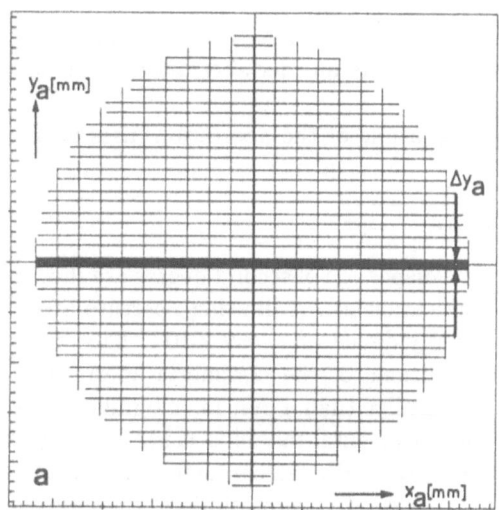

Fig. 2.38. Penumbral seam produced by the β–wobble ($\Delta\beta = \pm 7 \times 10^{-4}$) and by the residual chromatic aberration coefficient $C_{\beta\beta\kappa}$ after changing the nominal energy by $E = 20$ eV [2.49]; (a) computer simulated, (b) experimental.

$$\Delta x_o = \alpha_o \kappa C_{\alpha\alpha\kappa}/M_i^2 \ , \tag{2.192}$$

$$\Delta y_o = \beta_o \kappa C_{\beta\beta\kappa}/M_i^2 \ . \tag{2.193}$$

The chromatic displacement

$$\Delta x_o = \kappa^2 C_{\alpha\kappa\kappa}/M_i \tag{2.194}$$

resulting from the second-degree dispersion also depends on the magnification M_i of the intermediate image at the plane z_i in front of the filter. On the other hand the chromatic aberration of magnification

$$\Delta x_o = x_o \kappa C_{\alpha\gamma\kappa}/L \ , \tag{2.195}$$

$$\Delta y_o = y_o \kappa C_{\beta\delta\kappa}/L \tag{2.196}$$

does not depend on the location of the filter. Fortunately, the coefficients $C_{\alpha\gamma\kappa}$ and $C_{\beta\delta\kappa}$ are very small after all second-order aberrations have been corrected by the sextupoles, as depicted in Table 2.4. The fact that these coefficients are not zero indicates that the second-order aberrations have not been completely eliminated.

An appropriate measure of the magnitude of the aberration discs is the mean quadratic diameter referred to the object plane

$$d_o = \sqrt{2}\sqrt{\Delta x_o^2 + \Delta y_o^2} \quad . \tag{2.197}$$

The axial chromatic aberration and the dispersion of second-degree are shown in Fig. 2.39 as functions of the magnification M_i. A limiting aperture angle $\alpha_o = \beta_o = 10^{-2}$ rad and a relative energy deviation $\kappa = 2 \times 10^{-4}$ have been assumed. As can be seen, the contributions of the chromatic aberrations introduced by the filter itself remain smaller than 1 nm for magnifications $M_i \geq 2 \times 10^2$. Thus, even in the case of elemental mapping using an energy window of 20 eV at a nominal energy of 100 keV, the chromatic aberrations of the filter do not affect the spatial resolution. The resolution is limited in this case by the chromatic aberration of the objective lens.

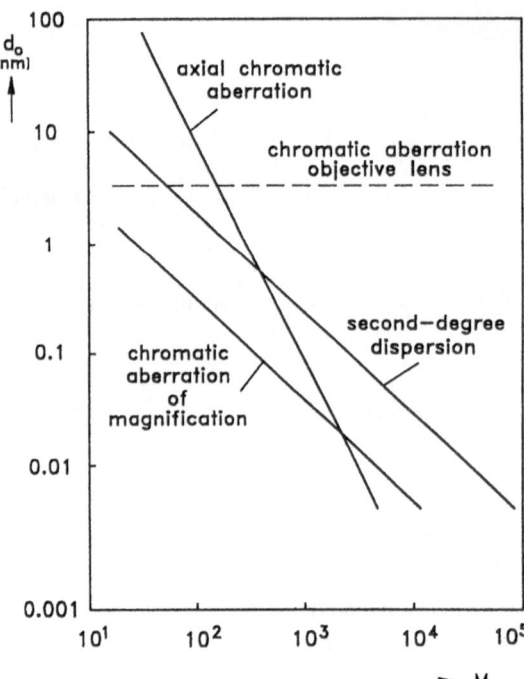

Fig. 2.39. Averaged diameters d_o of the residual chromatic aberrations of the second-order-corrected FHI filter referred to the object plane as functions of the magnification M_i of the intermediate image in front of the filter.

The usable image field at the plane z_i in front of the FHI-filter must be limited by an aperture to a radius $r_i = 0.5$ mm in order to avoid the influence of aberrations higher than second rank. In this case the chromatic aberrations determine the number of equally well resolved object points per diameter of the image field:

$$N = \frac{\sqrt{2} r_i M_i}{[C_{\alpha\alpha\kappa}^2 + C_{\beta\beta\kappa}^2 + (M_i\kappa/\alpha_o) \cdot C_{\alpha\kappa\kappa}^2]^{1/2} \alpha_o \kappa}. \qquad (2.198)$$

The results are depicted in Fig. 2.40. Two curves have been calculated for objective apertures $\alpha_o = \beta_o = 10^{-2}$ rad and relative energy deviations $\kappa = 2 \times 10^{-5}$ and $\kappa = 2 \times 10^{-4}$, taking into account the chromatic aberration coefficients of the FHI-filter listed in Table 2.4. The curves demonstrate that even with relatively large energy windows of 20 eV ($\kappa = 2 \times 10^{-4}$) as used for elemental mapping, more than $N = 10^3$ image points can be resolved for magnification $M_i \geq 10^2$. For contrast enhancement the energy window is limited to 2 eV ($\kappa = 2 \times 10^{-5}$) in order to filter out all inelastically scattered electrons. In this case, the effect of the chromatic terms is reduced to such an extent that they can be completely neglected. Hence, for elemental mapping and zero-loss filtering the remaining chromatic aberrations of the second-order-corrected energy filter do not affect the achievable lateral resolution.

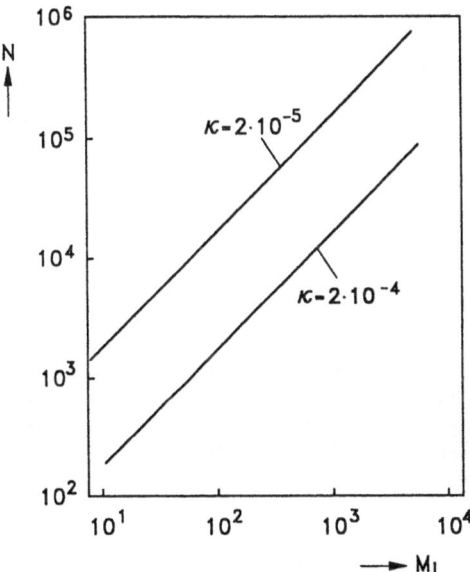

Fig. 2.40. Number N of equally well resolved object points along a diameter $d_a = 1$ mm of the achromatic image of the second-order-corrected FHI filter as a function of the corresponding magnification M_i assuming $\alpha_o = \beta_o = 10^{-2}$ for two energy windows (a) 2 eV ($\kappa = 2 \times 10^{-5}$) and (b) 20 eV ($\kappa = 2 \times 10^{-4}$), respectively.

First- and Second-Order Correction at the Energy-Selection Plane.
The dominant aberrations at the energy-selection plane are the second-order aperture aberration ($A_{\gamma\gamma\gamma}, B_{\gamma\delta\delta}$) and the first-order axial astigmatism ($A_{\gamma\gamma}, B_{\delta\delta}$). The latter aberration results from misalignment and machining imperfections.

As outlined in Sect.2.6.2, the aperture aberrations at the energy-selection plane are responsible for the non-isochromatic imaging. In Fig. 2.41 this effect is demonstrated by means of a plasmon-loss-filtered image of Al_3Li precipitates at an energy loss of 13.5 eV [2.50]. Before we took this micrograph, the coefficient $A_{\gamma\gamma}$ has been completely compensated and the aberration coefficient $A_{\gamma\gamma\gamma}$ was partially corrected to a value $A_{\gamma\gamma\gamma} = -17R_0$. In this case the energy shift at the final image plane has been found to be

$$\delta E = -0.288 \cdot r_I^2 \text{ eV} \quad , \tag{2.199}$$

where r_I in units of cm is the off-axial distance of the image point. Setting an energy window of 4 eV, all 13.5 eV loss electrons, which would contribute to the image field with $r_I > 2.7$ cm, are stopped by the slit aperture at the energy-selection plane so that the precipitates outside this radius disappear, as shown in Fig. 2.41a and explained in Fig. 2.18.

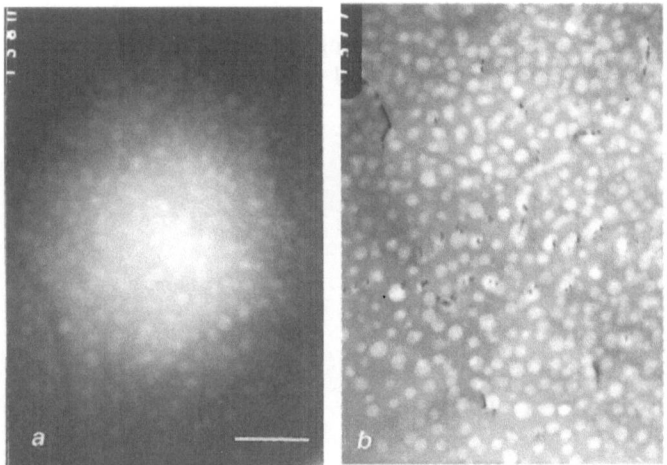

Fig. 2.41. Plasmon-loss-filtered images of Al$_3$Li precipitates using an energy window of 4 eV centred about the plasmon energy loss of 13.5 eV; (**a**) non-isochromatic filtering caused by the second-order aperture aberrations at the energy-selection plane $(A_{\gamma\gamma\gamma} = -17R_0,\ B_{\gamma\delta\delta} = -34R_0)$, (**b**) isochromatic imaging of the Al$_3$Li precipitates after second-order correction, $A_{\gamma\gamma\gamma} = B_{\gamma\delta\delta} = 0$ (bar = 200 nm).

In order to compensate for the two linearly independent second-order coefficients $A_{\gamma\gamma\gamma}$ and $B_{\gamma\delta\delta}$, three sextupoles are required. The first (S1) is located very close to the image plane z_i and the second (last of all) (S7) close to the conjugate achromatic image plane z_a. The third sextupole (S4) is placed as a single corrector at the central plane z_m of the filter. Owing to these special locations the sextupoles S1, S4, and S7 do not influence the second-order coefficients at the achromatic image plane because the fundamental rays x_α and y_β intersect the optic axis of the filter at the image planes. The quadrupole field, which is superposed onto the sextupole field at the mid–plane z_m, acts preferentially on the coefficient $A_{\gamma\gamma}$ because a strongly distorted image is formed in this plane $(x_{\gamma m} \gg y_{\delta m})$. The coefficient $B_{\delta\delta}$ is negligibly small for the FHI-filter.

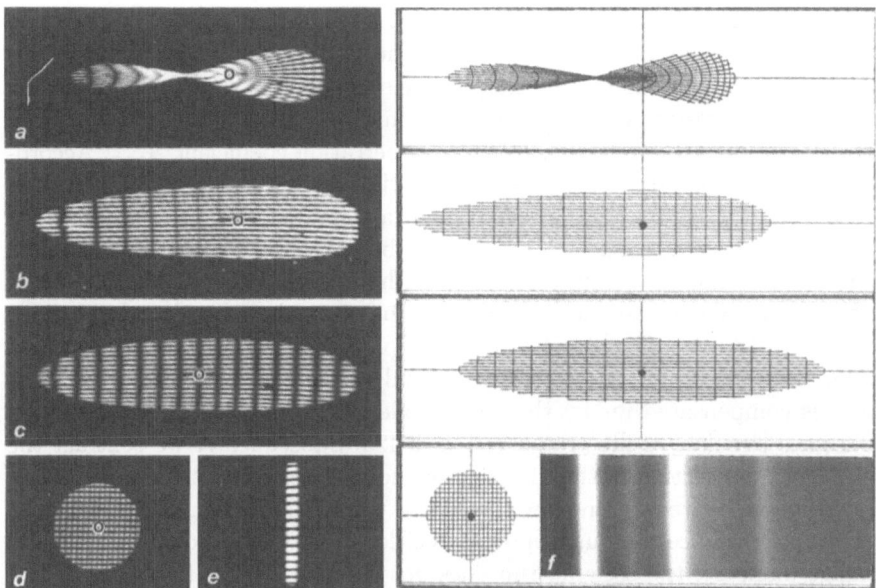

Fig. 2.42. Procedure to compensate for the first- and second-order aberrations at the energy-selection plane supported by computer-simulated images [2.49]; (a) caustic figure produced by the aberration coefficients $A_{\gamma\gamma\gamma}, B_{\gamma\delta\delta}$, and $A_{\gamma\gamma}$; (b) correction of $B_{\gamma\delta\delta}$ by exciting the sextupole pair S1/S7 indicated by the disappearance of the curvature of the grid bars; (c) elimination of the aberration coefficient $A_{\gamma\gamma\gamma}$ by symmetrization of the caustic by means of the sextupole S4; (d) correction of the first-order term $A_{\gamma\gamma}$ indicated by a circular caustic achieved by activating the quadrupole located at the central plane of the filter; (e) elimination of the extension of the caustic in the energy-dispersive direction by the quadrupole located behind the filter; (f) spectrum of an Al film obtained after performing the correction steps (b)–(f).

In the correction procedure, we observe the bars of the shadow image of the grid at a plane located somewhat above the energy-selection plane z_E. No additional procedure is required for the correction. In Fig. 2.42 the correction procedure is outlined in steps. The second-order coefficients and the first-order astigmatism cause the displacements

$$\Delta x = \gamma^2 A_{\gamma\gamma\gamma} + 0.5\,\delta^2 B_{\gamma\delta\delta} + \gamma A_{\gamma\gamma} \quad , \tag{2.200}$$

$$\Delta y = \gamma\delta B_{\gamma\delta\delta} + \delta B_{\delta\delta} \quad , \tag{2.201}$$

which produce the caustic figure shown in Fig. 2.42a. The actual compensation of the coefficient $B_{\gamma\delta\delta}$ is achieved by properly adjusting the sextupole pair S1/S7. The relations (2.200) and (2.201) indicate that this coefficient produces a curvature of the grid bars; the disappearance of this curvature indicates that $B_{\gamma\delta\delta}$ has been corrected (Fig. 2.42 b). The remaining aberration coefficient $A_{\gamma\gamma\gamma}$ causes a shift of the entire figure in the x direction, which depends quadratically on the parameter γ. As shown in Fig. 2.42b, the caustic remains asymmetric with respect to the marked centre of the grid. This imperfection is eliminated by exciting the sextupole S4. Owing to the strong first-order distortion of the intermediate image located at the mid-plane z_m, this sextupole acts preferentially on the aberration coefficient $A_{\gamma\gamma\gamma}$. Accordingly, the figure can be rendered symmetric, as shown in Fig. 2.42c. In this case $A_{\gamma\gamma\gamma}$ is also corrected. The first-order astigmatism term $A_{\gamma\gamma}$ enlarges the caustic in the x direction. This term can be cancelled by a quadrupole field superposed onto the sextupole field S4 without introducing any second-order aberrations. Owing to the limited space between the second and the third magnet a twelve-pole corrector has been placed at the central plane, which allows one to excite pure quadrupole and sextupole fields. The coefficient $A_{\gamma\gamma}$ is compensated for by the quadrupole field if the caustic cross section becomes circular, as illustrated in Fig. 2.42d. Subsequently we contract the circular cross-section in the direction of the dispersion by means of a second quadrupole located behind the filter until the smallest diameter is achieved (Fig. 2.42e). Since the quadrupole is positioned very close to the achromatic image plane z_a, the astigmatism does not show up at the final image. The associated defocus in the y direction does not disturb the filtering conditions because the slit aperture in the energy-selection plane is extended in this direction. By this measure we obtain a band-like energy spectrum, as can be seen from the energy spectrum of an Al film shown in Fig. 2.42f.

Once the energy filter has been corrected to first and second order, it is capable of performing isochromatic energy-filtering, as demonstrated in Fig. 2.41b, which shows the image of Al_3Li precipitates taken with 13.5 eV energy-loss electrons. No energy shift occurs in the final image in contrast to the image (Fig. 2.41a) transferred by the uncorrected energy filter.

Finally, we will discuss the effect of the remaining chromatic aberrations on the energy spectrum. Only the aberration coefficients that cause a displacement in the direction of the dispersion need be considered . The residual value of the chromatic aberration coefficient $C_{\alpha\gamma\kappa}$ is negligibly small after the second-order correction. This behaviour does not hold for the second-degree dispersion, which causes a quadratic distortion of the energy-loss spectrum. The resulting shift

$$\delta E_s = \kappa^{*2} C_{\gamma\kappa\kappa}/\Delta \tag{2.202}$$

is inversely proportional to the dispersion $\Delta = C_{\gamma\kappa}/E_0$ of the energy filter. For the FHI-filter this dispersion is $\Delta = 6 \times 10^{-3}$ mm/eV at a nominal energy of $E_0 = 10^5$ eV. The quadratic energy displacement does not disturb the energy resolution because this shift is independent of the sign of the energy deviation and can be corrected by suitably displaying the spectrum. On the other hand, the axial chromatic term $C_{\gamma\gamma\kappa}$ which is composed of an energy-dependent defocus and a chromatic axial astigmatism (see Sect.2.6.2) causes a broadening

$$\delta E_B = \gamma\kappa^* C_{\gamma\gamma\kappa}/\Delta \tag{2.203}$$

of the individual energy losses. Therefore, this aberration affects the energy resolution of the recorded energy loss spectrum.

Both the nonlinear shift δE_s and the broadening δE_B can be neglected if the energy-loss spectra are recorded sequentially by steadily changing the acceleration voltage. The mean energy of the selected energy loss then stays constant, so that the spread δE_B depends on the width of the chosen energy window. For example, if we select an energy range of 2 eV and choose an image field with radius $r_i = 0.5$ mm the resulting deviation $\delta E_B = 3 \times 10^{-2}$ eV can be neglected.

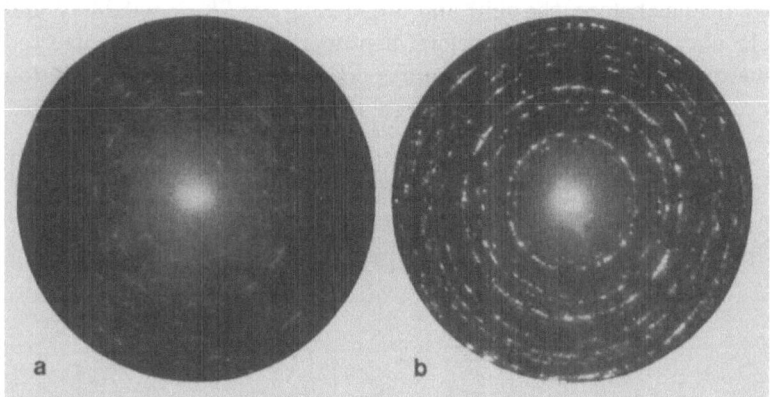

Fig. 2.43. Non-filtered (**a**) and zero-loss-filtered (**b**) diffraction diagrams of an Al_2O_3 film (H. Boersch [2.51]).

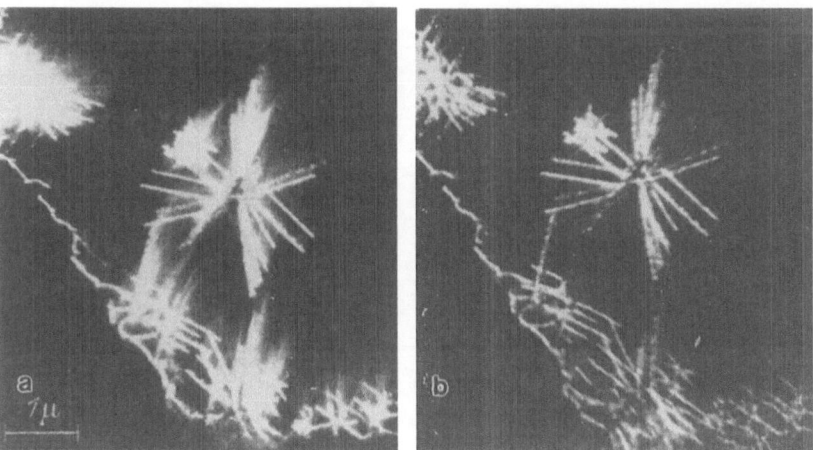

Fig. 2.44. Dark-field images of ZnO needles; (**a**) non-filtered and (**b**) zero-loss-filtered (G. Möllenstedt and O. Rang [2.52]).

The situation is quite different if the spectra are registered by a parallel-recording detector. Assuming a recorded energy range of 500 eV, taken from a limited image field with radius $r_i = 5 \times 10^{-2}$ mm, the nonlinear shift and the broadening of the 500 eV loss electrons at the far end of the spectrum become $\delta E_s = -3.8$ eV and $\delta E_B = 1.5$ eV. The recording of such a wide energy spectrum by the detectors currently used for parallel-recording, for example CCD-converters, is far from optimal because the energy resolution is determined by the pixel size of the detector rather than by the broadening δE_B. Hence, this aberration does not impair the energy resolution. Limiting the recorded energy range of the spectrum to about 50 eV and taking the same image field as before, the shift and broadening are $\delta E_s = -0.28$ eV and $\delta E_B = 0.15$ eV, respectively. Therefore, a properly designed and corrected imaging energy filter enables one to record spectra with high energy resolution which is needed, for example, to investigate near-edge (ELNES) structures.

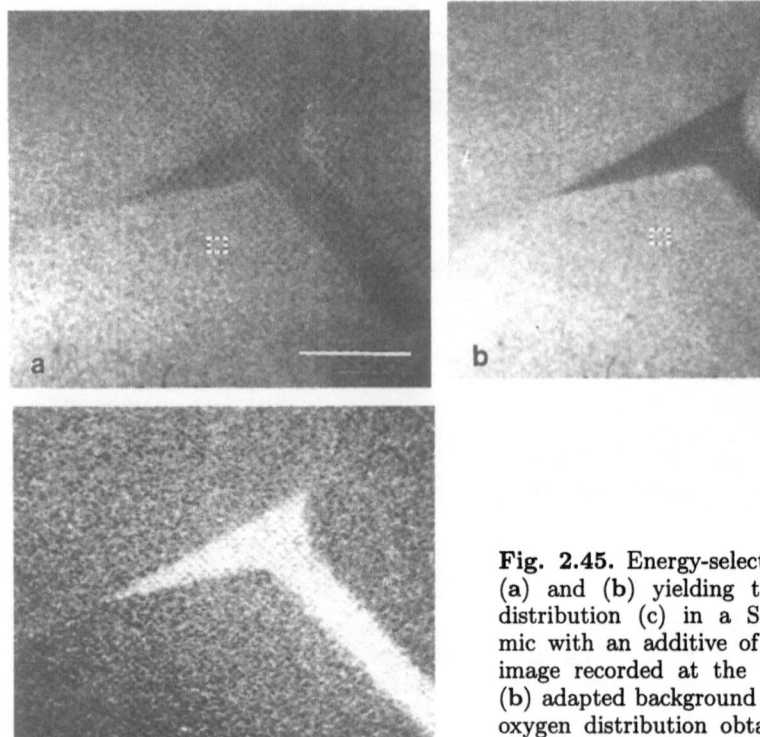

Fig. 2.45. Energy-selected images (a) and (b) yielding the oxygen distribution (c) in a Si_3N_4 ceramic with an additive of MgO; (a) image recorded at the O K-edge, (b) adapted background image, (c) oxygen distribution obtained after background subtraction (bar = 200 nm).

2.11 Conclusion

In the preceding sections we have outlined in detail both the theoretical principles and the methods for calculating and designing high-performance imaging energy filters and the experimental procedures necessary for accurately aligning the filter and for eliminating unwanted aberrations. Moreover, we have shown that symmetry conditions imposed on the filter reduce significantly the number of aberrations and, in addition, simplify considerably the actual alignment of the filter, because symmetry violations can be detected much more easily than deviations from a given nominal value.

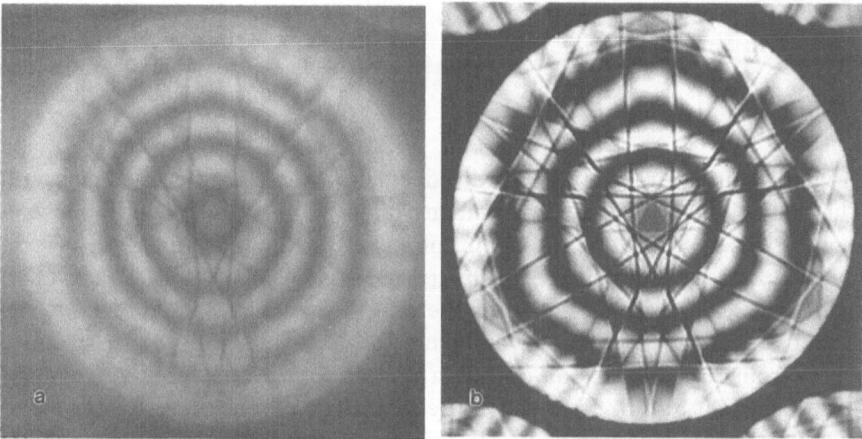

Fig. 2.46. Convergent-beam diffraction diagrams of an approximately 200 nm thick [111] oriented Si film taken at 100 kV; (**a**) unfiltered, (**b**) zero-loss filtered.

Here we have considered only imaging energy filters that allow the selection of an arbitrary energy loss. However, it should be noted that the idea of enhancing the contrast and the specimen resolution of relatively thick objects by filtering out the inelastically scattered electrons was proposed and performed long before the first imaging energy filter was developed. For this purpose *Boersch* [2.51] employed a retarding grid filter, while *Möllenstedt* and *Rang* [2.52] utilized the filtering properties of a strongly excited electrostatic retarding projector lens. Both filters act as high-pass filters which reflect all electrons whose velocities are too small to surmount the potential barrier, typically about 10 to 20 V above the cathode potential.

The retarding grid was used by Boersch to obtain a largely homogeneous electric field, thus minimizing the aberrations. Moreover, by placing the grid at an intermediate image of the object plane Boersch was able to filter large diffraction patterns without introducing third-order distortions and chromatic aberration of magnification at the recording plane. In addition

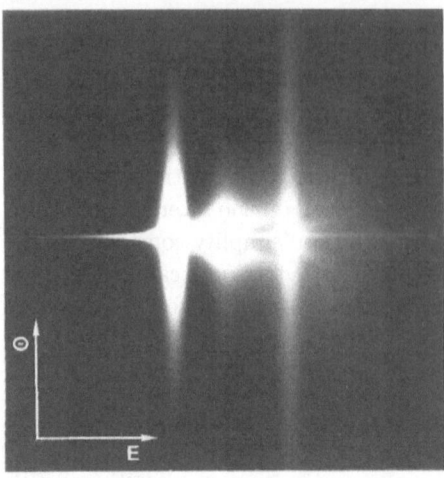

Fig. 2.47. Angularly resolved energy spectrum of a 15 nm thick Al film showing the surface-plasmon dispersion. The straight line corresponds to the volume plasmon at 13.5 eV.

the image curvature and the third-order field astigmatism were kept small. This behaviour is convincingly demonstrated in Fig. 2.43b, which shows a zero-loss-filtered diffraction diagram of an Al_2O_3 film obtained by Boersch. For comparison, the unfiltered diffraction pattern is shown on the left-hand side (Fig. 2.43a). The insertion of the grid creates a major problem as the mesh bars scatter the electrons and emit secondary electrons, which produce a background, preventing an optimum increase of the contrast.

Möllenstedt and Rang operated the retarding projector lens at the second maximum of the refraction power. In this case an image of the diffraction plane is formed at the centre of the lens. As a result, the integrands of the aberration coefficients which contain odd powers of the field rays $x_\gamma = y_\delta$ become antisymmetric functions if the field is symmetric with respect to the central plane of the lens. In this case, the third-order distortion and the chromatic aberration of magnification vanish. Thus, already these authors intuitively used symmetry conditions for eliminating aberrations, although this powerful principle was not yet known. Unfortunately, the image curvature and the field astigmatism of their lens were rather high, limiting the number of equally well resolved image points per diameter to about 150. Nevertheless, a remarkable improvement in image quality was obtained by Möllenstedt and Rang, as illustrated in Fig. 2.44 which shows the unfiltered (Fig. 2.44a) and the filtered (Fig. 2.44b) image of ZnO needles. Despite this success, these high-pass filters were not pursued further, primarily because they did not allow energy-selection, and because they were not applicable for high-resolution imaging.

The results of *Henry* and *Castaing* [2.5] and the increasing interest in utilizing the chemical information contained in the inelastically scattered electrons stimulated our investigations of magnetic imaging energy filters. This chapter presents the results of our work over a period of 20 years in

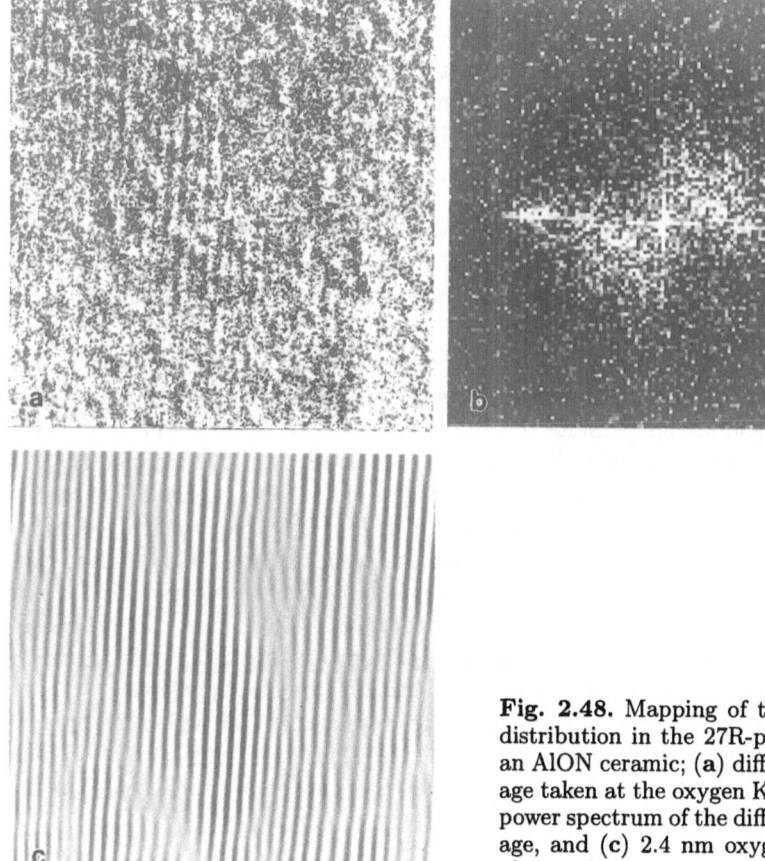

Fig. 2.48. Mapping of the oxygen distribution in the 27R-polytype of an AlON ceramic; (a) difference image taken at the oxygen K-edge, (b) power spectrum of the difference image, and (c) 2.4 nm oxygen lattice planes after image processing

condensed form, starting from some crude ideas without any experimental experience with non-rotationally symmetric electron optical elements. Many problems, which at first seemed unsurmountable, were gradually be overcome until it was possible to calculate, design, and precisely align a corrected high-performance imaging energy filter, which can be used routinely in a high-resolution electron microscope. The incorporation of such an energy filter adds a new dimension to the information obtainable with a conventional electron microscope.

   The progress made since the early days of energy filtering is best demonstrated by four characteristic experimental results which have been obtained with the FHI filter. The first example concerns elemental mapping. In order to obtain the spatial distribution of a distinct element, images of inelastically scattered electrons at and in front of the characteristic ionization edge are recorded. This procedure is illustrated in Fig. 2.45, which shows the image taken at the ionization edge (a), the calculated background image (b), and the background-subtracted image (c) of a Si_3N_4 ceramic. The last image, (a)–(b), represents the oxygen distribution. The second example shown in Fig. 2.46 demonstrates the contrast enhancement resulting from zero-loss filtering of convergent beam diffraction patterns. Fine structures can be observed in the filtered diagram (Fig. 2.46); they are completely masked by the inelastic background in the unfiltered diagram (Fig. 2.46a). The imaging filter technique also offers the possibility of investigating the scattering process itself by recording the angularly resolved energy spectrum. To elucidate this method we have depicted in Fig. 2.47 the angularly resolved spectrum of a 15 nm thick aluminum film showing the two branches of the surface-plasmon dispersion.

The spatial resolution that can be obtained with the mapping technique is primarily limited by the noise. However, by optimizing the recording procedure it should be possible to obtain a resolution limit in the order of a nanometre. To prove that this estimate is realistic, we have mapped the 2.4 nm oxygen lattice planes of the 27R-polytype of an AlON ceramic. The resulting background-subtracted image shown in Fig. 2.48a convincingly demonstrates that the oxygen lattice planes are well resolved. This finding can also be derived from the power spectrum (Fig. 2.48b), where the 2.4 nm spots are clearly distinguishable from the noise. This noise can be subsequently removed by means of standard image processing techniques as shown in Fig. 2.48c. In this filtered image, dislocations are clearly visible.

Summarizing, we can state that the development of high-performance imaging energy filters and their accurate alignment and correction has reached such a degree of perfection that they are likely to become an integral part of any future transmission electron microscope.

Acknowledgements
We want to thank S. Uhlemann for the numerical calculation of the secondary fundamental rays (Fig. 2.22 and 2.23) of the FHI filter and S. Kujawa for placing Figs. 2.31–34, 2.42 and 2.48 at our disposal. One of the authors (H. R.) is grateful to E. Zeitler for the opportunity to perform part of this work at the Fritz-Haber-Institut in Berlin. Thanks are due to Mrs. J. Reiffel for valuable assistance in editing.

References

2.1 C. Deininger, J. Mayer: Omega energy-filtered convergent beam electron diffraction. In *Electron Microscopy 1992*, ed. by A. Rios, J.M. Aria, L. Megias-Megias, A. López-Gallindo (Secretariado Publ., Univ. Granada 1992) Vol.I, pp.181–182

2.2 O. Scherzer: Vorschläge zur Terminologie unrunder Elektronenlinsen. Optik **22**, 314-318 (1965)

2.3 E. Plies: Proposal for an electrostatic energy filter and a monochromator. In *Electron Microscopy 1978*, ed. by J.M. Sturgess (Microscopical Soc. of Canada, Toronto 1978) Vol.1, pp.50–51.

2.4 H. Rose: Electrostatic energy filter as monochromator of a highly coherent electron source. Optik **85**, 95–98 (1990)

2.5 R. Castaing, L. Henry: Filtrage magnétique de vitesses en microscopie électronique. J. Microscopie **3**, 133–152 (1964)

2.6 M. Cotte: Recherches sur l'optique électronique. Ann. Phys. (Paris) **10**, 333–405 (1938)

2.7 R. M. Henkelman, F. P. Ottensmeyer: An energy filter for biological electron microscopy. J. Micr. **102**, 79–94 (1979)

2.8 S. Senoussi: Etude d'un dispositif de filtrage des vitesses purement magnétique adaptable à un microscope électronique à très haute tension. Thése de 3e Cycle, Univ. Paris-Orsay (1971)

2.9 H. Rose, E. Plies: Entwurf eines fehlerarmen magnetischen Energie-Analysators. Optik **40**, 336–341 (1974)

2.10 G. Zanchi, J. Ph. Pérez, J. Sevely: Adaption of a magnetic filtering device on a one megavolt electron microscope. Optik **43**, 495–501 (1975)

2.11 J. Ph. Pérez, J. Sirvin, A. Séguéla, J. C. Lacaze: Étude, au premier ordre, d'une systéme dispersif, magnétique, symmétric, de type alpha. J. Physique **45**, C 2, Suppl. 2, 171–179 (1984)

2.12 S. Lanio: High-resolution imaging magnetic energy filter with simple structure. Optik **73**, 99–107 (1986)

2.13 W. Pejas, H. Rose: Outline of an imaging magnetic energy filter free of second-order aberrations. In *Electron Microscopy 1978*, ed. by J.M. Sturgess (Microscopical Soc. of Canada, Toronto 1978) Vol.1, pp.44–45

2.14 H. Rose, W. Pejas: Optimisation of imaging magnetic energy filters free of second-order aberration. Optik **54**, 235–250 (1979)

2.15 D. Krahl, K.H. Herrmann, E. Zeitler: Experiments with an imaging filter in a CTEM. *Proc. 39th Ann. Mtg. EMSA* (San Francisco Press, San Francisco 1981) pp.366–367

2.16 S. Lanio, H. Rose, D. Krahl: Test and improved design of a corrected imaging magnetic energy filter. Optik **73**, 56–68 (1986)

2.17 W. Legler: Ein modifiziertes Wiensches Filter als Elektronenmonochromator. Z. Physik **171**, 424–435 (1963)

2.18 H. Boersch, J. Geiger, W. Stickel: Das Auflösungsvermögen des elektrostatisch-magnetischen Energieanalysators für schnelle Elektronen. Z. Physik **180**, 415–429 (1964)

2.19 W. H. J. Andersen, J. Kramer: A double-focusing Wien filter as a full-image energy analyser for the electron microscope. *5th Europ. Congr. on Electron Microscopy* (The Institute of Physics, Bristol 1972) pp.146–147

2.20 H. Rose: The retarding Wien filter as a high performance imaging filter. Optik **77**, 26–34 (1987)

2.21 R. L. Seliger: E×B mass-separator design. J. Appl. Phys. **43**, 2352–2357 (1972)

2.22 R. E. Collins: The design of double focussing Wien filters. J. Vac. Sci. Technol. **10**, 1106–1109 (1973)

2.23 M. Kato, K. Tsuno: Numerical analysis of trajectories and aberrations of a Wien filter including the effect of fringing fields. Nucl. Instr. Meth. Phys. Res. A **298**, 296–320 (1990)

2.24 K. Tsuno: Aberration analysis of a Wien filter for electrons. Optik **89**, 31–40 (1991)

2.25 M. Haider, W. Bernhardt, H. Rose: Design and test of an electric and magnetic dodecapole lens. Optik **63**, 9–23 (1982)

2.26 O. L. Krivanek, A. J. Gubbens, N. Dellby: Developments in EELS instrumentation for spectroscopy and imaging. Microsc. Microanal. Microstruct. **2**, 315–332 (1991)

2.27 W. Glaser: Über geometrisch-optische Abbildungen durch Elektronenstrahlen. Z. Physik **80**, 451–464 (1933)

2.28 O. Scherzer: Berechnung der Bildfehler dritter Ordnung nach der Bahnmethode. In *Beiträge zur Elektronenoptik*, ed. by H. Busch and E. Brüche (Barth, Leipzig 1937) pp. 33–41

2.29 H. Rose: Hamiltonian magnetic optics. Nucl. Instr. Meth. Phys. Res. A **258**, 374–401 (1987)

2.30 P. A. Sturrock: Perturbation characteristic functions and their application to electron optics. Proc. Roy. Soc. (London) A **210**, 269–289 (1952)

2.31 E. Zeitler: Analysis of an imaging magnetic energy filter. Nucl. Inst. Meth. Phys. Res. A **298**, 234–246 (1990)

2.32 A. J. Dragt: Elementary and advanced Lie algebra methods with applications to accelerator design, electron microscopes, and light optics. Nucl. Instr. Meth. Phys. Res. A *258*, 339–354 (1987)

2.33 M. Herzberger: *Modern Geometrical Optics* (Interscience, New York 1958)

2.34 E. H. Linfoot: *Recent Advances in Optics* (Oxford Univ. Press, Oxford 1955)

2.35 P. A. Sturrock: *Static and Dynamic Electron Optics* (Cambridge Univ. Press, Cambridge 1955) p. 60

2.36 H. Rose, E. Plies: Correction of aberrations in electron optical systems with curved axes. In *Image Processing and Computer-aided Design in Electron Optics*, ed. by P. W. Hawkes (Academic, London 1973) pp.344–369

2.37 H. Rose: Aberration correction of homogeneous magnetic deflection fields. Optik **51**, 15–38 (1978)

2.38 R. Degenhardt, H. Rose: A compact aberration-free imaging filter with inside energy selection. Nucl. Instr. Meth. Phys. Res. A **298**, 171–178 (1990)

2.39 E. Plies, H. Rose: Über die axialen Bildfehler magnetischer Ablenksysteme mit krummer Achse. Optik **34**, 171–190 (1971)

2.40 G. Hoffstätter, H. Rose: Gauge invariance in the eikonal method. Nucl. Instr. Meth. Phys. Res. A **328**, 398–401 (1993)

2.41 S. Uhlemann, H. Rose: The MANDOLINE filter – a new high-performance imaging filter for sub-eV EFTEM. Optik **96**, 163–178 (1994)

2.42 R. Herzog: Ablenkung von Kathoden- und Kanalstrahlen am Rande eines Kondensators, dessen Streufeld durch eine Blende begrenzt ist. Z. Physik **97**, 596–602 (1935)

2.43 E. Plies: Korrektur der Öffnungsfehler elektronenoptischer Systeme mit krummer Achse und durchgehend astigmatismusfreien Gaußschen Bahnen. Optik **40**, 141–160 (1974)

2.44 P. W. Hawkes, E. Kasper: *Principles of Electron Optics* (Academic, London 1989) Vol.2

2.45 H. Wollnik: *Optics of Charged Particles* (Academic, London 1987)

2.46 A. Abramowitz, A. Stegun: *Handbook of Mathematical Functions* (Dover, New York 1972) p.1004

2.47 A. J. Dragt: Lectures on Nonlinear Orbit Dynamics. In *Physics of High Energy Particle Accelerators*, AIP Conf. Proc. **87** (1982).

2.48 S. Lanio: Optimierung abbildender Energiefilter für die analytische Elektronenmikroskopie. Dissertation D 17, TH Darmstadt (1986).

2.49 S. Kujawa, D. Krahl, H. Niedrig, E. Zeitler: Second-rank aberrations of a magnetic imaging filter: measurement and correction. Optik **86**, 39–46 (1990)

2.50 I. Fromm, L. Reimer, R. Rennekamp: Investigation and use of plasmon losses in energy-filtering transmission electron microscopy. J. Micr. **166**, 257–271 (1992)

2.51 H. Boersch: Ein Elektronenfilter für Elektronenmikroskopie und Elektronenbeugung. Optik **5**, 436–450 (1949)

2.52 G. Möllenstedt, O. Rang: Die elektrostatische Linse als hochauflösendes Geschwindigkeitsfilter. Z. angew. Physik **3**, 187–189 (1951)

3. Plasmons and Related Excitations

Peter Schattschneider and *Bernard Jouffrey*

3.1 Basic Theory

3.1.1 The Continuum Theory of Low-Energy Losses

The Inelastic Scattering Cross-Section. There is a clear distinction between energy loss to inner-shell electrons and to valence or conduction electrons. The reason for the different treatment of these two processes is essentially that, in the former case, the initial state has a sharp energy while in the latter, there is a range of energy within the valence or conduction band. Orbitals of inner shell electrons show almost no overlap between neighbouring sites, so the exchange integral is negligibly small, leading to an extremely small bandwidth. The K-band in Na has a width of 2×10^{-19} eV, and a K-electron in Na jumps roughly once a week to a neighbouring site [3.1]. Those electrons – loyal to their atoms – are well described within an atomic model, which means that energy loss to inner-shell electrons can be treated within atomic theory. (Strictly speaking, this is true only for final states far above the Fermi level. When the excited inner shell electron, after interaction with the fast beam electron, occupies states slightly above the Fermi energy, the density of unoccupied states is mirrored as near edge structure, a typical solid state effect.)

The strong interaction of neighbouring valence electrons in a solid, on the other hand, is manifest in the very existence of energy bands. Energy loss to valence or conduction electrons can hence not be treated in an atomic picture. Let us then define low-energy losses as *losses to valence or conduction electrons.*[1]

In low-energy losses, the interaction between valence or conduction electrons is not merely a perturbation of an otherwise atom-like behaviour. Rather, the theory of those losses should start from the solid regarded as a many-body system. It need not be stressed that quantum mechanics still has problems with many-body systems with the result that the low-loss region

---

[1] That a particular value of energy loss cannot be used as an upper bound for the definition of low losses is evident in many examples of energy loss spectra. The N-edge in Sn is at $\simeq 25$ eV whereas Cd shows low-loss structure (plasmon and interband transitions) up to 45 eV [3.2].

Springer Series in Optical Sciences, Vol. 71
Energy-Filtering Transmission Electron Microscopy
Editor: Ludwig Reimer   ©Springer-Verlag Berlin Heidelberg 1995

in EELS is not yet understood in detail. Low-energy losses provide an excellent tool for testing the validity of quantum mechanical approaches, posing a number of exciting problems.

But low-energy losses are not only a playground for theoreticians. Important information on the solid can be extracted from the spectra as we shall see in Sect. 3.3. A natural way in proceeding, given to the quantum mechanical difficulties mentioned, is to use a phenomenological theory, which does not necessarily rely on quantum mechanics. Since we are dealing with the interaction between charges in EELS, classical electrodynamics can quite well describe the observed phenomena. Of course, this theory does not predict an energy-loss spectrum from the atomic numbers of the constituents of a specimen – it does not work *ab initio* – but speaking of oscillators and eigenmodes in a dielectric medium helps a great deal in characterizing the electronic properties of that medium.

The central quantity that describes the dielectric behaviour of a medium in electrodynamics is the dielectric function ε, that is the relative dielectric permittivity $\varepsilon = \varepsilon_t \varepsilon_0$, as a function of frequency ω and, possibly, wave vector q. Here, ε_t is the total permittivity of the medium, and ε_0 is the vacuum permittivity. The dielectric function, in turn, determines the dielectric response of a medium to an external perturbation, such as an incident electromagnetic wave or an impinging electron. Both probes allow one to obtain $\varepsilon(\omega, q)$ as a function of frequency ω and wave vector q. (These two independent variables relate to the energy E and the momentum p transferred from the probe to the medium as $E = \hbar\omega$ and $p = \hbar q$.)

At this point, one could ask "Why should I use EELS when, in principle, the same information can be obtained by the much simpler and cheaper optical spectroscopies?" The inherent advantage of EELS over optical methods is that by angle-resolved measurements, one has access to the q-dependent dielectric function $\varepsilon(\omega, q)$, whereas optical methods are, in principle, restricted to $|q| \simeq 0$ (see Fig. 3.1 for the scattering geometry). The wave number of the excitation is given by

$$q = \sqrt{q_\parallel^2 + q_\perp^2} \simeq \sqrt{q_E^2 + k_0^2 \theta^2}. \qquad (3.1)$$

The last approximate equality holds for typical scattering angles encountered in transmission microscopy. The quantity q_E is related to the energy loss E, $q_E/k_0 = E/2E_0$. For large scattering angles ($q_\perp \gg q_E$), (3.1) simplifies to $q \simeq \theta k_0$. Note that the smallest possible wave number of an excitation is q_E, corresponding to a scattering angle of $\theta = 0$. To illustrate this, assume a low-energy loss of 20 eV, and a primary voltage of 200 kV. Then, from the above, $q_E = 0.125$ nm^{-1}, corresponding to a wavelength of the excitation of 50 nm. We note in passing that the scattering cross-section is proportional to q^{-2}; see (3.12) below. Consequently, the angular halfwidth of an excitation is approximately given by the condition $q_\perp \simeq q_E$. In the above example, this gives an angular halfwidth of 0.05 mrad. Therefore, scattering by excitation

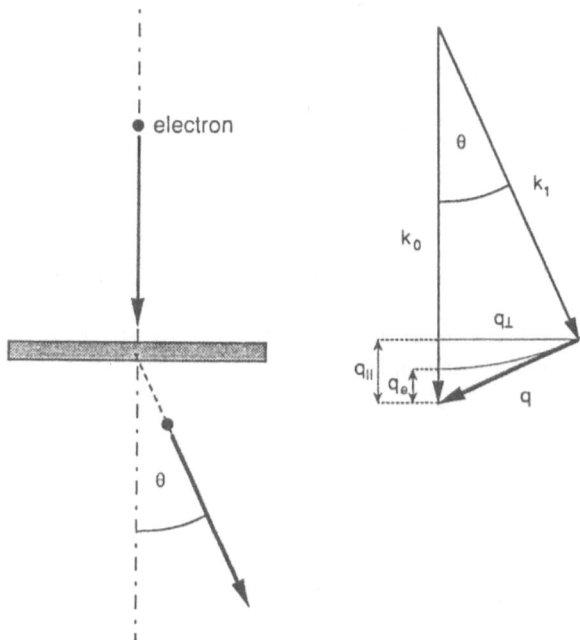

Fig. 3.1. Scattering geometry for transmission EELS. The scattering angles are exaggerated in the figure.

of plasmons is sometimes denoted as "forward scattering". Even with a (not too small) contrast aperture inserted into the microscope, almost all electrons that have excited a low-energy loss are collected by the post-specimen lenses.

The more practical advantages are that EELS covers a much larger energy range than optics, so not only does $\mathrm{Im}\{1/\varepsilon\}$ contain more information but also Kramers–Kronig analysis can be performed much more accurately, and that, in transmission geometry, it probes the bulk of the specimen. Not to mention the other possibilities of modern electron microscopes, such as forming sub-nanometre-sized probes and the simultaneous use of other electron microscopical techniques.

Spectra are often interpreted in a qualitative manner, borrowed from optical absorption spectroscopy, by assigning spectral structure to electronic transitions. We shall see that one has to be extremely cautious when doing this. The reason is that electronic oscillators such as interband transitions are well visible in the dielectric function, its imaginary part having local maxima at transition frequencies, but in EELS, ε is not accessible directly. Rather, the differential energy loss probability as a function of energy loss $E = \hbar\omega$ and scattering angle $\theta = q/k_0$ (isotropy assumed) is proportional to the *inverse* dielectric function

$$\frac{\mathrm{d}p}{\mathrm{d}E} \propto \mathrm{Im}\left\{-\frac{1}{\varepsilon(\omega,q)}\right\} . \tag{3.2}$$

The wave number of the incident electron is k_0.

In order to see how the loss probability relates to the dielectric function, we offer the following plausibility argument. The mean energy lost by an electron per unit path when it penetrates a foil is the integral of the energy loss $E(p)$ over its probability p

$$\bar{E} = \int E \mathrm{d}p = \int E \frac{\mathrm{d}p}{\mathrm{d}E} \, \mathrm{d}E \tag{3.3}$$

where $\mathrm{d}p/\mathrm{d}E$ is the differential energy-loss probability of the probe electron per unit path, and we have discarded the q-dependence for the moment. $\bar{E}$ can likewise be expressed as the power UI of the moving electron integrated over time, where U is the voltage induced in the medium by the moving electron which corresponds to the current I:

$$\bar{E} = \int UI \, \mathrm{d}t. \tag{3.4}$$

From the theory of Fourier transforms, we know that the inner product is constant irrespective of the representation, that is

$$2\pi \int U(t)I^*(t)\mathrm{d}t = \int \tilde{U}(\omega)\tilde{I}^*(\omega) \, \mathrm{d}\omega. \tag{3.5}$$

The tilde on top of a function symbol ($\tilde{f}$) denotes Fourier transformed quantities [$\tilde{f}(\omega) = \int_{-\infty}^{+\infty} \exp(i\omega t)\mathrm{d}t$]. We change to the energy variable $E = \hbar\omega$ and use $I = I^*$, whereupon (3.4) becomes

$$\bar{E} = \frac{1}{\hbar}\mathrm{Re}\int \tilde{U}\tilde{I}^* \, \mathrm{d}E. \tag{3.6}$$

Here, the real part was taken since the mean energy loss $\bar{E}$ is real. Comparison of (3.3) and (3.6) yields

$$\frac{\mathrm{d}p}{\mathrm{d}E} \propto \frac{1}{E}[\tilde{U}(\omega) \cdot \tilde{I}^*(\omega) + \tilde{U}^*(\omega) \cdot \tilde{I}(\omega)]. \tag{3.7}$$

The solution of the Poisson equation for the potential induced in the medium by the electron is[2]

$$\tilde{U} = \frac{\tilde{\rho}}{\varepsilon_0\varepsilon(\omega)q^2}, \tag{3.8}$$

---

[2] In using the Poisson equation, one has discarded the transverse fields set up by the moving electron. Since they are of relativistic origin, they become important when the electron is fast. In particular, when the electron moves faster than the velocity of light in the medium, the transverse field components create a sharp resonance in the loss spectrum, which is accompanied by emission of Cerenkov radiation. This effect is strong in transparent media with a high refractive index, for which the light velocity becomes small and the Cerenkov emission is not severely damped. But even in this case, the Cerenkov peak in the energy loss function is in the low-frequency range, so it does not much influence the energy-loss spectrum above some eV.

and the current $I = d\rho/dt$ is, in frequency representation

$$\tilde{I} = -\mathrm{i}w\tilde{\rho}. \tag{3.9}$$

Inserting these two relations into (3.7) immediately yields

$$\frac{dp}{dE} \propto \tilde{\rho}\tilde{\rho}^* \left(\frac{1}{\mathrm{i}\varepsilon} - \frac{1}{\mathrm{i}\varepsilon^*}\right) \propto \mathrm{Im}\left\{-\frac{1}{\varepsilon}\right\}. \tag{3.10}$$

The quantity on the left is the energy-loss spectrum. The quantity on the right is termed the "loss function". Equation (3.10) is the central relation in low-energy-loss spectrometry. It relates the measured energy-loss probability to the dielectric function ε.

In anisotropic media, ε is a second-rank tensor, and with the proportionality factor [3.3] $4\pi\varepsilon_0/(e\pi a_H)^2$, the exact relationship between the differential scattering probability and the loss function is[3]

$$\frac{\partial^3 p}{\partial E \partial q^2} = \frac{4\pi\varepsilon_0}{(e\pi a_H)^2} \mathrm{Im}\left\{-\frac{1}{q\varepsilon q}\right\}. \tag{3.11}$$

For isotropic media this simplifies to

$$\frac{\partial^3 p}{\partial E \partial q^2} = \frac{4\pi\varepsilon_0}{(e\pi a_H q)^2} \mathrm{Im}\left\{-\frac{1}{\varepsilon}\right\}. \tag{3.12}$$

Equation (3.11) or (3.12) is the basic formula for the interpretation of electron-energy loss spectra in the low and medium energy range. Owing to the causality of the dielectric response to an external perturbation, the complex function $1/\varepsilon(\omega)$ is determined solely by its imaginary part. Once $\mathrm{Im}\{1/\varepsilon\}$ is known, the Kramers–Kronig formula [3.4] yields

$$\mathrm{Re}\left\{\frac{1}{\varepsilon(\omega)}\right\} = 1 - \frac{2}{\pi}\int \frac{\mathrm{Im}\{-1/\varepsilon(\omega')\}}{\omega'^2 - \omega^2}\omega'd\omega' \tag{3.13}$$

and this gives

$$\varepsilon(\omega) = \frac{\mathrm{Re}\{1/\varepsilon(\omega)\} - \mathrm{i}\,\mathrm{Im}\{1/\varepsilon(\omega)\}}{|1/\varepsilon(\omega)|^2}. \tag{3.14}$$

The relationship between the loss function and ε is exemplified in Fig. 3.2, taken from an energy loss experiment on TiC and VC [3.5]. It is important to note that features in ε do not coincide with features in the loss function – a consequence of the occurrence of $\varepsilon\varepsilon^*$ in the denominator of the latter. This fact is referred to as screening (of ε_2 by $\varepsilon\varepsilon^*$). One may distinguish two limiting cases:

$$\mathrm{Im}\left\{-\frac{1}{\varepsilon}\right\} = \frac{\varepsilon_2}{\varepsilon\varepsilon^*} = \begin{cases} 1/\varepsilon_2 & |\varepsilon_1| \ll \varepsilon_2 \quad \text{anti-screening} \\ \varepsilon_2/|\varepsilon_1|^2 & |\varepsilon_1| \gg \varepsilon_2 \quad \text{screening.} \end{cases} \tag{3.15}$$

---

[3] p is a scattering probability per path length of the electron traversing the medium.

Owing to the screening effect, maxima in ε_2 are not visible directly in the loss function; when $|\varepsilon_1|$ is large, they are strongly attenuated. In the figure, this effect can be observed between 5 eV and 10 eV. When $|\varepsilon_1|$ is small – that is, in the vicinity of the plasma frequency, see Longitudial Modes in this Sect. – the situation is reversed: local maxima in ε_2 cause local minima in the loss function – see the energy range between 10 eV and 15 eV in the figure. This can be observed in media with strong interband transitions, for instance in the carbides and nitrides as exemplified in Fig. 3.2, and is also the case in the ceramic superconductors [3.6]. These examples show that one should be cautious when interpreting energy-loss spectra. Assignment of structure in an energy-loss spectrum to (alleged) interband transitions without having performed a Kramers–Kronig analysis is wrong.[4]

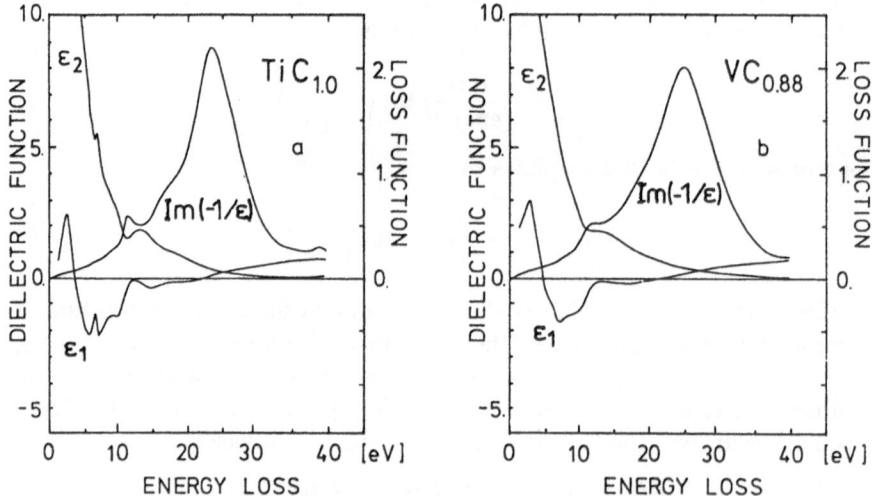

Fig. 3.2. Dielectric function $(\varepsilon_1, \varepsilon_2)$ and loss function $\mathrm{Im}\{-1/\varepsilon\}$ of TiC and VC. Note the different positions of the local maximum (at $\simeq 10$ eV) in ε_2 and in the loss function [3.5].

After proper Kramers–Kronig analysis of data, ε_2 can be interpreted phenomenologically, in terms of electronic oscillators at certain frequencies. Quantum mechanically, these oscillators are identified with electronic inter- or intraband transitions.

The Drude Model. The electronic properties of a medium, the possible eigenmodes, the response to an external longitudinal or transverse perturbation are all described by the frequency-dependent permittivity ε. Various models have been proposed for the permittivity. A widely used and conceptually simple one is the Drude model, proposed in 1900 in order to explain the

---

[4] From the above derivation, this it at once clear. It is virtually comparable to the statement that $\varepsilon = \varepsilon^{-1}$.

optical properties of metals [3.7]. It is a good approximation for the dielectric behaviour of free-electron-like metals.

We assume in the following that the conduction electrons can move freely in the metal. The equation of motion for the conduction electrons of mass m in an electric field $\boldsymbol{E}$ is[5]

$$m\frac{\partial \boldsymbol{v}}{\partial t} + m\frac{\boldsymbol{v}}{\tau} = e\boldsymbol{E} \tag{3.16}$$

where e is the charge of an electron, $\boldsymbol{v}$ its velocity and τ a characteristic relaxation time due to friction in the electron gas. Note that $\tau \to \infty$ means vanishing friction. On multiplying (3.16) by $\rho_0/m = ne/m$, we find

$$\frac{\partial \boldsymbol{j}}{\partial t} + \frac{\boldsymbol{j}}{\tau} = \frac{ne^2}{m}\boldsymbol{E}. \tag{3.17}$$

Here,

$$\boldsymbol{j} = \rho_0 \boldsymbol{v} = ne\boldsymbol{v} \tag{3.18}$$

is a polarization current (i.e. the response of the system) when $\boldsymbol{E}$ is caused by an external perturbation;. ρ_0 is the equilibrium charge density and n the equilibrium electron density. The polarization current $\boldsymbol{j}$ is related to the dipole moment density $\boldsymbol{P} = ne[\boldsymbol{r}(t) - \boldsymbol{r}_0]$, where $\boldsymbol{r}_0$ is the equilibrium position of an electron, by (3.18) as

$$\boldsymbol{j} = ne\frac{\partial \boldsymbol{r}(t)}{\partial t} = \frac{\partial}{\partial t}\boldsymbol{P}(t). \tag{3.19}$$

When we relate the electric field via the dielectric displacement

$$D = \varepsilon_0 \varepsilon \mid \boldsymbol{E} \mid \tag{3.20}$$

to the dipole moment density

$$\mid \boldsymbol{E} \mid = \frac{1}{\varepsilon_0(\varepsilon - 1)} P, \tag{3.21}$$

(3.17) becomes

$$\frac{\partial^2 P}{\partial t^2} + \frac{1}{\tau}\frac{\partial P}{\partial t} = \frac{ne^2}{m\varepsilon_0(\varepsilon - 1)}P. \tag{3.22}$$

This is a linear, homogeneous differential equation for $P(t)$. The solution is oscillatory, $P \propto \exp(-i\omega t)$, with the frequency ω satisfying the characteristic equation

---

[5] Note that the effective mass m^* used for the description of electrons or holes in conduction bands should not be used here since it is essentially a static mass whereas the effective mass used in (3.16) is for high frequency. As we shall see, the longitudinal eigenfrequency (that is the frequency at which the electrons oscillate) is normally much larger than the frequencies of interband transitions in terms of which m^* is defined, see also (3.52).

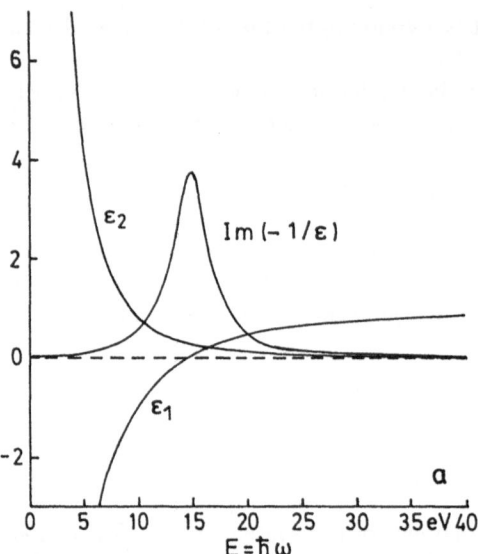

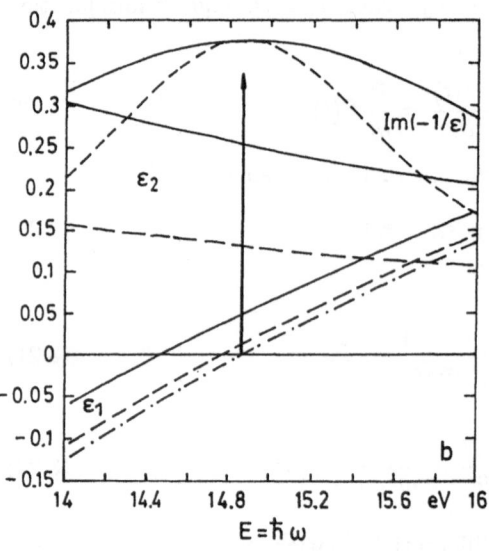

Fig. 3.3. (a) The real part ε_1 (full line) and imaginary part ε_2 (dashed line) of the Drude model for real frequency. The loss function $\mathrm{Im}\{-1/\varepsilon\}$ peaks approximately at the zero of ε_1: $\hbar\omega_p=15$ eV, $\hbar/\tau=4$ eV. **(b)** Enlargement in the vicinity of the plasma frequency. Heavy and thin full lines are the real and imaginary parts of the dielectric function for real frequency. Note that the zero of ε_1 does not coincide with the maximum of the loss function, which is at 14.87 eV. Dashed lines: same for $\mathrm{Im}\{\omega\}=-1/4\tau$. The peak in the loss function is now narrower. For $\mathrm{Im}\{\omega\}=-1/2\tau$, the zero of ε_1 (dash-dotted) is at the maximum of the loss function whereas ε_2 has vanished. The loss function becomes infinitely narrow, indicated by an arrowed vertical line.

$$-\omega^2 - \frac{i\omega}{\tau} = \frac{ne^2}{m\varepsilon_0(\varepsilon - 1)} \tag{3.23}$$

or, solving for $\varepsilon(\omega)$,

$$\varepsilon(\omega) = 1 - \frac{ne^2}{m\varepsilon_0} \cdot \frac{1}{\omega^2 + i\omega/\tau}. \tag{3.24}$$

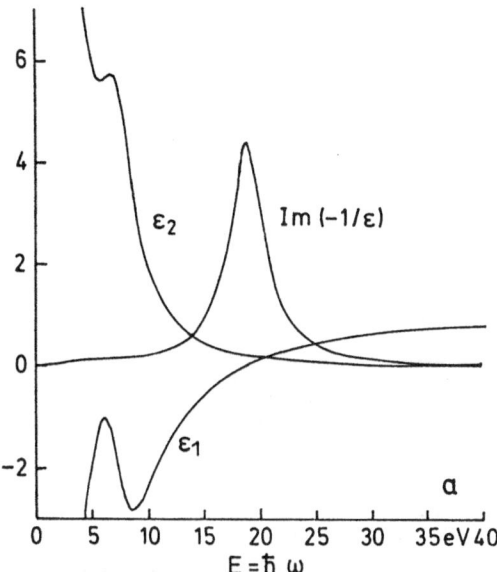

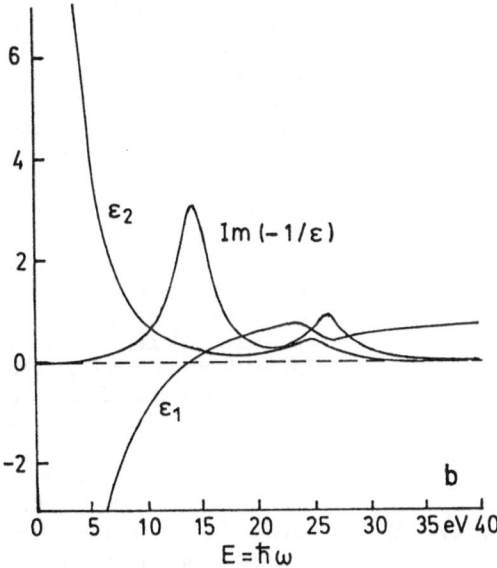

Fig. 3.4. Influence of the superposition of oscillators on the free-electron response. (a) Real part ε_1, imaginary part ε_2 of the dielectric function $\varepsilon(\omega)$ and loss function in the Drude–Lorentz model for a real frequency. Free electron parameters as on Fig. 3.3. An oscillator at $\hbar\omega_j = 7.5$ eV with strength $n_j = 0.5\, n_{free}$ is superposed. The plasma frequency shows blue-shift and the oscillator is severely screened. (b) Same for $\hbar\omega_j = 25$ eV and with strength $n_j = 0.2\, n_{free}$. The plasma frequency is red-shifted, the anti-screened oscillator frequency is blue-shifted.

It should be noted that, in the derivation given above, the system was assumed to be isotropic (otherwise $\chi = P/|\boldsymbol{E}|$ would depend on the direction of $\boldsymbol{E}$, so the susceptibility χ would be a tensor) and homogeneous. Equation (3.16) hence did not contain any space-dependent part and ε was independent of wave vector.

For reasons which will soon become clear, the quantity

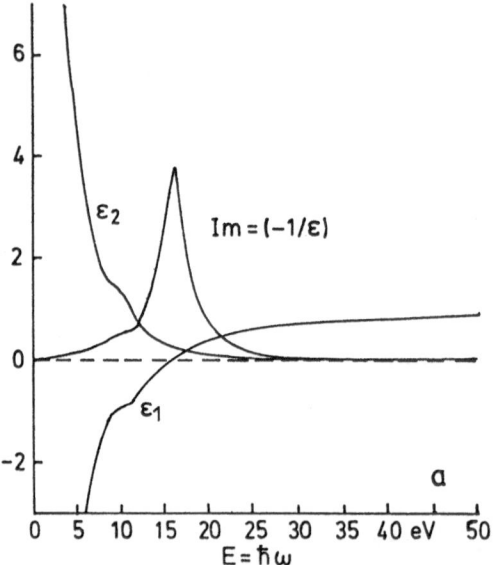

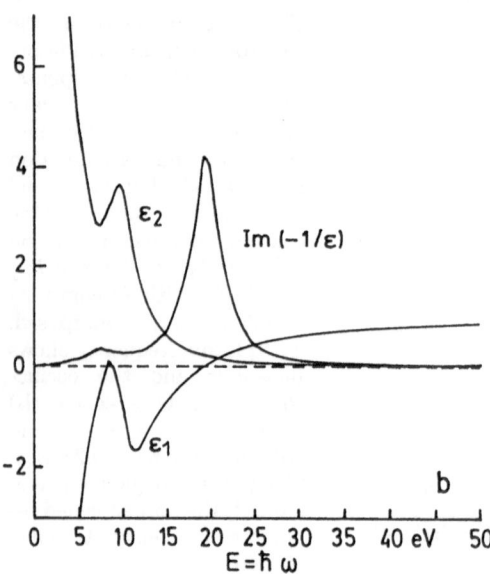

Fig. 3.5. Influence of the strength of superimposed oscillators on the free-electron response. (a) Real part ε_1, imaginary part ε_2 of the dielectric function $\varepsilon(\omega)$ and loss function in the Drude–Lorentz model for a real frequency. Free electron parameters as in Fig. 3.3. An oscillator at $\hbar\omega_j = 10$ eV with strength $n_j = 0.1\,n_{free}$ is superimposed. The plasma frequency shows blue-shift, and the oscillator is severely screened. (b) The same with strength $n_i = 0.5\,n_{free}$. The subsidiary peak at $E < 10$ eV gets stronger with increasing n_j and moves to lower energy. The zero crossing of ε_1 at $\simeq 8$ eV does not qualitatively change the loss function, as compared to (a).

$$\omega_{\mathrm{p}} = \sqrt{\frac{ne^2}{m\varepsilon_0}} \tag{3.25}$$

appearing in (3.24) is called the *plasma frequency*. The real and imaginary parts of the dielectric function ε (3.24), are given by

$$\varepsilon_1 := \mathrm{Re}\{\varepsilon\} = 1 - \frac{\omega_p^2}{\omega^2 + 1/\tau^2}, \tag{3.26}$$

$$\varepsilon_2 := \mathrm{Im}\{\varepsilon\} = \frac{1}{\omega\tau}\frac{\omega_p^2}{\omega^2 + 1/\tau^2}. \tag{3.27}$$

These "Drude expressions" for the dielectric function of a metal are displayed, as a function of *real* ω, in Fig. 3.3 together with the loss function. The parameters chosen are: $\hbar\omega_p = 15$ eV and $\hbar/\tau = 4$ eV. An imaginary part of ε appears due to the friction term j/τ which causes dissipation of energy in a system. (The imaginary part of ε is a measure of energy dissipation.) Note that the loss function peaks roughly at a frequency where $\varepsilon_1 = 0$. From the enlarged part in Fig. 3.3b, it can be seen that ε_1 crosses the abscissa at 14.46 eV whereas the loss function has its maximum at 14.87 eV. The dashed line shows ε_1 for *complex* ω. The imaginary part of the frequency was chosen to be $-1/2\tau$. Now ε_1 is zero at $\hbar\sqrt{\omega_p^2 - 1/4\tau^2} = 14.87$ eV. This is exactly where the loss function peaks. The imaginary part of ε vanishes altogether for $\mathrm{Im}\{\omega\} = 1/2\tau$. That is to say, inclusion of a friction term $1/\tau$ shifts the zero of ε from ω_p to the complex frequency

$$\omega_p' = \sqrt{\omega_p^2 - 1/4\tau^2} - \frac{i}{2\tau}. \tag{3.28}$$

It can be seen that the loss function $\mathrm{Im}\{-1/\varepsilon\}$ has a singularity at ω_p' where

$$\varepsilon(\omega_p') = 0. \tag{3.29}$$

The shift of the real part of the zero of ε is the same as the shift in the maximum of the loss function. In Fig. 3.3b, this is shown as a vertical line representing the δ-function. We note that $\mathrm{Re}\{\omega_p'\}$ is given by the position of the maximum in the loss function, and $2\,\mathrm{Im}\{\omega_p'\}$ is given by its FWHM. It is often carelessly said that the zero of ε_1 defines the plasma frequency. As demonstrated above, this is not true. Restricting attention to real ω, the difference can be quite large (0.41 eV in the given example). The correct statement is that *the maximum of the loss function coincides with the real part of the zero of ε.*

Band Transitions. The Drude model describes the dielectric behaviour of free electrons. But even in metals, the closest realization of a free-electron gas in a solid, this is an approximation since the conduction electrons in a metal are not really free but move under the influence of the lattice potential causing interband transitions. When the unoccupied bands overlap or the gaps are small, the model is justified. In insulators, the Drude model is certainly not applicable; there, one should consider the valence electrons as bound charges having certain eigenfrequencies. The same is true in semiconductors.

The question is then: how can we incorporate interband transitions into a phenomenological model?

This can be accomplished by description of a bound charge by its equation of motion, similar to (3.16):

$$m\left(\frac{\partial v}{\partial t} + \frac{v}{\tau} + \omega_j^2 x\right) = e \tag{3.30}$$

where ω_j is the oscillator frequency of the bound charge; $E_j = \hbar\omega_j$ is the excitation energy of the interband transition of the bound electron. Equation (3.24) now becomes

$$\varepsilon(\omega) = 1 + \frac{ne^2}{m\varepsilon_0} \cdot \frac{1}{\omega_j^2 - \omega^2 - i\omega/\tau}. \tag{3.31}$$

This model dielectric function is called the Lorentz model. When there is more than one interband transition, this can be generalized to

$$\varepsilon(\omega) = 1 + \frac{e^2}{m\varepsilon_0} \sum_j \frac{n_j}{\omega_j^2 - \omega^2 - i\omega/\tau}. \tag{3.32}$$

with n_j the number of electrons involved into transition with energy $E_j = \hbar\omega_j$.

We can now model any medium by proper choice of transition energies and number n_j of oscillating electrons. The Drude model is recovered by setting $E_j = 0$, and metals and semiconductors can be described by a superposition of free and bound charges. This is then called Drude–Lorentz model. In order to see how the oscillator frequency of bound charges influences the loss function we have modelled two systems in Figs. 3.4a,b. Both have free electrons with $\hbar\omega_p = 15$ eV, exactly as in Fig. 3.3a. In Fig. 3.4a, an oscillator at 7.5 eV is superimposed. The oscillator is rather strong (oscillator strength half that of the free electrons). Three effects are visible in the figure: first, ω_p' is shifted to a *higher* value than the free electron value; second, the oscillator appears as a small peak only in the loss function. This is due to screening since, at about 5 eV, the modulus of ε is large. And third, the oscillator appears shifted in the loss function; its maximum coincides roughly with the local minimum in ε_2.

In Fig. 3.4b, a weak oscillator at 25 eV is added to the free-electron model. Here, the plasma frequency is lowered. In the loss spectrum, the oscillator appears slightly enhanced due to anti-screening and, as in the previous example, it is shifted. This example confirms our earlier warning: it is, in general, not possible to deduce the transition frequencies from the loss function.

Another quite general aspect can also be discussed in this context, namely that ε can be negative in some frequency range. From (3.20), we see that this implies that D and E have different signs. Since the displacement field is the gradient of the Coulomb potential (of an external electron) and the electric field is the gradient of the *effective* potential that acts between the charge carriers in the medium, we see from (3.20) that

$$V_{\text{eff}} = \frac{V_{\text{Coul}}}{\varepsilon} > 0, \tag{3.33}$$

the effective interaction between electrons in the medium has become attractive by overscreening. This has consequences for the possibility of pair formation in the electron gas, a question which is under discussion in connection with the ceramic superconductors (Sect. 3.2.3).

The Random-Phase Approximation. The models discussed above are phenomenological models, in a way. One can construct the dielectric function when the strength and frequency of oscillators are known. These oscillators are, as already mentioned, inter- or intraband transitions. A model that predicts from first principles the dielectric behaviour due to electronic transitions would be much better. Such a model is necessarily based on quantum mechanics.

Despite the impressive successes of quantum mechanics, it has been having its problems with the interacting electron gas. In 1954, *Lindhard* [3.8] gave a formula for the dielectric function of the free electron gas for the first time. Explicitly, and written as a function of the dimensionless variables $Q = |q/2k_F|$ and $W = \hbar\omega/4QE_F$, the permittivity is

$$\varepsilon(W,Q) = 1 + \frac{3}{128\gamma^2 Q} \cdot Z(W,Q) \tag{3.34}$$

with

$$Z(W,Q) = 4Q + [1 - (W-Q)^2]\ln\frac{W-Q-1}{W-Q+1} - [1 - (W+Q)^2]\ln\frac{W+Q-1}{W+Q+1}. \tag{3.35}$$

and

$$\gamma = E_F/E_p = \frac{(3\pi^2)^{2/3}}{4\sqrt{\pi}} \cdot \left(\frac{n}{a_H^3}\right)^{1/6}. \tag{3.36}$$

E_F, E_P are the Fermi and plasmon energy. Equation (3.34) is valid for complex frequency.

The Lindhard dielectric function, as it is called, is wave number-dependent, unlike the macroscopic, classical models described above. That is to say, ε is nonlocal, i. e., fields in the vicinity of a chosen point contribute to the polarization. This is certainly a closer representation of a physical system than the preceding models.

A graph of the function is given in Fig. 3.6. It will be noticed that the microscopic model has a range of oscillators between zero frequency and a cutoff, which increases with wave number. These are intraband transitions. It is intuitively clear that for small wave numbers of an excitation – that is, for small momentum transfer between the ground state and the excited state – the Pauli exclusion principle significantly restricts the phase space of available final states, thus reducing the number of possible transitions. For $q = 0$, there are no intraband transitions. For small q, (3.34) becomes the Drude formula without damping.

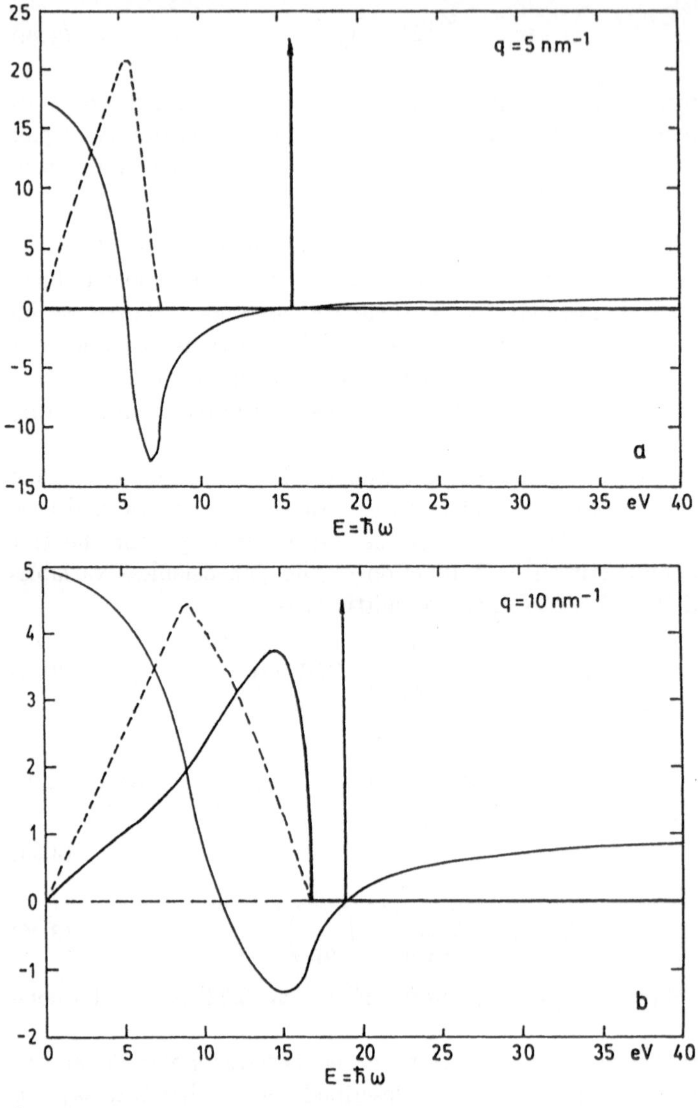

Fig.3.6. Caption see opposite page.

As already mentioned, the dielectric function describes the response of a medium to an external perturbation. It likewise allows the frequencies and wave vectors of the electromagnetic eigenmodes (solutions of the homogeneous Maxwell equations) in the medium to be determined. After having set up various models for the dielectric function of a medium, we are in a position to discuss the possible modes.

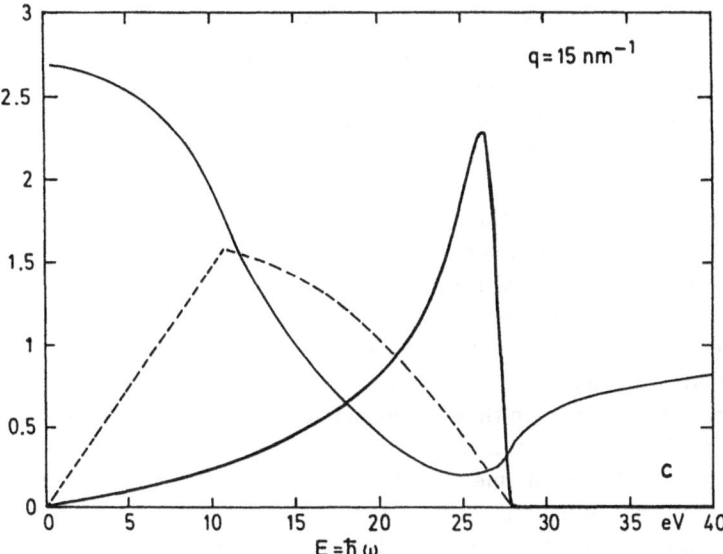

Fig. 3.6. The Lindhard dielectric function (3.34) ε_2 (dashed) and the loss function (full line) for small and large q (where ε_1 no longer has a zero). The loss function still has a well-defined maximum at the minimum of ε_1. $q = 5$, 10, 15 nm^{-1} in (a), (b) and (c), respectively. The free-electron parameters resemble those for aluminium.

Transverse Modes. For a description of transverse volume modes, it is convenient to start with the definition of the phase velocity [3.9] for electro-magnetic transverse waves $v_{\mathrm{ph}} = \omega/q$. On the other hand, the phase velocity is given by $v_{\mathrm{ph}} = c/n$ where c is the vacuum velocity of light, and $n = \sqrt{\varepsilon\mu}$ is the refractive index of the medium.[6] Hence

$$\frac{\omega}{q} = \frac{c}{\sqrt{\varepsilon\mu}}.\tag{3.37}$$

In the simplest model discussed above – the Drude model for a metal (3.26,27) – the general behaviour can be easily understood. Simplifying things by neglect of damping ($\tau \to \infty$) makes the dielectric function real, $\varepsilon = \varepsilon_1$. Additionally, we assume that the medium is non-magnetic, $\mu = 1$. Then from (3.26,37)

$$q = \frac{\omega_{\mathrm{p}}}{c}\sqrt{\varepsilon_1\mu} = \frac{\omega_{\mathrm{p}}}{c}\sqrt{\left(\frac{\omega}{\omega_{\mathrm{p}}}\right)^2 - 1} \quad \begin{cases} q \in R & \omega > \omega_{\mathrm{p}} \\ q\,\text{imag.} & \omega < \omega_{\mathrm{p}} \end{cases} \tag{3.38}$$

---

[6] Strictly, ε is the *transverse* dielectric function which defines the response of the medium to a transverse perturbation (a photon). This is not necessarily the same as the longitudinal permittivity used above. In isotropic media, the two functions are the same [3.10] except in the static limit. In the limit $q \to 0$ both functions coincide for any system since the distinction between transverse and longitudinal becomes meaningless. This fact allows comparison of optical resonant absorption experiments with EELS.

two cases can be distinguished, as shown in (3.38): for $\varepsilon_1 > 0$ or $\omega > \omega_p$ the wave number q is a real function of real ω, whereas for $\varepsilon_1 < 0$ it is purely imaginary. The real branch describes propagating, undamped waves. For $\varepsilon_1 < 0$, i.e. $\omega < \omega_p$, the transverse modes are damped and do not propagate (purely imaginary q). That leads to the important conclusion that a metal is transparent for frequencies above the plasma frequency, and opaque below.

When $\varepsilon_2 \neq 0$, (3.37) shows that for the transverse eigenmode q^2 has an imaginary part; hence q is complex even for $\omega > \omega_p$, i.e., transverse waves are always damped in a metal with finite conductivity.

An insulator can be described by the Drude–Lorentz-model (3.31) with only bound electrons, represented by a single strong oscillator at ω_j. In this case, (3.38) yields propagating transverse waves for $\omega \epsilon [0, \omega_j]$ and $\omega \epsilon [\omega_p, \infty]$. Such a medium is transparent in these frequency intervals. In Fig. 3.7 the longitudinal and transverse eigenmodes are drawn in a $(k_\perp, \omega)$-diagram (heavy lines). Also shown is the light line $\omega = k_\perp c$, which is simply the existence condition for the electromagnetic vacuum mode, and the coupled modes $\omega_s^\pm$ that can be excited on the surfaces of a thin slab (surface plasmons). The diagram is partitioned into 6 regions with qualitatively different behaviour as to incident light. In regions 1 and 2 the medium is transparent. In 3, light cannot propagate in the medium, 4 is the region of total reflection. In 5, 6 only surface modes can exist. The dotted lines in the lower left part of the figure are guided light modes in a thin slab.

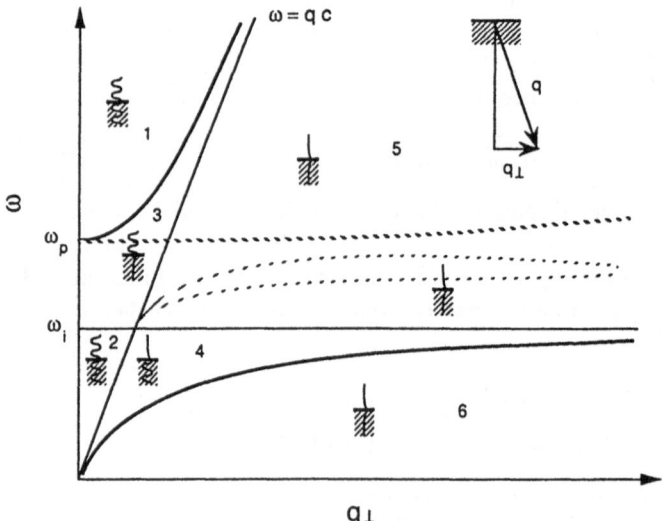

Fig. 3.7. The longitudinal and transverse eigenmodes of a medium with one Lorentz oscillator at ω_j. The $(q_\perp, \omega)$-plane can be classified according to the optical response of the medium. In the bulk (replace $q_\perp$ by q), the thick full lines are transverse modes (the polaritons) and the thick dashed line is the longitudinal mode (the plasmon). Thin dashed lines are surface modes.

For $\omega_j = 0$, the Drude dielectric function is recovered. In this case, only the high-energy branch of the transverse mode survives. Such is the case in a free-electron gas.

Figure 3.7 shows that the light line $\omega = ck$ (which defines the light velocity in vacuum) is strongly altered at its intersections with the transverse and longitudinal eigenfrequencies of the medium. The effect is in complete analogy with the hybridisation of the "free" photon as it would propagate in vacuum with phonons in a medium. The hybrid mode is called a polariton.[7]

The polariton can be imagined as a crossover of two dispersion lines, i.e. *two* excitations are effective at the same frequency and momentum. The state of the system would be degenerate at this particular frequency. As is well known from the physics of oscillations, degenerate states split when there is interaction between the states, as with a coupled pendulum, Bragg reflection or atomic orbital overlap. The number of branches in the dispersion graph remains constant when interaction is activated. In fact, Fig. 3.7 reveals that there are two polariton branches: the lower one is a hybrid with the transverse oscillator (usually an interband transition) at ω_t, the upper one starts off the *longitudinal* resonant frequency ω_p of the medium but is purely transverse. The existence of a transverse mode at ω_p is a consequence of negative ε for frequencies below the plasma frequency. The upper polariton branch is pushed out of the forbidden region between the transverse and the longitudinal resonant frequency.[8]

Longitudinal Modes. It is commonly accepted that the solutions of the Maxwell equations are transverse ($q \perp E \perp H$) in a particular case, namely in a source-free, infinitely extended, homogeneous and isotropic medium. Strictly, this is true for vacuum but not for an arbitrary medium. To see how this comes about, it suffices to write the first Maxwell equation in frequency representation:

$$\nabla \varepsilon E = 0. \qquad (3.39)$$

It is well known that a vector field with vanishing divergence must be transverse [3.8], $q \perp E$ and so there is apparently no component of the electric field parallel to q. That is, longitudinal solutions are seemingly forbidden. Note that this reasoning implicitly assumes that $\varepsilon \neq 0$.

If it is possible to have

$$\varepsilon = 0, \qquad (3.40)$$

there would be no restriction on the direction of E from (3.39) and longitudinal components of E would be allowed.

---

[7] When hybridizing with an interband transition as depicted in the figure, the mode is sometimes called a plasmon polariton, an expression borrowed from the very-low energy spectroscopies where phonon polaritons are well known.

[8] At the starting point of the upper branch, the photon interacting with the medium is at rest since $\lim_{\omega \to \omega_p} \varepsilon = 0$, so $\lim_{\omega \to \omega_p} v_{ph} = \infty$, and for the group velocity $v_g = c^2/v_{ph}$ it follows that $\lim_{\omega \to \omega_p} v_g = 0$. That is to say, the photon has acquired a rest mass by interaction with the medium.

At a first glance, it does not make sense to put $\varepsilon = 0$. This is because in the standard treatment, the dielectric response of a medium is assumed to be linear, local, and immediate:

$$\boldsymbol{D}(t, \boldsymbol{r}) = \varepsilon_{\text{loc}} \boldsymbol{E}(\text{t}, \boldsymbol{r}), \tag{3.41}$$

where ε_{loc} does not depend on t or $\boldsymbol{r}$. That means that the induced field at position $\boldsymbol{r}$ and at time t will be determined only by the (external) perturbation at the very same location and instant, the surroundings or history having no effect. Of course, such a medium cannot have $\varepsilon_{\text{loc}} \equiv 0$. Otherwise, *any* induced field would diverge: local, immediate response does not allow longitudinal wave propagation.

It is intuitively clear that the medium will respond to a perturbation within a certain spatial range since the Coulomb field is extended, and that the response will depend on the perturbation at earlier times since the responding charges have inertia. The local, immediate model (3.41) is hence unrealistic [3.3]. Retaining linearity, (3.41) should be replaced by

$$\boldsymbol{D}(t, \boldsymbol{r}) = \int \mathrm{d}t' \int \mathrm{d}^3 r' \varepsilon_{\text{nloc}}(t - t', \boldsymbol{r} - \boldsymbol{r}') \boldsymbol{E}(t', \boldsymbol{r}'). \tag{3.42}$$

for any homogeneous medium. The displacement field $\boldsymbol{D}$ is the convolution integral of the electric field with the non-local dielectric function $\varepsilon_{\text{nloc}}$. When Fourier transformed to frequency and wave vector[9], the convolution simply becomes a product:

$$\tilde{\boldsymbol{D}}(\omega, \boldsymbol{q}) = \varepsilon(\omega, \boldsymbol{q}) \tilde{\boldsymbol{E}}(\omega, \boldsymbol{q}). \tag{3.43}$$

Note that ε is frequency- and wave vector-dependent when the response is non-immediate and non-local, respectively.[10]

Condition (3.40) is now

$$\varepsilon(\omega, \boldsymbol{q}) = 0 \tag{3.44}$$

defining a dispersion relation

$$\omega_{\text{p}} = \omega_{\text{p}}(\boldsymbol{q}) \tag{3.45}$$

for longitudinal fields. These eigenmodes are well-defined excitations in plasmas (plasma waves). Therefore, the frequency ω_{p} is called the plasma frequency. In the quantum mechanical description, the quantum of plasma oscillation is called a plasmon.[11] We have already defined the plasma frequency for the Drude-model (3.25).

---

[9] We omit the tilde over the function symbol ε since everywhere else in this text ε is given in frequency-momentum representation

[10] The Drude-model implicitly assumes local response yielding a wave vector-independent ε. This follows directly from (3.20)

[11] A plasmon can be imagined as a particular coherent superposition of wave functions of moving electrons, just as coherently moving ions set up a phonon [3.11].

The very existence of plasma oscillations in a medium is proof that the dielectric response of that medium is *not immediate*, i. e., the medium has some inertia.

For completeness we mention that all possible electromagnetic eigenmodes in a medium can be obtained from the roots of the characteristic equation. In the homogeneous, isotropic, source-free case this equation is

$$\omega^2 \varepsilon \mu \left(\frac{\omega^2 \varepsilon \mu}{c^2} - q^2 \right) = 0 . \tag{3.46}$$

There are four eigenmodes, each defined by a root of the characteristic equation (3.46), namely

$$\varepsilon(\omega, q) \ = 0 \qquad \textit{longitudinal electric} \tag{3.47}$$

$$\mu(\omega, k) \ = 0 \qquad \textit{longitudinal magnetic} \tag{3.48}$$

$$\omega \sqrt{\varepsilon \mu} \ = \pm qc \qquad \textit{transverse electromagnetic} . \tag{3.49}$$

There are pure longitudinal electric and magnetic waves besides the well-known transverse electromagnetic ones. The transverse/longitudinal solutions are no longer decoupled when the system is not isotropic. For instance, a boundary induces coupling between both types of mode. This is the reason why a photon can excite plasmons, e.g. in Attenuated (or Frustrated) Total Reflection (ATR or FTR) experiments [3.3,12]. A formal treatment of the general case of an infinite bulk crystal shows that the modes couple, but depending on the crystal symmetry, there are still high-symmetry directions where transverse and longitudinal modes remain decoupled [3.13].

An important difference between longitudinal (electric) and transverse waves is seen by applying the continuity equation to the polarization current j_{pol} which is parallel to E in an isotropic continuum. Since $\nabla E = 0$ for transverse modes,

$$\frac{\partial \rho_{\text{pol}}}{\partial t} = -\nabla j_{\text{pol}} \begin{cases} \neq 0 & \textit{longitudinal} \\ \\ = 0 & \textit{transverse} . \end{cases} \tag{3.50}$$

The charge density remains constant for transverse waves whereas longitudinal waves cause periodic charge density fluctuations in the medium.

3.1.2 Plasmon Spectrometry

We have seen that plasma oscillations exist in metals (which can be considered as solid state plasmas) as well as the better known transverse eigenmodes (3.38). These transverse excitations can be detected by resonant absorption or emission of photons, by means of ellipsometry or optical transmission spectroscopy. From the measured optical absorption coefficient κ and the index of refraction n, $\varepsilon_2 = 2n\kappa$ can be calculated. When n is only slightly frequency-dependent, as is normally the case in the range of optical frequencies, ε_2 is a measure of the excitation strength of transverse resonant modes.[12]

The important question of how the longitudinal modes can be detected has been answered in Sect. 3.1.1. Poles of the loss function, (3.10), coincide with zeros of ε, that is, they define the longitudinal eigenmodes of the medium. It is immediately evident that electron-energy loss spectrometry is the method of choice for detection and analysis of longitudinal modes. In much the same way as optical or X-ray techniques are used to detect transverse modes, by resonant absorption or emission of photons, electron-energy loss spectrometry directly detects the longitudinal modes, by resonant exchange of quanta of the Coulomb field.

In order to discuss this point in more detail, let us examine Fig. 3.3b. In the Drude model, $1/\varepsilon(\omega)$ has a single pole at the complex frequency ω_p', given by (3.28), which is the zero of ε. The maximum of the loss function $\mathrm{Im}\{-1/\varepsilon\}$ is also at $\mathrm{Re}\{\omega_p'\}$. It is seen that, in the Drude model, the maximum in the loss function is at the longitudinal eigenfrequency, $\mathrm{Re}\{\omega_p'\}$. (Note that this value is different from the zero of ε_1. The smaller ε_2 at the zero, the higher will the maximum be, and the smaller is its FWHM, $\mathrm{Im}\{\omega_p'\}$. Only for vanishing friction will both values coincide with one another, and with the plasma frequency ω_p.) The maxima of the energy loss function coincide with the longitudinal eigenfrequency. The incident fast electron excites resonant longitudinal oscillations in the solid.

This is a very important result: The loss function directly yields the longitudinal resonance. This is likewise true for the Drude–Lorentz model where a superposition of longitudinal excitations shows up in the loss function. For more complex models the question of what are the exact values of the longitudinal resonant frequencies is only meaningful in the case of discrete oscillators. For a continuum of oscillators, (see the Lindhard model, for example) the poles become maxima in the complex ω-plane. The search for those maxima necessitates in general analytic continuation of the dielectric function into the complex plane. Note however that there is no need for calculation of the dielectric function from energy-loss measurements when one is interested in the *longitudinal* response of a medium; the loss function mirrors the longitudinal excitations in the most direct way, including their finite lifetime.[13]

---

[12] In the isotropic continuum theory adopted here, a photon can excite transverse modes only. Microscopically, absorption of a photon corresponds to an interband transition, with $\partial\rho/\partial t \neq 0$ around the excited atoms. By (3.50), there are now longitudinal contributions. These are local; the divergence of the macroscopic polarization current still vanishes when the excitation is along a high-symmetry direction. As will be discussed later, the local fields may change the macroscopic response function (Sect. 3.2.1).

[13] The same is of course true for spectroscopy of transverse excitations. As mentioned above, $\mathrm{Im}\{\varepsilon\}$ is usually interpreted in terms of inter- or intraband tran-

From this viewpoint, it can be questioned whether it is useful to give conditions for the existence of zeros of ε_1 in order to decide whether a structure in the loss function is a plasmon [3.14]. Comparison of Figs. 3.5a and b shows that the zero crossing at about 8 eV does not qualitatively change the spectrum of longitudinal excitations.

The question of plasmons in insulators deserves mention. Originally, a plasmon was believed to exist only when there are free electrons [3.8], i.e., in a metal. But as early as 1955, *Kanazawa* cautiously suggested that *it would seem possible that plasma oscillations exist in insulators* [3.15]. *Horie* [3.16] defined plasmons as excitations above the band gap of insulators, whereas excitons are below. As *Egri* [3.17] pointed out, the plasmons as well as the *longitudinal* excitons occur at zeros of the dielectric function. The close relation between plasmons and excitons has been noticed early. *Ferrell* showed that a plasmon in a metal is an "exciton in momentum space" [3.18,19]. Both the regular exciton and the plasmon are coherent superpositions of electron-hole excitations.

The excited state can be described by the probability amplitude of finding a hole at a distance r from an excited electron. This has been called the wave function of the plasmon. Its envelope is a measure of the extension, or localisation, of the plasmon. It has been calculated for metals [3.11,20] and insulators [3.21]. Figure 3.8 shows this wave function for a free-electron gas, in the direction of propagation, for the two- and three-dimensional case. Loosely speaking, the first zero gives the extension of the plasmon. In metals, this is of the order of tenth of nanometre. In aluminium, this is 0.22 nm, slightly less than the distance between neighbouring lattice sites. *Rogan* and *Inglesfield* [3.23] also obtain a plasmon extension over only a few lattice sites in metals, in an equation-of-motion approach to tight-binding systems. Their plasmon looks very similar to a Frenkel exciton in a tight-binding insulator.

We shall now turn to the discussion of plasmons in the original sense, that is, we assume that at a certain energy loss, a distinct and narrow maximum is observed. The plasma resonance can quite well be described by three quantities: the energy $\hbar\omega_p$, the halfwidth and the dispersion coefficient.

In the following, we give the classical values, which are valid for a large class of metals and semiconductors.[14]

The plasma frequency has already been defined in terms of the electron density-to-mass ratio, as in (3.25) for the Drude model. The Lindhard dielectric function gives the same value. This can be generalized as

---

sitions for real ω although the transverse eigenmodes are given as the poles of ε, which occur at complex values of the frequency.

[14] In metals and semiconductors, $\hbar\omega_p$ is of order (1–10) eV. Instead of electrons, ions may oscillate longitudinally. In this case, the longitudinal optical phonon formally replaces the plasmon. Instead of ω_p, the resonance is now at the longitudinal optical phonon (LO) frequency, of order (0.01–0.1) eV, due to the much higher inertia of ions.

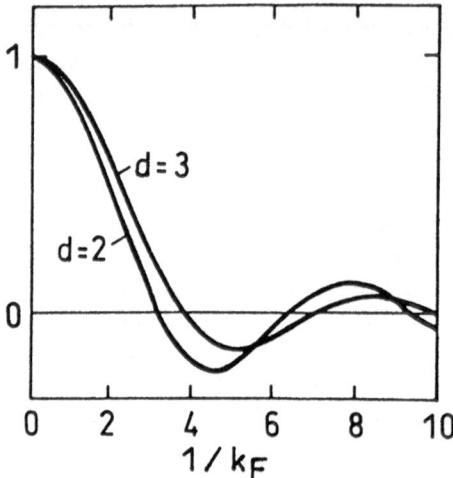

Fig. 3.8. Wave function of the plasmon in a free-electron gas in the direction of propagation, for the two-dimensional case ($d=2$) and the three-dimensional case ($d=3$). Abscissa in units of k_F^{-1} [3.21].

$$\omega_{\mathrm{p}} = \sqrt{\frac{ne^2}{\hat{m}\varepsilon_0}} \; . \tag{3.51}$$

Here, $\hat{m} = m(1 + \chi_{\mathrm{b}})$ is a masslike parameter describing the influence of bound charges on the otherwise nearly free electrons of mass m and density n that set up the plasma oscillation. The polarizability χ_{b} of bound electrons tends to change the plasma frequency of the free electrons [3.24]. In a semiconductor, n would be the density of conduction electrons, and χ_{b} would be the polarizability of the valence electrons. It should be mentioned however, that the notion of an effective mass is sensible only when $|\chi_{\mathrm{b}}(\omega)| \ll 1$ in the vicinity of ω_{p}, i.e., when bound electrons act as a *small* disturbance of the movement of "free" electrons: $n_{\mathrm{b}} \ll n$ and the eigenfrequency of the bound electrons $\omega_{\mathrm{b}} \ll \omega_{\mathrm{p}}$. The factor $1 + \chi_{\mathrm{b}} = \varepsilon_{\mathrm{b}}$ is the relative dielectric permittivity of the bound electrons. Sometimes the medium is considered as a background with ε_{b} into which the free electrons are embedded. The background screening seems to change the mass of the free electrons that participate in the plasma oscillation. Note however that

$$\frac{n}{\hat{m}} \approx \frac{n_{\mathrm{t}}}{m} \tag{3.52}$$

when the bandgap is small, where n_{t} denotes the total density of free and bound electrons. Instead of n conduction electrons with effective mass $\hat{m}$, one can likewise think of *all* electrons with mass m participating in the plasma oscillation. We shall see that introduction of an effective mass is not sufficient to understand the plasmon in the solid state. Electron correlation and exchange as well as renormalization of the plasmon energy by virtual phonons and by core contributions must be treated on a quantum mechanical footing.

The halfwidth of the resonance [3.24] is

$$\triangle E = 2\hbar\varepsilon_2(\omega_p)\left[\frac{\partial\varepsilon_1}{\partial\omega}|_{\omega_p}\right]^{-1}. \tag{3.53}$$

In the Drude model, this is

$$\triangle E = \hbar\omega_p \cdot \varepsilon_2(\omega_p). \tag{3.54}$$

3.1.3 Plasmon Dispersion

The dispersion coefficient α is a measure of the change of the plasma frequency with wave number. It is defined by the following equation:

$$E_p(q) = \hbar\omega_p + \frac{\hbar^2}{m}\alpha q^2 + O(q^4). \tag{3.55}$$

In the Drude model, longitudinal electric waves exist for any q at ω_p. This is the frequency dispersion for the collective oscillation. Note that the phase velocity of the longitudinal mode, $c_{ph} = \omega/k$ decreases from ∞ to 0 as k increases, and its group velocity $\partial\omega/\partial k$ vanishes. This result implies that the longitudinal mode cannot transport any energy. This unphysical behaviour is caused by the constituting equation of motion (1.16), which did not contain any space-dependent term. This immediately implies that ε is also independent of position and, therefore, that its Fourier transform does not depend on wave number q. In order to improve the model, a diffusion term due to a gradient of pressure p in the electron gas can be added to the equation of motion (3.16)

$$m\frac{\partial v}{\partial t} + m\frac{v}{\tau} + \frac{\nabla p}{n_0} = eE \tag{3.56}$$

Here, n_0 is the average density of electrons. It can be shown [3.3] that the dispersion relation for the longitudinal mode is now

$$\omega^2 = \omega_p^2 + \frac{3}{5}v_F^2 q^2 \tag{3.57}$$

with the Fermi velocity v_F. For the dispersion coefficient as defined by (3.55), we find

$$\alpha = \frac{3}{5}\frac{E_F}{\hbar\omega_p}. \tag{3.58}$$

with the Fermi energy E_F. The asymptote of expression (3.57) is $\omega = v_F\sqrt{3/5}$.

For the dispersion of (3.57), we have $v_g v_p = 3/5v_F^2$. As an aside, we mention that the Lorentz transform of a wave train leads to a similar result for photons: $v_g v_p = c^2$. The plasmon behaves like an electromagnetic wave train in vacuum, with $v_F\sqrt{3/5}$ as the velocity limit instead of the velocity of light.

A plasmon "mass" m_p, similar to the mass of a quasiparticle, can be defined as

$$m_p := \frac{\hbar}{\partial^2\omega/\partial q^2} = m_0\frac{5E_p}{6E_F}.$$ (3.59)

This mass[15] is very close to the electron mass m_0.

The simplest quantum mechanical model capable of predicting plasmons – the RPA, short for random phase approximation – leads to the same dispersion coefficient (3.55) as the above classical model but only for the small-q expansion of the plasma frequency. For higher values of q, quartic terms become important as already calculated in 1957 [3.18,28]. In the RPA, the next higher-order term in the plasmon dispersion equation (3.55) is

$$O(q^4) = E_F\left[\frac{E_F}{2E_p} - \frac{6}{35}\left(\frac{E_F}{E_p}\right)^3\right]\left(\frac{q}{k_F}\right)^4.$$ (3.60)

This term is relatively unimportant at wave numbers much smaller than the Fermi wave number k_F and was neglected in almost all interpretations of experiments aiming at determination of plasmon dispersion.[16] As we shall see in Sect. 3.2.1, it is in the large-q region that the basic theory fails, so the quartic-term correction in the RPA is not really at stake.

3.2 Beyond the Continuum

3.2.1 Refinements of the Basic Model

The random phase approximation (RPA) was the first quantum mechanical approach that was able to predict important properties of the electron gas such as the ground state energy, the band width, or the effective mass. It also succeeded in predicting the existence of plasmons. The expression (3.34) given by Lindhard for the dielectric function of the electron gas in the RPA model describes the plasma frequency quite well in the case of simple metals. The dispersion is only moderately well described. For the halfwidth of the plasmon it fails completely since (3.53) gives zero halfwidth of the plasmon for the Lindhard function. Only in higher order perturbation theory does the

---

[15] Note however that the last equality is valid in the random phase approximation only. In fact, the dispersion can become negative – see later – that is, the mass can become negative; in that case, acceleration would reduce the energy of the quasiparticle. This can lead to spontaneous creation of a collective movement of charges, giving rise to a spontaneous charge density wave instability (Peierls transition) [3.25–27]

[16] In aluminium, the quartic term is 3.5 eV at $q = k_F$, but only 0.2 eV at $q = 0.5 k_F$. When fitting experimentally obtained values to a quadratic dispersion curve up to this wave number, α would be overestimated by only 5%. Nevertheless, the quartic term is the explanation for the alleged discontinuous slope change that has caused some confusion in the past (Sect. 3.2.2).

plasmon acquire a finite lifetime, caused by multipair excitations. This results in a quadratic dependence of linewidth on the wave number. At $q = 0$, the plasmon linewidth vanishes. There are several calculations of plasmon decay for jellium beyond the RPA, starting in 1959 [3.29] – see also *Bachlechner* et al. [3.30] and references therein. A recent recalculation [3.30] revealed some inconsistencies in the earlier theoretical work. These authors give a general expression for the plasmon halfwidth, valid for arbitrary electron densities in the jellium.

The linewidth of the plasma excitation has been measured for a number of materials, for $q = 0$ and as a function of wave number. For a survey, see *Raether* [3.24], and the more recent measurements on the alkali metals [3.31]. It was soon realized that the jellium theory is not able to explain the finite, and sometimes considerable, halfwidth found at $q = 0$. Moreover, in some cases the q-dependence deviates markedly from the prediction, as in the alkali metals [3.31,32] or in some III-V semiconductors [3.33]. Important contributions to the linewidth come from interband transitions and in particular cases from the core polarizability. In the search for improvements, attempts have long been made to include the effects of electron correlation (phase relations between individual electrons are random in the RPA) and exchange in refined theories. By now it is clear that correlational corrections are important but not sufficient. There are other effects due to the periodic lattice potential, to interband transitions and to the ion cores. The relative importance of these effects depends on the material. Moreover, these effects influence the plasma frequency, the dispersion and the halfwidth in different ways. Whereas the lattice periodicity leads to strong damping, thus giving a fairly good description of the halfwidth, it has little influence on the dispersion. The core polarizability acts so as to change the plasma frequency, again with little influence on the dispersion and almost no influence on halfwidth. Correlational effects, on the contrary, may markedly change the dispersion. In the alkali metals, these effects are so strong that the plasmon dispersion becomes negative for Cs. So the state of affairs is rather subtle, and there is no simple and general theory of the plasmon in solids.

In the following, the more recent investigations on plasmons are surveyed.

Correlation in the Electron Gas. The first point to be examined is the limited applicability of the RPA, which is good for weakly coupled plasmas. The coupling strength Γ in a classical plasma is given by the ratio of the potential energy to the mean thermal energy of a particle. For strong coupling, that is for $\Gamma \gg 1$, correlational effects become important. It has been suggested that the coupling constant Γ in the degenerate electron gas can be obtained by the replacement $3kT/2 \rightarrow 3E_F/5$, which yields[17] $\Gamma = 1.36\,r_s$. But since electrons in the vicinity of E_F primarily contribute to the collective movement it is also reasonable to use the replacement $3kT/2 \rightarrow E_F$ that is

---

[17] Here the mean distance between conduction electrons is $2r_s a_H$, and $r_s = (4\pi n/3)^{-1/3} a_H^{-1}$ is a dimensionless length. $a_H = 0.053$ nm is the Bohr radius.

normally used to obtain the Thomas–Fermi screening length from the Debye screening length [3.3]. Then, $\Gamma = 0.81 r_s$.

For a plausibility argument, we can put

$$\Gamma \approx r_s. \tag{3.61}$$

It can be seen that high electron density corresponds to weak coupling. The RPA is good for $r_s < 1$. In simple metals ($2 < r_s < 6$), one is in the intermediate to strong coupling regime and the RPA may fail.

For the electron densities found in simple metals, exchange and correlation effects neglected in the RPA may be important in the calculation of ε. These are usually treated as follows. In a homogeneous, isotropic system, the dielectric function is the ratio of the external potential V to the total potential V_t in the medium

$$\varepsilon = \frac{\tilde{D}}{\varepsilon_0 \tilde{E}} = \frac{\tilde{V}}{\tilde{V}_t} = \frac{\tilde{V}}{\tilde{V} + \tilde{V}_i} \tag{3.62}$$

since $\tilde{D}/\varepsilon_0$ and $\tilde{E}$ are the gradients of $\tilde{V}$ and $\tilde{V}_t$, respectively. In the last equality, the total potential was replaced by the sum of the external and the induced potential $\tilde{V}_i$. The latter can be expressed in terms of the Coulomb potential of the induced charge density $\tilde{\rho}$:

$$\tilde{V}_i = \frac{\tilde{\rho}}{\varepsilon_0 q^2}. \tag{3.63}$$

On the other hand, the induced charge density $\tilde{\rho}$ is given by the density response function χ as

$$\tilde{\rho} = e^2 \chi \tilde{V}. \tag{3.64}$$

From (3.62-64), it follows that

$$\varepsilon = \frac{1}{1 + v\chi}, \tag{3.65}$$

where we have introduced the Coulomb potential energy

$$v = \frac{e^2}{\varepsilon_0 q^2}. \tag{3.66}$$

The problem of calculating the dielectric function is the problem of calculating the density response χ. In the random phase approximation, one assumes that the charges in the medium, normally in equilibrium, respond to a perturbation as if they were independent of one another. The density response function can then be calculated from the Lindhard polarizability χ^L for non-interacting electrons

$$\chi^L(\omega, q) = 2 \sum_k \frac{f(E_k) - f(E_{k+q})}{\hbar\omega + i\delta + E_k - E_{k+q}} \tag{3.67}$$

where f is the Fermi occupation function and $E_{\boldsymbol{k}}$ is the energy of free electrons with wave vector $\boldsymbol{k}$. Assuming that the charges react as if they were free, but to the *total* potential, the induced charge density reads[18]

$$\tilde{\rho} = e^2 \chi^{\mathrm{L}} \tilde{V}_t. \tag{3.68}$$

Combining (3.63), (3.64) and (3.68), we find

$$\chi_{\mathrm{RPA}} = \frac{\chi^{\mathrm{L}}}{1 - v\chi^{\mathrm{L}}}. \tag{3.69}$$

Inserting this into (3.65), we have

$$\varepsilon_{\mathrm{RPA}} = 1 - v\chi^{\mathrm{L}}. \tag{3.70}$$

On calculating χ^{L} explicitly, (3.70) becomes the Lindhard function (3.34). It is valid for a homogeneous electron gas, for which the fields and potentials are assumed not to vary throughout the medium.

The tendency of electrons to avoid each other, due to the repulsive Coulomb fields (correlation) and to the Pauli exclusion principle (exchange), invalidates this assumption. This effect can be taken into acount by assuming that each electron carries a virtual exchange-correlation potential $\tilde{V}_{\mathrm{ec}}$ and that, as before, the charges respond to this potential as if they were independent. Since each electron carries "its own hole", $\tilde{V}_{\mathrm{ec}}$ has to be multiplied by the particle density in order to obtain the exchange-correlation term for the whole arrangement of charges. The exchange-correlation corrected response function χ_{ec} is given by

$$e^2 \chi_{\mathrm{ec}} \tilde{V} = \tilde{\rho} = e^2 \chi^{\mathrm{L}} \left(\tilde{V}_t + \frac{\tilde{\rho}}{e} \tilde{V}_{\mathrm{ec}} \right). \tag{3.71}$$

Proceeding as above using (3.64) we have

$$\chi_{\mathrm{ec}} = \frac{\chi^{\mathrm{L}}}{1 - (v + e\tilde{V}_{\mathrm{ec}})\chi^{\mathrm{L}}}. \tag{3.72}$$

instead of (3.69). The exchange-correlation potential can be pictured as being caused by a virtual particle density $G(\boldsymbol{r})$, the *exchange-correlation hole*. The Poisson equation for the hole is

$$\nabla^2 V_{\mathrm{ec}}(\boldsymbol{r}) = \frac{eG(\boldsymbol{r})}{\varepsilon_0} \tag{3.73}$$

which is immediately solved by Fourier transform

---

[18] Although both χ_{RPA} and χ^{L} have the same dimension, they have different meaning. The former is a response function (density response to an *external* potential $\tilde{V}$), the latter relates to the polarization of the medium, $\tilde{P} = -v\chi^{\mathrm{L}}\tilde{E}$, i.e., it describes a response to the *total* potential. Hence the naming "polarizability".

$$\tilde{V}_{ec}(q) = -\frac{e}{\varepsilon_0 q^2}\tilde{G}(q). \tag{3.74}$$

From (3.65), (3.71) and (3.74), the expression

$$\varepsilon_{ec} = 1 - \frac{v\chi^L}{1 + v\tilde{G}\chi^L} \tag{3.75}$$

is eventually obtained.

The factor $\tilde{G}(q)$ is the Fourier transform of the virtual exchange-correlation hole that builds up around each electron. Numerical Fourier–Bessel transform of $\tilde{G}(q)$ gives the charge distribution of the exchange-correlation hole in configuration space [3.34-36].

There are several approximations to model such a hole. The models given by *Hubbard* [3.37], *Singwi* [3.35] and *Utsumi* and *Ichimaru* [3.38] are widely used.

Iwamoto et al. [3.39] calculate the different effects of the Pauli exclusion principle (q_e) and the Coulomb repulsion (q_c). They find hole radii $r_{e,c} = 1/q_{e,c}$ for metallic densities, $r_e \approx 0.7\,r_s$ whereas r_c increases from $0.3\,r_s$ to $0.42\,r_s$ with decreasing density. Obviously, the Pauli principle is very effective in keeping electrons apart. For the low density metals, correlation effects create a hole of almost half the interparticle distance around each electron.

The plasmon dispersion can be calculated from the plasma defining condition $\varepsilon_{ec} = 0$. The corrected dispersion coefficient α_{ec} is well approximated by

$$\alpha_{ec} = \alpha_{RPA}(1 - Ar_s - Br_s^2) \tag{3.76}$$

where A and B are constants depending on the expression for the exchange-correlation hole $\tilde{G}$. A is always positive, and the constant B is zero or very small and can be neglected for metallic electron densities. Thus, the effect of the exchange-correlation hole is to reduce the plasmon dispersion coefficient.

There has been a wealth of work on the exchange-correlation hole in the electron gas, with the following values obtained for the constant A: 0.026 [3.40], 0.028 [3.41], 0.055 [3.42], 0.089 [3.35], 0.107 [3.43], 0.109 [3.44], 0.125 [3.45], 0.181 [3.37], 0.198 [3.34].

For aluminium, a range of values for α (from 0.27 to 0.41) is obtained. Experimental values of the dispersion coefficient[19] agree only roughly with this model prediction – see Table 3.1. The upper part of the table lists some theoretical values, the lower part refers to experiment.

The scatter of predictions is only slightly worse than that of experiments when only the three most recent experimental results are considered. The first dispersion experiment reported by *Watanabe* [3.46] in 1956 is also given in the table. The high value obtained by *Batson* and *Silcox* [3.47] is due to the fact that these authors used a linear fit up to high momentum transfer where

---

[19] From an experimentally determined dispersion $E_p(q) = \hbar\omega_p + Cq^2$, the value of α may be readily found as $\alpha = 13.12$ C eV^{-1} nm^{-2}.

Table 3.1. Plasmon dispersion coefficient in aluminium

Source	$\alpha_{ec}/\alpha_{RPA} = 1 - Ar_s$	α_{ec}
Hubbard [3.37]	0.625	0.27
Nozières and *Pines* [3.42]	0.82	0.36
Vashishta and *Singwi* [3.44]	0.77	0.34
Suehiro [3.40]	0.946	0.416
Fink et al. [3.48]	0.68	0.30
Schattschneider et al. [3.49]	0.79	0.35
Batson and *Silcox* [3.47]	0.86	0.38
Watanabe [3.46]	1.13	0.50

the quartic term in the dispersion becomes important.[20] Since the predicted values correspond to the initial slope of the dispersion line, the lower values are probably more reliable.

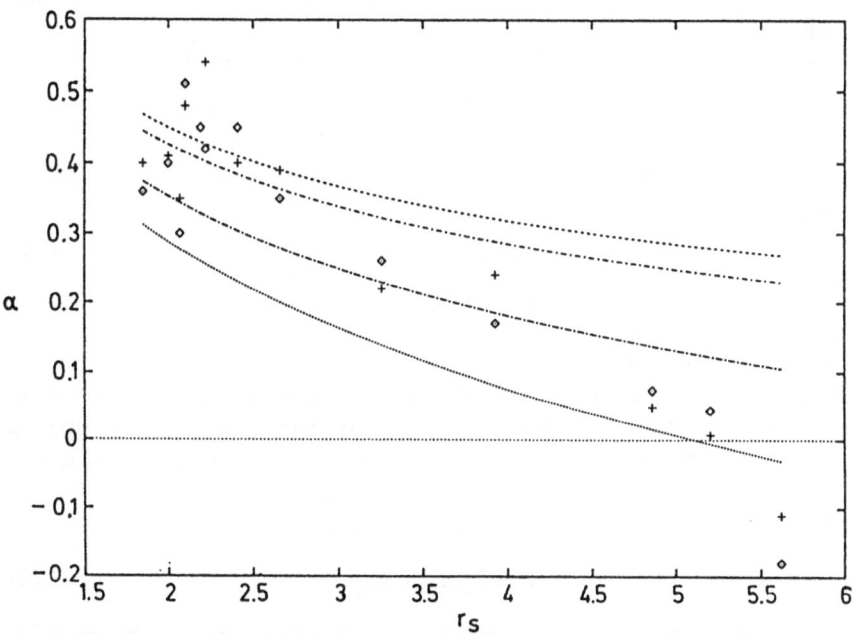

Fig. 3.9. Plasmon dispersion coefficient for metals. Symbols are the most recent (◊) and next most recent (+) experimental results available. Lines: RPA (dashed), lowest [3.40], intermediate [3.43] and highest [3.34] (dotted) reported exchange-correlation correction to RPA. The intermediate curve obeys several sum rules. $2r_s a_H$ is the mean distance between electrons.

---

[20] This is also the case in other simple metals and has led to considerable uncertainty and discussion. See also the discussion of the "discontinuous slope change" below.

Figure 3.9 shows the theoretical dispersion coefficients in the range of metallic densities, compared with experimental results. In order to avoid confusion, not all of the proposed corrections are given. We restrict presentation to the minimal correction taking into account exchange only [3.40], the maximal correction (that is, the minimal dispersion coefficient given in the theoretical literature [3.34]) and an intermediate correction that has been shown to satisfy several sum rules [3.43].[21] Evidently, the corrections improve the RPA but the agreement with experiment cannot be considered satisfactory. The graph shows that "turning the knob" labelled correlation correction does not solve the problem of plasmon dispersion in metals. There are other effects, leading to the very strong negative dispersion in the alkalis, and to deviations for small r_s – where the RPA should be excellent – that do not depend only on the electron density.

Local Fields. Another important cause of deviation from the RPA is the periodic lattice potential in a crystal. When translational symmetry is broken, the dielectric function, in momentum representation, becomes a second-rank tensor

$$
\bar{\varepsilon} = \begin{pmatrix}
\varepsilon_{00}(\omega,) & \varepsilon_{10}(\omega,\boldsymbol{q}) & \cdots & \varepsilon_{G0}(\omega,\boldsymbol{q}) \\
\varepsilon_{01}(\omega,\boldsymbol{q}) & \varepsilon_{11}(\omega,\boldsymbol{q}) & \cdots & \cdots \\
\vdots & \vdots & \ddots & \vdots \\
\cdots & \cdots & \cdots & \varepsilon_{GG}(\omega,\boldsymbol{q})
\end{pmatrix}
$$

With this notation

$$\tilde{D}(\boldsymbol{q}) = \bar{\varepsilon}\tilde{E}(\boldsymbol{q}). \tag{3.77}$$

The off-diagonal elements couple field components that differ by reciprocal lattice vectors $\boldsymbol{G}$. The off-diagonal elements depend on the Fourier coefficients of the lattice potential.

Owing to the presence of off-diagonal elements of ε, an external long-wavelength perturbation, given by $\tilde{D}_0(\boldsymbol{q})$, say, will create rapidly varying electric field components $\tilde{E}_G(\boldsymbol{q})$, often referred to as local fields.[22] That is, the local fields change the macroscopic, long-wavelength response of the medium.

The plasmon-defining condition (3.40) is now replaced by

$$\mathrm{Det}[\bar{\varepsilon}(\omega,\boldsymbol{q})] = 0. \tag{3.78}$$

The zeros of Eq.(3.78) are the poles of the loss function. They can be regarded as being caused by umklapp processes in the lattice, much as the band structure in crystals is caused by a superposition of free electron parabolas with

---

[21] A recent calculation based on the Fermi hypernetted-chain theory has a steeper slope with $A = 0.074$ and $B = 0.012$. It seems to fit the experiments slightly better [3.50].

[22] These must not be confused with the exchange-correlation fields discussed above. Those are in the vicinity of an electron in the homogeneous electron gas (jellium), caused by correlation and exchange (*exchange-correlation hole*) which are sometimes also denoted *local fields*.

relieved degeneracy at the crossing points (the Brillouin zone boundary). This results in plasmon bands (Fig. 3.10). Unlike higher electronic band excitations, however, the higher plasmon branches are much damped because of the plasmon decay mechanism [3.51,52] and are unlikely to be observed directly.[23]

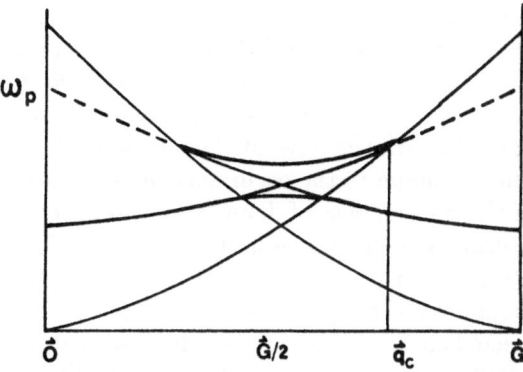

Fig. 3.10. Coupling causes plasmons at the Brillouin zone boundary $G/2$ to split into two branches, this is quite similar to the formation of energy bands. When $q_c > G/2$, the two branches should be observable.

The high-energy branches have little effect on the low-energy branch except next to the Brillouin zone boundary. Accordingly, the off-diagonal elements can be treated as a small perturbation when q is small. In this case, the (lowest) plasmon branch is given by the *macroscopic* dielectric function ε_M according to (3.40) [3.3,51]. It is important to note that, in general, the diagonal element of the inverse dielectric function differs from the inverse of the diagonal element

$$\varepsilon_M^{-1} = (\bar{\varepsilon}^{-1})_{00} \neq (\bar{\varepsilon}_{00})^{-1}. \tag{3.79}$$

The off-diagonal elements are always small, so one can write, after extracting the diagonal part of the matrix,

$$\bar{\varepsilon} = \mathrm{Diag}(\varepsilon)[\bar{1} + \bar{\Delta}] \tag{3.80}$$

where $\bar{1}$ is the unity tensor with elements $\delta_{GG'}$ and $\bar{\Delta}$ is a second rank tensor with elements $\Delta_{GG'} = (\varepsilon_{GG'} - \varepsilon_{GG})/\varepsilon_{GG}$. We have $\Delta_{GG'} \ll 1$ for $G \neq G'$. We can then expand the inverse

$$\bar{\varepsilon}^{-1} = \mathrm{Diag}(\varepsilon)^{-1}[\bar{1} - \bar{\Delta} + \bar{\Delta}^2 - \ldots] \tag{3.81}$$

noting that $\Delta_{GG} = 0$:

$$\varepsilon_M(\omega, q) = \varepsilon_{00}(\omega, q) - \varepsilon_{00}(\omega, q)(\bar{\Delta}^2)_{00}. \tag{3.82}$$

---

[23] Indirect observation of plasmon bands by use of coherent inelastic scattering has been reported recently [3.53] (Sect.3.2.2.).

We see that the local fields give a small *second-order* contribution to the macroscopic dielectric function.[24] This results in damping of the plasmon, caused by the non-vanishing imaginary parts of ε_{0G} when G lies in the single-particle excitation continuum. The reactive (real) part of the correction term shifts the zero of $\mathrm{Re}\{\varepsilon\}$, and so changes the plasma frequency. The direction of shift depends on the sign of the additional term; in general, it is positive, thus lowering the plasma frequency.

The local-field theory was developed in the early sixties [3.55,56]. *Paasch* [3.57] calculated the plasma frequency and the halfwidth of the alkali metals in this approach. The dielectric response of Si including local field effects showed a lowering of the excitation maxima to lower energies, and a broadening [3.58]. A similar calculation for covalent crystals [3.59] assigned the redshift of the optical absorption maximum to the occurrence of an exciton. *Sturm* [3.51,60] calculated the dielectric response of diamond and zincblende crystals in this approximation, using a pseudopotential to obtain the Δ_{0G}. The effect is a slight shift of the plasmon to lower energy, and a distinct broadening. A similar calculation for aluminium [3.61] yields only small corrections to the plasma frequency, as expected since Al is a paradigm for free-electron behaviour. The most prominent effect is on the linewidth, giving a FWHM of 0.39 eV at $q = 0$ compared to about 0.5 eV from experiment. Whereas the linewidth for $q = 0$ can quite well be described by an RPA-calculation including local field effects from the lattice, the situation is more complicated for $q > 0$. Quadratic dependence of wave number was found in the small-q range [3.51], as it is for the dispersion of the plasmon, but with noticeable differences between experiments and disagreement with theoretical predictions.

Local fields have only little effect on the dispersion [3.62]. For aluminium, this effect increases α by 5.5 % [3.63], or according to another calculation, by 7 % [3.61].

Attempts habe been made to model the coupling of the plasmon to lattice waves [3.64]. This leads to a dependence on the Debye wave number.

Core Polarizability. The core polarizability also contributes a fraction $\Delta\varepsilon$ to the dielectric function ε. Its influence on the plasmon energy has been treated phenomenologically by adding an oscillator with known strength and energy to the dielectric response. Since core excitations are in general higher than the plasmon, this will lower the plasmon energy, see Sect.3.1.1 (Band Transitions). It has, however, been pointed out recently that there are two shortcomings in this approach. First, the core polarization is definitely a local effect and should thus be modelled as such; and second, the oscillator strength of core electrons is smaller than suggested by the available number of core electrons since transitions to occupied valence states are forbidden. Conversely, the oscillator strength of the valence electrons is increased in

---

[24] A more detuilled yet short derivation of the local-field effects can be found in the review by *Sturm* [3.53].

order to satisfy the f-sum rule [3.65]. Thus, the medium has a higher *effective* valence electron density. This makes the downward shift of plasmon energy much smaller than believed earlier. For aluminium, *Sturm* [3.65] obtained E_p = 15.6 eV. (The free-electron value is 15.8 eV.) This is probably the most accurate prediction available including any and all feasible corrections to the RPA.

Core polarization contributes to the plasmon linewidth via its imaginary part, as given in (3.53). The core electron transition frequency as well as the dipole form factor of the ion cores are necessary for the calculation. This effect is important when transitions can be found in the vicinity of the plasma frequency. In indium, this contribution is of the same order of magnitude as the local-field effect discussed above [3.65]. In Al, the effect is very small since the core excitation edge is at 72 eV, far above the plasmon energy, and the damping term at 15 eV caused by the polarization of L-shell electrons is very small.

The influence of core excitations on the plasmon dispersion is not known at present. A simple argument, assuming a q-independent core contribution to the real part of ε, shows that the dispersion is slightly increased. In aluminium, the RPA dispersion coefficient is increased by slightly less than 1 %, and is thus smaller than any other effect. An accurate calculation would necessitate knowledge of the q-dependent core polarizability.

In conclusion, it can be said that the agreement of the most refined available predictions with experimental values is not what one would call overwhelming. The standard model substance, aluminium, steadfastly resists theory: the best prediction is still 0.6 eV higher than the measured plasmon energy (15 eV).

3.2.2 Recent Developments

The Discontinuous Slope Change. In the standard experimental system, aluminium, there seemed to be a discontinuous change of slope in the dispersion diagram $\omega_p(q)$; different authors reported different dispersion coefficients α for both the high- and low-q domain. Measured values ranged from 0.13 to 0.38 for α_{low} and from 0.30 to 0.46 for α_{high}. A similar slope change was reported for indium [3.66]. *Schattschneider* et al. [3.49] and *Sprösser-Prou* et al. [3.48] recently repeated the measurement of the Al-bulk plasmon with high energy and momentum resolution, motivated by the fact that the equipment used in the earlier experiments was probably not capable of giving reliable results. They were unable to find any discontinuous slope change in $\omega_p(q)$. This is in accord with the earlier experimental results of *Batson* and *Silcox* [3.47]. Unlike, *Batson* and *Silcox*, however, *Sprösser-Prou* et al. [3.48] did find deviations from the quadratic dependence (3.57), attributed by them to higher order terms in the RPA-series expansion.

In a recent experiment, to determine the plasmon dispersion in indium, no discontinuous slope change was found but, as in aluminium, a quartic term appeared in the dispersion [3.67].

It is almost certain, therefore, that the discontinuous slope change does not exist.[25]

Negative Dispersion. Experiments on the alkali metals have made it clear that the deviations from the RPA increase as r_s increases; they thus seem to be caused by strong correlational effects, which cannot be easily described by a correction to the RPA. In Cs, the dispersion is negative [3.31] (Fig. 3.9). The reason for this is that the coupling constant in a plasma, Γ, is of the same order as r_s, as already mentioned, so the heavy alkalis cease to resemble metals with weak coupling and the RPA, which is a weak-coupling theory, breaks down. It has been speculated that in Cs, a crystallisation of the Fermi liquid into a Wigner lattice could be incipient [3.69,70]. In the Wigner crystal, there should be optical phonons with a negative dispersion, as in a crystal.[26] But since the Wigner crystallisation should occur at $\Gamma \simeq 178$ in a nondegenerate electron gas [3.69], or after another calculation at $r_s \approx 100$ [3.72], Wigner crystallization is highly improbable for the heavy alkalis. It has however been shown that the plasmon dispersion in a strongly coupled classical plasma already becomes negative [3.73] for coupling constants equivalent to between $r_s \simeq 5$ [3.74], and $r_s \simeq 10$ [3.73]. An RPA calculation with a correlation correction yields negative dispersion for $r_s > 6$ [3.31], in qualitative agreement with the independent classical calculation. The quantum mechanical hypernetted chain (HNC) approach gives also a negative dispersion [3.50]. *Taut* [3.25] showed that negative dispersion for high r_s is forced by an anomaly in the structure factor at $2k_F$. The behaviour of plasmons in the heavy alkalis can be explained qualitatively in this way [3.75] but none of the theories seems to be in quantitative agreement with experiment.[27]

High-q Plasmons and Plasmon Bands. There was early evidence that the plasmon dispersion flattens when $q > q_c$, i. e., when the plasmon can decay into electron-hole pairs [3.77]. The flattening could be explained qualitatively by a takeover of oscillator strength by the electron-hole continuum

---

[25] The "softening" of the plasmon, i.e. the flattening of the plasmon dispersion at higher momentum transfer, reported for III-V semiconductors, such as InSb [3.33] and for some ternary semiconductors [3.68] is probably a real effect of unknown origin.

[26] As early as 1974, an incipient Wigner lattice was proposed in order to explain the anomaly of the dynamical structure factor at large q [3.71]. Today, it seems certain that the effect is due to both correlational effects and to the lattice potential.

[27] Recently, a calculation taking into account the combined effect of exchange-correlation and the local fields reduced the discrepancy with the measured dispersion coefficients in the alkali metals. The best results were obtained using a local density approximation [3.76]. It must however be conceded that the theory can only reproduce the initial slope; in the heavy alkalis in particular, it fails for $q > 0.5\,q_F$.

as shown in Fig. 3.11. Some experimenters claimed to observe two maxima in the high-q spectrum. It has now been confirmed that the inelastic scattering cross-section does indeed exhibit structure in the high-q regime in many materials [3.71,78-81]. Most of the experiments have been performed with X-ray or synchrotron radiation rather than with electrons because of intensity considerations.

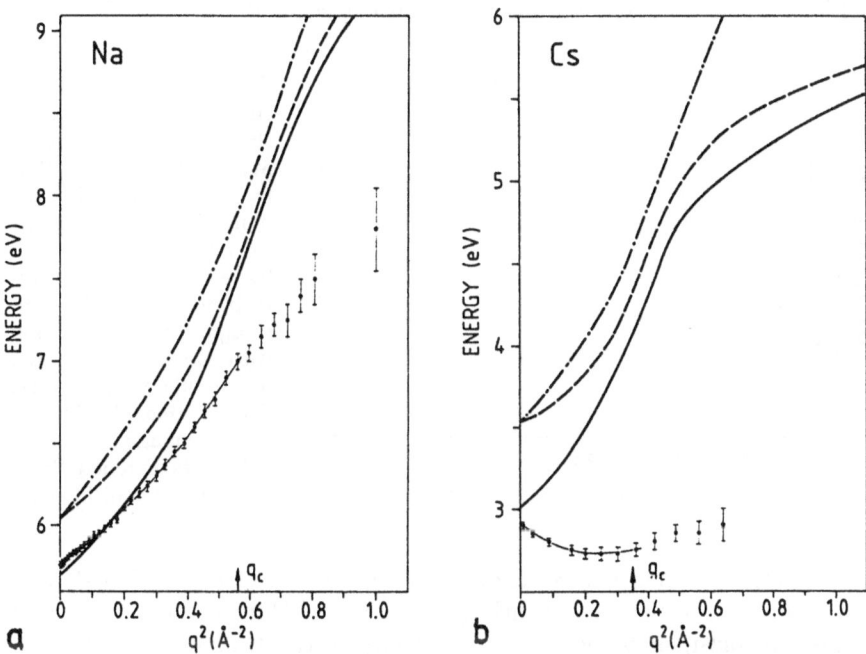

Fig. 3.11. Plasmon dispersion for (a) Na and (b) Cs. Dash-dotted: RPA; dashed: local field corrections [3.44]; full line: core polarization included. Dots: measurements. Note the softening of the plasmon at $q > q_c$ [2.192].

There are several explanations for this behaviour, which qualitatively agree with experiment. Besides the plasmon band model discussed above, exchange effects [3.82,83] have been proposed as explanation for the structure at high q. *Schülke* has given a qualitative explanation of his findings of a double-peak structure in Li [3.78,84] and in Be [3.79,80] by energy gaps at the Brillouin zone boundaries. These gaps cause a modulation in the dynamical structure factor. It is essentially the same effect that leads to zone boundary collective states except that, in the present case, the real part of the dielectric function does not vanish within the gap – see the discussion on zone boundary collective states in Sect. 3.3.4. Recently *Sturm* and *Schülke* [3.85] have given a quantitative explanation for the structure visible in Si. These authors conclude that the structure visible beyond cutoff is a resonance, or better, an antiresonance between plasmons and the single-particle excitation continuum at large q. It has been argued above that the plasmon at $q \simeq 0$ has finite width because it can decay into the electron-hole continuum by umklapp processes in the lattice. Conversely, the electron-hole excitations that have $E = E_p$ can decay into a plasmon. This process removes intensity from the continuum, which will consequently have a local minimum[28] at $E \approx E_p$. Although it is essentially caused by the local fields, exchange corrections enhance the effect. Figure 3.12 shows this minimum at the plasma frequency in silicon.

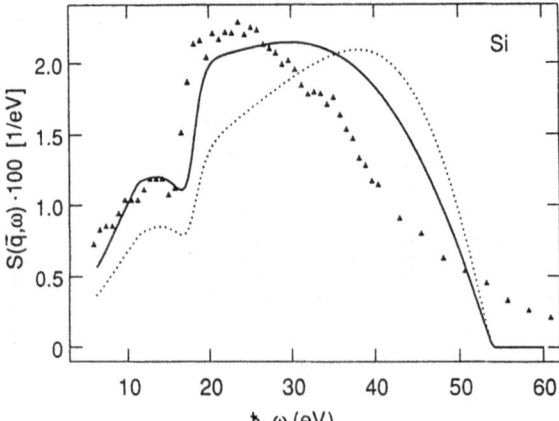

Fig. 3.12. Plasmon antiresonance in Si next to the Brillouin zone boundary (q=23.6 nm^{-1}). The local minimum in the dynamical structure factor $S(q,\omega)$ is caused by a decay of the intraband transition continuum into a plasmon with $q \simeq 0$, $E = E_p$, by umklapp processes. Dotted line: RPA; full line: exchange-correlational effects included. Symbols: experiment [3.53].

Recently, plasmon bands, as discussed in Sect. 3.2.1 (Local fields) were found in an experiment with synchrotron radiation. Normally, the separation of the two plasmon branches in the vicinity of the Brillouin zone boundary is so much smaller than their width that the peaks are not separable experimentally, even in the semiconductors that have $q_c > G/2$. Using a two-beam geometry (only two Bloch waves are well excited), the diagonal and the nondiagonal elements of the inverse dielectric matrix can be separately determined, by tuning the excitation strength of the two Bloch waves. This allowed for a separation of the plasmon maximum into two branches (Fig. 3.13).

Coherent Double Plasmon Excitation. Double plasmon excitations are commonly found in electron energy loss spectra obtained in transmission experiments on thin metallic films.

The relative intensities of the double-to-single plasmon excitation range from some ten to some hundred percent, depending on the thickness of the specimen [3.4]. This is because the fast electron traversing the specimen in

---

[28] This is analogous to the Fano-Breit-Wigner interference that draws intensity from a broad background by excitation of phonons, as observed in infrared-reflectivity measurements [3.86].

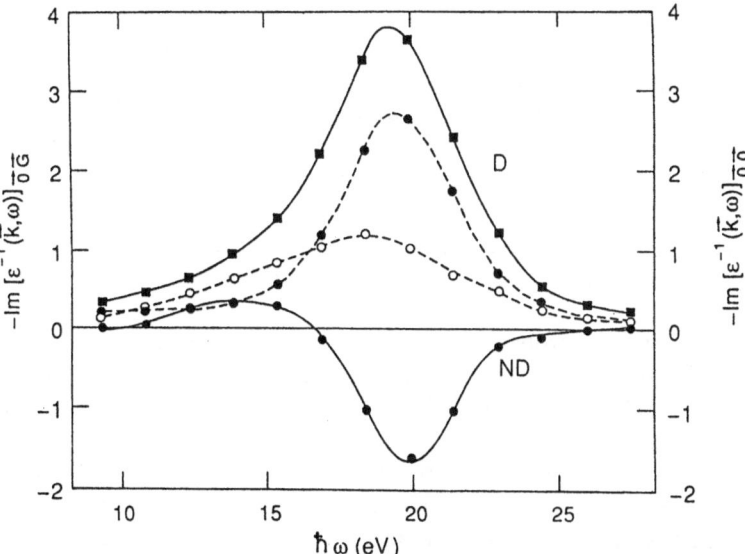

Fig. 3.13. Experimental proof of plasmon branches at G_{111} in Si. Diagonal (D) and non-diagonal (ND) elements of the inverse dielectric tensor, as determined from inelastic X-ray scattering (full lines). From these, the two plasmon branches were calculated (dashed lines) [3.53].

an EELS experiment interacts a number of times with the solid state plasma; in each interaction along its trajectory it will loose the plasmon excitation energy E_p with a high probability and hence the EELS will exhibit a number of peaks at the multiples of the single plasmon excitation energy. The peak intensities obey a Poisson distribution [3.4].

Ashley and *Ritchie* [3.87] considered the possibility of excitation of *two* plasmons in a single interaction of the fast probe electron with the solid state plasma. This process should cause a subsidiary maximum in the EELS at the same energy as the subsequent excitation of two single plasmons, namely at $2\,E_p$, so that it cannot be observed directly. Since the latter interaction is, essentially, an incoherent one whereas the simultaneous excitation of two quanta of collective oscillation in one event corresponds to inclusion of second order terms in a perturbation theoretical treatment of the solid state plasma, these cases are referred to as *coherent* and *incoherent* excitations.

Ashley and *Ritchie* originally calculated the relative probability F_2 for the coherent double plasmon loss in the random phase approximation. They found, assuming a plasmon cutoff wave number q_c of 11 nm^{-1} in aluminum, a relative intensity $F_2 = 0.04$. This value is very sensitive to the choice of q_c. A value of $q_c = 15$ nm^{-1} yields [3.88] $F_2 \simeq 0.17$. *Srivastava* et al. [3.89] predicted $F_2 = 0.024$ from a quantum mechanical calculation based upon a canonical transform of the many–body Hamiltonian including second-order terms.

Not only the theoretical predictions but also the experimental results disagree by an order of magnitude. The main problem here is to single out the coherent contribution from the much larger incoherent double plasmon peak. *Spence* and *Spargo* [3.88] reported an experimental value of $F_2 = 0.13$ for coherent double plasmon excitation in energy loss spectra of an aluminium single crystal, a value higher than but not inconsistent with the prediction of *Ashley* and *Ritchie*.

A straightforward way of obtaining the coherent contribution experimentally is to process the spectra so as to remove plural incoherent events from measurements.

The single scattering probability f_1 is masked by plural losses even for thicknesses less than one mean-free-path length. A scattering experiment always yields the plural scattering probability p. The problem is how to retrieve f_1 from measurement of p.

A number of methods have been reported in the literature for retrieval of the single-loss probability. Some of them are suited for EELS obtained in the electron microscope in the image mode, i. e., when all electrons are collected independent of their angle of scattering [3.47]. More recently, methods have been used which work in the diffraction mode [3.90]. In the latter case, the scattering probability $p = p(E, \theta)$ is measured in the focal plane of the objective lens as a function of energy loss E and scattering angle θ.

For image mode spectra of aluminium, F_2 was found to be less than 0.02 after correcting for incoherent double losses by deconvolution [3.91]. This value is close to the theoretical value reported by *Srivastava* et al. [3.89]. According to a recent investigation on the influence of the objective aperture of the microscope on deconvolution [3.92], there can be errors up to some percent of the original (incoherent) double plasmon peak, due to truncation of the angular profile by the objective aperture. This could explain the fact that experimentally, F_2 was found to decrease with thinner specimens.

Batson and *Silcox* [3.47] found $F_2 \simeq 0.07$ after removal of plural incoherent losses in diffraction mode energy-loss spectra of aluminium. In a more recent diffraction mode experiment on aluminium, an upper limit of $F_2 \leq 0.005$ was obtained [3.93]. This limit is far below the predicted values. For processing, a deconvolution procedure for radially symmetric angular profiles, based on Fourier–Bessel transforms was used [3.94]. Figure 3.14 shows diffraction mode spectra, before and after deconvolution. Note that the triple plasmon peak is completely removed after deconvolution. The dispersion of the plasmon can also be seen clearly. The small structure remaining at $E = 2\,E_p$ in Fig. 3.14b is attributed to the coherent excitation of two plasmons. It should be noted that the noise appears exaggerated in the figure since the intensity is on a logarithmic scale.

Double Bragg inelastic scattering causes a small amount of incoherent double plasmon intensity to remain at high scattering angles [3.47,96] with the result that the coherent contribution is smaller than the observed values.

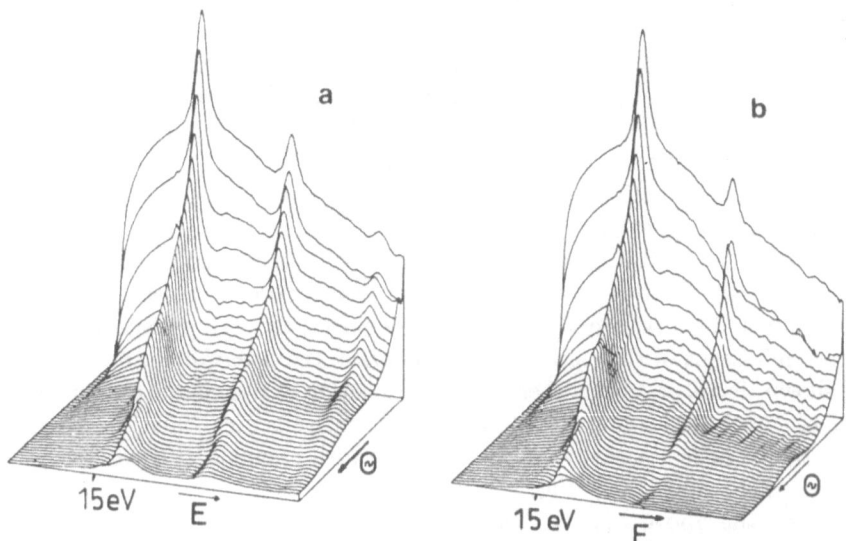

Fig. 3.14. Angularly resolved plasmon spectrum of aluminium (intensity on a logarithmic scale): (a) as measured, the plasmon dispersion is clearly visible; (b) after removal of momentum-dependent multiple scattering. The remaining small intensity at the double plasmon may be due to coherent two-plasmon excitation [3.95].

This effect is the more prominent the thicker the specimen. The results are thus most reliable for very thin specimens at small angles.

In an experiment performed in image mode [3.93] a value of $F_2 \le 0.006$ was found for a 24 nm thick Al film, in agreement with angle-resolved measurements. The higher values for F_2 previously reported[29] may be due to the use of less sophisticated processing, of thicker films in which Bragg inelastic scattering is a serious problem, and to investigation of single crystals in which dynamical effects may enhance the probability for coherent double plasmon excitation.

3.2.3 Acoustic-Like Plasmons and Layered Structures

Acoustic-Like Plasmons. As discussed in Sect. 3.1.3, the plasma frequency (3.55) is finite for zero wave number, with only weak dependence on q and vanishing group velocity $\partial \omega / \partial q$. This dispersion is similar to that of an optical phonon and the (regular) plasmon is therefore sometimes termed optical. It should be noted that this is the only reason for this appelation, it is definitely not implied that the plasmon can be excited optically as is the case

---

[29] *Ingram* et al. [3.97] observe remaining oscillator strength at the double plasma frequency, after deconvolution and Kramers–Kronig analysis of reflection EEL spectra in aluminium, without giving F_2 values.

for optical phonons. With this nomenclature in mind, it is natural to ask whether there are acoustic plasmons. This question has been discussed since the fifties [3.98]. Acoustic plasmons should obey the same dispersion relation as their phonon analogues, $\omega \propto q$, which occur in a lattice with basis and two constituents of different masses. This reasoning can be extended by assuming a two-component solid state plasma, with two different effective masses of electrons. Since the Coulomb interaction is repulsive, an out-of-phase motion of the two charge carriers would reduce the potential energy of the charge fluctuation, necessitating less energy than an in-phase motion at the same wave number. That is to say, the out-of-phase motion in a two-component plasma would constitute a new low-energy excitation with a dispersion relation reminiscent of that of an acoustic phonon, the *acoustic plasmon*. It can be shown that $\omega \propto q$ when there is a superposition of two charge carriers with different masses, leading to two different Fermi velocities [3.99]. Note again that the name is not really appropiate since with phonons, the two branches change their role [3.100] (Fig. 3.15).

From the above remarks, one would expect that semiconductors with two types of charge carrier, such as the III-V compounds, should regularly exhibit acoustic plasmons. But in spite of great efforts the search for acoustic plasmons in the semiconductors has not been successful.

Ousaka et al. [3.101] realized that in homogeneous, isotropic solid state plasmas, there is always band mixing between charge carriers of different mass. It is not meaningful to assume two Fermi distributions. Rather, the system should be described by a unique dielectric function that mirrors the possible intra- and interband transitions. In doing so, it is straightforward to show that the dielectric function is in principle of the Drude–Lorentz type [3.101] (see Sect. 3.1.1). For particular cases, the low-energy longitudinal excitation is almost linear in q for intermediate wave numbers, but it inevitably tends to the interband transition frequency for $q \to 0$. When there is no bandgap the plasmon merges into the edge of the electron-hole continuum and thus ceases to exist as a well-defined excitation. This is proof that in a homogeneous, isotropic system, acoustic bulk plasmons do not exist.

On the other hand, when the above restrictions of homogeneity and isotropy are lifted, the situation changes dramatically. Soon after the calculation of the static dielectric response of an electron gas of layered structure by *Visscher* and *Valicov* [3.102], *Grecu* [3.103] and *Fetter* [3.104] analysed the plasmon modes in a model system of equally spaced two-dimensional layers of electron gas. There has been a vast amount of theoretical work on such systems, spurred by the technological ability to produce artificial heterostructures such as GaAs/GaAlAs.

These studies differ widely in approach, approximations and results, and the dielectric properties depend strongly on the particular system chosen (ideal two-dimensional electron gas in an insulating matrix [3.105–107]; semiconductor well structure with finite bandwidth where Fermi levels, distance

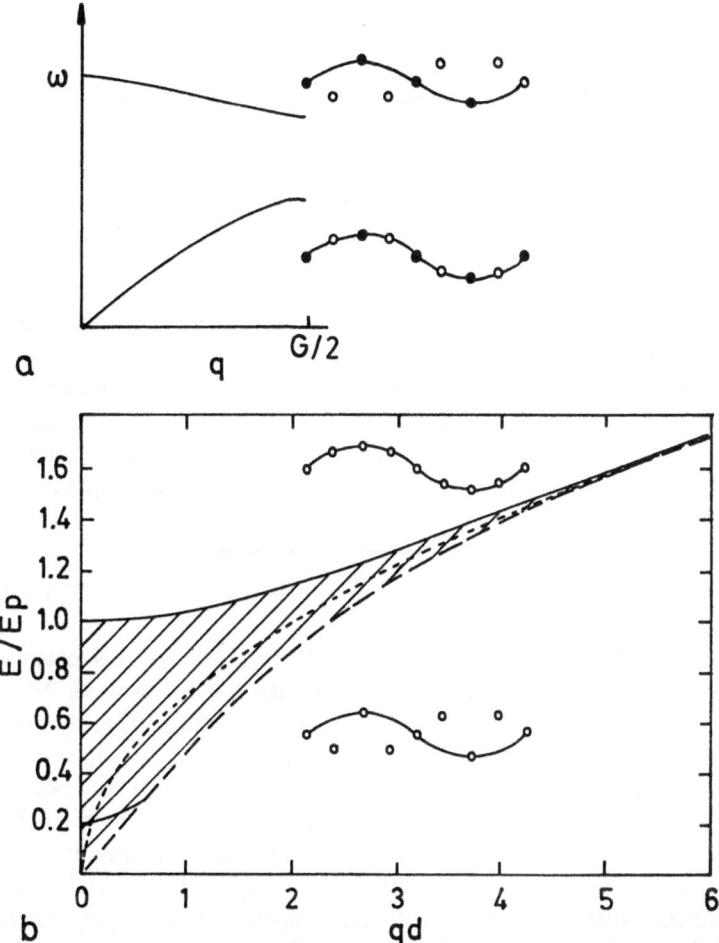

Fig. 3.15. (a) Schematic representation of the acoustic (lower) and optical phonon branch in a crystal. Next to the branches, the positions of positive (shaded circles) and negative ions (white circles) are sketched. (b) plasmon continuum in a layered two-dimensional electron gas. Abscissa in dimensionless units qd (d is the layer distance). Unlike the phonon case, the acoustic (lower) branch corresponds to out-of-phase motion of electrons on adjacent planes (dashed line). The dotted line is the surface plasmon dispersion $\omega \propto q^{1/2}$. The continuum is generated by the coupling of a large number of surface plasmons from individual layers with removal of the degeneracy. The lowermost full line indicates the asymptotic behaviour of the "almost acoustic" plasmon in a homogeneous system with two different charge carriers [3.101]. The branch tends to the interband transition frequency for $q \to 0$.

and thickness of the wells can be varied [3.108–112]; electron-hole systems [3.113] or metallic superlattices [3.114–116]; and metal/insulator systems [3.117]). The question of coupling between plasmons and polaritons with respect to metal optics and emphasis of surface modes has also been discussed [3.118–122].

One general feature occurs in all of those calculations: in layered structures, there is a continuum (or, to be exact, a quasi-continuum) of modes, spanned by the coupling between layers with relieve of degeneracy. The upper bound of the continuum emerges from the high-energy bulk plasmon and, as $q \to \infty$, it tends to the plasma frequency given by the average charge density in the system. The lower bound emerges from the lowest-energy longitudinal oscillator of one of the constituents; it tends to the same value as the upper bound. When the system consists of layers of two-dimensional electron gas interspersed by an insulator, the lower bound is proportional to the wave number for low wavenumber, i. e., it is *acoustic*.

In order to visualize this behaviour, we discuss the plasmon dispersion for the idealized case of an infinite stack of two-dimensional layers of electron gas. As a function of the in-layer component q and the component q_z of the wave vector perpendicular to the conducting planes, the plasma frequency is (e. g. [3.106,123])

$$\omega(q, q_z) = \omega_p \sqrt{\frac{qd}{2} \frac{\sinh(qd)}{\cosh(qd) - \cos(q_z d)}} \qquad (3.83)$$

where d is the layer distance. The factor ω_p is the well known bulk plasma frequency of (3.25), where the *mean* electron density has to be used. Note that the mean density is given by $n = n_s/d$ when n_s is the surface density in the conducting plane. Graphs of this function for $q_z = 0$ and $q_z = \pi/d$ are given in Fig.3.15. These are the upper ("optical") and lower ("acoustic") limits of the quasi-continuum of plasma excitations. The appearence of a continuum can be understood when one considers the system as a large number N of stacked, two-dimensional electron gas layers. Each layer in itself supports a propagating surface plasmon [3.3] that disperses as $\omega \propto \sqrt{q}$, shown as the dashed line in the figure. These modes are N-fold degenerate without coupling. The Coulomb interaction strongly couples the layers, so the degeneracy is relieved with a rather large splitting between the N modes. The upper branch is a plasmon with $q_z = 0$, the charge carriers in adjacent planes oscillating in phase, whereas the lower bound is due to a charge oscillation with $q_z = \pi/d$. This is the highest wave number the layered structure can support. Charge carriers in adjacent planes move out of phase in this mode – this is exactly the definition of an acoustic plasmon given above. We see that layered structures do in fact feature acoustic plasmons.[30]

---

[30] The acoustic plasmon, as well as any mode in the continuum except the optical mode, is not strictly longitudinal. This is at once clear since the electrons can only move in the conducting planes whereas the wave vector of the mode points in an arbitrary direction. But even when movement out of the planes is allowed, there are in general transverse components because the layered structure breaks symmetry, and this causes coupling of transverse and longitudinal modes. See the discussion in Sect. 3.1.2.

The q-dependence in the square-root shows that the modes behave originally as surface plasmons. The second factor in the square-root is the form factor of the superlattice, which modifies the planar behaviour by coupling between modes. In the limit $qd \to 0$, the strong-coupling limit, it follows from (3.83) that the upper bound of the continuum is the bulk plasma frequency, and the lower bound is linear in q

$$\omega(qd \ll 1) = \omega_{\mathrm{p}} \frac{qd}{2} \tag{3.84}$$

showing "acoustic" behaviour. The group velocity is $\omega_p d/2$. When $qd \to \infty$, the weak-coupling limit, corresponding to large layer spacing compared to the wavelength of charge fluctuations, one expects re-emergence of the planar mode. In fact,

$$\omega(qd \gg 1) \propto \sqrt{q}, \tag{3.85}$$

independent of q_z as can be seen from the figure. In general, this behaviour is also found in more complex systems (see literature cited above) although the results differ in detail.

When the layer distance d tends to zero, one would expect recovery of regular bulk behaviour. In fact, for very small d, correlation effects lead to an enhanced reduction of the phase velocity of the acoustic plasmon until it finally merges with the electron-hole continuum. As an existence condition for acoustic plasmons, *Ivliev* and *Sobakin* [3.124] give

$$d > r_{\mathrm{s}} \gamma(r_{\mathrm{s}}) \tag{3.86}$$

where γ is inversely proportional to r_{s} and is of the order of 0.1 for typical values of r_{s}, the mean electron distance in the layers. This is a rather favourable condition, since only for layer spacings much below the interelectron distance do the acoustic plasmons cease to exist. It is not clear whether this condition really holds or whether, at much larger values of d, a transition to 3-dimensional behaviour already takes place as other calculations show [3.112]. This seems to be the case in GaAs/GaAlAs heterostructures [3.125] at a layer thickness of $\simeq 10$ nm. EELS experiments on Mo/V superlattices show that at a layer thickness of 40 monolayers or less, the dielectric function deviates markedly from the linear combination of the single bulk dielectric functions [3.126]. This indicates a transition from the weak to the strong coupling regime.

Acoustic plasmons have been detected in heterostructures by the use of inelastic photon scattering in the meV-energy range [3.125,127–129]. EELS has not been so successful, and this is probably due to the fact that spectrometer resolution is too low for the very low-loss region of acoustic plasmons. Another reason is certainly the low excitation probability in the plasmon continuum. The question of excitation strength within the plasmon band has also been addressed [3.109,130–134]. The results agree in that the excitation strength is largest at the upper energy bound, decreasing rapidly to lower energies, with possibly a submaximum at the lower acoustic bound. Moreover, the excitation strength is concentrated at the high energy bound (the regular bulk plasmon) more and more when the wavenumber is decreased [3.109,130]. This indicates that in general it will be difficult to detect acoustic modes with EELS.[31]

An EELS study of intercalated graphite showed evidence for a low-loss plasmon band, but the effect was too small for the results to be conclusive [3.131]. In potassium intercalated graphite, a peak splitting of each of the two graphite plasmon peaks at 7 eV and 27 eV was observed [3.136]. The authors ascribe the lower peak to charge carriers donated by potassium, and the higher peak to a broad, unidentified interband transition at about 26 eV. The lower peak splitting is too large to have been caused by an acoustic plasmon band but possibly the higher band, showing a broad, momentum-dependent plateau with distinct edges, is an indication of an acoustic-like plasmon band. Asymmetries in the plasmon profile can also be seen in two other experiments [3.137,138], but these authors did not try to analyse the spectra in terms of plasmon bands.[32]

In conclusion, there is at present no definite proof of a plasmon continuum in the energy range above 1 eV.

Ceramic Superconductors. Low-loss spectra of high-temperature super-conductors have been published since 1988 [3.6,137,140–150]. The standard treatment using the Kramers–Kronig transform was applied to the spectra [3.6,140] in order to assign the structure in ε_2 to particular interband transitions. Apart from the well-known problem of visual structure assignment,[33] the experimental results differ considerably. This is partly due to early problems in specimen preparation as *Bozovic* et al. [3.151] pointed out in a review of experimental work, but also to the experimental setup and data processing. For instance, the plasma frequency reported at 13.3 eV [3.140] is due to a zero crossing of ε_1 after Kramers–Kronig analysis. More recent data show only a minimum at this energy loss.

Although several authors have argued that EELS in the low-loss regime has a number of advantages over optical and X-ray methods, such as probing the bulk [3.140], probing the ground state [3.6], or probing small volumes

---

[31] A recent experiment in reflection EELS of misfit-layer compounds was negative [3.135], but this may be due to the strong contribution of surface excitations masking the acoustic bulk plasmon.

[32] *Ritsko* et al. [3.138] report an anomalous decrease in excitation strength with increasing in-plane wave vector. Such a decrease has been predicted for super-lattices, and has been related to the onset of localisation [3.106].

[33] For instance, in $YBa_2Cu_3O_7$, the maximum in ε_2 has been found and assigned as follows: 6.6 eV: $O2p \rightarrow Y4d$, $Ba5d$ [3.140], 8 eV: $Cu3d,O2p \rightarrow Ba4f$ [3.6], 7.5 eV: $Cu3d$, $O2p \rightarrow Ba$, Y [3.137]. The paper of *Ramsey* et al. [3.143] which assigns structure in REEL spectra between 6.4 and 9.4 eV to "interband transitions and surface plasmon", is unreliable because no Kramers–Kronig analysis was performed.

[3.152], the standard treatment described above has merely confirmed the coarse structure of electronic transitions already known from optical experiments, and added some uncertainty to the finer details – for a review of experiments, see [3.151] – without having contributed to an understanding of the mechanism of superconductivity. This is because the dielectric properties in the energy range above 1 eV or so are probably not related to superconductivity. The more recent EELS-measurements between some eV and some 10 eV agree reasonable well with one another and with optical measurements [3.148] which, in turn, seem to depart not too far from calculations in the conventional band approach, as was shown for the two examples $Bi_2Sr_2CaCu_2O_8$ [3.153] and $Ba_{1-x}K_xBiO_3$ [3.154].

More interesting insights into the electronic properties of the ceramic superconductors and possible clues to the mechanism of superconductivity can be obtained by high resolution EELS in the range up to 1 or 2 eV. This brings us back to acoustic plasmons.

Among other more arcane candidates for a pairing interaction in superconductivity at high temperature has been the plasmon. In 1981, *Ruvalds* [3.155] in a rather speculative paper suggested acoustic plasmons as a pairing mechanism for superconductivity at room temperature. There are several papers on the possibility of plasmon-assisted Cooper pairing, e.g. [3.100,132,156,157]. *Mahan* and *Wu* [3.158] discuss the case for regarding the plasmon as an intermediate boson for superconductivity. They conclude that the gap equation cannot be satisfied when the plasmon acts as a pairing boson. As they state, this result is well known, *since plasmon oscillations do not make aluminium into a high-temperature superconductor.* Their result was obtained by inclusion of all wave number components of the alleged boson that mediates pairing. The problem is however more complicated. Recently, it has been shown [3.123,159] that, under certain assumptions, the gap equation can be satisfied with surprisingly good agreement between calculated and measured values of T_c. Moreover, in both papers, an increase of T_c with the number of CuO_2-planes in the unit cell is predicted. Although different approaches were chosen, both authors argue that plasmons in layered media would cause superconductivity only under the assumption of strong localisation. This has been suggested independently [3.26], see also the remark on localisation above.[34]

By momentum-resolved EELS, the Karlsruhe group [3.161,162] showed that the peak visible at about 1 eV for a number of ceramic superconductors has in fact a dispersion $E \propto q^2$ suggesting a bulk plasmon (Fig. 3.16). These findings were confirmed later on with slightly different values for the

---

[34] Recently, it has been shown that electronic pairing mechanisms, when they act in a two-dimensional electron gas, are very efficient when the Fermi level lies at or next to a van Hove singularity in the density of states [3.160]. The cuprate superconductors fulfil both requirements. As in the plasmon model, it is the low-momentum, low-frequency electronic interactions that mediate the pairing, thus also pointing to an acoustic or almost acoustic collective excitation.

plasma frequency and the dispersion coefficient [3.149]. The authors specu-
late that this is due to differences in the electron density caused by differences
in stoichiometry, and argue qualitatively that the effective mass of the charge
carriers m^* scales as the hole concentration n, thus leaving the plasma frequency almost unchanged when n is changed, see (3.51). This is an interesting
speculation since it would allow $m^* \to 0$ for a half-filled band.

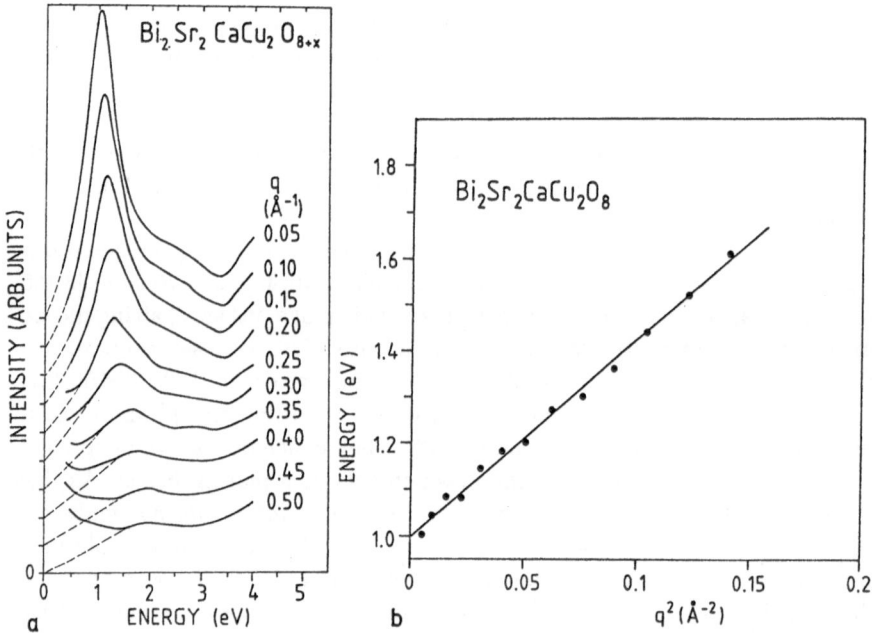

Fig. 3.16. Plasmon dispersion in the ceramic superconductor $Bi_2Sr_2CaCu_2O_8$; (a)
Energy loss spectra; (b) position of maxima [3.162].

Fink et al. [3.137,163] argue that the dispersion coefficient in $YBa_2Cu_3O_7$
and in $Bi_2Sr_2CaCu_2O_8$ should be calculated on the assumption that the system is two-dimensional (collective movement of charge carriers in the planes
containing O_2). This claim must be considered with caution since the classical
model underlying this argument neglects all the facts that have been known to
influence seriously the plasmon energy and dispersion in solids (Sect. 3.2.1);
the experimental value differs only by about 20% from the expected bulk
plasmon dispersion, so the surprisingly good agreement of the Fermi velocity
with predictions from band structure calculations, derived from the assumption of a two-dimensional electron gas[35] is not really conclusive.

---

[35] Moreover, in a strictly two-dimensional electron gas the dispersion is $\hbar\omega \propto \sqrt{q}$
[3.164].

Plasmon spectrometry would be important for deciding if acoustic-like plasmons, such as predicted, exist in the ceramic superconductors. Condition (3.86) yields a spacing of less than 0.25 nm, favourable for acoustic plasmons. *Griffin* and *Pindor* [3.165] give even better conditions. They find $n-1$ additional modes below the continuum where n is the number of CuO_2 layers in a unit cell. An experimental proof of these predictions would support theories of superconductivity based on the plasmon-exchange mechanism.

There is only one experiment that indirectly confirms these predictions: in optical reflectance/ellipsometry measurements on two ceramic superconductors, the calculated loss function was shown to vary as E^2 as $E \rightarrow 0$, in contradiction to the expected Drude behaviour [3.166]. The same E^2 dependence was predicted for a layered electron gas model with parameters that fit reasonably well [3.167]. Unfortunately, this feature shows up in the limit of very small energies only. EELS cannot probe this energy region because of the zero-loss contribution. A test of the theory by EELS should rely on more distinctive predictions such as an anomaly in dispersion of the plasmon,[36] its temperature dependence [3.169] or the quantitative contribution of acoustic modes to the loss spectrum.

A recent attempt to explain plasmon dispersion in $YBa_2Cu_3O_7$ by a quantum mechanical model featuring acoustic plasmons [3.170] is not convincing. The prediction deviates so slightly from the standard bulk plasmon model that it is not possible within the given experimental uncertainty to rule out or confirm the existence of acoustic plasmons. More accurate measurements are necessary, as well as more detailed calculations, which could be compared directly with experiment.

3.3 Standard Interpretation of Low-Energy Losses

The previous sections were largely devoted to the theory of longitudinal excitations in the solid, and to experimentals tests of those theories. Admittedly, the more practical applications of plasmon spectroscopy are rare. This is primarily due to the complexity of this many-electron phenomenon, as discussed above, which makes it difficult to interpret plasmon spectra in material-specific, easy-to-understand terms. One can however use (3.12) to calculate the dielectric function from the loss function. Its imaginary part is essentially a measure of the strength of intra- and interband transitions, and can be compared with predictions from band calculations, as in optical spectrometry.

---

[36] Recently, a line splitting of the plasmon when it enters the superconducivity-gap has been predicted [3.168]. This argument would however become invalid when the gap is opened by the very existence of plasmons, such as in a charge density wave instability [3.25].

This may be considered the standard application of low-loss EELS. In the light of the preceding chapters, it amounts to determination of transverse eigenmodes from the measurement of longitudinal ones.

It is hard to see why one should buy a TEM and a spectrometer, struggle with multiple scattering and surface modes, and then apply demanding data processing such as Kramers–Kronig analysis instead of directly measuring the reflection coefficient for light!

The advantages are to be found in all the capabilities of modern electron microscopes, that is, a vast palette of additional techniques, including energy-dispersive X-ray microanalysis and high-resolution facilities, in a single instrument. This implies investigation of specimens on the nanometre or subnanometre scale. Moreover, for intermediate specimen thickness, surface modes and surface overlayers do not pose too difficult a problem whereas in optical methods, surface overlayers may be an obstacle. Lastly, the energy (or frequency) range is practically unlimited.

Attempts have been made to calculate the dielectric function of almost every material from EELS data, in general in reasonable agreement with results from optics. For a review up to 1980, see the book of *Raether* [3.23]. *Fink* [3.171] summarizes more recent work, primarily of his group.

It is important to note that (3.89) shows that EELS in the image mode does not furnish a quantity that can be compared with optical data. It is rather EELS in the diffraction mode with $q = 0$ that is the proper experiment, though image-mode EELS has mostly been used. This is justified by the approximation $\varepsilon(\omega, q) = \varepsilon(\omega, 0)$. The Lorentzian q-dependent prefactor in the scattering cross-section (3.12) can be integrated out, and one has

$$Im\left\{-\frac{1}{\varepsilon}\right\} = \frac{(e\pi a_H)^2}{8\pi^2\varepsilon_0} \frac{k_0^2}{\ln(q_m/q_E)} \frac{dp}{dE} \tag{3.87}$$

where the last factor is the image mode spectrum

$$\frac{dp}{dE} = \int_{q_E}^{q_m} \frac{\partial^3 p}{\partial E \partial q^2} d^2q. \tag{3.88}$$

Referring to Fig. 3.1, the logarithm in (3.87) can be written in terms of scattering angle, as $0.5 \ln[1+(\beta/\theta_E)^2]$ where β is the collection half-angle corrresponding to q_m. Here, q_m is the maximum wave number allowed (e. g. by the contrast aperture), and $q_E = k_0 E/2E_0$ is the momentum halfwidth of the scattering cross section. See the comment after (3.1) in Sect. 3.1.1.

Before performing the Kramers–Kronig analysis (3.13), surface and multiple scattering contributions to the spectrum must be removed. Surface effects can in principle be removed by a self-consistent method originated by *Wehenkel* and *Gauthé* [3.172]. For removal of multiple scattering, several deconvolution methods based on Fourier transform of the spectrum or on matrix algebra have been devised. The former methods are described in [3.4], for the latter, as well as for a comparison of techniques, see [3.49,172][37]. For a more detailed survey, the reader should refer to the books of *Egerton* [3.4] and *Raether* [3.24].

Apart from the calculation of the dielectric function by means of Kramers–Kronig analysis as described above, and apart from the more theoretical investigations into dispersion, width and high-q behaviour, there are some methods of low-loss spectrometry that are applicable in particular cases. In the following, a survey is given.

3.4 Band-Gap Analysis

The imaginary part of ε_2 measures photon absorption.[38] It is a weighted integral over the dipole oscillator strength f_{if} between initial and final states [3.174]

$$\varepsilon_2(E) = \frac{\pi e^2 \hbar^2}{2mE\varepsilon_0} \frac{1}{(2\pi)^3} \int\limits_{E_f - E_i = const.} \frac{f_{if}}{|\nabla_k(E_f - E_i)|} \, dS \qquad (3.89)$$

where E_f and E_i are the energies of the final and initial states in the interband transition, and the integral is over the surface in momentum space with constant energy difference. Assuming f_{if} independent of energy for the transitions involved,

$$\varepsilon_2(E) \propto \frac{1}{E} \int\limits_{E_f - E_i = const.} \frac{dS}{|\nabla_k(E_f - E_i)|} . \qquad (3.90)$$

The integral is known as the joint density of states (JDOS). It can be calculated directly from the band structure.

When both the initial and the final bands are parabolic (which is a good approximation in the vicinity of high-symmetry points in the reciprocal lattice) the gradient is

$$\nabla_k(E_f - E_i) = \frac{\hbar^2(m_v + m_c)}{m_v m_c} k \qquad (3.91)$$

and the surface area $\int dS = 4\pi k^2$. m_v, m_c are the effective masses in the valence and conduction band. Neglecting prefactors,

---

[37] Surface effects amount to some percent of the total loss probability when the specimen thickness is in its optimal range of slightly less than one mean free path for inelastic scattering. This is of the order of 100 nm for 200 kV beam voltage. Surface correction is thus not important for the standard case. *Egerton* [3.4] has given a FORTRAN source code including all the necessary steps for the computation of ε.

[38] In this and the next section, we write $\varepsilon(E)$ instead of $\varepsilon(E/\hbar)$.

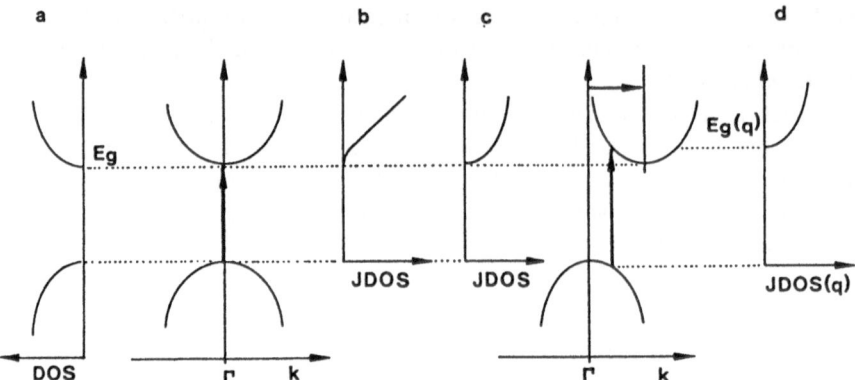

Fig. 3.17. Vertical and non-vertical transitions for parabolic bands near the Γ-point in the Brillouin zone. (**a**) Density of states, (**b**) joint density of states for angle-integrated EELS-experiments, (**c**) joint density of states for optical experiments or angle-resolved EELS. (**d**) Joint density of states for momentum transfer $q \neq 0$ (non-vertical transitions). E_g is the gap energy. The thick arrows indicate onset energy of transitions. In this example $E_g(q) > E_g$.

$$JDOS \propto \sqrt{E - E_g}, \tag{3.92}$$

where E_g is the band gap energy [3.175][39] (Fig. 3.17).

When ε_1 is large in the energy range in question, these transitions will be screened in the EEL spectrum. From (3.15), we see that the loss function is proportional to ε_2 when screening is important. Equation (3.90) shows that one measures JDOS/E. Such is the case in the 1 eV-range in semiconductors.[40] The JDOS thus obtained can be compared with band calculations. This was done for titanium and vanadium carbides and nitrides with good agreement between theory and experiment [3.5].

By fitting the energy dependence of JDOS, *Batson* [3.180] could detect changes in the gap energy at a single misfit dislocation (Fig. 3.18). Owing to screening and to the high JDOS for transitions from localized states in the dislocation, as few as five localized states on a GaAs/GaInAs-interface could be detected within a 0.5 nm probe. It was pointed out recently [3.181] that excitonic effects which invalidate the simple JDOS-proportionality of (3.90) for optical spectra may not be as difficult a problem in EELS. In that paper, it was shown that the square root-dependence of (3.92) can be fitted to low-

---

[39] A steep rise (or decrease) is found for transitions next to points of high symmetry in the reciprocal lattice [3.174]. This can be used for qualitative interpretation of optical spectra.

[40] The equation $E^2\varepsilon_2 \propto$ JDOS is often used [3.176–178]. As *Liang* [3.179] has pointed out, this is based on the ad hoc assumption that the matrix element G_{if} of the momentum operator between initial and final states is independent of energy. A more natural choice is to set the oscillator strength $f_{if} = 2G_{if}/mE$ independent on frequency. This leads to (3.90).

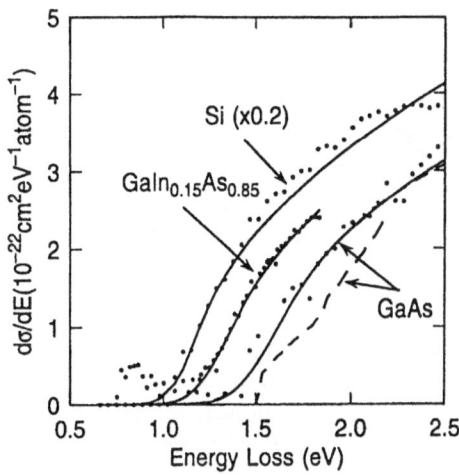

Fig. 3.18. Interband absorption for Si, GaAs and GaInAs compared to optical results for GaAs (dashed line). The solid lines show the joint density of states according to Eq.(3.92) [3.139].

loss spectra of GaInAs, GaAs and Si. Such is not the case with the optical results (Fig. 3.18).

The same formula has been used to analyse low-loss transitions in EELS of ceramic superconductors [3.150]. The authors conclude that the sharp rise in JDOS at 3.6 eV is due to transitions in the CuO_2 layers.

When taking spectra in the microscope image mode the pre-spectrometer optics performs an integration over scattering angle. This corresponds to integration over k-space for both occupied and empty bands. It can be shown that this leads to a convolution of the density of states (DOS) of both bands [3.182] when calculating ε

$$\varepsilon_2(E) \propto \int \rho_i(E - E')\rho_f(E')\mathrm{d}E'. \tag{3.93}$$

For parabolic bands, this gives

$$\varepsilon_2(E) \propto \frac{1}{E^2} \cdot (E - E_g)^2. \tag{3.94}$$

In an EELS study of hydrogenated amorphous carbon [3.183], this relationship was used to derive the band-gap energy. Strong dependence of the gap energy for $\pi - \pi^*$ transitions on the hydrogen content was found (Fig. 3.19).

An EELS analysis of boron nitride along the same lines yielded an optical band-gap of 4 ± 0.5 eV [3.183].

When comparing EELS results on JDOS with optical experiments, it must be recalled that, according to the limited momentum resolution of the incident beam and the spectrometer, neither of the two cases described above will be seen in reality. The second remarkable difference is that in EELS, the dipole selection rule is no longer valid [3.3], that is, more transitions than in optical spectra may occur at $k \neq 0$.

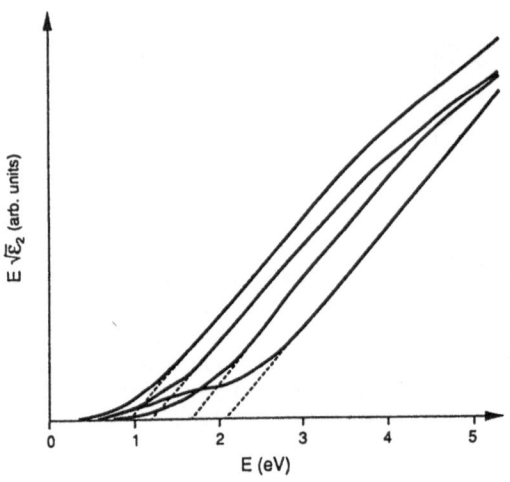

Fig. 3.19. Band gap energies (zeros of extrapolated straight lines, dashed) in hydrogenated amorphous carbon. Curves move to the right with decreasing hydrogen content [3.139].

3.5 Momentum Dependence

One of the advantages of EELS over photon-related methods for probing the solid is its ability to determine the wave vector-dependent dielectric response, simply by fixing the scattering angle. This would give information on the band shape. Equation (3.89) for the optical JDOS can be generalized when ε is momentum-dependent as follows [3.184,185]:

$$\varepsilon_2(E, q) \propto \frac{1}{E} \int\limits_{E=E_f-E_i} \frac{\mathrm{d}S}{|\nabla_k[E_f(k+q) - E_i(k)]|}. \tag{3.95}$$

The integral can be visualized as a JDOS between bands shifted by q with respect to one another.

In 1971, *Zeppenfeld* [3.176] analysed graphite with momentum-resolved EELS, exploiting the above idea of non-vertical transitions. He was able to determine the valence band shape in the basal plane of graphite, in good agreement with theory. Momentum-resolved energy loss spectra of Si were interpreted in terms of nonvertical interband transitions [3.186].

Fink et al. [3.187] used momentum-resolved EELS to analyse the band shape in oriented polyacetylene. In the one-dimensional polymers, the band structure can be described by the simple model

$$E(q) = \pm\sqrt{b_1^2 + b_2^2 + 2b_1 b_2 \cos(qa)}, \tag{3.96}$$

and (3.24) can easily be evaluated. There are only two parameters $b_{1,2}$ related to the band gap

$$E_g = 2(b_1 - b_2) \tag{3.97}$$

and to the combined band width

$$\Delta = 2\,(b_1 + b_2).\tag{3.98}$$

The parameters b_1, b_2 can be obtained from a fit of (3.96) to the measurements.

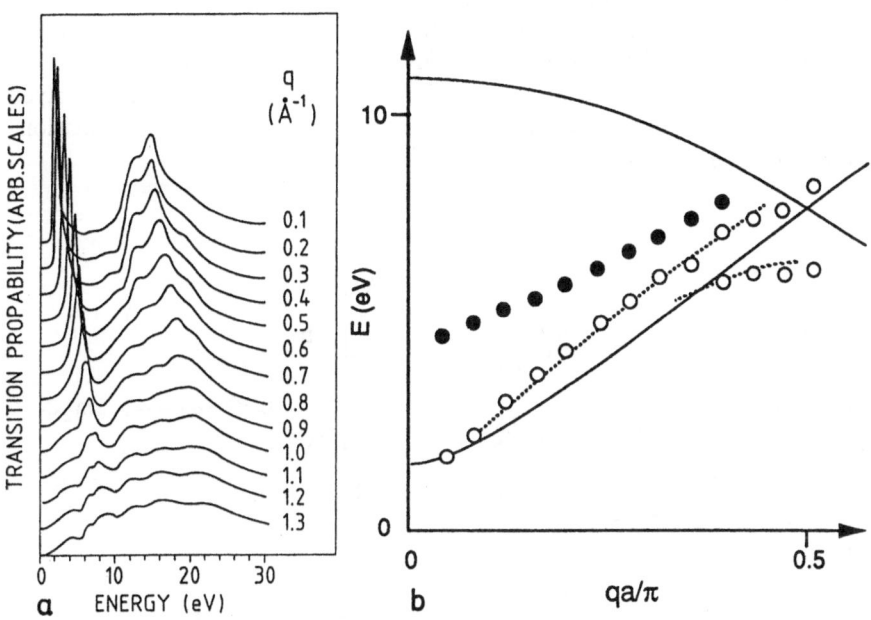

Fig. 3.20. (a) Optical conductivity $\sigma = \omega\varepsilon_2$ for trans-polyacetylene, for various momentum transfers q parallel to the c-axis. (b) Loci of the maxima in (a) ($\circ$) and maxima in the correspondig loss function ($\bullet$). The combined bandwidth as a function of q (full lines) and a calculation of the maxima assuming a bandgap of 1.8 eV and a total bandwidth of 11 eV, with local field corrections (dotted line), are also shown [3.187].

Figure 3.20a shows maxima in JDOS (3.95), caused at the lower band edge due to van Hove singularities. The positions of maxima of the loss function are shown in Fig. 3.20b. The dotted lines are, according to the authors, calculations based on a model including local-field corrections. The non-vanishing oscillator strength below the bandgap of 1.8 eV has been ascribed to excitonic transitions or to soliton-antisoliton pair excitation. There remains some unexplained structure in the JDOS below the van Hove singularities. It may be speculated whether this is caused by a gap separating intra- and interband transitions, similar to the zone boundary gap in metals (see later). This would possibly have consequences on the localisation of these excited states. Note that a fit to the maxima in EELS, that is the longitudinal modes, denoted by circles in the figure, would result in a wrong band gap. Only the *dispersion* of maxima in $\text{Im}\{1/\varepsilon\}$ is an approximate measure of the combined bandwidth.

This fact can be used for qualitative interpretation of features in the loss function. Figure 3.21 is a loss spectrum of Co_5Sm. In order to test whether the broad peaks resemble plasmons (which should disperse according to (3.55)) or screened interband transitions, the q-dependence was investigated [3.189]. It was found that the features show practically no dispersion, or even a slightly negative one. The combined bandwidth is less than 1 eV, even for the high-energy maximum. Obviously, the features in the spectrum are caused by interband transitions between narrow bands, that is, the electrons involved in the transitions are rather localized. It is suggested that these transitions are due to d-electrons.

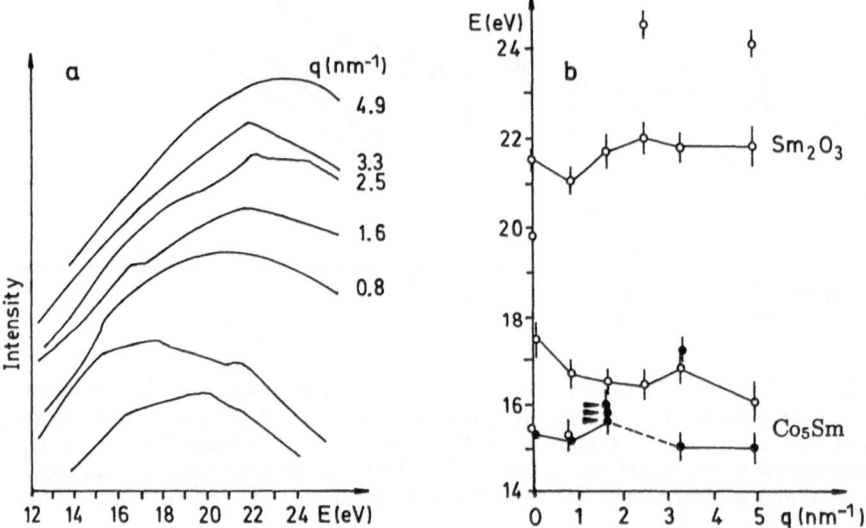

Fig. 3.21. Dispersion of faint features in Co_5Sm indicates transitions between flat bands. (a) Series of spectra for momentum transfers from 0 to 4.9 nm^{-1}. (b) Positions of discernible structures in (a). Arrows indicate change of maxima during electron irradiation [3.189].

3.6 Zone Boundary Collective States

The JDOS as defined above is conceptually simple but elaborate computations are needed to obtain numerical results. A different approach to the same topic of momentum-dependent JDOS offers some interesting insights that are not directly available from (3.95) for the momentum dependent JDOS. This approach has come to be known as the theory of zone boundary collective states (ZBCS).

One starts from the model of a free electron gas, with the lowest energy levels filled within the Fermi sphere. This gives the well-known Lindhard dielectric function. Its imaginary part mirrors the possible intraband transitions in the electron gas, the momentum transfer given. When a crystal potential is present there exist energy gaps at the Brillouin zone boundaries. At their edges, the density of states has van Hove singularities. They can be pictured as the result of pushing states out of the band gap, thereby creating new features in the dielectric function, and hence in the energy loss spectrum. The effect can be understood in a simple model, assuming a one-dimensional lattice potential. We consider the two lowest bands in the extended zone scheme. The energy–momentum relation of the free electron acquires a band gap of magnitude $2V_1$, the first harmonic Fourier coefficient of the lattice potential. The momentum transfer q is assumed to be perpendicular to the zone boundary. The possible final momentum states allowed by the Pauli principle are within the hatched area in Fig. 3.22. These transitions give rise to absorption within an energy band.[41] This follows immediately from the geometrical construction as in the figure. Switching on the lattice potential pulls those final states out of the energy gap that lie on the Brillouin zone boundary. (In the figure denoted as k'). Then, the initial states must have been on the dashed line, $k = k'-q$. The transition energy of these states is

$$E = E_f - E_i = \frac{\hbar^2}{2m}(k'^2 - k^2) \pm \frac{V_G}{2} = \frac{\hbar^2}{2m}q(g - q) \pm \frac{V_G}{2}. \qquad (3.99)$$

The NFE-transition band will acquire a gap at this energy, as depicted schematically in the figure. The pushing out of states from the gap forces an increase of oscillator strength at the gap edges, by the f-sum rule. The q-dependent gap position (3.99) is drawn in Fig. 3.22 in the electron-hole excitation continuum. The continuum of electron-hole excitations falls apart into an intraband region and an interband region. In addition, this continuum will be folded back by an umklapp process at $\pm G/2$. In these areas, the lower edges of which are also given in the figure, there are interband transitions too.

In optical spectra, the band gap cannot be seen as is evident from the figure: it is at zero energy at $q = 0$. The gap can be detected by q-dependent spectroscopies – inelastic X-ray or EELS. In reality, the bands will not have the simple parabolic shape assumed above. This will change the gap dispersion. Monitoring the gap dispersion would thus be a way of probing the band structure. In practice, the band gap which is of the order of the lattice potential will be smeared out since other transitions (higher bands as sketched

---

[41] A calculation of the available phase space for transitions as a function of energy, based on the given geometry, yields exactly the oscillator strength distribution over energy as given by the imaginary part of the Lindhard ε (3.34).

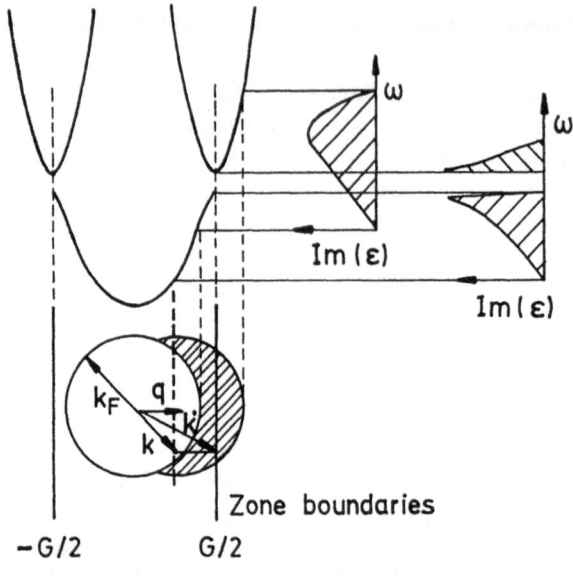

Zone boundaries

−G/2   G/2

Fig. 3.22. Explanation of zone boundary states in a medium with one-dimensional periodicity. The first Brillouin zone is not completely filled. The shaded area in the lower part shows the final states for a fixed momentum transfer q. Without a zone boundary, this would yield the imaginary part of the Lindhard dielectric function, simply by counting the states corresponding to a certain energy transfer (projection onto the free electron parabola). The final states on the zone boundary are pushed out of the energy gap, causing a frequency interval where $\varepsilon_2 = 0$. The position of this interval depends on momentum.

in the figure, as well as other zone boundaries) make a background. Another difficulty is the faint intensity at larger momentum transfer in EELS. In principle, the band gap can be measured with inelastic X-ray or synchrotron scattering since it is visible in the dynamical structure factor. Likewise, it causes an oscillation in the real part ε_1 of the dielectric function as discussed in Sect. 3.1.1. When the push-out effect of states is strong, ε_1 may even have a zero crossing, that is, a well defined longitudinal mode will be seen in EELS. This defines an undamped longitudinal mode in the gap, the zone boundary collective mode. It can be questioned whether this mode should be called a plasmon [3.51] especially when ε_1 does not cross zero. Since it is in an energy gap, it is perhaps better to refer to it as a longitudinal zone boundary exciton. Anyway, as discussed in Sect. 3.1.1, each interband transition has a corresponding longitudinal mode which is visible in the EELS spectrum. The ZBCS are close in energy to the transverse modes at the zone boundary gap edges.

ZBCS were predicted by *Foo* and *Hopfield* [3.190]. Later, doubts were raised as to their detectability [3.191]. They have been observed with inelastic X-ray spectrometry in Li and Be [3.79,85,80] and with EELS in Al [3.3.192,193] and Na [3.31]. *Sturm* presented detailed calculations, in excellent agreement with experimental findings for Al [3.194] and Na [3.195] (Fig. 3.23).

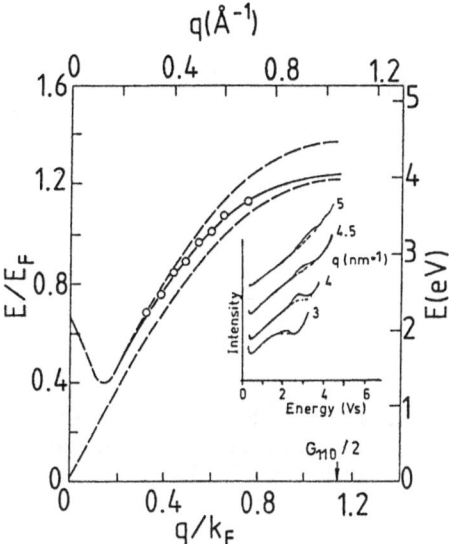

Fig. 3.23. Zone boundary collective states in sodium, $q \parallel < 110 >$. Dashed lines denote edges of the intra- and interband continuum (3.99). The circles are the positions of the zone boundary collective states as retrieved from the loss function (see insert) [3.171].

The case of graphite deserves special mention. Graphite has a filled zone. There are no intraband transitions, and consequently the oscillator strength has a sharp onset. This behaviour was found by *Zeppenfeld* [3.176] and in a later experiment [3.196] with slightly different results (Fig. 3.24), where the momentum-dependent onset of interband transitions is given. The dispersion of this onset obeys (3.99). Additionally, the momentum-dependent local maximum in the loss function is given; this is the longitudinal resonance corresponding to the van Hove singularity of the density of states at the band edge. There is no zone boundary collective state in the strict sense (i. e. an excitation in the gap) when the band is filled.

3.7 Plasmons in Small Particles

Plasmons in small particles have been studied extensively. The central feature of this type of experiments is that the volume-to-surface ratio is given by the size of the particle, approximately. Hence, the smaller the particle, the more will surface modes contribute to the loss spectrum. Non-local effects (i. e. the non-abrupt change of the dielectric function next to the boundary) increase the excitation probability of surface plasmons in small spheres [3.197]. A detailed treatment of surface modes can be found in [3.198].

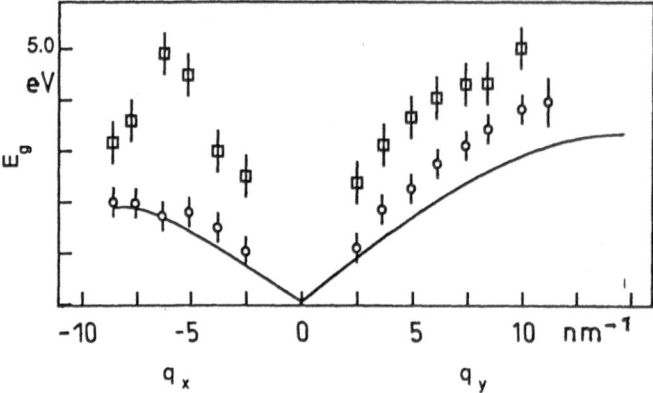

Fig. 3.24. Momentum-dependent band gap in the basal plane of graphite. Circles: from experiment, lines: theory. The squares indicate the positions of local maxima in the loss function, resembling longitudinal resonances. Unlike the ZBCS, they are above the gap and are thus heavily screened [3.176].

Surface plasmons have been theoretically described for a plane surface [3.192–195], for spherical geometry [3.203–211], and for cylindrical pores [3.212,213]. *Lucas* [3.214] calculated the attractive force between voids, arising from the interaction of surface plasmons. This effect is similar to the Van der Waals forces between small spheres.

In spheres and cylinders, the basic fact is that when the surface excitation is expanded into eigenfunctions on the surface, its highest oscillator strengths occur at the low-multipole indices; a finite expansion is therefore sufficient. For a sphere in vacuum, the l-th mode is given by [3.24]

$$\omega_l = \omega_p \sqrt{\frac{l}{2l+1}} \tag{3.100}$$

whereas for a spherical void

$$\omega_l = \omega_p \sqrt{\frac{l+1}{2l+1}}. \tag{3.101}$$

The bulk plasmon is the monopole ("breathing") mode and the eigenfrequencies converge to $\omega_p/\sqrt{2}$, the surface mode for a plane.

Surface plasmons in metal spheres were found quite early [3.203,215–217]. *Batson* [3.218] carried out more detailed experiments and his findings are in overall agreement with theory. The finer features of spectra not predicted by the single-sphere model were explained as a coupling effect between neighbouring particles [3.219].

Owing to the evanescent fields, the surface modes are well localised at the surface (or interface) and can thus be used to image small particles in the STEM [3.220–222]. An interesting aspect is that the surface modes draw oscillator strength from the bulk mode, thus introducing a correction term to the latter. The same is, to a lesser extent, true for EELS in the regular plane film geometry (Fig. 3.25). The correction term (*Begrenzungsterm*) has been calculated by *Ritchie* [3.199], and discussed in detail by *Geiger* [3.223]. Later it was shown to oscillate both for the plane [3.201,202] and spherical geometry [3.224]. The correction is strong for $qr < 1$. This was experimentally proven in EELS of potassium clusters [3.225].

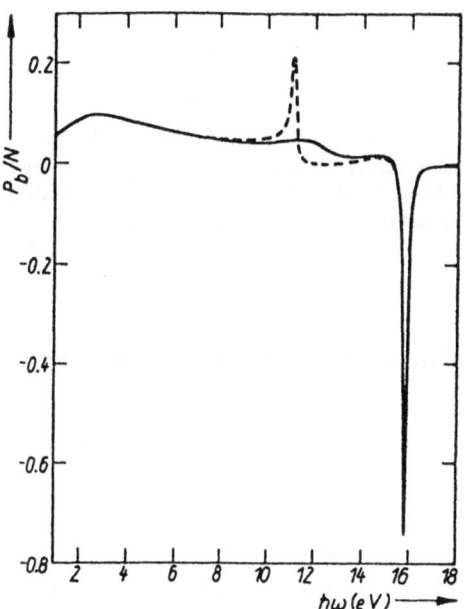

Fig. 3.25. Surface-induced correction (*Begrenzungsterm*) to the bulk energy-loss probability. Calculation for a 2.5 nm thick Al foil, with (full line) and without (dashed line) dispersion. Incident energy 100 keV. The surface mode draws almost half of the intensity from the volume mode [3.202].

Walsh [3.226] has measured plasmon modes in beam-drilled microholes in Al and AlF$_3$ in a STEM. There is qualitative agreement with theory, with an unexplained structure in the spectra attributed to surface plasmons on microspheres created during drilling.

The possibility of combining highly sensitive spectrometers with imaging methods, such as a PEELS with a STEM, allows the as yet unexplained details of plasmon spectra in small particles to be explained. A faint excitation at about 4 eV in Si spheres of about 20 nm diameter was shown to be a surface mode originating in the outermost layer of a Si sphere covered by an inner shell of SiO$_2$ and a very thin amorphous Si outer shell [3.227]. Figure 3.26 is a comparison of predictions from dielectric theory, based upon the three-component model, with experiment.

Surface plasmons of gas precipitates (bubbles) were experimentally found in 1970 [3.228]. They were studied by *Manzke* [3.229] and by *vom Felde* [3.230,231] who derived the internal gas pressure from the change in plasmon energy. Pressures up to 500 kbar were found for He-bubbles in Ni [3.171].

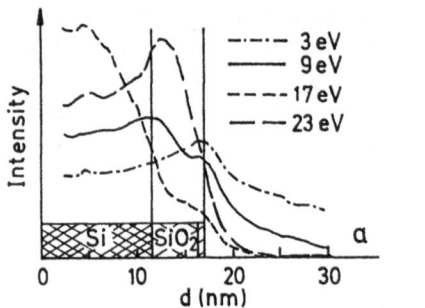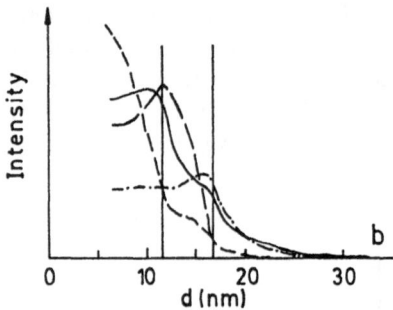

Fig. 3.26. (a) Energy-loss line scans across the edge of a small Si-particle, for E = 3, 9, 17 and 23 eV. (b) Dielectric theory based upon a three-component model assuming amorphous Si as the outermost layer (as depicted on the abscissa), gives a very similar prediction. Note the narrow maximum of the 3 eV loss at the outermost edge. This feature could not be explained by a two-component model (Si and SiO_2) [3.227].

These experimental findings were confirmed recently by calculations of *Serra* et al. [3.211].[42]

3.8 Fingerprinting Methods

Normally, quantification in EELS is done with element-characteristic ionisation edges. The plasmon would have the advantage of much higher intensity (by a factor 1000 and more) but it is not element-specific. However, E_p depends on density, according to (3.25). This has been used for concentration determination in alloys [3.233,234] and for comparison of electron density in amorphous and crystalline FeB-alloys [3.235]. *Wang* tried to analyse the mass-density in diamond-like carbon films from plasmon position [3.236]. It must however be remarked that the plasma frequency in fact depends on the density-to-mass ratio, and that this mass is simply a parameter that describes the complex influence of the valence band shape and interband transitions on plasmon energy. The change in electronic structure during a phase transition in VO_2 has been analysed by this method [3.237].

When the dependence of plasmon energy on concentration is known from some alloy standard, microanalysis by "fingerprinting" is possible. This has been used for the detection of hydrogen[43] in metals [3.239,240]. Identification of hydride phases in Zr by accurate measurement of the differences in plasmon

---

[42] EELS experiments with a STEM show that the helium $1S \rightarrow 2P$ transition that can be seen as a faint structure in the low-loss spectrum shifts to higher energy in small bubbles, in accord with increasing pressure [3.232].

[43] The occurrence of a collective mode at 7 eV in hydrogen-loaded Nb and V was also used as a fingerprint technique [3.238].

energy (of the order of 0.2 eV) has also been reported [3.241]. Lithium-rich precipitates in a Al-10.5 % Li alloy could thus be detected on a scale of better than 50 nm [3.242]. In a similar way, metallic Al and Al^{3+}-ions in spinel could be discriminated [3.243].

The halfwidth of the plasmon is a measure of the inverse lifetime of the excited state. Since the plasmon is composed of correlated electron-hole pairs, one can in principle measure the collision time of conduction electrons in a metal. The collision times τ thus obtained are different from the electronic collision times responsible for the (d.c.) conductivity. This is because τ is frequency-dependent [3.244]. Plasmon-lifetime analysis was also used in studies of icosahedral phases. In icosahedral Al_6Mn, the plasmon is broader than in the amorphous phase [3.245]. This is not the case in the PdUSi and in the AlCuV-system. These findings suggest that the change in lifetimes stem from disorder in the icosahedral phase and not from differences in electronic structure [3.246].

The temperature dependent change in carrier density, and hence plasma frequency was used to determine the sample temperature to within ± 20 K in a reflection EELS experiment on Sn [3.247].

3.9 Coupling of Plasmons to Light

Since the plasmon is, by definition, a longitudinal mode it seems to be impossible to excite it by resonant absorption of a photon.[44] The decoupling of the longitudinal and transverse modes discussed in Sect. 3.1.1 was due to the assumption of isotropy. Strictly, this is only true in a liquid, or in the amorphous phase. In crystals, the modes decouple only in certain high symmetry directions.[45] of the wave vector [3.13,248]. In an arbitrary direction in a crystal, a normal mode has both transverse and longitudinal components.

The second aspect that makes possible the excitation of plasmons by light is the presence of boundaries. By destroying the translational invariance in one direction, they necessarily break the isotropy of the system, and again – even for isotropic bulk material – the modes are coupled. Plasmons excited by light were detected in 1970 [3.249,250]. This coupling has consequences for the reflection coefficient; in fact, it can be shown experimentally and theoretically that the Fresnel equations in their standard form are violated [3.3,12].

---

[44] This must not be confused with inelastic photon scattering. Both the inelastic scattering cross-sections of electrons and photons are proportional to the dynamical structure factor $S(q, \omega)$.

[45] This is only true for the macroscopic response (restriction of the momentum to the first Brillouin zone) As discussed in Sect. 3.2.1, the local fields in a crystal always generate longitudinal components of the fields on a microscopic scale.

3.10 Reflection EELS

Until recently, transmission EELS experiments and reflection (REELS) studies have been different fields of activity. There was almost no connection between these two methods[46] although they both measure the inelastic scattering cross-section of electrons. There are, essentially, three reasons for the different treatment: surface excitations, multiple scattering, and of course the different physics of surfaces.

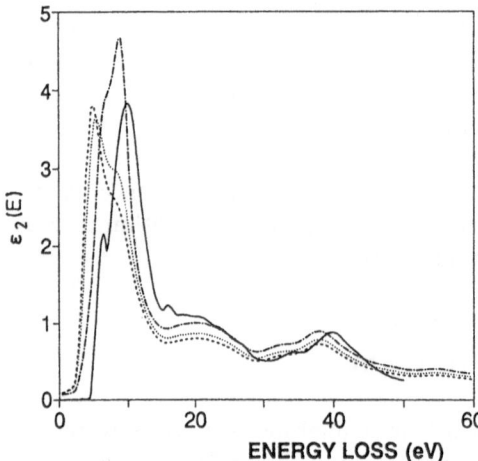

Fig. 3.27. Imaginary part of the dielectric function of ZrO_2. The amount of surface contribution that has been subtracted tentatively from the REELS spectra before Kramers–Kronig analysis is 20% (dashed); 40% (dotted); and 60% (thin full line). The heavy full line is a result from transmission (standard) EELS. The unknown surface contribution in REELS is strongest on the low-energy side of the main peak [3.252].

The primary electrons used in REELS have penetration depths of some nanometres into the specimen. This should suffice to excite bulk modes, but the dwell time in the specimen[47] is much shorter than the dwell time in the evanescent field of the surface modes. Thus, the latter dominate the loss spectrum.

Multiple scattering, including elastic processes, makes the scattering geometry much more involved than in the transmission (EELS) case. The situation is so complex that REELS spectra have usually been interpreted in a more qualitative manner, assigning maxima in the loss spectra to electronic transitions. Plasmons (surface and bulk) were seen and identified as additional excitations. In principle, this is wrong. Determination of electronic transitions, in the sense of the optical methods, requires Kramers–Kronig analysis of the loss function, which shows the *longitudinal* excitations. The presence of surface excitations is a serious problem here. Recently, attempts

---

[46] Common citations are very rare in the two fields. Note also that the REELS community plots energy-loss spectra in an energy–coordinate system.

[47] The electric field of the bulk modes is restricted to the volume whereas the field of the surface modes penetrates into the vacuum.

have been made to apply more accurate data analysis such as Kramers–Kronig analysis [3.6,97,251,252], multiple scattering corrections [3.253,254] or the use of model dielectric functions [3.255–258]. In spite of the above-mentioned difficulties arising from mixing of surface and volume modes and the scattering geometry in REELS, the results are encouraging. As an example, ε_2 for ZrO_2, obtained from EELS and REELS, is given in Fig. 3.27. Except on the low-energy side of the main peak where surface modes of unknown magnitude introduce errors in the Kramers–Kronig analysis of REEL spectra, the agreement is good.

These recent attempts suggest that two hitherto mutually ignored relatives are becoming aware of each other, a development from which both could profit. In view of the fundamental desire for unification in physics, and probably not only in physics, we hope that this tendency continues.

References

3.1 A. Sommerfeld, H. Bethe: *Elektronentheorie der Metalle* (Springer, Berlin 1967)

3.2 C.C. Ahn, O.L. Krivanek: *EELS Atlas* (Arizona State Univ., Tempe 1983)

3.3 P. Schattschneider: *Fundamentals of Inelastic Electron Scattering* (Springer, Wien, NewYork 1986)

3.4 R.F. Egerton: *Electron Energy-Loss Spectroscopy in the Electron Microscope* (Plenum, NewYork 1986)

3.5 J. Pflüger, J. Fink, W. Weber, K.P. Bohnen: Dielectric properties of TiC_x, VC_x, and VN_x from 1.5 to 40 eV determined by electron energy-loss spectroscopy. Phys. Rev. B **30** 1155–1163 (1984)

3.6 A. Balzarotti, M. DeCrescenci, N. Motta, F. Patella, A. Sgarlata: Energy loss study of the electronic structure of $YBa_2Cu_3O_{7-\delta}$ high T_c superconductors. Sol. Stat. Comm. **68**, 381–386 (1988)

3.7 P. Drude: Zur Elektronentheorie der Metalle. Phys. Zeitschr. **14**, 161–165 (1900)

3.8 J. Lindhard: On the properties of a gas of charged particles. Danske Vidensk. Selsk. Mat.-Fys. Medd. **28**, 1–57 (1954)

3.9 D. Jackson: *Classical Electrodynamics* (Wiley, NewYork 1962)

3.10 D. Pines: *Elementary Excitations in Solids* (W.A. Benjamin, NewYork 1964)

3.11 I. Egri: Excitons and plasmons in metals, semiconductors and in insulators: A unified approach. Phys. Reports **119**, 363–402 (1985)

3.12 F. Forstmann, R.R. Gerhardts: Metal optics near the plasma frequency. In *Festkörperprobleme*, ed. by J. Trensch, Adv. in Solid State Physics **22** (Vieweg, Braunschweig 1982) pp.291–323

3.13 W. Hanke: Dielectric theory of elementary excitations in crystals. Adv. Phys. **27**, 287–341 (1978)

3.14 T. Miyakawa: Excitons and plasmons in insulators. J. Phys. Soc. Jpn. **24**, 768–786 (1968)

3.15 H.K. Kanazawa: On the theory of plasma oscillations in metals. Prog. Theor. Phys. **13**, 227–242 (1955)

3.16 C. Horie: Exciton and plasmon in insulating crystals. Prog. Theor. Phys. **21**, 113–134 (1959)

3.17 I. Egri: Plasmons in semiconductors and insulators: A simple formula. Sol. Stat. Commun. **44**, 563–566 (1982)

3.18 R.A. Ferrell: Characteristic energy loss of electrons passing through metal foils. II. Dispersion relation and short wavelength cutoff for plasma oscillations. Phys. Rev. **107**, 450–462 (1957)

3.19 R.A. Ferrell, J.J. Quinn: Characteristic energy losses of electrons passing through metal foils: momentum-exciton model of plasma oscillations. Phys. Rev. **108**, 570–575 (1957)

3.20 K. Sawada: Correlation energy of an electron gas at high density. Phys. Rev. **106**, 372–383 (1957)

3.21 I. Egri: Electronic interband transitions: plasmons, Frenkel- and Wannier-excitons. Z. Physik B **42**, 99–106 (1981)

3.22 I. Egri: The internal structure of plasmons. Z. Physik B **53**, 183–189 (1983)

3.23 J. Rogan, J.E. Inglesfield: Electronic excitations in tight-binding systems. J. Phys. C **14**, 3585–3602 (1981)

3.24 H. Raether: *Excitation of Plasmons and Interband Transitions by Electrons.* Springer Tracts Mod. Phys. Vol. 88 (Springer, Berlin, NewYork 1988)

3.25 M. Taut: Plasmon dispersion in the weighted density approximation: correlation-induced anomaly. J. Phys. C **21**, 899–909 (1988)

3.26 A. Gold: Instability of layered quantum liquids. 1. Local-field correlation in superlattices. Z. Physik B **86**, 193–206 (1992)

3.27 A. Gold: Local-field correction of the charged Bose condensate for two and three dimensions. Z. Physik B **89**, 1–10 (1992)

3.28 Y.I. Ichikawa: Theory of collective oscillations of electrons in solids. Prog. Theor. Phys. **18**, 247–263 (1957)

3.29 D.F. DuBois: Electron interactions. II. Properties of a dense electron gas. Ann. Phys. (N.Y) **8**, 24–77 (1959)

3.30 M.E. Bachlechner, W. Macke, H.M. Miesenböck, A. Schinner: Perturbational analysis of plasmon decay in jellium. Physica B **168**, 104–114 (1991)

3.31 A. vom Felde, J. Sprösser-Prou, J. Fink: Valence-electron excitation in the alkali metals. Phys. Rev. B **40**, 10181–10193 (1989)

3.32 P.C. Gibbons, S.E. Schnatterly, J.J. Ritsko, J.R. Fields: Line shape of the plasma resonance in simple metals. Phys. Rev. B **13**, 2451–2460 (1976)

3.33 R. Manzke: Wavevector dependence of the volume plasmon of GaAs and InSb. J. Phys. C **13**, 911–917 (1980)

3.34 K.S. Singwi, M.P. Tosi, R.H. Land, A. Sjölander: Electron correlations at metallic densities. Phys. Rev. **176**, 589–599 (1968)

3.35 K.S. Singwi, A. Sjölander, M.P. Tosi, R.H. Land: Electron correlations at metallic densities IV. Phys. Rev. B **1**, 1044–1053 (1970)

3.36 G.D. Mahan: *Many-Particle Physics* (Plenum, New York 1981)

3.37 J. Hubbard: The desription of collective motions in terms of many-body perturbation theory. II. The correlation energy of a free-electron gas. Proc. Roy. Soc. (London) A **243**, 336–352 (1957)

3.38 K. Utsumi, S. Ichimaru: Dielectric formulation of strongly coupled electron liquids at metallic densities. IV. Analytical expression for the local-field correction. Phys. Rev. A **26**, 603–610 (1982)

3.39 N. Iwamoto, E. Krotscheck, D. Piners: Theory of electron liquids. II. Static and dynamic form factors, correlation energy, and plasmon dispersion. Phys. Rev. B **29**, 3939–3951 (1984)

3.40 H. Suehiro, Y. Ousakas, H. Yasuharas: A note on the plasmon dispersion coefficient. J. Phys. C **18**, 6007–6010 (1985)

3.41 A.K. Gupta, P.K. Aravind, K.S. Singwi: Plasmon dispersion in electron liquid. Sol. State Commun. **26**, 49–52 (1978)

3.42 P. Nozieres, D. Pines: Electron interaction in solids. Characteristic energy loss spectrum. Phys. Rev. **113**, 1254–1267 (1959)

3.43 N. Iwamoto, D. Pines: Theory of electron liquids. I. Electron-hole pseudopotentials. Phys. Rev. B **29**, 3924–3935 (1984)

3.44 P. Vashishta, K.S. Singwi: Electron correlations at metallic densities. Phys. Rev. B **6**, 875–887 (1972)

3.45 K.N. Pathak, P. Vashishta: Electron correlations and momentum sum rules. Phys. Rev. B **7**, 3649–3656 (1973)

3.46 H. Watanabe: Experimental evidence for the collective nature of characteristic energy loss of electrons in solids – studies on the dispersion relation of plasma frequency. J. Phys. Soc. Jpn. **11**, 112–119 (1956)

3.47 P.E. Batson, J. Silcox: Experimental energy-loss function $Im[-1/\varepsilon(q,\omega)]$. Phys. Rev. B **27**, 5224–5239 (1983)

3.48 J. Sprösser-Prou, A. vom Felde, J. Fink: Aluminium bulk-plasmon dispersion and its anisotropy. Phys. Rev. B **40**, 5799–5801 (1989)

3.49 P. Schattschneider, F. Födermayr, D.S. Su: Deconvolution of plasmon spectra. Scanning Microscopy, Suppl.2, 255–269 (1988)

3.50 E. Krotscheck: Private communication (1993)

3.51 K. Sturm: Electron energy loss in simple metals and semiconductors. Adv. Phys. **31**, 1–64 (1982)

3.52 L.E. Oliveira, K. Sturm: High-frequency dielectric properties of covalent semiconductors within the nearly-free-electron approximation. II. The two-plasmon-band model. Phys. Rev. B **22**, 6283–6293 (1980)

3.53 K. Sturm, W. Schülke: Shape of plasmon bands near the Brillouin zone boundary. Phys. Rev. B **46**, 7193–7195 (1992)

3.54 K. Sturm: Dynamic structure factor: an introduction. Z. Naturforschg. A **48**, 233–242 (1993)

3.55 S.L. Adler: Quantum theory of the dielctric constant in real solids. Phys. Rev. **126**, 413–420 (1962)

3.56 N. Wiser: Dielectric constant with local field effects included. Phys. Rev. **129**, 62–68 (1963)

3.57 G. Paasch: Influence of interband transitions on plasmons in the alkali metals: Pseudopotential calculation. Phys. Stat. Sol. **38**, K123–K126 (1970)

3.58 S.G. Louie, J.R. Chelikowsky, M.L. Cohen: Local-field effects in the optical spectrum of silicon. Phys. Rev. Lett. **34**, 155–158 (1975)

3.59 W. Hanke, L.J. Sham: Local-field and excitonic effects in the optical spectrum of a covalent crystal. Phys. Rev. B **12**, 4501–4511 (1975)

3.60 K. Sturm: Local field effects in the plasmon line shape of semiconductors of the diamond and zinc-blende structures. Phys. Rev. Lett. **40**, 1599–1902 (1978)

3.61 G. Sölkner: Plasmonen in einfachen Metallen. Dissertation, Technische Universität Wien 1986

3.62 K. Sturm: Band structure effects on the plasmon dispersion in simple metals. Z. Physik B **29**, 27–32 (1978)

3.63 H. Bross: Pseudopotential theory of the dielectric function of Al – the volume plasmon dispersion. J. Phys. F **8**, 2631–2649 (1978)

3.64 J.L. Parish: Charge renormalization describing plasmon dispersion in metals. Phys. Rev. B **42**, 10940–10944 (1990)

3.65 K. Sturm, E. Zaremba, K. Nuroh: Core polarization and the dielectric response of simple metals. Phys. Rev. B **42**, 6973–6992 (1990)

3.66 K.J. Krane: Dispersion and damping of volume plasmons in polycrystalline aluminium and indium. J. Phys. F **8**, 2133–2137 (1978)

3.67 P. Schattschneider, D.S. Su: Hyperquadratic plasmon dispersion in indium. J. Electron. Spectrosc. Rel. Phenom. **58**, R13–R18 (1992)

3.68 R. Gründler, T. Stöcker, H. Boudriot, K. Deus, H.A. Schneider: Volume plasmon dispersion of polycrystalline films of the ternary semiconductors $ZnSnAs_2$ and $ZnSiAs_2$. Phys. Stat. Sol. (b) **140**, K19–K22 (1987)

3.69 G. Kalman, K.I. Golden: Response function and plasmon dispersion for strongly coupled liquids. Phys. Rev. A **41**, 5516–5527 (1990)

3.70 A. vom Felde, J. Fink, Th. Büche, B. Scheerer, N. Nücker: Plasmons in the heavy alkali metals. Strong deviation from RPA. Europhys. Lett. **4**, 1037–1042 (1987)

3.71 P.M. Platzman, P. Eisenberger: Presence of an incipient electron lattice in solid-state electron gases. Phys. Rev. Lett. **33**, 152–154 (1974)

3.72 D.M. Ceperley, B.J. Alder: Ground state of the electron gas by a stochastic method. Phys. Rev. Lett. **45**, 566–569 (1980)

3.73 J.P. Hansen: Plasmon dispersion of the strongly coupled one component plasma in two and three dimensions. J. Phys. Lett. **42**, 397–400 (1981)

3.74 P. Carini, G. Kalman, J.I. Golden: Plasmon dispersion for strong coupling. Phys. Lett. A **78**, 450–453 (1980)

3.75 M. Taut: Anomaly in the plasmon dispersion of alkaline metals. Sol. State Commun. **65**, 905–909 (1988)

3.76 M. Taut, K. Sturm: Plasmon dispersion constant of the alkali metals. Solid State Commun. **82**, 295–299 (1992)

3.77 P. Zacharias: Behaviour of collective volume excitations in aluminium near the critical wavevector: transition into single-particle excitation. J. Phys. C **7**, L26–L28 (1974)

3.78 W. Schülke, H. Nagasawa, S. Mourikis, P. Lanzki: Dynamic structure of electrons in Li metal: Inelastic synchrotron x-ray scattering results and interpretation beyond the random-phase approximation. Phys. Rev. B **33**, 6744–6757 (1986)

3.79 W. Schülke, U. Bonse, H. Nagasawa, S. Mourkis, A. Kaprolat: Lattice-induced double peak in the dielectric response of Be metal. Phys. Rev. Lett. **59**, 1361–1364 (1987)

3.80 W. Schülke, H. Nagasawa, S. Mourkis, A. Kaprolat: Dynamic structure of electrons in Be metal by inelastic x-ray scattering spectroscopy. Phys. Rev. B **40**, 12215–12228 (1989)

3.81 A. Vradis, G.D. Priftis: Dynamic structure factor $S(k, \omega)$ of beryllium by x-ray inelastic scattering experiments. Phys. Rev. B **32**, 3556–3561 (1985)

3.82 F. Brosens, J.T. Devreese, L.F. Lemmens: Dielectric function of the electron gas with dynamical-exchange decoupling. II. Discussion and results. Phys. Rev. B **21**, 1363–1370 (1980)

3.83 N.M. Glezos: Influence of exchange effects on the plasmon excitation spectrum of metals: Application in the case of beryllium. Phys. Rev. B **43**, 7538–7545 (1991)

3.84 W. Schülke, H. Nagasawa, S. Mourikis: Dynamic structure factor of electrons in Li by inelastic synchrotron x-ray scattering. Phys. Rev. Lett. **52**, 2065–2068 (1984)

3.85 K. Sturm, W. Schülke, J.R. Schmitz: Plasmon-Fano resonance inside the particle-hole excitation spectrum of simple metals and semiconductors. Phys. Rev. Lett. **68**, 228–231 (1992)

3.86 J.M. Bassat, P. Odier, F. Gervais: Two-dimensional plasmons in nonstoichiometric La_2NiO_4. Phys. Rev. B **35**, 7126–7128 (1987)

3.87 J.C. Ashley, R.H. Ritchie: Double-plasmon excitation in a free-electron gas. Phys. Stat. Sol. **38**, 425–434 (1970)

3.88 J.C.H. Spence, A.E.C. Spargo: Observation of double-plasmon excitation in aluminium. Phys. Rev. Lett. **26**, 895–897 (1971)

3.89 K.S. Srivastava, S. Singh, P. Gupta, O.K. Harsh: Low-energy double plasmon satellites in the x-ray spectra of metals. J. Electron. Spectrosc. Rel. Phenom. **25**, 211–217 (1982)

3.90 P. Schattschneider, F. Födermayr, D.S. Su: Coherent double-plasmon excitation in aluminium. Phys. Rev. Lett. **59**, 724–727 (1987)

3.91 D.L. Misell, A.J. Atkins: An attempt to observe the double plasmon loss by electron spectroscopy. J. Phys. C **4**, L81–L84 (1971)

3.92 D.S. Su, P. Schattschneider, P. Pongratz: Aperture effects and the multiple-scattering problem of fast electrons in electron energy-loss spectroscopy. Phys. Rev. B **46**, 2775–2780 (1992)

3.93 P. Schattschneider, P. Pongratz: Coherence in energy loss spectra of plasmons. Scanning Microscopy **2**, 1971–1978 (1988)

3.94 P. Schattschneider, M. Zapfl, P. Skalicky: Hybrid deconvolution for small-angle inelastic multiple scattering. Inverse Problems **1**, 381–391 (1985)

3.95 P. Schattschneider: The dielectric description of inelastic electron scattering. Ultramicroscopy **28**, 1–15 (1989)

3.96 P. Schattschneider, D.S. Su, P. Pongratz: Influence of Bragg scattering on plasmon spectra in aluminium. Scanning Microscopy **6**, 123–128 (1992)

3.97 J.C. Ingram, K.W. Nebesny, J.E. Pemberton: Optical properties of metal surfaces from electron energy loss spectroscopy in the reflection mode. Appl. Surf. Sci. **44**, 279–291 (1990)

3.98 V.P. Silin: O spektre vozbuzdijenij systemy elektronov i ionov. Zh. Eksp. teor. Fiz. **23**, 649–659 (1952)

3.99 P.M. Platzman, P.A. Wolff: Waves and interactions in solid state plasmas. In *Solid State Physics*, Suppl.13 (Academic, New York 1973)

3.100 V.Z. Kresin, H. Morawitz: Plasmon and phonon mechanisms of superconductivity in the layered high-T_c copper-oxides. Physica C **153–155**, 1327–1328 (1988)

3.101 Y. Ousaka, H. Yasuhara, T. Nagashima: A possible mechanism of a low-energy plasmon excitation in metals. J. Phys. Soc. Jpn. **60**, 640–650 (1991)

3.102 P.B. Visscher, L.M. Falicov: Dielectric screening in a layered electron gas. Phys. Rev. B **3**, 2541–2547 (1971)

3.103 D. Grecu: Plasma frequency of the electron gas in layered structures. Phys. Rev. B **8**, 1958–1961 (1973)

3.104 A.L. Fetter: Electrodynamics of a layered electron gas. II. Periodic array. Ann. Phys. (New York) **88**, 1–25 (1974)

3.105 J.K. Jain, P.B. Allen: Plasmons in layered films. Phys. Rev. Lett. **54**, 2437–2440 (1985)

3.106 J.K. Jain, S. Das-Sarma: Elementary collective excitations in multilayered two-dimensional systems. Surf. Sci. **196**, 466–475 (1988)

3.107 P. Hawrylak: Plasmon and electron-hole-pair damping of excited vibrational and electronic states in quasi-two-dimensional electron systems. Phys. Rev. B **35**, 3818–3822 (1987)

3.108 A.C. Tselis, J.J. Quinn: Theory of collective excitations in semiconductor superlattice structures. Phys. Rev. B **26**, 3318–3385 (1984)

3.109 P. Hawrylak, J.W. Wu, J.J. Quinn: Inelastic electron scattering by collective charge-density excitations at the surface of a semiconductor superlattice. Phys. Rev. B **32**, 4272–4274 (1985)

3.110 D. Heitmann: Two-dimensional plasmons in homogenous and laterally microstructure space charge layers. Surf. Sci. **170**, 332–345 (1986)

3.111 G. Eliasson, G.F. Guiliani, J.J. Quinn, R.F. Wallis: Plasmon bands in periodic conducting heterostructures. Phys. Rev. B **33**, 1405–1407 (1986)

3.112 H. Ishida: Dimensional crossover of plasmon in semiconductor superlattice. J. Phys. Soc. Jpn. **55**, 4396–4407 (1986)

3.113 S. Das-Sarma, J.J. Quinn: Collewctive excitatiuons in semicinductor superlattices. Phys. Rev. B **25**, 7603–7618 (1982)

3.114 P. Apell, C. Holmberg: Bulk and surface collective modes in metal superlattices. Superlatt. Microstruct. **2**, 297–301 (1986)

3.115 M. Babiker: Electron energy-loss spectroscopy on metallic superlattices. J. Phys. C **20**, 3321–3335 (1987)

3.116 V.Z. Kresin, H. Morawitz: Plasmon spectrum in layered conductors. Phys. Lett. A **145** 368–370 (1990)

3.117 M. Babiker, N.C. Constantinou, M.G. Cottam: General linear responce theory of polaritons in binary superlattices. J. Phys. C **20** 4581–4596 (1987)

3.118 P. Apell, O. Hunderi, R. Monreal: Superlattice optics. Physica Scripta **34**, 348–352 (1986)

3.119 Ph. Lambin, J.P. Vigneron, A.A. Lucas, A. Dereux: Electrodynamics of a phase-stratified medium, with applications to electron energy-loss spectroscopy, infrared reflectivity measurements and attenuated total reflection. Physica Scripta **35**, 343–353 (1987)

3.120 J.-P. Vigneron, P. Lambin, A.A. Lucas, H. Morawitz: Collective polarization waves in high T_c superconductors. Physica C **153-155**, 1313–1314 (1988)

3.121 M. Babiker, N.C. Constaninou, M.G. Cottam: Raman intensities and polarisation selection rules of plasmon-polaritons in superlattice structures. Sol. State Commun. **68**, 967–970 (1988)

3.122 G.A. Faris, M.M. Auto, E.L. Albuquerque, P. Fulco: Effect of a charge layer on the plasmon-polariton dispersion curve in doped semiconductor superlattices. Z. Physik B **80**, 207–211 (1990)

3.123 M. Cui, H. Tsai: Plasmon theory of high-T_c superconductivity.Phys. Rev. B **44**, 12500–12510 (1991)

3.124 S.V. Ivliev, V.N. Sobakin: Spectra of normal oscillations of a system of interacting electrons in a multilayer 2d structure. Soviet Phys. Solid State **33**, 1840–1844 (1991)

3.125 D. Kirillov, C. Webb, J. Eckstein: Propagation of plasmons across layers of GaAs-Ga$_{1-x}$Al$_x$As superlattices. Phys. Rev. Lett. **49**, 1366–1368 (1986)

3.126 N. Zaluzek: Electron energy loss spectroscopy in advanced materials. In *Transmission electron energy loss spectrometry in materials science*, ed. by M.M. Disko, C.C. Ahn, B. Fultz (The Minerals, Metals and Materials Soc., Warrendale 1992) pp.241–266

3.127 D. Olego, A. Pinczuk. A.C. Gossard, W. Wiegmann: Plasma dispersion in a layered electron gas: a determination in GaAs-(AlGa)As heterostructures. Phys. Rev. B **25**, 78677–7870 (1982)

3.128 G. Fasol, H.P. Hughes, K. Ploog: Resonance Raman scattering by intrasubband excitations of GaAs multi-quantum well structures. Surf. Sci. **170**, 497–500 (1986)

3.129 A. Pinczuk, M.G. Lamont, A.C. Gossard: Discrete plasmons in finite semiconductor multilayers. Phys. Rev. Lett. **56**, 2092–2095 (1986)

3.130 M.V. Krasheninnikov. A.V. Chaplik: Theory of electromagnetic excitation of two-dimensional plasma waves in multilayer superlattices. Sov. Phys. JETP **57**, 614–628 (1983)

3.131 K.W.K. Shung: Dielectric function and plasmon structure of stage-1 intercalated graphite. Phys. Rev. B **34**, 979–993 (1986)

3.132 V.Z. Kresin, H. Morawitz: Layer plasmons and high-T_c superconductivity. Phys. Rev. B **37**, 7854–7857 (1988)

3.133 J. Barnas, M. Zimpel, P. Grunberg: Interface magnetic and collective electronic modes in randomly layered metallic structures. Vacuum **41**, 1414–1415 (1990)

3.134 E.L. Albuquerque, G.A. Farias, M.M. Auto: Electron energy-loss spectroscopy of a n-i-p-i semiconductor superlattice. Phys. Stat. Sol. (b) **164**, 463–468 (1991)

3.135 Y. Ohno: Plasma excitation in misfit-layer compounds. Phys. Rev. B **46**, 1664–1674 (1992)

3.136 D.M. Hwang, M. Utlaut, S.A. Solin: Electron energy loss studies of potassium intercalated graphite. Synth. Met. **3**, 81–88 (1981)

3.137 H. Romberg, N. Nücker, J. Fink, T. Wolf, X.X. Xi, B. Koch, H.P. Geserich, M. Durrler, W. Assmus, B. Gegenheimer: Dielectric function of $YBa_2Cu_3O_{7-\delta}$ between 50 meV and 50 eV. Z. Physik B **78**, 367–380 (1990)

3.138 J.J. Ritsko, M.J. Rice: Plasmon spectra of ferric-chloride-intercalated graphite. Phys. Rev. Lett. **42**, 666–669 (1979)

3.139 J. Fink, T. Müller-Heinzerling, J. Pflüger, B. Scheerer, B. Dischler: Investigation of hydrocarbon plasma generated carbon films by electron energy loss spectroscopy. Phys. Rev. B **30**, 4713–4718 (1984)

3.140 J. Yuan, L.M. Brown, W.Y. Liang: Electron energy-loss spectroscopy of the high-temperature superconductor $Ba_2YCu_3O_7$. J. Phys. C **21**, 517–526 (1988)

3.141 Y. Chang, M. Onellion, D.W. Niles, R. Joynt, J.M. Margaritondo: High-temperature superconductor $Ba_2YCu_3O_{7-x}$: Plasmon and ultraviolet optical transition studies. Sol. State Commun. **63**, 721–720 (1987)

3.142 A. Ondo, K. Saiki, K. Ueno, A. Koma: Low-energy electron energy loss spectroscopy on $YBa_2Cu_3O_{7-y}$. Jpn. J. Appl. Phys. **27**, L304–L307 (1988)

3.143 M.G. Ramsey, F.P. Netzer: Monitoring of the surface decomposition of Y-Ba-Cu-O and La-Sr-Cu-O by electron spectroscopy. Mater. Sci. Engineering B **2**, 269–276 (1989)

3.144 C. Tarrio, S.E. Schnatterly: Inelastic electron scattering in the high-T_c compound $YBa_2Cu_3O_{7-x}$. Phys. Rev. B **38**, 921–924 (1988)

3.145 N. Nücker, H. Romberg, S. Nakai, B. Scheerer, J. Zink, Y.F. Yan, Z.X. Zhao: Plasmons and interband transitions in $Bi_2Sr_2CaCu_2O_8$. Phys. Rev. B **39**, 12379–12381 (1989)

3.146 N. Nücker, U. Eckern, J. Fink, P. Müller: Long-wavelength collective excitatioins of charge carriers in high-T_c superconductors. Phys. Rev. B **44**, 7155–7158 (1991)

3.147 P. Batson, M.F. Chisholm: Preliminary study of electron energy-loss spectra of $YBa_2Cu_3O_{7-\delta}$. J. Electron Microsc. Tech. **8**, 311–315 (1988)

3.148 I. Terasaki, T. Nakahashi, S. Takebayashi, A. Maeda, K. Uchinokura. The dielectric functions of high-T_c cuprates obtained from the optical and electron energy loss measurements. Supercond. Sci. Techn. **4**, 397–399 (1991)

3.149 Y.Y. Wang, G. Feng, A. Ritter: Electron energy-loss and optical transmittance investigation of $Bi_2Sr_2CaCu_2O_8$. Phys. Rev. B **42**, 420–425 (1990)

3.150 Y.Y. Wang, A.L. Ritter: Optical excitations in $Bi_2Sr_2CuO_6$ and $Bi_2Sr_2CaCu_2O_8$: Evidence for localized (excitonic) and delocalized charge-transfer gaps. Phys. Rev. B **43**, 1241–1244 (1991)

3.151 I. Bozovic: Plasmons in cuprate superconductors. Phys. Rev. B **42**, 1969–1984 (1990)

3.152 B.N.J. Persson, J.E. Demuth: High-resolution electron energy-loss study of the surface and energy gaps of cleaved high-temperature superconductors. Phys. Rev. B **42**, 8057–8071 (1990)

3.153 Y.A. Uspenskii, S.N. Rashkeev: Microscopical calculations of the optical properties of $Bi_2Sr_2CaCu_2O_8$. Phys. Lett. A **153**, 373–376 (1991)

3.154 A.M. Bratkovsky,, S.N. Rashkeev: The electronic structure and optical properties of $Ba_{1-x}K_xBiO_3$ and $BaPb_{1-x}Bi_xO_3$ superconducting systems. Phys. Lett. A **142**, 172–178 (1989)

3.155 J. Ruvalds: Are there acoustic plasmons . Adv. Phys. **30**, 677–695 (1981)

3.156 M.T. Tachiki, S. Takahashi: Charge fluctuation mechanism of high-T_c superconductivity and the isotope effect in oxide superconductors. Phys. Rev. B **39**, 293–299 (1989)

3.157 J. Ruvalds: Plasmons and high-temperature superconductivity in alloys of copper oxides. Phys. Rev. B **35**, 8869–8872 (1987)

3.158 G.D. Mahan, J.W. Wu: Plasmons and high-temperature superconductivity. Phys. Rev. B **39**, 265–273 (1989)

3.159 P. Longe, S.M. Bose: Acoustic plasmon exchange in multilayered systems: II. Application to high-T_c superconductors. J. Phys C **4**, 1811–1818 (1992)

3.160 D.M. Newns, H.R. Krishnamurthy, P.C. Pattniak, C.C. Tsuei, C.L. Kane: Saddle-point pairing: an electronic mechanism for superconductivity. Phys. Rev. Lett. **69**, 1264–1267 (1992)

3.161 J. Fink, N. Nücker, H. Romberg, S. Nakai: Electronic structure studies of high-T_c superconductors by valence and core electron excitation. In *Proc. Intl. Symp. Electronic Structure of High T_c Superconductors*, ed. by A. Bianconi, A. Marcelli (Pergamon, Oxford 1988) pp.293–312

3.162 S. Nakai, N. Nücker, H. Romberg, M. Alexander, J. Fink: Electron energy-loss studies of high-T_c superconductors $YBa_2Cu_3O_{7-x}$ and $Bi_2Sr_2CaCu_2O_8$. Physica Scripta **41**, 596–600 (1990)

3.163 J. Fink, J. Pflüger, Th. Müller-Heinzerling, H. Nücker, B. Scheerer: The electronic structure of previous and present high-T_c superconductors investigations with high-energy spectroscopies. In *Earlier and Recent Aspects of Superconductivity*, ed. by J.G. Bednorz, K.A. Müller (Springer, Berlin, NewYork 1990) pp.377–406

3.164 F. Stern: Polarizability of a two-dimensional electron gas. Phys. Rev. Lett. **18**, 546–548 (1967)

3.165 A. Griffin, A.J. Pindor: Plasmon dispersion relations and the induced electron interaction in oxide superconductors: Numerical results. Phys. Rev. B **39**, 11503–11513 (1989)

3.166 I. Bozovic, J.H. Kim, Jr. Harris, W.Y. Lee: Optical study of plasmons in $Tl_2Ba_2Ca_2Cu_3O_{10}$. Phys. Rev. B **43**, 1169–1172 (1991)

3.167 A.C. Sharma, I. Kulshrestha: Inverse dielectric response function for copper oxide superconductors. Phys. Rev. B **46**, 6472–6476 (1992)

3.168 H.A. Fertig, S. Das: Collective excitations and mode coupling in layered superconductors. Phys. Rev. B **44**, 4480–4494 (1991)

3.169 V.Z. Kresin, H. Morawitz: Electron energy-loss spectroscopy of layered systems. Phys. Rev. B **43**, 2691–2695 (1991)

3.170 F. Tsay, Y. Wang, T.J. Yang: The dispersion relation of plasmons in a simplified system of CuO_2 layers and Cu-O chains. Z. Physik B **88**, 255–260 (1992)

3.171 J. Fink: Recent developments in energy-loss spectroscopy. Adv. Electron. Electron Phys. **75**, 121–232 (1989)

3.172 C. Wehenkel, B. Gauthé: Electron energy loss spectra and optical constants for the first series from 2 to 120 V. Phys. Stat. Sol. (b) **64**, 515–525 (1974)

3.173 P. Schattschneider, G. Sölkner: A comparison of techniques for the removal of plural scattering in energy loss spectroscopy. J. Micr. **134**, 73–87 (1984)

3.174 H. Ibach, H. Lüth: *Solid State Physics* (Springer, Berlin, Heidelberg, NewYork 1990)

3.175 J.C. Phillips: *Bonds and Bands in Semiconductors* . (Academic, New York 1973)

3.176 K. Zeppenfeld: Nichtsenkrechte Interbandübergänge in Graphit durch unelastische Elektronenstreuung. Z. Physik **243**, 229–243 (1971)

3.177 G. Leveque, S. Robin-Kandare, L. Martin: Band structure of layer crystals NbSe$_2$ and MoSe$_2$ and an interpretation of their optical spectra. Phys. Stat. Sol. (b) **63**, 679–690 (1974)

3.178 M.D. Rechtin, B.L. Averbach: Atomic arrangements and fast electron energy losses in sputtered thin films of vitreous Se-As alloys with 0-24 As. J. Non-Cryst. Solids **12**, 391–421 (1973)

3.179 W.Y. Liang, A.R. Beal: A study of the optical joint density-of states function. J. Phys. C **9**, 2823–2832 (1976)

3.180 P.E. Batson, K.L. Kavanagh, J.M. Woodall, J.W. Mayer: Observation of defect electronic states associated with misfit dislocations at the GaAs/GaInAs interface. Phys. Rev. Lett. **57**, 2729–2732 (1986)

3.181 P.E. Batson: Electron energy loss studies in semiconductors. In *Transmission Electron Energy Loss Spectrometry in Materials Science*, ed. by M.M. Disko, C.C. Ahn, B. Fultz (The Minerals, Metals and Materials Soc., Warrendale 1992) pp.217–240

3.182 J. Tauc, R. Grigorovici, A. Vancu: Optical properties and electronic structure of amorphous germanium. Phys. Stat. Sol. **15**, 627–637 (1966)

3.183 W.G. Sainty, P.J. Martin, R.P. Netterfield, D.R. McKenzie, D.M. Cockayne: The structure and properties of ion beam synthesized boron nitride films. J. Appl. Phys. **64**, 3980–3986 (1988)

3.184 J. Fink, B. Scheerer, W. Wernet, M. Monkenbusch, G. Wegner: Electronic structure of pyrrole-based conducting polymers: an electron energy-loss spectroscopy study. Phys. Rev. B **34**, 1101–1115 (1986)

3.185 J. Fink, B. Scheerer, W. Wernet, M. Monkenbusch, G. Wegner: Electronic structure of pyrrole-based conducting polymers. Synth. Met. **18**, 71–76 (1987)

3.186 C.H. Chen: Electron energy loss studies of direct nonvertical interband transitions in silicon. Phys. Stat. Sol. (b) **83**, 347–351 (1977)

3.187 H. Fritzsche, J. Fink, N. Nücker, B. Scheerer, G. Leising: Momentum-dependent dielectric function of the cis-transoid conformation of cis-polyacetylene. Phys. Rev. B **40**, 8033–8036 (1989)

3.188 J. Fink, G. Leising: Momentum-dependent dielectric functions of oriented trans-polyacetylene. Phys. Rev. B **34**, 5320–5328 (1986)

3.189 P. Schattschneider, J. Fidler, V. Chopov: Wavenumber-resolved energy loss spectra of Co$_5$Sm. J. Electron. Spectr. Rel. Phenom. **31**, 25–32 (1983)

3.190 E.N. Foo, J.J. Hopfield: Optical absorption and energy loss in sodium in the Hartree approximation. Phys. Rev. **173**, 635–644 (1968)

3.191 K.C. Pandey, P.M. Platzman, P. Eisenberger: Plasmons in periodic solids. Phys. Rev. **9**, 5046–5055 (1974)

3.192 E. Petri, A. Otto: Direct nonvertical interband and intraband transitions in Al. Phys. Rev. Lett. **34**, 1283–1286 (1975)

3.193 M. Urner-Wille: Wavevector-dependent fine structure in the electron energy-loss spectrum of aluminium. J. Phys. D **10**, 49–53 (1977)

3.194 K. Sturm, L.E. Oliveira: Theory of a zone-boundary collective state in Al: a model calculation. Phys. Rev. B **30**, 4351–4365 (1984)

3.195 K. Sturm, L.E. Oliveira: Zone boundary collective states in lithium and sodium. Europhys. Lett. **9**, 257–262 (1989)

3.196 H. Venghaus: Redetermination of the dielectric function of graphite. Phys. Stat. Sol. (b) **71**, 609–614 (1975)

3.197 D.R. Penn, P. Apell: Anomalous electron energy loss in small spheres. J. Phys. C **16**, 5729–5743 (1983)

3.198 H. Raether: Surface plasmons on smooth and rough surfaces and on gratings. Springer Tracts Mod. Phys. Vol. 111 (Springer, Berlin, New York 1988)

3.199 R.H. Ritchie: Plasma losses by fast electrons in thin films. Phys. Rev. **106**, 874–881 (1957)

3.200 M. Scheinfein, A. Muray, M. Isaacson: Electron energy loss spectroscopy across a metal-insulator interface at sub-nanometer spatial resolution. Ultramicroscopy **16**, 233–240 (1985)

3.201 P.M. Echenique, J. Bausells, A. Rivacoba: Energy-loss probability in electron microscopy. Phys. Rev. B **35**, 1521–1524 (1987)

3.202 D.B. Tran-Thoai, E. Zeitler: Inelastic scattering of fast electrons by thin metal slabs. Phys. Stat. Sol. (a) **120**, 467–474 (1990)

3.203 F. Fujimoto, K. Komaki: Plasma oscillations excited by a fast electron in a metallic particle. J. Phys. Soc. Jpn. **25**, 1679–1687 (1968)

3.204 J.C. Ashley, T.L. Ferrell: Excitation by fast electrons of surface plasmons on spherical voids in metals. Phys. Rev. B **14**, 3277–3281 (1976)

3.205 J.C. Ashley, T.L. Ferrell, R.H. Ritchie: X-ray excitation of surface plasmons in metallic spheres. Phys. Rev. B **10**, 554–558 (1973)

3.206 N. Barberan, J. Bausells: Plasmon excitation in cavities. Solid State Commun. **73**, 651–655 (1990)

3.207 H. Kohl: Image formation by inelastically scattered electrons: image of a surface plasmon. Ultramicroscopy **11**, 53–66 (1983)

3.208 H. Kohl, C. Colliex: Theory of image formation by plasmon loss electrons in small spheres. In *Electron Microscopy and Analysis 1985*, ed. by J. Tatlock, Inst. Phys Conf. Ser. 78 (Hilger, Boston 1986) pp.87–90

3.209 W. Ekardt: Collective multipole excitations in small metal particles: critical angular momentum for the existence of collective surface modes. Phys. Rev. B **32**, 1961–1970 (1985)

3.210 N. Zabala, A. Rivacoba: Support effects on the surface plasmon modes of small particles. Ultramicroscopy **35**, 145–150 (1991)

3.211 L. Serra, F. Garcias, J. Navarro, N. Barberan, M. Barranco: Electronic surface excitations of cavities in metals. Phys. Rev. B **46**, 9369–9379 (1992)

3.212 M. Schmeits: Inelastic scattering of fast electrons by spherical surfaces. J. Phys. C **14**, 1203–1216 (1981)

3.213 C.A. Walsh: An analytical expression for the energy loss of fast electrons travelling parallel to the axis of cylindrical interface. Phil. Mag. B **63**, 1063–1078 (1991)

3.214 A.A. Lucas: Plasmon cohesive energy of voids and void lattices in irradiated metals. Phys. Rev. B **7**, 3527-3537 (1973)

3.215 F. Fujimoto, K. Komaki, K. Ishida: Surface plasma oscillations in aluminium fine particles. J. Phys. Soc. Jpn. **23**, 1186 (1967)

3.216 M. Creuzberg: Entstehung von Alkalimetallen bei der Elektronenbestrahlung von Alkalihalogeniden. Z. Physik **194**, 211–218 (1966)

3.217 U. Kreibig, P. Zacharias: Surface plasma resonance in small spherical silver and gold particles. Z. Physik **231**, 128–143 (1970)

3.218 P.E. Batson: A new surface plasmon resonance in clusters of small aluminium spheres. Ultramicroscopy **9**, 277–282 (1982)

3.219 P.E. Batson: Surface plasmon coupling in clusters of small spheres. Phys. Rev. Lett. **49**, 936–940 (1982)

3.220 C. Colliex: An illustrated review of various factors governing the high spatial resolution capabilities in EELS microanalysis. Ultramicroscopy **18**, 131–150 (1985)

3.221 M. Achèche, C. Colliex, P. Trebbia: Characterization of small metallic clusters by electron energy loss spectroscopy. Scanning Electron Microscopy 1986/I (SEM Inc., AMF O'Hare) pp.25–32

3.222 M. Achèche, C. Colliex, H. Kohl, A. Nourtier, P. Trebbia: Theoretical and experimental study of plasmon excitations in small metallic spheres. Ultramicroscopy **20**, 99–106 (1986)

3.223 J. Geiger: Energieverluste schneller Elektronen in Festkörpern. Habilitationsschrift, Technische Universität Berlin 1967

3.224 D.B. Tran-Thoai, E. Zeitler: Inelastic scattering of fast electrons by small metal particles. Phys. Stat. Sol. (a) **107**, 791–797 (1988)

3.225 A. vom Felde, J. Fink, W. Ekardt: Quantum size effects in excitation of potassium clusters. Phys. Rev. Lett. **61**, 2249–2252 (1988)

3.226 C.A. Walsh: Analysis of electron energy-loss spectra from electron-beam-damaged amorphous AlF$_3$. Phil. Mag. A **59**, 227–246 (1989)

3.227 D. Ugarte, C. Colliex, P. Trebbia: Surface- and interface-plasmon modes on small semiconducting spheres. Phys. Rev. B **45**, 4332–4343 (1992)

3.228 P. Henoc, L. Henry: Observation des oscillations de plasma a l'interface d'inclusions gazeuses dans une matrice cristalline. J. Phys. (Paris) **31**, C1, 55–57 (1970)

3.229 R. Manzke, M. Campagna: Study of He bubbles in Al by electron energy loss spectroscopy. Sol. State Commun. **39**, 313–317 (1981)

3.230 A. vom Felde, J. Fink, Th. Müller-Heinzerling, J. Pflüger, B. Scheerer, G. Linker, D. Kaletta: Pressure of argon, neon and xenon bubbles in aluminium. Phys. Rev. Lett. **53**, 922–925 (1984)

3.231 A. vom Felde, J. Fink: Excitation of bubble surface plasmons in rare-gas-irradiated aluminium films. Phys. Rev. B **31**, 6917–6920 (1985)

3.232 A.J. McGibbon et al.: Microscopy in solid state science. Microsc. Res. Techn. **24**, 299–315 (1993)

3.233 G. Hibbert, J. Edington: Experimental errors in combined electron microscopy and energy analysis. J. Phys. D **5**, 1780–1786 (1972)

3.234 D.B. Williams, J.W. Edington: High resolution microanalysis in materials science using electron energy-loss measurements. J. Micr. **108**, 113–145 (1976)

3.235 H. Yasuda, H. Nakayama, H. Fujita, K. Ueda, T. Ishida: Structural analysis of amorphous Fe-B alloys on the basis of plasma losses. Mater. Trans. JIM **30**, 717 (1989)

3.236 Y. Wang, H. Chen, R.W. Hoffman: Structural analysis of hydrogenated diamond-like carbon films from electron energy loss spectroscopy. J. Mater. Res. **5**, 2378–2386 (1990)

3.237 K. Breuer, W. Tews, E. Steinbeiss: Energy loss measurements on thin vanadium dioxide films. Phys. Stat. Sol. (b) **73**, K95 (1976)

3.238 L.M. Brown, A.P. Stephens: Characterisation of niobium and vanadium hydrides by electron energy loss spectroscopy and by STEM. Acta Met. **33**, 827–833 (1985)

3.239 C. Colliex, P. Trebbia: Energy loss spectroscopy in the electron microscope as a tool for microanalysis. In *Developments in Electron Microscopy and Microanalysis*, ed. by J.A. Venables (Academic, London 1976) pp.123–128

3.240 N.J. Zaluzec, T. Schober, B.W. Veal, D.G. Westlahe: EELS study of metal-hydrogen systems. In *Analytical Electron Microscopy 1981* (San Francisco Press, San Francisco 1981) pp.191–192

3.241 O.T. Woo, G.J.C. Carpenter: EELS characterisation of zirconium hydrides. Microsc. Microanal. Microstruct. **3**, 35–44 (1992)

3.242 J.A. Hunt, D.B. Williams: Electron energy-loss spectrum-imaging. Ultramicroscopy **33**, 47–73 (1991)

3.243 N.D. Evans, S.J. Zinkley, J. Bentley, E.A. Kenik: Quantification of metallic Al-profiles in Al$^+$ implanted MgAl$_2$O$_4$ spinel by electron energy loss spectroscopy. In *Proc. 49th Ann. Meeting EMSA*, ed. by G.W. Bailey (San Francisco Press, San Francisco 1991) pp.728–729

3.244 P. Schattschneider, G. Sölkner, M.Q. Lü, J. Hafner, K. Riedl, H. Sassik: Electron energy loss spectroscopy of plasmons in a glassy metal. Scanning Microscopy, Suppl.1, 151–160 (1987)

3.245 C.H. Chen, D.C. Joy, H.S. Chen, J.J. Hauser: Observation of anomalous plasmon linewidth in the icosahedral Al-Mn quasicrystals. Phys. Rev. Lett. **57**, 743 (1986)

3.246 L.E. Levine, P.C. Gibbons, K.F. Kelton: Electron energy-loss spectroscopy studies of icosahedral-phase plasmons. Phys. Rev. B **40**, 9338–9341 (1989)

3.247 M.P. Seah, G.C. Smith: Plasmons: quanta for micro-region temperature measurements. J. Mat. Sci. **21**, 1305–1309 (1986)

3.248 D.L. Johnson: Local fields effects, x-ray diffraction, and the possibility of observing the optical Borrmann effect: solutions to Maxwell equations. Phys. Rev. B **12**, 3428–3437 (1975)

3.249 I. Lindau, P.O. Nilsson: Experimental evidence for excitation of longitudinal plasmons by photons. Phys. Lett. A **31**, 352–353 (1970)

3.250 M. Anderegg, B. Feuerbacher, B. Fitton: Optically excited longitudinal plasmon in potassium. Phys. Rev. Lett. **27**, 1565–1568 (1971)

3.251 J.C. Ingram, K.W. Nebersny, J.E. Pemberton: Optical constants of the noble metals determined by reflection electron energy loss spectroscopy. Appl. Surf. Sci. **44**, 293–300 (1990)

3.252 F. Yubero, J.M. Sanz, E. Elizalde, L. Galan: Kramers-Kronig analysis of reflection electron energy loss spectra of Zr and ZrO$_2$. Surf. Sci. **237**, 173–180 (1990)

3.253 M. De Crescenzi, E. Colavita, L. Papagno, G. Chiarello, R. Scarmozzino, L.S. Caputi, R. Rosei: Electronic properties of Fe$_{80}$B$_{20}$ alloys.: ordering and disordering effects. J. Phys. F **13**, 895–907 (1983)

3.254 Z.L. Wang: Electron multiple inelastic scattering in the geometry of RHEED. Ultramicroscopy **26**, 321–326 (1988)

3.255 P.A. Cox, J.P. Kemp: Theoretical treatment of screening in the EELS of metalic oxide systems. Surf. Sci. **210**, 225–237 (1989)

3.256 S. Tougaard, I. Chorkendorff: Quantitative analysis of reflection electron energy-loss spectra of aluminium. Sol. State. Commun. **57**, 77–79 (1986)

3.257 S. Tougaard, J. Kraaer: Inelastic electron scattering cross sections for Si, Cu, Ag, Au, Ti, Fe and Pd. Phys. Rev. B **43**, 1651–1661 (1991)

3.258 G. Gensterblum, J.J. Pireaux, P.A. Thiry, R. Caudano, A.A. Vigneron: High-resolution electron energy loss spectroscopy of thin films of C60 on Si(100). Phys. Rev. Lett. **67**, 2171–2174 (1991)

4. Inner-Shell Ionization

Ferdinand Hofer

The electron energy-loss spectrum contains, at higher energy losses (above 50 eV), characteristic edges due to ionization of inner atomic shells. These edges are of considerable interest to the electron microscopist because they permit the qualitative and quantitative microanalysis of nearly all chemical elements on a nanometre scale. However, additional insight into the chemical and structural properties associated with the atom being excited may be gained by examination of the fine structure of these edges.

In this chapter, the basic physics of inner-shell ionization and the various edge types and edge fine structure features occurring in EELS are described. Since the inner-shell ionizations probed by EELS resemble a close relationship to the core level absorption edges of x-ray absorption spectroscopy (XAS), we have to consider this spectroscopic method too. Furthermore it is necessary to accent that we are mainly dealing with edges which correspond to single scattering distribution, i.e. spectra which have been measured using very thin specimens or which have been deconvoluted.

4.1 Theory of Inner-Shell Excitations

This theory, first developed by Bethe [4.1] and later expanded [4.2], describes the dependence of the inelastic scattering by an isolated atom on energy loss and scattering angle. Although originally formulated only for single atoms or gaseous targets, the theory is applicable also to the inelastic scattering from the inner atomic shells in solids. The inelastic scattering of an incident electron in the initial state (energy $E_0 = eU$ and wave vector $\boldsymbol{k}_0$) by an atom generates an outgoing electron in the final state (energy $E_0 - E$ and wave vector $\boldsymbol{k}_n$, E = energy loss). Such a process is schematically presented in Fig. 4.1. The scattering centre (atom) suffers a change in energy and reaches an excited state of energy $E_1 = E$ above its ground state and receives a momentum transfer $\boldsymbol{q}'$ given by

$$h\boldsymbol{q}' = h(\boldsymbol{k}_0 - \boldsymbol{k}_n).\tag{4.1}$$

Since in practice the scattering angle θ is measured, it is useful to relate this to the momentum transfer. For $\theta \ll 1$ rad and $E_0 \ll m_0 c^2$, it is a good approximation to take $k_n/k_0 = 1$ and to write

Springer Series in Optical Sciences, Vol. 71
Energy-Filtering Transmission Electron Microscopy
Editor: Ludwig Reimer   ©Springer-Verlag Berlin Heidelberg 1995

$$q'^2 = k_0^2(\theta^2 + \theta_E^2) \tag{4.2}$$

where $k_0 = 1/\lambda$ is the magnitude of the incident-electron wave vector and $\theta_E = E/2E_0$ is the characteristic angle of inelastic scattering with an energy loss E. The probability of an inelastic scattering process is described by a double differential cross-section with respect to E and q'. This can be obtained by using the first Born approximation, which assumes that the energy loss and momentum transfer are small compared with the energy and momentum of the incident beam (weak interaction). The cross-section is written as [4.1]:

$$\frac{d^2\sigma}{d\Omega dE} = \frac{4}{a_H^2 q'^4} \frac{k_n}{k_0} S(q', E) \tag{4.3}$$

where $d\Omega$ is the solid angle around the final wave vector and $a_H = 0.0569$ nm is the Bohr radius of the hydrogen atom. The first term in (4.3) is essentially the Rutherford cross-section for scattering from a point charge. The function $S(q', E)$ is the form factor for inelastic scattering, which modifies the Rutherford scattering that would occur if the atomic electrons were free. The form factor is a property of the atom, dimensionless and independent of the incident electron energy. It is normally expressed in two different ways, one appropiate for valence excitations (see previous chapter) and the other for core excitations.

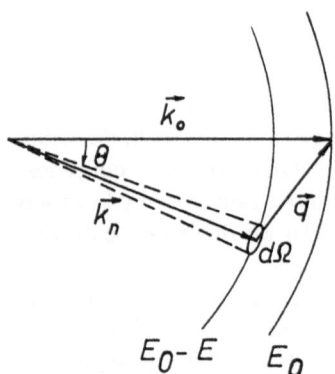

Fig. 4.1. The energy and momentum parameters involved in an inelastic scattering event.

The probability of inner-shell excitation may be expressed in terms of a one-electron transition matrix element between an initial core state $|0\rangle$ and a final unoccupied state $|n\rangle$. Both incident and scattered electrons can normally be treated as plane waves. Within the Born approximation, the inelastic form factor is related to the square of the absolute value of the transition matrix element in the following way:

$$S(q', E) = |\langle n|\exp(2\pi i q' \cdot r)|0\rangle|^2 \tag{4.4}$$

where the operator $\exp(2\pi i q' \cdot r)$ arises from the interaction between the incident electron and the atomic electrons (r = coordinate of atomic electrons). Since this interaction is primarily electrostatic (below 300 keV), it can be described by a Coulomb potential. Inner-shell electrons (1s, 2s, 2p, 3s, 3p, 3d etc.) are excited to final states that mostly lie in the continuum and are taken to be normalized per unit energy range (Fig. 4.2) [4.3].

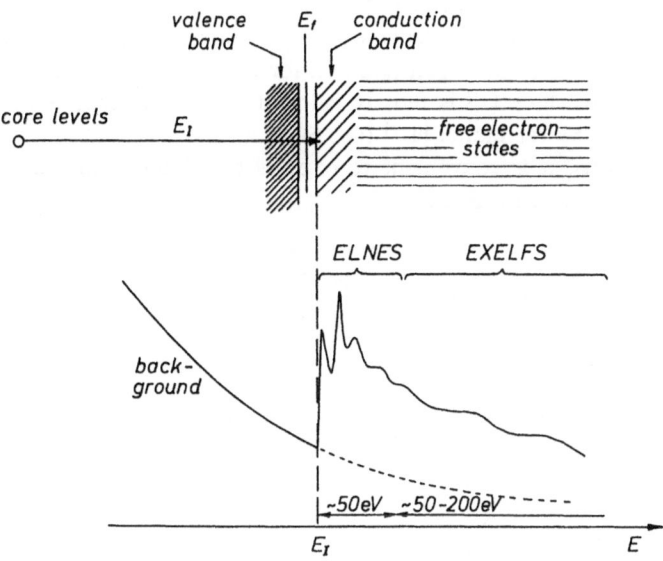

Fig. 4.2. Schematic representation of a core-loss excitation within the band stucture model and its relationship with the energy-loss spectrum (including edge fine structures). An electron initially on an atomic core level is promoted to a vacant state above the Fermi level.

Out of this theory arises the very useful concept of the generalized oscillator strength (GOS), which is directly related to the transition matrix element and to the inelastic form factor [4.2]:

$$\frac{df(q,E)}{dE} = \frac{2mE}{h^2 q^2} |\langle n | \exp(2\pi i \, q' \cdot r) | 0 \rangle|^2 \qquad (4.5)$$

The GOS represents the probability of a given electron undergoing a transition from its initial state to a particular state in the continuum. In the limit $q' \to 0$, the GOS collapses to the optical oscillator strength, which characterizes the response of an atom to incident photons (optical absorption). It is very important to emphasize that for high incident energies and small q' the GOS probed by optical techniques (x-ray absorption for instance) and by EELS is almost identical. Since the inner-shell levels of an atom are highly localized, the GOS for core ionization can be considered as basically

an atomic quantity, with the effects of the solid added as a refinement after-
wards [4.4]. A three-dimensional plot of the GOS in the coordinate system
(E, q') is known as the Bethe surface (Fig. 4.3). This surface represents all
the information concerning the inelastic scattering of charged particles by
atoms and can be used for the calculation of cross-sections. Many experimental investigations and theoretical calculations of the GOS have therefore
been performed: Following the original work of Bethe [4.1], several models
have been developed for the calculations of photo-ionization cross-sections
[4.5,6], which have improved our knowledge of the atomic effects involved in
such processes. Some of them have been adapted for typical electron microscopical situations, where the GOS is mostly calculated using a hydrogenic
model [4.7] or a Hartree–Slater model [4.7] (see Sect. 4.2.1).

For many applications, it is convenient to express the differential cross-
section in terms of the quantities E and θ :

$$\frac{\mathrm{d}^2\sigma}{\mathrm{d}\Omega\mathrm{d}E} = \frac{e^4}{(4\pi\epsilon_0)^2 E_0 E} \frac{1}{\theta^2 + \theta_E^2} \frac{\mathrm{d}f(q', E)}{\mathrm{d}E} . \qquad (4.6)$$

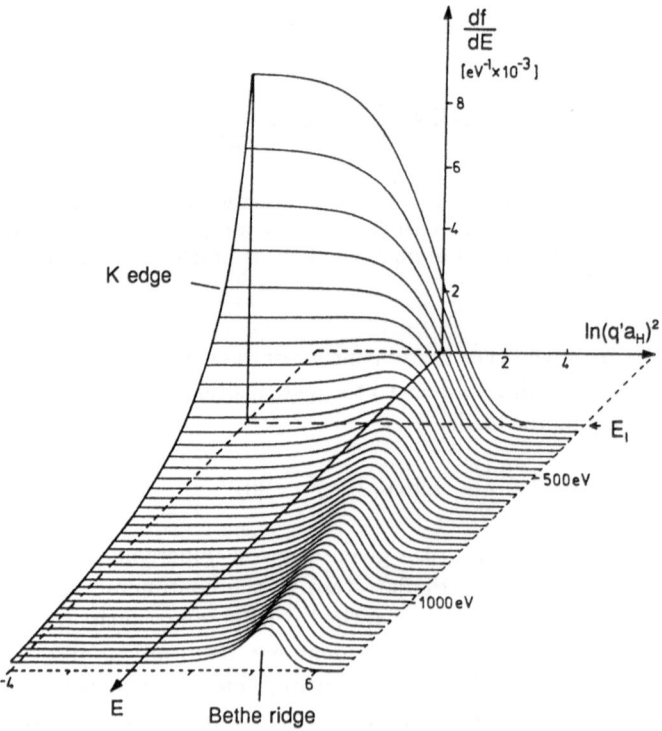

Fig. 4.3. Bethe surface for K-shell ionization, calculated using a hydrogenic model;
from [4.7]. The GOS is zero for energy losses below the ionization energy E_I. The
horizontal coordinate is related to the scattering angle θ.

Angular Distribution. As already mentioned, a convenient representation of the angular dependence of the cross-sections is the Bethe surface. Figure 4.3 shows the GOS of a K ionization calculated using the hydrogenic model [4.7] as a function of E and $\ln(q'a_H^2)$. The individual curves show the angular dependence of K-shell ionization for different amounts of energy loss. Just above the ionization energy E_I , this scattering is forward peaked whereas for large energy losses (several times E_I) the scattered electrons are concentrated into the Bethe ridge with a maximum at the Compton angle $\theta_C = (E/E_0)^{1/2}$. The scattering into this angle can be described by classical scattering at a quasi-free electron and it can be used to obtain a measure of the ground-state momentum wave function of the atomic electrons [4.9].

In order to see the relation betweeen the energy loss spectrum and the x-ray absorption spectrum it is convenient to expand the operator in the matrix element as

$$\exp(2\pi i \mathbf{q}' \cdot \mathbf{r}) = 1 + 2\pi i q'(\mathbf{u} \cdot \mathbf{r}) - \frac{1}{2} 4\pi^2 q'^2 (\mathbf{u} \cdot \mathbf{r})^2 + ... \qquad (4.7)$$

where $\mathbf{u}$ is a unit vector in the direction of $\mathbf{q}$'. The first term in (4.7) vanishes since the initial and final state wave functions are orthogonal. The second term is equivalent to the dipole matrix element for absorption of x-rays where $\mathbf{u}$ is replaced by the polarization vector of the electric field. This dipole term gives rise to transitions involving a change in angular momentum quantum number $\Delta l = \pm 1$. It is normally the most important contribution to the transition.

At small scattering angles, the main angular dependence comes from the Lorentzian factor $1/(\theta^2 + \theta_E^2)$ in (4.6). In this case, the GOS is almost constant (independent of q' and θ) and this regime is known as the dipole region. The half-width of the angular distribution is represented by θ_E. In the small-angle scattering limit, the oscillator strengths probed by optical techniques are identical with those given by EELS. Cross-sections for inelastic scattering of fast electrons can therefore be compared with photoabsorption data [4.10,11].

At larger scattering angles, deviations from the Lorentzian profile are introduced by the q' dependence of the GOS. In this case, the dipole selection rules do not apply and dipole forbidden transitions (e.g. quadrupole) may then be observed. The q' vector dependence of inelastic scattering can thus be investigated by EELS, which is not possible with optical absorption.

Energy Distribution. The number of electrons scattered as a result of inner-shell excitations into angles less than the aperture α and with an energy loss E can be expressed as an energy differential cross-section $d\sigma(\alpha)/dE$, which can be compared with experiments (EELS-spectrum). It may be shown that the approximate behaviour of the energy differential cross-section obeys the relationship [4.3]

$$\frac{d\sigma(\alpha)}{dE} \propto E^{-r} \qquad (4.8)$$

where the value of r is the downward slope in a log-log transformed spectrum. The parameter r depends on α, the collection semi-angle and varies from about 4, for large α, to about 6 for small α [4.12]. However, this is only the case for K edges about some 10 eV above threshold and to some extent for L edges, but not for M, N or O edges.

For thin specimens in which multiple inelastic scattering is negligible, the inner-shell contribution to the EELS spectrum, the intensity of the edge $I(\alpha, E)$, is related to the energy differential cross-section by

$$I(\alpha, E) = N I_t(\alpha) \frac{d\sigma(\alpha)}{dE} \qquad (4.9)$$

where N is the number of atoms per unit specimen area contributing to the ionization edge and $I_t(\alpha)$ is the total intensity transmitted through the collection aperture of semi-angle α.

Partially Integrated Cross-Sections. Quantitative analysis by EELS requires knowledge of the partial cross-sections of inner-shell excitations [4.13], which can be obtained by integrating the energy differential cross-sections over a range Δ above E_I and for scattering inside an aperture α:

$$\sigma_I(\alpha, \Delta) = \int\limits_{E_I}^{E_I + \Delta} \frac{d\sigma(\alpha)}{dE} dE \qquad (4.10)$$

Total Cross-Sections. For large collection angles and energy windows the partial cross-section becomes the total cross-section for producing an inner-shell vacancy, a quantity of considerable interest in both x-ray and Auger spectroscopy [4.11,14].

4.2 Determination of Ionization Cross-Sections

Ionization cross-sections are not only useful for their practical application to spectral quantification procedures but also provide a necessary guide to the development and refinement of theoretical models for electron-solid interactions. The ionization cross-sections can be either calculated using theoretical models or determined experimentally using thin-film standards. Up to now calculated cross-sections have been mainly used in practical microanalysis, although it has been long known that experimental cross-sections can provide more accurate quantifications.

4.2.1 Calculation of Cross-Sections

The central quantity in the Bethe theory is the generalized oscillator strength (GOS) (4.5). In order to calculate the GOS and the ionization cross-sections, one has to know the initial and final state wave functions of the inner-shell electron, which can be calculated by the following methods, principally based on atomic models.

Hydrogenic Model. This method is widely used for the calculation of inner-shell cross-sections and it was the first to be developed [4.1]. It is based on the wave mechanics of the hydrogen atom. Since analytical expressions are available for the wave functions of the hydrogen atom, which can be obtained from an exact solution of the Schrödinger equation, this method is easily applied. To make the hydrogen wave functions applicable to inner-shell electrons of atoms with higher Z, the nuclear charge $+Ze$ and the screening of the nuclear field by the remaining $Z-1$ electrons have to be taken into account. The wave functions remain hydrogen-like in form and can be used for the calculation of the GOS by using standard methods [4.1,15]. These hydrogenic formulations have been incorporated into a short computer program called SIGMAK, which enables the GOS and the partial cross-sections for K shells of the elements Li to Si to be calculated [4.7]. This program has subsequently been revised to include both retardation and exact relativistic kinematics [4.3]. The hydrogenic model predicts the saw tooth profile of the K edges and the cross-sections are in good agreement with experimental data obtained under conditions relevant to EELS in AEM.

A similar model has been developed for L shells (SIGMAL1) but here, the simple treatment of screening proves inadequate and an empirical factor has to be inserted to match experimentally determined edge shapes [4.16]. This SIGMAL1 program evaluates partial cross-sections for L shell ionizations of the elements Al to Zn, using relativistic kinematics and the hydrogenic expression for the GOS [4.17] with an empirical correction based on photoabsorption data [4.18]. Severe discrepancies are found when this model is applied to transition metal compounds [4.19–21]. For this reason, Egerton readjusted the model to correspond more closely to photoabsorption data and to take into account energy loss measurements [4.22]. Recent measurements reveal that this version of the hydrogenic model provides quite good partial cross-sections for EELS quantification (see next chapter). Recently a hydrogenic model for both M_{45} and M_{23} shells has also been derived, using a correction based on experimental photoabsorption data [4.23].

Hartree–Slater Model. The initial and final wave functions can be also calculated from an atomic Hartree–Slater model [4.8,24]. The initial state is taken to be a one-electron wave function solved with a self-consistent potential and with averaged exchange by assuming a central (spherically symmetric) field within the atom. The final states are computed by solving the radial Schrödinger equation using the same central fields for the continuum energies above threshold. An advantage of this approach is that it can be used for all ionization edges. It also predicts a number of shape effects, such as the sawtooth shape of the K edges or the delayed edge onsets occurring 20–30 eV above the thresholds of L and M edges. However, transitions to unoccupied bound states (white lines), which can occur in the case of M and L edges, are not included in the calculations [4.4,25]. Recently, oscillator strengths for the white-line components of L and M edges have been calculated, which can be used to generate partial cross-sections including both bound and continuum states [4.26].

4.2.2 Experimental Determination of Cross-Sections

Since many questions concerning the accuracy of calculated cross-sections have arisen in the last decade, several groups have determined partial cross-sections by EELS in a TEM. Such measurements provide a test of the accuracy of the atomic models and hence lead to more accurate EELS quantification.

Absolute Measurements. This method is generally applicable and can be used for many elements ranging from Li to U [4.27]. However, a difficulty associated with measuring absolute cross-sections is that a thin film standard of known thickness, known composition and density must be manufactured for each element of interest. The main problem of this method is the thickness measurement, which can be only made with sufficient accuracy by means of convergent-beam electron diffraction (CBED) [4.28] (Sect. 6.4). Since thickness measurement by CBED can be a time consuming procedure, which is also restricted to thick, crystalline specimens, only very few measurements of absolute ionization cross-sections have been published: many of the measurements were performed in the seventies, during the period when electron microscopes were first fitted with spectrometers [4.27,29-33]. In a more recent investigation, Crozier used thin evaporated films of C, Al, Fe, Cu, Ag and Au, the thicknesses of which were determined by weighing [4.34,35]. The experimental problems were such that the agreement of these measurements with the calculated cross-sections varied considerably. This may partly be due to the fact that, in very thin metal films, thin oxide layers develop due to air oxidation and contamination films may introduce considerable systematic errors in the determination of cross-sections. If, however, thicker samples are used, the effect of thin oxidation layers decreases but the spectrum has to be corrected for multiple scattering.

k-Factor Determination. The difficulties of the absolute measurements can be partially avoided if ratios of cross-sections are determined by means of simple compound standards [4.22,36]. In this method, a thin film standard is used which must contain one light element (B), which gives rise to a K edge in the EELS spectrum and the element (A), the cross-section of which is sought. The cross-section ratio (or k-factor) can then be determined according to the following equation:

$$k_{AB} = \frac{\sigma_B(\alpha, \Delta)}{\sigma_A(\alpha, \Delta)} = \frac{I_B(\alpha, \Delta)}{I_A(\alpha, \Delta)} \frac{N_A}{N_B} \tag{4.11}$$

where N is the number of atoms per unit area, $I_i(\alpha, \Delta)$ is the core-loss intensity integrated up to an energy region of width Δ starting at the edge onset, $\sigma_i(\alpha, \Delta)$ is the partial cross-section integrated over a collection angle α and an energy region Δ. The measured cross-section ratios can be viewed as EELS k factors in analogy to the Cliff–Lorimer method in x-ray spectroscopy of thin films [4.37]. Absolute cross-section values can be determined by using a calculated cross-section value for the light element. This is possible because the K edges of light elements can be accurately calculated [4.38]. Preferably, k factor determination should be performed with oxide compounds [4.36], although borides or carbides have also been used [4.22,39]

At the present state of development, the k factor approach has some essential advantages: all elements and all ionization edges (K, L, M, N and O) can be quantified with good accuracy and an accurate knowledge of specimen thickness is not necessary. Additionally, contamination layers or thin amorphous layers do not disturb the measurement of intensity ratios and cross-section ratios show a smaller dependence on multiple scattering, the effect of which on each edge cancels [4.40]. Furthermore, k factors can be viewed as effective cross-sections that allow a quantification of EELS also under particular experimental situations [4.41,42]. Experimental k factors can be easily compared with calculated cross-section ratios, thus providing a test of the models under the conditions of interest to the electron microscopist.

Experimental k factors have been determined for K, L_{23}, M_{45}, M_{23} and N_{45} shells [4.22,41,43–45]. Several other measurements on light and medium element materials have been also reported [4.16,46,47]

4.2.3 Parametrization of Cross-Sections

The comparison of cross-sections or k factors measured in different laboratories is complicated by the fact that different values of incident energy and collection angle have been employed. The compilation and parametrization of cross-section values has been proposed several times [4.25,48]. One efficient way of comparing values determined in different experimental conditions is to represent $\sigma(\alpha, \Delta)$ in terms of an integrated oscillator strength $f(\Delta)$, which is independent of α and E. This can be easily done with a short computer program [4.48] and the $f(\Delta)$ data have already been tabulated for several energy windows [4.38,49]. Furthermore, partial cross-sections for particular experimental situations can be derived from these tabulated $f(\Delta)$ values by using again a short computer program [4.48].

4.3 The Shape of the Ionization Edges

Since the core-electron wave functions change little when atoms aggregate to form a solid, the general shapes of ionization edges are determined mainly by atomic effects. At small scattering angles, the dominant transitions that are observed obey the dipole selection rules $\Delta l = \pm 1$ and $\Delta j = 0, \pm 1$ where l and j are the quantum numbers of the subshell from which the electron has been excited. Thus only allowed transitions from occupied states towards unoccupied states of the correct symmetry are observed: s→p, p→d or s, d→f or p, etc. These transitions are usually classified using the standard spectroscopy notation, e.g. K excitation for 1s electrons, L_1 for 2s, L_2 for $2p_{1/2}$, L_3 for $2p_{3/2}$, etc. (see Table 4.1 on page 259). The shape of the ionization edges varies with atomic number and ionized shell and the edge shapes can be classified into four main families [4.50] which are displayed in Fig. 4.4. The threshold energies of the ionization edges represent the difference in energy between the core level initial state and the lowest energy final state of the excited electron. They can be shifted with respect to the atomic energies and will be discussed in Sect. 4.4.1.

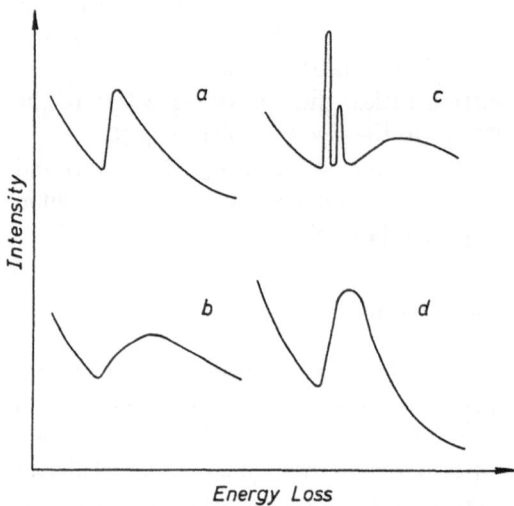

Fig. 4.4. Representation of the four main families of edge shape: (a) saw tooth profile, (b) delayed edge, (c) edge with white lines, (d) low-loss plasmon like edges, taken in revised form from [4.50].

In this section we describe the particular ionization edges that are assessible in the range of most EELS spectrometers (within 0–2000 eV) and usable for practical microanalysis. All these edges have been compiled in the Gatan EELS atlas [4.51]. We also compare experimental and calculated ionization cross-sections and integrated oscillator strengths.

4.3.1 The Background to the Inner-Shell Ionizations

The ionization edges are superimposed on a downward-sloping background, which can arise from valence electron scattering, from other ionization edges and, for thicker specimens, also from plural scattering. This background has to be subtracted when carrying out quantitative analysis of ionization edges. Since the background intensity is often comparable to or even larger than the core-edge intensity (e.g. in the low-loss region), accurate subtraction of the background is essential. Generally, the background can be modelled quite well by using least square fitting of the $A \cdot E^{-r}$ function (4.8) [4.12,30] although many other methods have been proposed, especially for the low-loss region, as discussed in Sect. 5.2.1.

4.3.2 K Ionizations

EELS is especially useful because it can detect the light elements using the K ionization edges. Generally, these edges exhibit a saw-tooth profile with a sharp rise at the threshold followed by a slow decay. In experimental spectra, this is often complicated by fine structure effects, although the overall edge-shape does not vary too much (Fig. 4.5). The K excitations can be relatively easily described by hydrogenic or Hartree–Slater wave functions. However, differences between the cross-sections calculated by these two models have been observed, which can extend up to 15% or 20% [4.38].

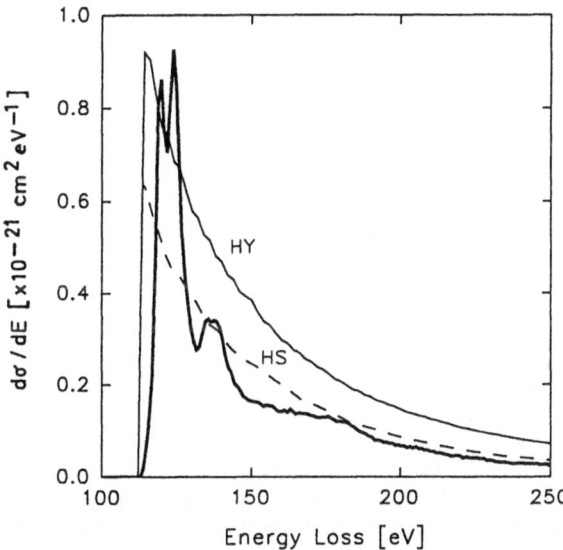

Fig. 4.5. The K edge of beryllium from the compound BeO (background subtracted). The experimental edge profile is presented as the differential cross-section, which is compared with the hydrogenic (Hy) and with the Hartree-Slater (HS) values; normalized relative to the K edge of oxygen [4.55].

H and He: The K edges of these elements have been measured with EELS only in gaseous samples [4.51]. Hydrogen can be indirectly detected in transition metal hydrides via an energy shift of the plasmon excitation [4.52,53]. In the case of He implanted into solids, weak peaks in the energy-loss range 21–23 eV have been found, arising from atomic-like 1s → 2p transitions within the He atoms [4.54].

Li – Si: EELS is a powerful method for the detection and quantitative analysis of the elements ranging from Li to Si. These edges are very well known

[4.51] and therefore routinely used. Figure 4.5 shows the Be K edge, which is a typical example for the K edges. In order to make comparisons with the predictions of the models, the experimental Be K edge is represented in terms of a differential cross-section (conversion by a modified k-factor approach [4.55]). Although the atomic theories do not take into account the fine structures at the edge threshold (solid state effects), they nevertheless describe the edge profile quite well. While the Hartree–Slater model shows good agreement with the experimental edge, the hydrogenic profile gives too high values, especially at the edge threshold, which is in agreement with previous observations [4.3].

P – Cl: Since the K edges of these elements lie in the energy-loss range above 2000 eV, they exhibit very low ionization cross-sections. These edges have been only rarely been used in EELS microanalysis, therefore.

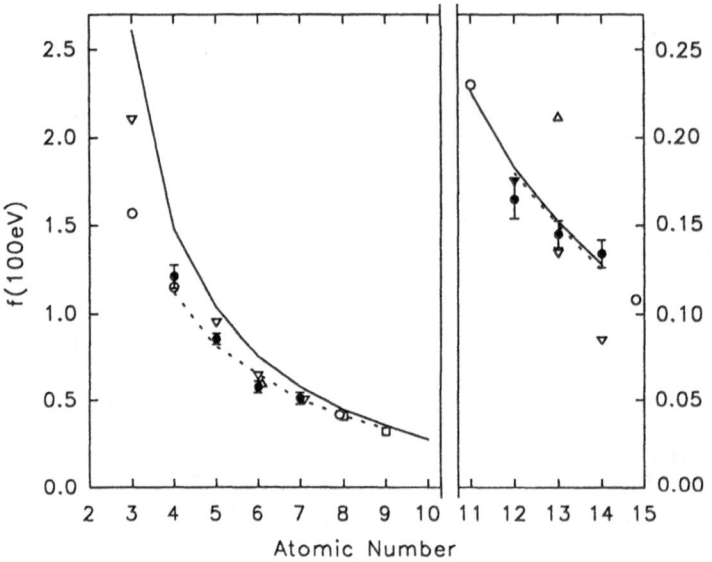

Fig. 4.6. Dipole oscillator strengths $f(100eV)$ for K transitions deduced from calculations and energy-loss experiments. Calculated values: —— Hydrogenic values (SIGMAK); – – – Hartree–Slater values. Experimental values: • *Hofer* et al.[4.38,60]; ▽ *Malis* and *Titchmarsh* [4.22]; △ *Crozier* [4.35]; photoabsorption: ○ *Henke* et al.[4.56]; □ *Saloman* et al.[4.57].

For efficient comparison of experimental and calculated k-factors and cross-sections, all EELS and photoabsorption data have been converted to $f(100eV)$ values; these are presented in Fig. 4.6 [4.49]. In the case of the elements B to Si, all the oscillator strengths for 100 eV do not differ by more than 10%, thus demonstrating that the use of the calculated cross-sections for these elements is reasonable. The Hartree–Slater values show even better agreement with the experimental ones in case of the elements Li and

Be. Since the experimental data have been normalized to the Hartree–Slater cross-section for the oxygen K edge, the differences between the experimental and the hydrogenic data are significantly emphasized. We have to state that a comparison of experimental k-factors with hydrogenic k-factors would yield a closer agreement [4.38]. Furthermore, it is worth mentioning that the photoabsorption data for K edges [4.56,57] agree quite well with the Hartree–Slater values and with the experimental EELS findings. However, if energy windows substantially smaller than 100 eV are to be used, quantification errors will increase because the theories cannot predict accurately enough the detailed edge profiles of all edges (Fig. 4.5). There will then be a clear advantage in using experimentally determined k-factors.

4.3.3 L_{23} Ionizations

For elements heavier than Al or Si, it is better to use the 2p excitations for microanalysis and the resulting L_{23} edges are routinely used for the detection of the elements Al to Zn. While the edge shapes for the K edges have much the same form for all elements, the edges for excitation of the 2p subshell show an interesting variation with Z. In general, the L_{23} edges are dominated by a "centrifugal barrier" effect, which leads to delayed edge maxima [4.58]. The 2s transition (L_1 edge) is also visible in the spectra but is much smaller than the 2p transition and is therefore not important for microanalysis.

Na – Ar: The L_{23} edges of the elements Na to Ar exhibit delayed maxima, which are typically 10–15 eV above threshold (Fig. 4.7a). The L edges of Na and Mg are difficult to use owing to the vicinity of the plasmon peak [4.59] and calculated cross-sections are not available. However, the elements ranging from Al to Ar are readily detectable in routine microanalysis using the L_{23} edges, which can be described by both the hydrogenic and Hartree–Slater model. In the case of insulating materials (e.g. oxides), the rounded L edges of these elements are sharpened at the edge threshold as can be seen for the Si L_{23} edge of SiO_2 in Fig. 4.7a [4.51,60]. In addition, the experimental Si L_{23} edge of Fig. 4.7a is compared with the calculated edge profiles, demonstrating that the theoretical models fit quite well.

K – Cu: In the case of the elements K to Cu, excitations of the 2p electrons are possible not only to the continuum but also to unoccupied bound d states. The transition to these states gives sharp lines at the threshold, often called "white lines" because of their appearance on photographic plates. There are two white lines due to spin orbit splitting from the two 2p subshells. Figure 4.7b shows the L_{23} edge of Ni in NiO, which is a typical example of such an edge consisting of both white lines and a continuum part. The ratio between the L_3 ($2p_{3/2}$) and the L_2 ($2p_{1/2}$) lines is generally not the 2:1 ratio expected from the occupation number of the inner shell states. In some cases, such as Mn, Cr and Fe, the ratio varies considerably according to the ionic charge (see Sect. 4.4.1). The L_{23} edges of these elements can be calculated by both models. This is shown for the L_{23} edge of Ni from the compound NiO (Fig. 4.7b).

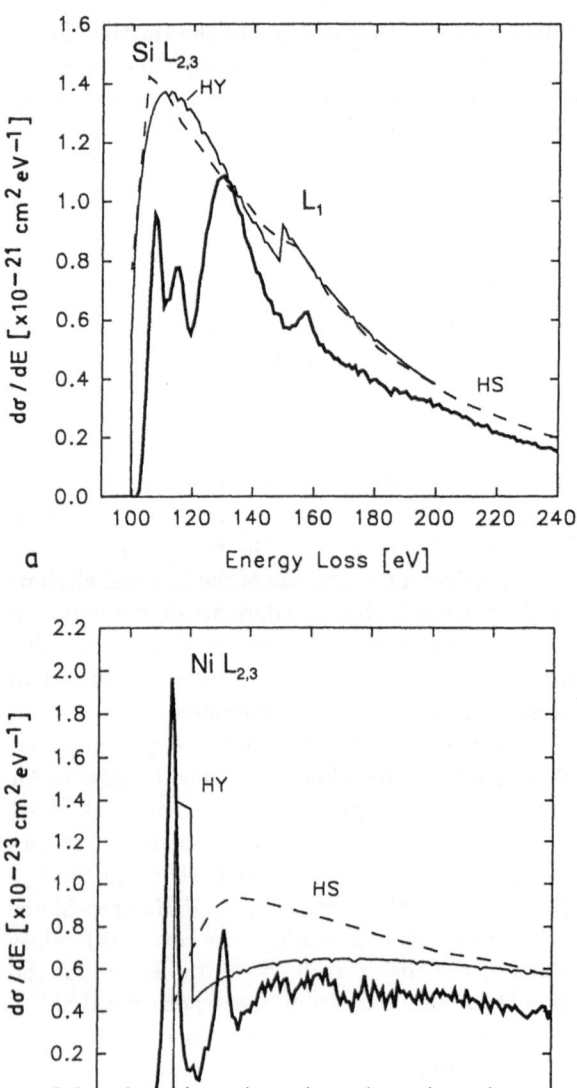

Fig. 4.7. The L_{23} edges of **(a)** silicon from the compound SiO_2 and of **(b)** nickel from the compound NiO. The experimental edge profiles are presented as differential cross-sections [4.55] which are compared with the hydrogenic (Hy) and with the Hartree–Slater (HS) values.

While the SIGMAL2 program takes partial account of the white lines, the Hartree–Slater theory cannot describe the near-edge region. Towards higher energy losses (some 10 eV above threshold) the Hartree–Slater model gives cross-sections that are appreciably higher than the experimental ones. This can lead to compensation of the white-line deficiency and hence to improved agreement of the partial cross-sections as they are obtained by integration over larger energy regions [4.61]. The intensity of the white lines decreases from K to Cu when viewed relative to the continuum part of the edges. The L_{23} edge of Cu is very interesting because in metallic copper the 3d band is filled and lies just below the Fermi level. As a consequence, the strong 2p–3d transitions do not occur. However, in Cu oxides, white lines can be found because electron transfer between Cu and O produces unfilled d states and the Fermi level is lowered into the Cu d band [4.62].

Cu – Br: For Cu metal and for the elements Zn to Br the d shell is filled and there are no white-line transitions. Therefore these edges exhibit only round delayed maxima which are caused by transitions to continuum states. They look similar to the L_{23} edges of the elements Na to Ar. While the hydrogenic model (SIGMAL2) only allows the L_{23} edges of Cu and Zn to be calculated, the Hartree–Slater model can be used for all the elements ranging from Cu to Br.

Rb – Pd: These edges exhibit white lines comparable to those of the first-row transition elements, which are caused by the excitation of 2p electrons to empty 4d states [4.51]. Although these edges exhibit a good signal-to-background ratio they have seldom been used in EELS analysis owing to the low count-rates in this spectral region (above 2000 eV) [4.63,64]. With the exception of the Y L_{23} edge [4.64], experimental or calculated cross-sections are not available.

For purposes of comparison, the partial cross-sections and k-factors have again been converted to $f(100\,\mathrm{eV})$ values, which are presented in Fig. 4.8. The experimental values for the elements S to Co agree quite well with the calculated values, but disagree for the elements Al, Si and Ni to Zn. In the case of Ni, Cu and Zn, however, the Hartree–Slater values deviate by more than 20% from the experimental and hydrogenic values. One explanation for this result is that the hydrogenic model (SIGMAL2) has been fitted to experimentally determined data and therefore white lines are included to some extent, which is not the case for the Hartree–Slater model. As in the case of the K edges, the experimental data are normalized to the Hartree–Slater oxygen K edge and hence the differences between experimental and hydrogenic values are emphasized. However, it has been found that the experimental k-factors of the elements Sc to Zn agree closely with the hydrogenic cross-section ratios [4.38].

4.3.4 M_{45} Ionizations

The M_{45} edges (3d shell) enable the elements Rb to Au to be detected. They exhibit complicated edge shapes, which can only dealt with by complicated theories and are not so strongly influenced by solid state effects as the K or L edges [4.65]. On the other hand, the cross-sections of these ionizations are very large and thus very suitable for microanalysis.

Rb – I: These edges show many of the same features that are seen with the filled d-shell L_{23} edges in the preceding two rows. For Rb to I, there is a marked delayed maximum caused by dominant transitions from 3d states to

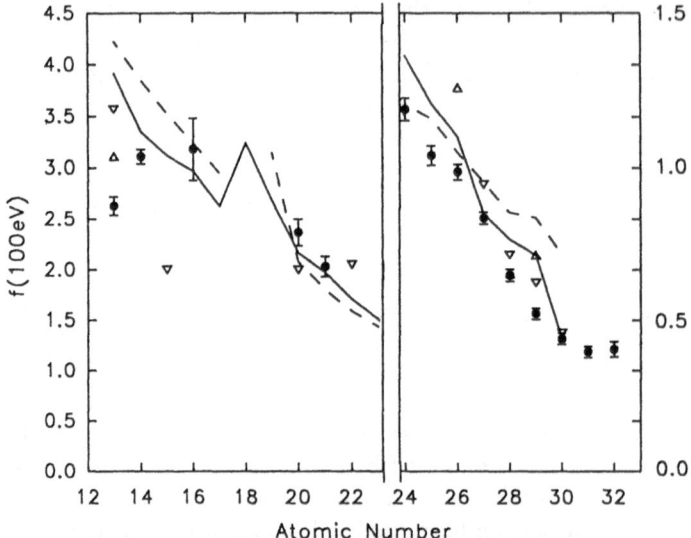

Fig. 4.8. Dipole oscillator strengths f(100eV) for L_{23} transitions deduced from calculations and energy-loss experiments. Calculated values: —— Hydrogenic values (SIGMAL2); – – – Hartree–Slater values [4.61]. Experimental values: • *Hofer* et al.[4.38,60]; ▽ *Malis* and *Titchmarsh* [4.22]; △ *Crozier* [4.35].

f-like continuum states. It can be seen from Fig. 4.9a for Zr that the maximum can occur as much as 60 to 80 eV beyond the threshold energy. On the M_{45} are superimposed the much less intense M_{23} (3p) and M_1 (3s) excitations. On examining the profile of the Zr M_{45} edge of Fig. 4.9a, we see large differences betweeen the experimental and calculated data. While the Hartree–Slater theory overestimates the cross-section, especially at thresholds, the hydrogenic model gives values that are too low over the entire energy range. For large momentum transfers, non-dipole transitions can make substantial contributions to the M_{45} edges of the elements Rb to Mo. If these spectra are recorded with large collection angles (e.g. 100 mrad), a sharp peak occures at the threshold, which has been attributed to monopole transitions to the 4d band [4.66].

Cs – Yb: In the rare earths and Cs and Ba, the f-shell becomes bound and there are possibilities of white-line transitions to unoccupied f states. Just as in the case of the L_3/L_2 ratios in the transitions metals, the ratio of the $3d_{5/2}$ (M_5) to the $3d_{3/2}$ (M_4) transitions does not exhibit the expected 3:2 ratio for the occupations of the d shell [4.67]. Figure 4.9b shows the M_{45} edge for La, where we can see the intense white lines at threshold, followed by the continuum part of the edge. The models overestimate the continuum portion slightly but the intense white lines are not predicted. For rare earths with higher atomic number, the white-line components are less important than the transitions to the continuum [4.55,58]. For Yb, the f shell is almost

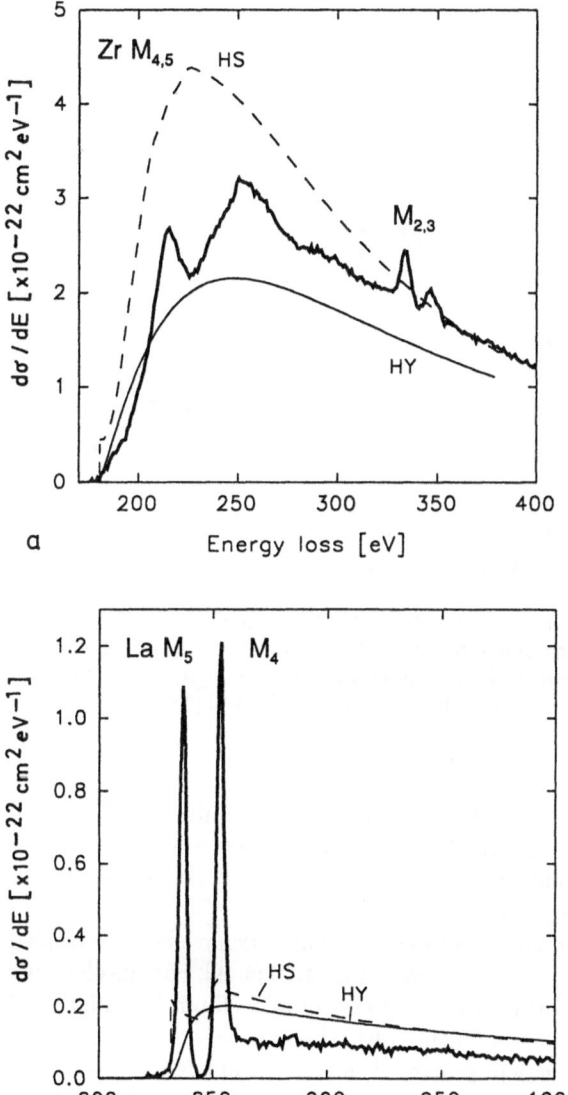

Fig. 4.9. The M_{45} edges of (a) zirconium from the compound ZrO_2 and of (b) lanthanum from La_2O_3. The experimental edge profiles are presented as differential cross-sections [4.25], which are compared with the hydrogenic (Hy) and with the Hartree–Slater (HS) values.

filled and the intensity of the second white line in the vicinity of the M_4 edge is weak. If the spectra are recorded with large momentum transfer (e.g. 100 mrad), non-dipole transitions can also be observed: sharp M_2 and M_3 peaks are then visible, representing $\Delta l = 2$ transitions from the 3p to unfilled 4f states [4.51]. The M_{45} edges of these elements are very well suited for elemental microanalysis since the intense white lines give very low detection limits for the rare earths.

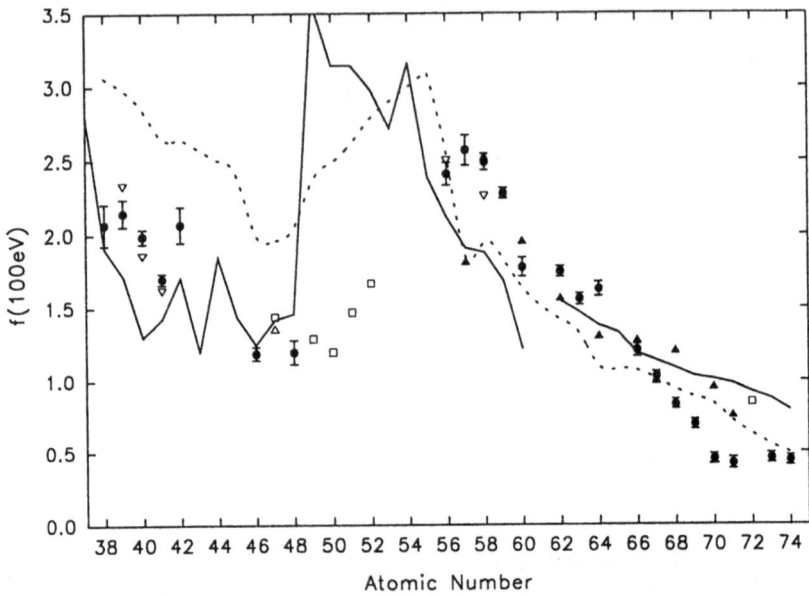

Fig. 4.10. Dipole oscillator strengths $f(100\text{eV})$ for M_{45} transitions deduced from calculations and energy-loss experiments. Calculated values: —— Hydrogenic values [23]; – – – Hartree–Slater values [4.61] Experimental values: • *Hofer* et al. [4.43,60]; ▽ *Chadwick* and *Malis* [4.64]; △ *Crozier* [4.35]; ▲ *Manoubi* et al.[4.67]; □ *Yang* and *Egerton* [4.68].

Lu – Au: For Yb and higher-Z elements, the f shell is completely filled and therefore white lines cannot occur. The M_{45} edges of these elements again show delayed maxima similar to the second row transition elements thus making them harder to recognize.

Calculated and measured cross-sections and k-factors have been converted to $f(100 \text{ eV})$ values, which are represented in Fig. 4.10 [4.49]. For the elements Sr to Te (Z=38–52), the experimental data agree quite well but the calculated ones deviate significantly. For the lanthanides, the experimental values from different workers are more widely scattered; and the models underestimate the f values of the elements Ba to Ho (Z=56–67) because these edges exhibit very intense white lines at the threshold, which are not included in the model. For Er (Z=68) and heavier elements the agreement between the theoretical and experimental f values is less good.

4.3.5 M_{23} Ionizations

In the case of the first row transition elements (K to Zn), two kinds of ionization edges can be found within the range of EELS spectrometers. Besides the L_{23} edges, these elements give rise to M_{23} edges between 30 and 100 eV. The excitation of 3p electrons results in plasmon-like peaks, which are su-

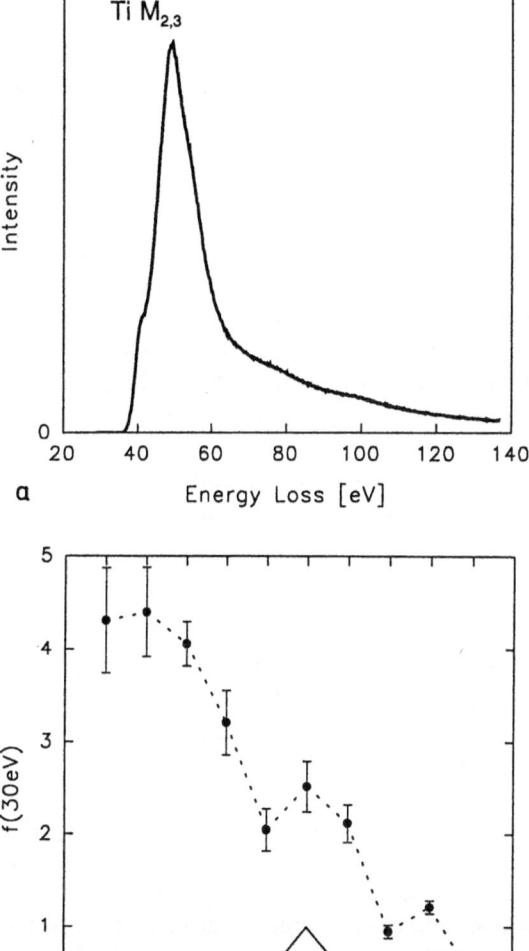

Fig. 4.11. M_{23} edges: (a) The M_{23} edges of titanium from the compound TiO_2 after background subtraction; (b) dipole oscillator strengths $f(30eV)$ for M_{23} ionizations deduced from calculations and energy-loss experiments. Calculated values: —— Hydrogenic model [55]; – – – Hartree–Slater model. Experimental values: • *Wilhelm* and *Hofer* [4.45].

perimposed on the rapidly falling valence-electron background (Fig. 4.11a). Background subtraction is therefore difficult and another complication is that their shapes are often influenced by a Fano resonance, which is an interaction between bound states and continuum. This can give a slight dip before the threshold, thereby further complicating background subtraction [4.58,69].

Although these ionizations were already investigated during the seventies using XAS [4.70,71] and EELS [4.31,72], they have only rarely been used in microanalysis [4.73,74]. M_{23} edges can exhibit considerable fine structural details, which have been observed in investigations on Cr compounds [4.75] and Mn oxides [4.76].

A comparison of the M_{23} and L_{23} edges of the first row transition elements has revealed that the advantage of the M_{23} edges provided by the larger intensities is offset by the worse peak-to-background ratios. The sensitivity of the M_{23} edges is comparable to that of the L_{23} edges [4.75]. The M_{23} edges are therefore interesting for creating element distribution images of the first-row transition metals, as has been demonstrated recently [4.77].

Although calculated partial cross-sections and oscillator strengths are available for the 3p excitations [4.8,23] these values are not expected to be very accurate. This can be seen from Fig.4.11b where calculated oscillator strengths $f(30eV)$ are compared with measured values [4.45]. The experimental values of $f(30eV)$ approach the number of electrons in the M_{23} subshell (6), which one would expect if the f sum rule were applied to the subshell. This emphasizes the reliability of the experimental data [4.78] which are for use in quantitative microanalysis provided that we remember that the spectra should always be deconvoluted to improve the background below the M_{23} edges.

4.3.6 N_{45} Ionizations

For the elements Cs, Ba and the lanthanides and in the energy-loss range 80 to 200 eV, a family of rather characteristic profiles, which can be attributed to the excitation of the $4d_{3/2}$ and $4d_{5/2}$ electrons to f states, is observed. These edges, which have seldom been used in EELS microanalysis, were characterized in the seventies by using EELS [4.79] and XAS [4.80,81]. The N_{45} edges show many of the features of the M_{23} edges: they are characterized by the existence of one main resonant absorption, which lies typically 10 or 20 eV above the edge onset (Fig. 4.12a). Moreover, they display a variety of fine structures, which do not follow a simple rule: discrete multiplet peaks occur at the threshold but these are only visible as small humps in EELS spectra recorded in electron microscopic systems because of the limited energy resolution. Fano resonances may also be found [4.69]. However, these edges present less difficulty with regard to background subtraction than M_{23} edges. An atomic model developed by Dehmer [4.82] and Sugar [4.83] provides a description of these edges, which considers transitions of the type $4d^{10}4f^n \rightarrow 4d^9\,4f^{n+1}$ in triply ionized atoms. Since the f shell is bound rather tightly to the core of the rare earth atoms, the f state is screened from most solid state effects. Therefore the edges depend little on whether one analyses an oxide or the metal. The effects of the Fano resonance, which depresses the intensity in the vicinity of the threshold, can be seen most strongly for the elements from Gd onwards. As we move over to Lu the effects of this resonance diminish, as the f shell is now full.

Since the N_{45} edges are very useful for microanalysis of the elements ranging from Ba to Tm, both calculated and experimental oscillator strengths

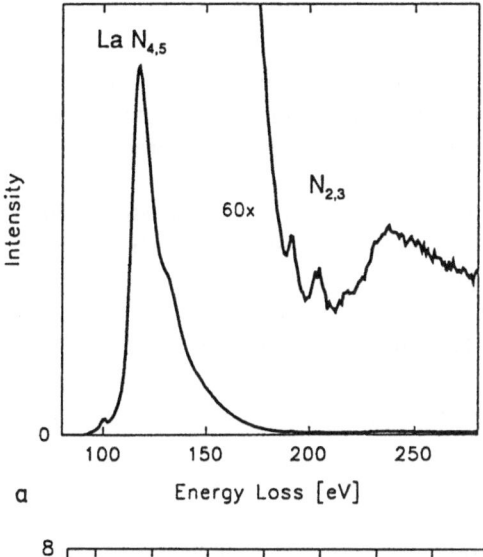

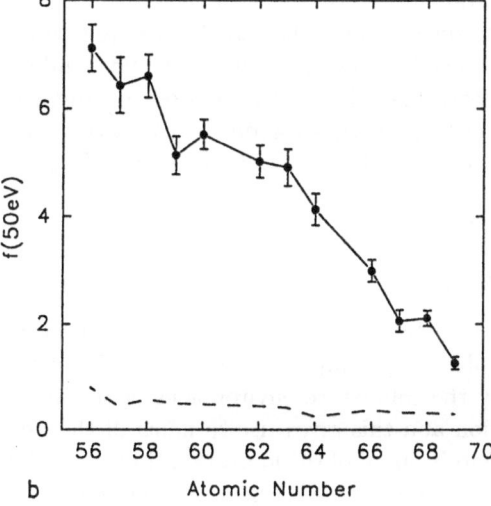

Fig. 4.12. N_{45} edges: (a) The N_{45} edge of lanthanum from the compound La_2O_3 after background subtraction; (b) dipole oscillator strengths $f(50eV)$ for N_{45} ionizations deduced from calculations and energy-loss experiments: – – – Hartree–Slater values; • *Hofer* [4.44].

[4.44] are presented in Fig.4.12b. Again the Hartree–Slater model predicts much lower values. This situation is believed to arise because one-electron theories can predict neither the detailed edge shapes nor the intensities of the N_{45} ionizations [4.58]. The experimental k-factors for the N_{45} edges have been successfully applied to the quantification of rare-earth compounds, as demonstrated recently [4.38].

For elements heavier than the rare earths, the N_{67} or 4f excitations have very large oscillator strengths [4.58] but they lie at very low energies and are dominated by multiple scattering.

4.3.7 O_{45} Ionizations

For Th and U the O_{45} or 5d excitation provides a useful means of identification and analysis because these edges are prominent as a double peak structure between 90 and 150 eV. The O_{45} edges have been characterized by both EELS [4.51,84] and XAS [4.85] and may be described by spin orbit and exchange splitting of the $5d^9 5f^1$ and $5d^9 5f^3$ final states. As the edges exhibit almost no fine structural details, the spectra have been interpreted in much the same atomic way as the N_{45} edges [4.85]. The partial cross-sections may be calculated by means of the Hartree–Slater method. It has recently been shown that the detection limit for Th and U using the O_{45} edges can be as little as one atom [4.86].

4.4 Edge Fine Structures

Edge fine structures arouse considerable interest in the application of EELS, especially in the field of materials science, because they can be used to extract information regarding local charge distributions, coordinations and bonding characteristics from nanometre areas. The edge fine structures are divided into the near-edge fine structures (ELNES) within about 50 eV of the edge onset and the extended fine structure (EXELFS) about some 100 eV above the edge onset (Fig. 4.2).

4.4.1 Near-Edge Fine Structures

Most of the edges that lie in the accessible energy-loss range contain a more complex structure than can be explained in simple atomic terms. The core-loss spectrum may be modified by the solid-state environment of the atom undergoing the inner-shell excitation and this gives rise to much of the fine structure that lies within approximately 50 eV of the ionization threshold, the so called energy-loss near-edge structure (ELNES). Since the edge intensity is proportional to the differential cross-section (4.9), it is possible to write the single scattering distribution of an inner shell edge as

$$\frac{d^2\sigma}{d\Omega dE} \propto |M(E)|^2 N(E) \tag{4.12}$$

where $M(E)$ is the atomic transition matrix governing the overlap between the initial and final state wave functions coupled by the dipole selection rule and $N(E)$ is the symmetry-projected unoccupied density of states (DOS) [4.87,88]. The function $M(E)$ represents the basic edge shape, which is primarily atomic; it is modulated by $N(E)$, which is dependent on the environment and bonding of the atom. Since $M(E)$ can be considered as a smoothly varying function of energy-loss, most of the near-edge structure therefore originates from changes in the DOS term. Thus the ELNES can be used to measure the energy distribution of unoccupied states directly. Equation (4.12) takes no account of the selection rules for inelastic scattering but for small collection angles the dipole approximation is quite acceptable. According to the dipole selection rule, a K excitation probes the unoccupied p-like DOS and an L_{23} excitation probes both the s-like and d-like unoccupied DOS. This fact can be put to use and to some extent it is possible to choose the symmetry of the final states by selecting a particular edge. Additionally, wherever possible, matrix element variation should always be included when comparing near-edge structures with the projected density of states. The finite lifetime of both the initial and final states, which will decay by de-excitation processes, can lead to a broadening of the near edge features (especially in case of metals). These arguments are based on a simple one-electron transition model, which neglects influences due to many body effects, notably relaxation due to the formation of a core hole, interaction between the excited inner-shell electron and the core hole as well as electron–electron interactions [4.89]. Core hole production may, especially in the case of insulators, give rise to a redistribution of the available unoccupied states, as well as to the formation of temporary bound states below the Fermi level, so-called core exciton levels [4.90].

Experimental Constraints. For ELNES studies an energy resolution of about 0.3 to 0.5 eV is desirable, which can only be obtained with microscopes operated with a field-emission gun and parallel EELS detection [4.94,95,96]. The accuracy of the energy scale is also very important (e.g. for identification of valencies) and sometimes more useful than very high energy resolution. In many practical applications of ELNES, LaB_6 sources and parallel-detection spectrometers are sufficient when used with optimized illumination conditions, as has been demonstrated for the ELNES of the carbon K edge (Fig. 4.13).

Interpretation of Near-Edge Structures. One approach to understanding near edge structures is to collect "fingerprints" from elements in similar compounds (or environments) and to try to identify common features (see fingerprints for the carbon K edge in Fig. 4.13) [4.97]. The other approach is to calculate the near-edge structures using a theory and to identify the features which show sensitivity to a property of interest. Various models for the interpretation of near-edge structures have been developed. In general, however, these methods neglect any effects due to the excitation process and for near-edge calculations these effects (e.g. core hole) have to be taken into account. Although the interpretation of near-edge structures is still a subject of debate, the technique has aroused considerable interest and many applications have already been described (e.g. [4.98,99]).

Band-Structure Calculations. The basic argument behind these methods is that the excited electron is probing the density of unoccupied states multiplied by the square of the matrix element. The calculations assume an

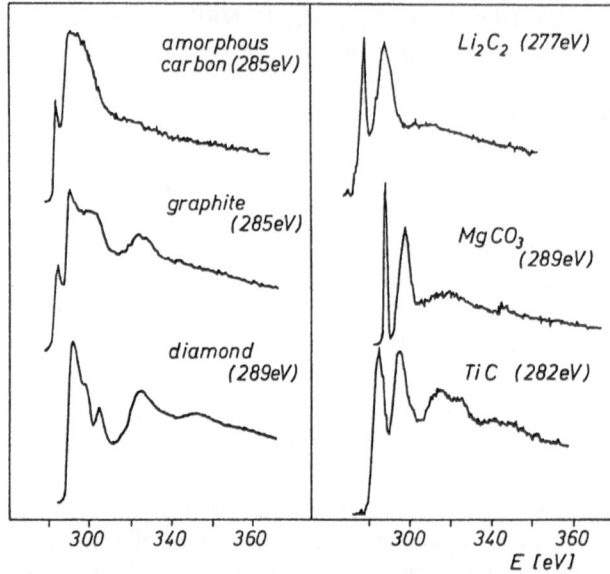

amorphous
carbon (285eV)

graphite
(285eV)

diamond
(289eV)

Li_2C_2 (277eV)

$MgCO_3$
(289eV)

TiC (282eV)

300   340   360   300   340   360
                                  E [eV]

Fig. 4.13. Near edge fine structures of the K edge of carbon in different carbon modifications and compounds (recorded with an energy resolution of 2 eV), from [4.97].

infinite periodic crystal lattice and are relatively complicated even for simple unit cells. Furthermore, previously published calculations do not include the higher unoccupied states as well as the site and symmetry projections necessary for comparison with experimental ionization edges. The unoccupied DOS needs to be resolved into the correct angular momentum component at the site of the excited atom, which can be successfully calculated by using the Augmented Plane Wave (APW) method [4.100]. This method has been used in the interpretation of near edge structures of transition metals and their carbides, oxides and nitrides [4.101,102]. Another method, which is better suited to describing the electronic properties of semiconductors, is the pseudopotential band theory. A new formulation of this method has been developed, which allows the various site and symmetry projected components of the total DOS to be determined [4.103]. This has been used with considerable success to interpret the ELNES of a range of semiconducting and ceramic materials [4.103–105]. A typical example of a comparison of calculation with experiment is given as Fig. 4.14a where the ELNES of the boron K edge in cubic boron nitride is presented [4.107].

Multiple-Scattering Calculations. The most common method, which has been extensively used in x-ray absorption, is the cluster calculation or XANES method [4.108], subsequently improved by Vvendensky [4.109]. This multiple-scattering (MS) theory is based on the interference between the outgoing electron wave of the ejected electron and the electron wave backscattered from the surrounding atoms. This, in turn, gives information about the relative positions of the neighbouring atoms (geometrical environment). A similar theory is applied to the interpretation of extended fine structures (see Sect. 4.4.2).

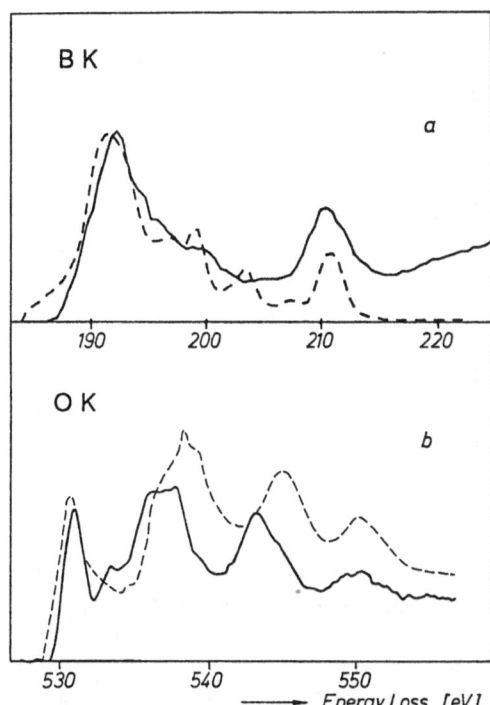

Fig. 4.14. Examples for the successful calculation of near-edge fine structures (solid line experimental results; dashed line calculated edge). **(a)** Boron K edge in cubic boron nitride, pseudo-atomic-orbital calculation; from [4.107]. **(b)** Oxygen K edge in $SrTiO_3$, ICXANES calculation with seven shells; from [4.126].

MS-calculations are performed in real space using a cluster approach. The multiple scattering within the cluster is then calculated using the ICXANES computer code of Vvendensky [4.109]. This method allows the cluster size (number of nearest neighbour shells) to be progressively increased and it is hence possible to identify the specific scatterers and scattering paths that give rise to various near-edge features. One advantage of this approach is that the core-hole effect may be modelled by the choice of a different potential for the central atom in the cluster. There have been attempts to incorporate the core hole using an atom of atomic number $Z+1$ for the excited atom [4.110] or by using special transition states [4.111]. This method has been very successful for systems with large unit cells (e.g. minerals such as rhodizite [4.112,113]) and for simple cubic oxides [4.114]. The ELNES of the oxygen K edges of titanium(IV) oxygen compounds could be modelled by MS-calculations for clusters consisting of up to seven shells surrounding the central oxygen atom [4.115] (see Fig. 4.14b).

Molecular Orbital Calculations. An alternative procedure is to look at the electronic structure in terms of bonds between atoms. In solids, ions or molecules interact with each other to form bands of orbitals, which extend throughout the structure. Therefore in many cases, the band structure of solids can be directly related to the molecular orbital structure (MO) of the individual ions or molecules. The MO-calculation of unoccupied orbitals can be performed by means of the SCF-Xα method [4.116], which has subsequently been refined [4.117]. This method has been used with considerable success for the interpretation of EELS spectra from gas phase molecules [4.118] and can be very useful for the interpretation of the ELNES of transition metal ions in solids [4.112,119].

Determination of Nearest-Neighbour Coordination. It has been found that the observed ELNES can exhibit a structure specific to the arrangement and type of atoms in the first coordination shell - a so-called "coordination fingerprint" [4.60,120]. This can be seen when the local DOS is dominated by the interaction of the central atom with the first coordination shell. In complex structures with large, complicated unit cells, there exists very little order outside the nearest neighbours and the ELNES of the central atom/ion is then dominated by the first coordination shell. Additionally, certain atoms/ions such as the O^{2-} ion are strong backscatterers [4.110] and where such species occur in the first coordination sphere this will give rise to a potential cage or barrier for electron scattering [4.121]. The mean free path of the excited electron is reduced at kinetic energies above 10 eV and this, together with the finite lifetime of the core hole created during the excitation, limits the size of the cluster probed by the excited electron. Furthermore, in insulators the attraction due to the core hole causes a localization of the unoccupied DOS and the electronic structure in this region may then be described in a simple MO-picture [4.89]. In such cases, the interpretation of coordination fingerprints is most readily achieved via MO-calculations and MS-calculations on clusters consisting of the central atom/ion and the first coordination sphere. Since the nature of the final state MOs are essentially governed by the symmetry of the polyhedral cluster and the number of valence electrons, isoelectronic species which possess the same symmetry exhibit very similar ELNES (e.g. the P, S, Cl L_{23} edges of PO_4^{3-} , SO_4^{2-} , ClO_4^- [4.60,122] or the C and N K edges of CO_3^{2-} and NO_3^- [4.123] (see Fig. 4.13). Experimentally, coordination fingerprints may be determined by the comparison of spectra obtained from a range of reference materials possessing the same coordination of a particular atom. Measurements will also be possible in the case of amorphous compounds since the lack of long-range order will minimize the contribution from outerlying shells. Finally, it has to be said that not all ionization edges provide coordination fingerprints [4.124]: spectra originating from deeper core levels (e.g. K or L edges) generally exhibit more characteristic differences in near-edge structures between different coordinations than those from higher lying initial states (e.g. M or N edges), which have a larger overlap with continuum wave functions.

Many examples of coordination fingerprints have been published, including both cation and anion fingerprints determined by XAS and EELS [4.125,126]. In the following we present some typical examples of EELS coordination fingerprints.

Several EELS [4.113,127] and XAS [4.128,129] measurements on boron compounds have revealed the existence of a boron coordination fingerprint, which permits us to distinguish between boron tetrahedrally coordinated by oxygen (BO_4 unit) and boron in trigonal planar oxygen coordination (BO_3 unit). Figure 4.15 shows the boron K edges measured from the borate minerals vonsenite and rhodizite. In rhodizite, boron is in tetrahedral coordination to oxygen while in vonsenite (Fe_3BO_5), the coordination is trigonal planar. The remarkable differences arise from the way the MO's containing boron 2p character are constructed in the two different symmetries [4.127]. The experimental findings have been confirmed by the results of multiple scattering calculations for BO_3 and BO_4 clusters [4.113].

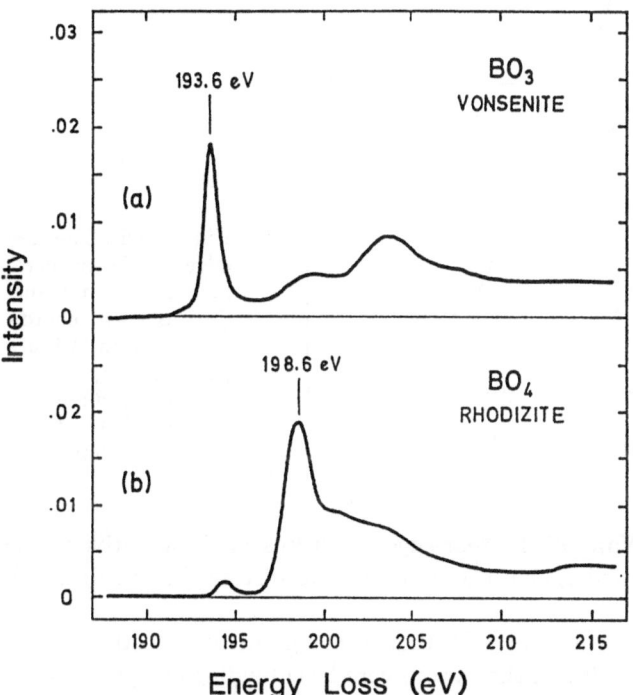

Fig. 4.15. The K edge of boron in triangular–planar and tetrahedral oxygen coordination: (a) vonsenite with [BO_3]-units; (b) rhodizite with [BO_4]-units, from [4.151].

In the case of the 3d transition elements, the L_{23} edges also show interesting coordination fingerprints. Much work has already be done on the high-energy K edges using XAS [4.130–132]. The L_{23} ELNES reflect the crystal field splitting of the metal 3d band [4.133] and this is considerably different in the octahedral and tetrahedral coordinations [4.134]. In Fig. 4.16 we show the titanium L_{23} edges for Ba_2TiO_4 (tetrahedral Ti) and $BaTiO_3$ (octahedral Ti), which are distinctly different and offer a means of distinguishing between the two coordinations [4.123,126]. Other interesting examples have been found in the case of the L_{23} edges of Al and Si [4.123,135,136] and in principle, coordination fingerprinting should be extendable to other 3d transition metal L_{23} edges.

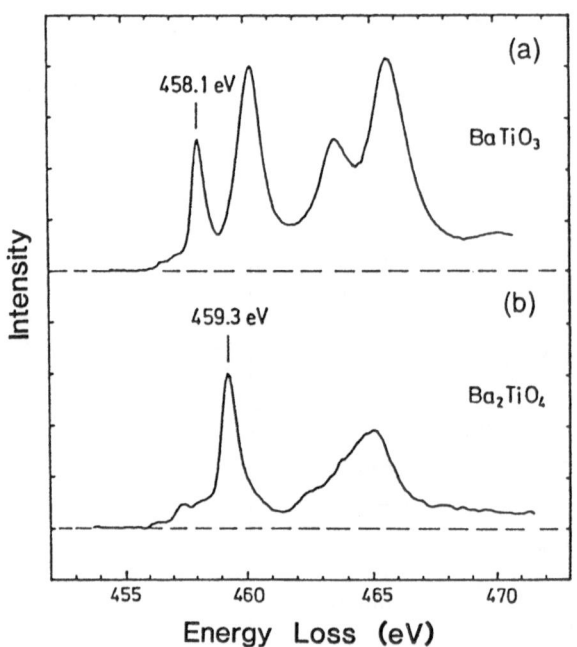

Fig. 4.16. The L_{23} edge of Ti in octahedral and tetrahedral oxygen coordination (from [4.125]). (a) $BaTiO_3$ (octahedral coordination); (b) Ba_2TiO_4 (tetrahedral coordination).

Determination of Valencies. Near-edge structures can be sensitive to the valency of the atom undergoing excitation in two distinct but intrinsically related ways:

First, it is well known from XPS that a change in the effective charge on an atom results in shifts of the various core-level binding energies [4.137]. The position of the EELS edge threshold corresponds to the difference in energy between the initial and final states of the transition. Since the energies of these states will be influenced by the charge state of the atom undergoing excitation, measurement of the threshold energy provides a means of determining charge distributions by EELS. These effects are described as a chemical shift of the edge [4.50] and are best observable when going from a metal to an insulator where the presence of a band gap will lead to a shift of the edge to higher energy loss in the case of the insulator. Such a chemical shift can be easily observed for the K edges of the light elements and for the L_{23} edges of the elements ranging from Mg to Cl. A typical example is shown in Fig. 4.17. The chemical shift of a first peak of the Si L_{23} ELNES of various amorphous Si compounds is linearly correlated to the electronegativity of the ligand and this gives a measure of the charge transfer from the silicon to the ligand [4.138]. Since in EELS occupied and unoccupied states are involved, the interpretation of chemical shifts is more complicated than in XPS, where only occupied states are probed. Chemical shifts have been documented in both EELS and XAS: EELS data have been collected recently by Brydson [4.125], while a survey of the XAS data has been prepared by Mande and Sapre [4.139].

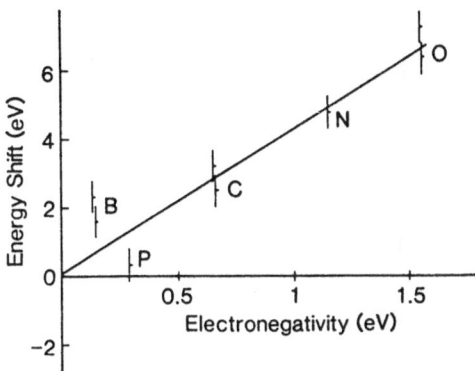

Fig. 4.17. Chemical shift of the first peak in the Si L_{23} ELNES of amorphous silicon alloys (relative to amorphous Si) as a function of the Pauling electronegativity of the ligand atom; from [4.138].

Second, the valence and, in some cases, the spin state of the excited atom can affect the intensity distribution in the near-edge structure. This often occurs in edges which are dominated by quasiatomic transitions; the white lines are an example. As already mentioned, EELS and XAS measurements of the L_{23} edges of the 3d transition elements showed that the ratio of the two spin orbit components from the $2p_{3/2}$ and $2p_{1/2}$ transitions did not follow the statistical 2:1 ratio expected from the ratio of the initial states [4.62,140,141]. A similar disagreement is observed for the M_{45} edges of the rare earths, where statistically the intensity ratio of the M_{45} ($3d_{5/2}$) and M_4 ($3d_{3/2}$) white lines should be 3:2 [4.65,67,142]. There are two methods of characterizing the valence of the excited atom:

1. The intensity ratios of the white lines can be used to determine the number of electrons in the final-state d or f band and whether they are in a high or low spin ground state. For the 3d transition metals and compounds, the L_3/L_2 ratio depends on the 3d state occupancy, although not in a simple fashion (Fig. 4.18a) [4.131]. For the rare earth metals and compounds, the M_5/M_4 ratio increases with atomic number (Fig. 4.18b)[4.67,142].

2. It has been shown that the sum of the L_3 and L_2 white-line intensities normalized to the continuum contribution does vary linearly with the number of d electrons [4.143]. The normalized sum of the L_{23} edges of the 4d transition metals exhibits a more complex behaviour, although it seems possible to use

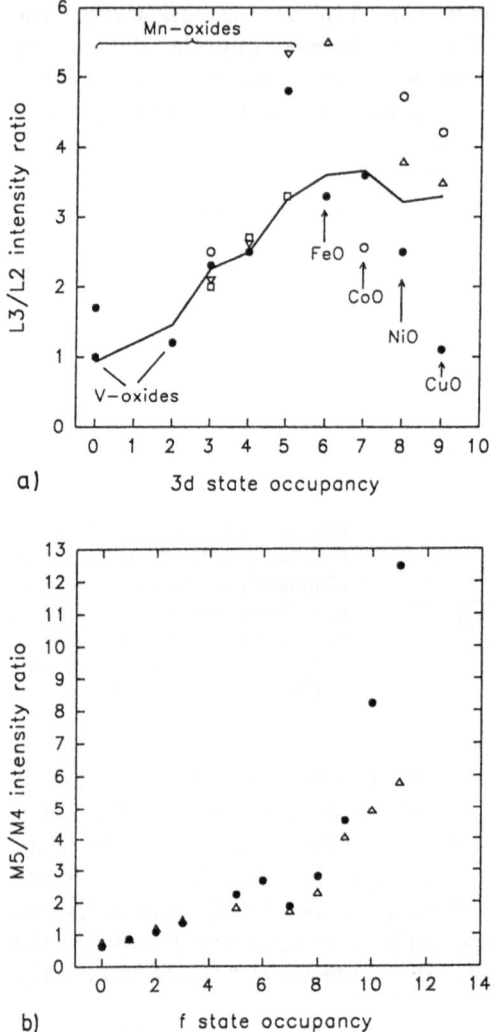

Fig. 4.18. White line intensity ratios for (a) L_{23} edges (L_3/L_2) and (b) M_{45} edges (M_5/M_4). (a) —— calculated and experimental (o *Waddington* et al. [4.145]; • *Sparrow* et al. [4.140]; △ *Leapman* et al. [4.62]; ▽ *Rask* and *Miner* [4.155]; *Paterson* and *Krivanek* [4.67]. (b) *Thole* et al. [4.65]; △ *Manoubi* et al. [4.67]).

it to study charge transfers for the second half of the 4d series where the relationship is linear [4.144].

Experimentally determined white-line ratios have been tabulated both for the 3d transition metal ions [4.145,146] and for the rare earths [4.65] and can be used to measure the valency of these elements.

Furthermore, it should be mentioned that the white lines can also provide information about the magnetic properties of the specimen [4.147] because the ratio of the L_3 and L_2 white lines is related to the local magnetic moment [4.148,149]. Another possibility to fingerprint the valency of a particular transition metal uses the fact that the detailed L_{23} white-line shapes observed at an energy resolution of about 0.5 eV depend specifically on the number of d electrons. The spectra are dominated by quasiatomic transitions from a $2p^6 3d^n$ initial state to $2p^5 3d^{n+1}$ final state, which are caused by the strong effect of the core-hole potential on the 3d band. These fine-structure features have been modelled theoretically using atomic multiplet theory including the crystal field caused by the ligands [4.150]. The M_{45} edges of the rare earth ions show similar fine structure, thus again allowing valency determination [4.65,151].

Determination of Bond Lengths and Bond Angles. Small variations in coordination fingerprints can often be related to differing degrees of departure from perfect coordination (i.e. variations in bond lengths and bond angles). Comparison of experimentally observed spectra with those derived from standards can yield an estimate of the distortion [4.125]. This has been demonstrated for the Ti L_{23} ELNES from a series of titanium (IV) oxides, where the broadening and splitting of near-edge features increased as the Ti site symmetry was lowered [4.152,153].

According to scattering theory, we may identify an edge region ranging from about 20 to 40 eV above threshold where multiple scattering in the continuum occurs giving rise to multiple scattering resonances (MSR), which can be viewed as quasibound states embedded in the continuum [4.89]. The energy positions of these MSR (relative to the edge onset) are strongly dependent on interatomic distances; they vary as $1/R^2$, where R is the bond length [4.154]. Although the shape resonances may permit the corresponding bond lengths to be determined semi-quantitatively [4.85], this has not been used in EELS so far.

Orientation Dependence of ELNES-Features. In crystalline material, the near edge is capable of providing additional information about chemical bonding. For example, in layered materials, the fine-structural features depend on momentum transfer and hence can be used to investigate anisotropy in the chemical bonding and band structure [4.91,92]. A typical example is the carbon K edge of graphite (Fig. 4.19). Here, the spectra have been recorded from a graphite sample oriented with the c-axis parallel to the incident electrons for different collection angles. When θ is small, q' is along the c-axis and the first peak is strongly enhanced. For large collection angles the first peak becomes weaker and the second one becomes dominant. We are thus able to identify the first peak as a $1s \rightarrow \pi^*$ transition and the second one as $1s \rightarrow \sigma^*$ transition, which is in agreement with previous band structure calculations. Similar effects on the ELNES can be obtained by varying the crystal orientation with respect to the incident beam while keeping the collection angle fixed [4.93].

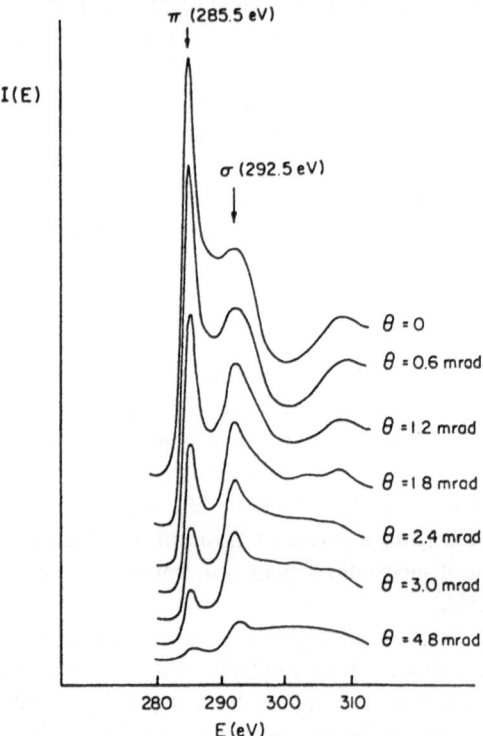

$I(E)$

π (285.5 eV)

σ (292.5 eV)

$\theta = 0$

$\theta = 0.6$ mrad

$\theta = 1.2$ mrad

$\theta = 1.8$ mrad

$\theta = 2.4$ mrad

$\theta = 3.0$ mrad

$\theta = 4.8$ mrad

280   290   300   310

E (eV)

Fig. 4.19. Dependence of fine structure features on momentum transfer: carbon K edge in graphite recorded at different collection angles and with the c-axis of graphite parallel to the incident beam (from [4.92]).

4.4.2 Extended Energy-Loss Fine Structures

Beyond the near-edge structure, extending over several hundred electronvolts, is the extended electron energy-loss fine structure (EXELFS, see Fig. 4.2), which can be analysed in the same way as the extended fine structures beyond x-ray absorption edges (EXAFS) [4.156]. It can be simply interpreted in the following way: as the ejected electron tries to escape from its parent atom, it is scattered by the neighbouring atoms. Depending on the wavelength of the electron and the distance to the nearest-neighbour atoms, the reflected wave will interfere destructively or constructively with that of the ejected electron, thus giving maxima and minima in the scattering cross-section. The crucial aspects of EXELFS are that the wavelengths of the various oscillatory components are related to the interatomic distances while the amplitudes correspond to the coordination numbers in each shell.

The oscillatory part of the spectrum above a core edge is given by the standard EXAFS-formula [4.156] which describes the interference between the outgoing spherical wave and the reflected waves. The probability that an energy loss will be modulated sinusoidally is given by the interference amplitude $\chi(k)$ in k space:

$$\chi(k) = -\sum_j \frac{N_j}{kR_j^2} e^{-2\sigma_j^2 k^2} e^{-2R_j/\lambda(k)} \left|F_j(k)\right| \sin[2kR_j + \phi(k)] \qquad (4.13)$$

This equation is written in terms of the wave vector $\boldsymbol{k}$ of the ejected electron which is related to its kinetic energy (difference between the energy loss E and the ionization energy $E_{\rm I}$):

$$k^2 = 2m(E - E_{\rm I})/\hbar^2 \qquad (4.14)$$

The summation in (4.13) is over the j different coordination shells at radius R_j each containing N_j atoms surrounding the central excited atom. The quantity $F_j(k)$ is the backscattering amplitude from each of the N_j neighbouring atoms of the j-th shell and depends on the type of atoms; σ_j is the Debye-Waller factor which accounts for the thermal vibration of the atoms j. The length $\lambda(k)$ is the range of the ejected electron, which is a function of its kinetic energy. The total phase shift $\phi(k)$ of the ejected electron is

$$\phi_j(k) = \phi_{\rm a}(k) + \phi_{\rm b}(k) - \pi, \qquad (4.15)$$

where $\phi_{\rm a}(k)$ is the phase shift due to the potential of the ecited atom and $\phi_{\rm b}(k)$ is the phase shift caused by the backscattering at the neighbouring atoms j.

The information on interatomic distances can be obtained by processing the EELS data with procedures that have already been extensively discussed [4.3,157,158]. First the background beyond the edge has to be subtracted and it is necessary to deconvolute the edge spectrum to remove multiple scattering contributions. Following the original EXAFS procedure, the interatomic distances can be obtained from the Fourier transform of the experimental spectrum. An alternative approach is to use (4.13) to calculate $\chi(k)$ iteratively, starting from an assumed structure model, and to fit the calculated function to the experimental data by varying the parameters $F_j(k), N_j$, and σ_j of the model.

The final result is the radial distribution function (N_j/R_j^2 or N_j as a function of R) (Fig. 4.20), which shows peaks corresponding to most probable interatomic distances. Since contributions from shells a few distances away are strongly attenuated in practice due to the term $\exp(-2R_j/\lambda)$ in (4.13), it is mainly the first and second neighbour peaks that can usually be distinguished. To convert the peak position into absolute distances, a knowledge of the phase shifts is needed, which can be obtained by measurements on similar compounds with well-known structure [4.168] or by calculation from first principles [4.169].

The use of EXELFS was first suggested during the seventies and was applied to nanometre-sized areas soon after [4.159]. One advantage of this method is that it is possible to obtain structural data not only from crystalline but also from amorphous samples: the extended fine structures give the interatomic distances around a particular atom, especially when the atom

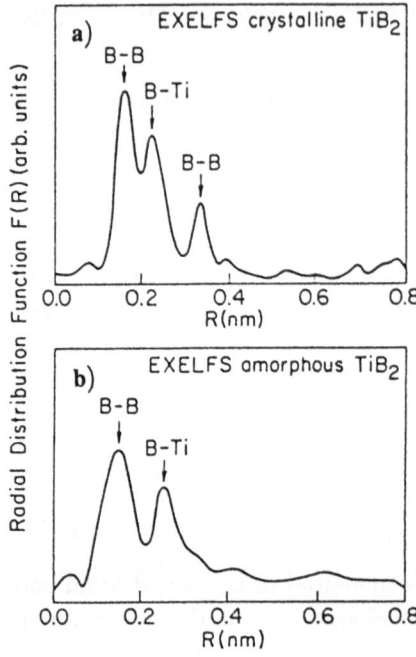

Fig. 4.20. An EXELFS investigation of titanium borides [4.161]): Fourier transform of EXELFS $\chi(k)$ data from the K edge of boron in (a) crystalline TiB_2 and (b) amorphous TiB_2 spectra recorded at 78 K.

is on specific sites in a crystal structure. Since EXELFS (and also EXAFS) is sensitive to short range order, aperiodic amorphous systems can be studied with the same ease as crystals. In amorphous materials one obtains a local radial distribution function for a particular atom. Although coordination numbers can also be derived (but with more elaborate techniques), they should be viewed with caution because they depend strongly on the accuracy of the data processing methods [4.170].

For example, EXELFS has been used to study amorphous and crystalline alumina [4.160,171], amorphous titanium borides [4.161] (Fig. 4.20), amorphous silicon carbides [4.162] and silicon nitride and glasses [4.163]. Many other applications have been published. Orientation dependent EXELFS spectra from layered materials have been measured [4.157,164] and it emerged that the spectra record preferentially the interatomic distances in directions perpendicular and parallel to the plane of the specimen. In addition to nearest neighbour distances, short-range order and thermal vibrations can be measured by EXELFS and its temperature dependence: recent work showed that an analysis of the temperature variation of the EXELFS amplitude yields a localized measurement of the Debye temperature [4.165]. EXELFS has been also used to study the short range order and local thermal vibrations of the Al and Fe atoms in chemically disordered and DO_3 ordered Fe_3Al [4.166].

It is difficult to obtain EXELFS data of the quality that one expects almost routinely using EXAFS because the EELS data suffer from the high background, multiple scattering and from overlapping of edges. However, despite all these obstacles, the great promise of EXELFS lies in the fact that one can study truly microscopic regions as well as edges of light elements [4.167].

Table 4.1. The most prominent ionization edges useable in EELS-microanalysis. The observed edge shapes refer to the notation of Fig. 4.4.

Edge	Excitation of	Elements	Edge shapes
K	1s	Li - Si	a
L_{23}	$2p_{1/2}, 2p_{3/2}$	Mg - Ar	b
		K - Ni	c
		Cu - Br	b
M_{23}	$3p_{1/2}, 3p_{3/2}$	K - Cu	d
M_{45}	$3d_{3/2}, 3d_{5/2}$	Se - Kr	d
		Rb - J	b
		Cs -Yb	c
		Lu - Au	b
N_{45}	$4d_{3/2}, 4d_{5/2}$	Cs - Yb	d
O_{45}	$5d_{3/2}, 5d_{5/2}$	U, Th	d

References

4.1 H. Bethe: Zur Theorie des Durchgangs schneller Korpuskularstrahlen durch Materie. Ann. Physik **5**, 325–400 (1930)
4.2 M. Inokuti: Inelastic collisions of fast charged particles with atoms and molecules: the Bethe theory revisited. Rev. Mod. Phys. **43**, 297–347 (1971)
4.3 R.F. Egerton: *Electron Energy-Loss Spectroscopy in the Electron Microscope* (Plenum, New York 1986)
4.4 P. Rez: Inner-shell spectroscopy: an atomic view. Ultramicroscopy **28**, 16–23 (1989).
4.5 U. Fano, J.W. Cooper: Spectral distribution of atomic oscillator strengths. Rev. Mod. Phys. **40**, 441–507 (1968)
4.6 S.T. Manson: Atomic photoelectron spectroscopy. Part I. Adv. Electr. Electron Phys. **42**, 73–111 (1976)
4.7 R.F. Egerton: K-shell ionization cross-sections for use in microanalysis. Ultramicroscopy **4**, 169–179 (1979)
4.8 R.D. Leapman, P. Rez, D.F. Mayers: K, L, and M shell generalized oscillator strengths and ionization cross-sections for fast electron collisions. J. Chem. Phys. **72**, 1232–1243 (1981)

4.9 B.G. Williams, T.G. Sparrow, R.F. Egerton: Electron Compton scattering from solids. Proc. Roy. Soc. (London) A **393**, 409–422 (1984)

4.10 C.J. Powell: Cross sections for ionization of inner-shell electrons by electrons. Rev. Mod. Phys. **48**, 33–47 (1976)

4.11 C.J. Powell: Cross sections for inelastic electron scattering in solids. Ultramicroscopy **28**, 24–31 (1989)

4.12 D.M. Maher, D.C. Joy, R.F. Egerton, P. Mochel: The functional form of energy-differential cross-sections for carbon using transmission electron energy-loss spectroscopy. J. Appl. Phys. **50**, 5105–5109 (1979)

4.13 R.F. Egerton: Formulae for light element microanalysis by electron energy-loss spectrometry. Ultramicroscopy **3**, 243–251 (1978)

4.14 J.N. Chapman, W.A.P. Nicholson, P.A.Crozier: Understanding thin film x-ray spectra. J. Micr. **136**, 179–191 (1984)

4.15 D.H. Madison, E. Merzbacher: Theory of charged-particle excitation. In *Atomic Inner-Shell Processes*, ed. by B. Crasemann (Academic, New York 1975) Vol.1, pp.1–72

4.16 R.F. Egerton: SIGMAL: a program for calculating L-shell ionization cross-sections. *Proc. 39th Ann. EMSA Meeting*, ed by G.W. Bailey (Claitor's Publishing, Baton Rouge 1981) pp.198–199

4.17 B.H. Choi, E.Merzbacher, G.S. Khandelwale: Tables for Born approximation calculations of L-subshell ionization by simple heavy charged particles. Atomic Data Tables **5**, 291–304 (1973).

4.18 W.J. Veigele: Photon cross-sections from 0.1 keV to 1 MeV for elements Z = 1 to Z = 94. Atomic Data Tables **5**, 51–111 (1973).

4.19 P.S. Sklad, J. Bentley, P. Angelini, G.L. Lehman: Reliability of the quantification of EELS measurements. In *Analytical Electron Microscopy 1984*, ed by D.B. Williams and D.C. Joy (San Francisco Press, San Francisco (1984) pp. 285–288

4.20 C. Allison, W.S. Williams, M.P. Hoffman. Quantitative electron energy loss spectroscopy of vanadium carbide. Ultramicroscopy **13**, 253–264 (1984)

4.21 N. Stenton, M.R. Notis, D.B. Williams: Quantitative electron energy-loss measurements on the system ZrO_2–CaO. J. Am. Ceram. Soc. **67**, 227–232 (1984)

4.22 T.F. Malis, J.M. Titchmarsh: A 'k factor' approach to EELS analysis. In *Electron Microscopy and Analysis 1985*, ed. by G.J. Tatlock, Inst. Phys. Conf. Ser. **78** (Inst. of Physics, Bristol 1985) pp.181–182

4.23 B.P. Luo, E. Zeitler: M-shell cross-sections for fast electron inelastic collisions based on photoabsorption data. J. Electr. Spectr. Rel. Phenom. **57**, 285–295 (1991)

4.24 S.T. Manson: Inelastic collisions of fast charged particles with atoms: ionization of the aluminium L shell. Phys. Rev. A **6**, 1013–1024 (1972)

4.25 P. Rez: Cross-sections for energy-loss spectrometry. Ultramicroscopy **9**, 283–288 (1982)

4.26 D. Rez, P. Rez: The contribution of discrete transitions to integrated inner shell ionization cross-sections. Microsc. Microanal. Microstruct. **3**, 433–443 (1992)

4.27 D.C. Joy, R.F. Egerton, D.M.Maher: Progress in the quantitation of electron energy-loss spectra. In *Scanning Electron Microscopy 1979/II*, ed. by O. Johari, (SEM, AMF O'Hare, Chicago 1979) pp.817–826

4.28 P.M. Kelly, A. Jostens, R.G. Blake, J.G. Napier: The determination of foil thickness by scanning transmission electron microscopy. Phys. Stat. Solidi (a) **31**, 771–780 (1975)

4.29 M. Isaacson: Interaction of 25 keV electrons with the nucleic acid bases, adenine, thymine, and uracil. II Inner-shell excitation and inelastic scattering cross-sections. J. Chem. Phys. **56**, 1813–1818 (1972)

4.30 R.F. Egerton: Inelastic scattering of 80 keV electrons in amorphous carbon. Phil. Mag. **31**, 199–215 (1975)

4.31 R.D. Leapman, V.E. Cosslett: Electron energy loss spectrometry: mean free paths for some characteristic x-ray excitations. Phil. Mag. **33**, 1–10 (1976)

4.32 P.Y. Kihn, J. Sevely, B. Jouffrey: Excitation des niveaux atomiques K du carbone, du magnesium et de l'aluminium par des electrons de 60 keV. Phil. Mag. **33**, 733–741 (1976)

4.33 G.W. Rossouw, M.J. Whelan: The K-shell cross-section for 80 kV electrons in single graphite and AlN. J. Phys. D **12**, 797–807 (1979)

4.34 P.A. Crozier: Absolute inner shell cross-section determination for EELS analysis. *Proc. 46th Ann. EMSA Meeting*, ed. by G.W. Bailey (San Francisco Press, San Francisco 1988) pp.534–535

4.35 P.A. Crozier: Measurement of inelastic scattering cross-sections by electron energy-loss spectroscopy. Phil. Mag. B **61**, 311–336 (1990)

4.36 F. Hofer: EELS quantification of M edges by using oxidic standards. Ultramicroscopy **21**, 63–68 (1987)

4.37 G.W. Lorimer: Quantitative x-ray microanalysis of thin specimens. In *Quantitative Electron Microscopy*, ed. by J.N. Chapman and A.J. Craven (Scottish Universities Summer School in Physics, Edinburgh 1984) pp.305–340

4.38 F. Hofer: Determination of inner-shell cross-sections for EELS quantification. Microsc. Microanal. Microstruct. **2**, 215–230 (1991)

4.39 K.M. Krishnan, C. Echer: Measurements of ionization cross-sections for electron energy-loss microanalysis under well defined scattering conditions. In *Microbeam Analysis*, ed. by D.G. Howitt (San Francisco Press, San Francisco 1991) pp.259–262

4.40 A.J. Bourdillon, W.M. Stobbs: Elastic scattering in EELS – Fundamental corrections to quantification. Ultramicroscopy **17**, 147–150 (1985)

4.41 F. Hofer, P. Golob: Quantification of electron energy-loss spectra with K and L shell ionization cross-sections. Micron and Microscopica Acta **19**, 73–86 (1988)

4.42 J.M. Titchmarsh, T.F. Malis: On the effect of objective lens chromatic aberration on quantitative electron energy-loss spectroscopy (EELS). Ultramicroscopy **28**, 277–282 (1989)

4.43 F. Hofer, P. Golob, A. Brunegger: EELS quantification of the elements Sr to W by means of M_{45} edges. Ultramicroscopy **25**, 81–84 (1988)

4.44 F. Hofer : EELS quantification of the elements Ba to Tm by means of N_{45} edges. J. Micr. **156**, 279–283 (1989)

4.45 P. Wilhelm , F. Hofer: EELS microanalysis of the elements Ca to Cu using M_{23} edges. *Proc. 10th Europ. Congr. on Electron Microscopy*, ed. by A. Ríos, J.M. Arias, L. Megias-Megias, and A. López-Galindo (Publ. de la Univ. Granada, Granada 1992) Vol.I, pp.281–282 (1992)

4.46 M. Grande, C.C. Ahn: Deconvolution and quantification of energy loss transition metal oxide spectra. *Inst. Phys. Conf. Ser.* **68**, 123–126 (Inst. of Physics, Bristol 1984)

4.47 P.A. Crozier, J.N. Chapman, A.J. Craven, J.M. Titchmarsh: On the determination of inner-shell cross-section ratios from NiO using EELS. J. Micr. **148**, 279–284 (1987)

4.48 R.F. Egerton: A simple parametrization scheme for inner-shell cross-sections. *Proc. 46th Ann. Meeting EMSA*, ed. by G.W. Bailey (San Francisco Press, San Francisco 1988) pp.532–533

4.49 F. Hofer: Measurement of partial scattering cross-sections for EELS-quanti-fication. In *Microbeam Analysis*, ed. by D.G. Howitt (San Francisco Press, San Francisco 1991) pp.255–258

4.50 C. Colliex: Electron energy loss spectroscopy in the electron microscope. In *Adv. Opt. and Electr. Micr. Vol.9*, ed. by R. Barer and V.E. Cosslett (Academic, London 1984) pp.65–177

4.51 C.C. Ahn, O.L. Krivanek: *EELS Atlas* (ASU Center for Solid State Science, Tempe AZ and Gatan, Inc., Warrendale, PA 1983)

4.52 N.J. Zaluzec, T. Schober, D.G. Westlake: Application of EELS to the study of metal-hydrogen systems. *Proc. 39th Ann. Meeting EMSA*, ed. by G.W.Bailey (Claitor's Publishing, Baton Rouge 1981) pp.194–195

4.53 O.T. Woo, G.J.C. Carpenter: EELS characterization of zirconium hydrides. Microsc. Microanal. Microstr. **3**, 35–44 (1992)

4.54 G.J. Thomas: Study of hydrogen and helium in metals by electron energy-loss spectroscopy. In *Analytical Electron Microscopy 1981*, ed. by R.H. Geiss (San Francisco Press, San Francisco 1981) pp.195–197

4.55 F. Hofer, B. Luo: Towards a practical method for EELS quantification. Ul-tramicroscopy **38**, 159–167 (1991)

4.56 B.L. Henke, P. Lee, J. Tanaka, R.L. Shimabukuro, B.K. Fujikawa: Low en-ergy x-ray attenuation coefficients: photoabsorption, scattering and reflec-tion: E = 100 – 2000 eV, Z = 1 – 94. Atomic and Nuclear Data Tables **27**, 1–144 (1982)

4.57 E.B. Saloman, J.H. Hubbell, J.H. Scofield: X-ray attenuation cross-sections for energies 100 eV to 100 keV and elements Z = 1 and Z = 92. Atomic Data and Nuclear Data Tables **38**, 1–97 (1988)

4.58 C.C. Ahn, P. Rez: Inner-shell edge profiles in electron energy-loss spec-troscopy. Ultramicroscopy **17**, 105–116 (1985)

4.59 R.D. Leapman, J.A. Hunt: Comparison of detection limits for EELS and EDXS. Microsc. Microanal. Microstr. **2**, 231–244 (1991)

4.60 F.Hofer, P. Golob: New examples for near-edge fine structures in electron energy-loss spectroscopy. Ultramicroscopy **21**, 379–384 (1987)

4.61 J. Auerhammer, P. Rez, F. Hofer: A comparison of theoretical and experi-mental L and M cross-sections. Ultramicroscopy **30**, 365–370 (1989)

4.62 R.D. Leapman, L.A. Grunes, P.L. Fejes: Study of the L_{23} edges in the 3d transition metals and their oxides by electron energy-loss spectroscopy with comparisons to theory. Phys. Rev. B **26**, 614–635 (1982)

4.63 E. Bischoff, M. Rühle: Quantitative EELS microanalysis of SiAlON ceramics. *Proc. 11th Int'l Congr. on Electron Microscopy*, ed. by T. Imura, S. Maruse, T. Suzuki (Jpn. Soc. Electron Microscopy, Tokyo 1986) Vol.I, pp.535–536

4.64 M.M. Chadwick, T.F. Malis: AEM-characterization of sintered silicon nitride with yttria and alumina additions. Ultramicroscopy **31**, 205–216 (1989)

4.65 B.T. Thole, G. van der Laan, J.C. Fuggle, G.A. Sawatzky, R.C. Karnatak, J.M. Esteva: 3d x-ray absorption lines and the $3d^9\ 4f^{n+1}$ multiplets of the lanthanides. Phys. Rev. B **32**, 5107–5118 (1985)

4.66 J.M. Auerhammer, P. Rez: Dipole-forbidden excitations in electron energy-loss spectroscopy. Phys. Rev. B **40**, 2024–2030 (1989)

4.67 T. Manoubi, P. Rez, C. Colliex: Quantitative electron energy loss spec-troscopy on M_{45} edges in rare earth oxides. J. Electr. Spectr. Relat. Phenom. **50**, 1–11 (1990)

4.68 Y.Y. Yang, R.F. Egerton: Measurements of oscillator strengths for use in elemental analysis. Phil. Mag. B **66**, 697–709 (1992)

4.69 U. Fano: Effects of configuration interaction on intensities and phase shifts. Phys.Rev. **124**, 1866–1878 (1961)

4.70 R. Bruhn, B. Sonntag, H.W. Wolff: 3p excitations of atomic and metallic Fe, Co, Ni and Cu. J. Phys. B **12**, 203–212 (1979)

4.71 L.C. Davis, L.A. Feldkamp: Interpretation of 3p core-excitation spectra in Cr, Mn, Fe, Co and Ni. Solid State Commun. **19**, 413–416 (1976)

4.72 P. Trebbia, C. Colliex: Developments in the study of the excitation of inner-shell electrons by energy analysis. *Proc. 8th Int'l Congr. Electron Microsc.*, ed. by J.V. Sanders and D.J. Goodchild (Australian Acad. of Sci., Canberra 1974) Vol.1, pp.382–383

4.73 S.A. Collett, L.M. Brown, M.H. Jacobs: Microanalytical electron microscopy on type 304 steel: Correlation between EDX and EELS results. In *Quantitative Microanalysis with High Spatial Resolution*, ed. by G.W. Lorimer, M.H. Jacobs, P. Doig (The Metals Society, London 1981) pp.159–164

4.74 J.C. Barbour, J.W. Mayer, L.A. Grunes: An investigation of electron energy-loss spectroscopy used for compositional analysis of crystalline and amorphous silicides. Ultramicroscopy **14**, 79–84 (1984)

4.75 F. Hofer, P. Wilhelm: EELS microanalysis of the elements Ca to Cu using M edges. Ultramicroscopy **49**, 189–197 (1993)

4.76 J.H. Paterson, O.L. Krivanek: ELNES of 3d transition metal oxides II. Variation with oxidation state and crystal structure. Ultramicroscopy **32**, 319–325 (1990)

4.77 J.A. Hunt, D.B. Williams: Electron energy-loss spectrum imaging. Ultramicroscopy **38**, 47–73 (1991)

4.78 R.F. Egerton: Oscillator strength parametrization of inner-shell cross-sections. Ultramicroscopy **50**, 13–28 (1993)

4.79 P. Trebbia, C. Colliex: Study of the excitation of 4d electrons in rare-earth metals by inelastic scattering of a high energy electron beam. Phys. Stat. Solidi (b) **58**, 523–532 (1973)

4.80 T.M. Zimkina, V.A. Fomichev, S.A. Gribovskii, I.I. Zhukova: Anomalies in the character of the x-ray absorption of rare earth elements of the lanthanide group. Sov. Phys. Solid State **9**, 1128–1130 (1967)

4.81 R. Haensel, P. Rabe, B. Sonntag: Optical absorption of cerium, cerium oxide, praseodymium, praseodymium oxide, neodymium oxide and samarium in the extreme ultraviolet. Solid State Commun. **8**, 1845–1848 (1970)

4.82 J.L. Dehmer, A.F. Starace, U. Fano, J. Sugar, J.W. Cooper: Raising of discrete levels into the far continuum. Phys. Rev. Lett. **26**, 1521–1525 (1971)

4.83 J. Sugar: Potential barrier effects in photoabsorption: II) Interpretation of photoabsorption resonances in lanthanide metals at the 4d electron threshold. Phys. Rev. B **5**, 1785–1792 (1972)

4.84 H.R. Moser, B.Delley, W.D. Schneider, Y. Baer: Characterization of f electrons in light lanthanide and actinide metals by electron energy-loss and x-ray photoelectron spectroscopy. Phys. Rev. B **29**, 2947–2955 (1984)

4.85 G. Kalkowski, G. Kaindl, W.D. Brewer, W. Krone: Near-edge x-ray absorption fine structure in uranium compounds. Phys. Rev. B **35**, 2667–2677 (1987)

4.86 O.L. Krivanek , C. Mory, M. Tence, C. Colliex: EELS quantification near the single-atom detection level. Microsc. Microanal. Microstruct. **2**, 257–267 (1991)

4.87 L.V. Lazaroff, D.M. Pease. *X-ray Spectroscopy* (McGraw-Hill, New York 1974)

4.88 S.T. Manson: The calculation of photoionization cross-sections: an atomic view. In *Photoemission in Solids I. General Principles* ed. by M. Cardona, L. Ley, Topics in Applied Physics, Vol.26 (Springer , Berlin , Heidelberg 1978) pp.135–163

4.89 R. Brydson: Interpretation of near-edge structure in the electron energy-loss spectrum. EMSA Bull. **21**, 57–67 (1991)

4.90 S.T. Pantelides: Electronic excitation energies and the soft-x-ray absorption spectra of alkali halides. Phys. Rev. B **11**, 2391–2402 (1975)

4.91 B.M. Kincaid, E.A. Meixner, P.M. Platzman: Carbon K edge in graphite measured using electron energy-loss spectroscopy. Phys. Rev. Lett. **40**, 1296–1299 (1978)

4.92 R.D. Leapman, J. Silcox: Orientation dependence of core edges in electron energy loss spectra from anisotropic materials. Phys. Rev. Lett. **42**, 1361–1364 (1979)

4.93 R.D. Leapman, P.L. Fejes, J. Silcox.: Orientation dependence from anisotropic materials determined by inelastic scattering of fast electrons. Phys. Rev. B **28**, 2361–2373 (1983)

4.94 W. Engel, H. Sauer, E. Zeitler, R. Brydson, B.G. Williams, E. Zeitler: EELS and the crystal chemistry of rhodizite. Part 1: Instrumentation and chemical analysis. J. Chem. Soc. Faraday Trans. 1, **84**, 617–630 (1988)

4.95 P.E. Batson, K.L. Kavanagh, C.Y. Wong, J.M. Wooodall: Local bonding and electronic structure obtained from electron energy loss scattering. Ultramicroscopy **22**, 89–102 (1987)

4.96 O.L. Krivanek, J.H. Paterson: ELNES of 3d transition metal oxides. I. Variations across the periodic table. Ultramicroscopy **32**, 313–318 (1990)

4.97 F. Hofer: Energy-loss spectroscopy and imaging. In *Procedures in Electron Microscopy*, ed. by A.W. Robards, A.J. Wilson, (Wiley, Chichester 1993) pp. 15:3.1–21

4.98 L.Dori, J. Bruley, D.J. DiMaria, P.E. Batson, J.Tornello, M. Arienzo: Thin-oxide dual electron-injector annealing studies using conductivity and electron energy-loss spectroscopy. J. Appl. Phys. **69**, 2317–2323 (1991)

4.99 D.F. Blake, F. Freund, K.M. Krishnan, C.J. Echer, R. Shipp, T.E. Bunch, A.G. Tielens, R.J. Lipari, C.J.D. Hetherington, S. Chang: The nature and origin of interstellar diamond. Nature **332**, 611–613 (1988)

4.100 Loucks T.L.: *Augmented Plane Wave Method* (Benjamin, New York 1967)

4.101 J.E. Muller, Jepsen O., Wilkins J.W.: X-ray absorption spectra: K-edges of 3d metals, edges of 3d and 4d metals and M-edges of palladium. Solid State Commun. **42**, 365–368 (1982)

4.102 P. Blaha, K. Schwarz: Electron densities in TiC, TiN, and TiO derived from energy band calculations. Int. J. Quantum Chem. **23**, 1535–1552 (1983)

4.103 X. Weng, P. Rez, O.F. Sankey: Pseudo-atomic-orbital band theory applied to electron energy loss near-edge structures. Phys. Rev. B **40**, 5694–5704 (1989)

4.104 X. Weng, H. Ma, P. Rez: Carbon K-shell near-edge structure: Multiple scattering and band theory calculations. Phys. Rev. B **40**, 4175–4178 (1989)

4.105 J.Pflüger, J. Fink, K. Schwarz: Electronic structure of unoccupied states of stoichiometric ZrN, NbC and NbN as determined by high resolution electron energy-loss spectroscopy. Solid State Commun. **55**, 675-677 (1985)

4.106 X. Weng, P. Rez, P.E. Batson: Single electron calculations for the Si L_{23} near edge structure. Solid State Commun. **74**, 1013–1015 (1990)

4.107 P. Rez, X. Weng, H. Ma: The interpretation of near edge structure. Microsc. Microanal. Microstruct. **2**, 143–151 (1991)

4.108 P.J. Durham, J.B. Pendry, C.H. Hodges: Calculation of x-ray absorption near-edge structure, XANES. Computer Phys. Commun. **25**, 193–205 (1982)

4.109 D.D. Vvendensky, D.K. Saldin, J.B. Pendry: An update of DLXANES, the calculation of x-ray absorption near-edge structure. Computer Phys. Commun. **40**, 421–440 (1986)

4.110 T. Lindner, H. Sauer, W. Engel, K. Kambe: Near-edge structure in electron energy-loss spectra of MgO. Phys. Rev. B **33**, 22–24 (1986)

4.111 R. Brydson, D.D. Vvendensky, W. Engel, H. Sauer, B.G. Williams, E. Zeitler, J.M. Thomas: Chemical information from electron energy-loss near-edge structure. Core hole effects in the beryllium and boron K-edges in rhodizite. J. Phys. Chem. **92**, 962–966 (1988).

4.112 R. Brydson, B.G. Williams, H. Sauer, W. Engel, R. Schlögl, M. Muller, E. Zeitler, J.M. Thomas: EELS and the crystal chemistry of rhodizite. Part 2: Near-edge structure. J. Chem. Soc. Faraday Trans. 1 **84**, 631–646 (1988)

4.113 H. Sauer, R. Brydson, W. Engel, P. Rowley: Coordination fingerprints of boron measured by EELS. In *Proc. 12th Int'l Congr. on Electron Microscopy*, ed. by P. Ingram, J.R. Michael, D.B. Williams (San Francisco Press, San Francisco 1990) Vol.2, pp.54–55

4.114 P. Rez, X. Weng: Multiple scattering approach to oxygen K near-edge structures in electron energy-loss spectroscopy of alkaline earths. Phys. Rev. B **39**, 7405–7412 (1989)

4.115 R. Brydson, H. Sauer, W. Engel, F. Hofer: Electron energy-loss near edge structures at the oxygen K edges of titanium(IV) oxygen compounds. J. Phys. Condens. Matter **4**, 3429–3437 (1992)

4.116 K.H. Johnson: Scattered-wave theory of the chemical bond. In *Advances in Quantum Chemistry, Vol.7*, ed. by P.O. Loewdin (Academic, New York 1973) pp.143-158

4.117 D. Dill, J.L. Dehmer: Electron-molecule scattering and molecular photoionization using the multiple-scattering model. J. Chem. Phys. **61**, 692–699 (1974)

4.118 W. Wurth, J. Stöhr: Model calculations for molecular photoabsorption spectra. Vacuum **41**, 237–239 (1990)

4.119 K.M. Krishnan: Iron L_{23} near-edge structure studies. Ultramicroscopy **32**, 309–311 (1990)

4.120 R. Brydson, H. Sauer, W. Engel, J.M. Thomas, E. Zeitler: Coordination fingerprints in electron loss near-edge structures: Determination of the local site symmetry of aluminium and beryllium in ultrafine minerals. J. Chem. Soc. Chem. Commun. **15**, 1010–1012 (1989)

4.121 A. Bianconi, M. Dell'Ariccia, P.J. Durham, J.B. Pendry: Multiple scattering resonances and structural effects in the x-ray absorption near edge spectra of Fe II and Fe III hexacyanide complexes. Phys. Rev. B **26**, 6502–6508 (1982)

4.122 Sekiyama H., Y. Kitajima, N. Kosugi, H. Kuroda, T. Ohta: K and L_{23} absorption spectra PO_4 , SO_4 and ClO_4. Photon Activity Report 1983/84, VI, 119 (1984)

4.123 F. Hofer: Beiträge zur Anwendung der Elektronenenergieverlustspektroskopie im Elektronenmikroskop. Habilitationsschrift, Technische Universität Graz (1988)

4.124 C. Colliex, T. Manoubi, M. Gasgnier, L.M. Brown: Near-edge fine structures on EELS core loss edges. In *Scanning Electron Microscopy 1985/II* (SEM Inc., AMF O'Hare 1985) pp.489–512

4.125 R. Brydson, H. Sauer, W. Engel: Electron energy loss near-edge structure as an analytical tool - the study of minerals. In *Transmission Electron Energy Loss Spectrometry in Materials Science*, ed. by M.M. Disko, C.C. Ahn, B. Fultz (The Minerals, Metals and Materials Society, Warrendale 1992) pp.131-154.

4.126 R. Brydson, H. Sauer, W. Engel, E. Zeitler: EELS as a fingerprint of the chemical coordination of light elements. Microsc. Microanal. Microstruct. **2**, 159–169 (1991)

4.127 P.N. Rowley, R. Brydson, J. Little, S.R.J. Saunders: Electron energy loss studies of Fe-Cr-Mn oxide films. Phil. Mag. B **62**, 229–241 (1990)

4.128 K.H. Hallmeier, R. Szargen, A. Meisel, E. Hartmann, E.S. Gluskin: Investigation of core excited quantum yield spectra of highly symmetric boron compounds. Spektrochim. Acta A **37**, 1049–1055 (1981)

4.129 W.H.E Schwarz, L. Mensching, K.H. Hallmeier, R. Szargen: K-shell excitation of BF_3 , CF_4 and MBF_4 compounds. J. Chem. Phys. **82**, 57–65 (1983).

4.130 F.W. Lytle, R.B. Greegor, A.J. Panson: Discussion of x-ray absorption near edge structure: application to Cu in the high T_c superconductors $La_{1.8}$ $Sr_{0.2}$ CuO_4 and $YBa_2Cu_3O_7$. Phys. Rev. B **37**, 1550–1562 (1988)

4.131 J. Wong, F.W. Lytle, R.P. Messmer, D.H. Maylotte: K-edge absorption spectra of selected vanadium compounds. Phys. Rev. B **30**, 5596–5610 (1984)

4.132 G.A. Waychunas, G.E. Brown: Application of EXAFS and XANES spectroscopy to problems in mineralogy and geochemistry, In *EXAFS and Near-Edge Structures III*, ed. by K.O. Hodgson, B. Hedman, J.E. Penner-Hahn, Springer Ser. in Chemical Physics, Vol. 31 (Springer, Berlin, Heidelberg 1984) pp.336–342

4.133 F.M.F. de Groot, J.C. Fuggle, B.T. Thole, G.A. Sawatzky: L x-ray-absorption edges of d^o compounds: K^+ , Ca^{2+} , Sc^{3+} , and Ti^{4+} in O_h (octahedral) symmetry. Phys. Rev. B **41**, 928–937 (1990)

4.134 G.J. Ballhausen: *Introduction to Ligand Field Theory* (McGraw Hill, New York 1962)

4.135 P.L. Hansen, D. McComb, R. Brydson: ELNES fingerprint of Al coordination in nesosilicates. Micron and Microsc.Acta **23**, 169–170 (1992)

4.136 D.W. McComb, P.L. Hansen, R. Brydson: A study of silicon ELNES in nesosilikates. Microsc. Microanal. Microstruct. **2**, 561–568 (1991)

4.137 P.K. Gosh: *Introduction to Photoelectron Spectroscopy.* (Wiley, New York 1983)

4.138 G.J. Auchterlonie, D.R. McKenzie, D.J.H. Cockayne: Using ELNES with parallel EELS for differentiating between a-Si:X thin films. Ultramicroscopy **31**, 217–222 (1989)

4.139 C. Mande, B. Sapre: Chemical shifts in x-ray absorption spectra. *Adv. X-ray Spectroscopy*, ed. by C. Bonelle, C. Mande (Pergamon, Oxford 1982) pp.287–301

4.140 T.G. Sparrow, B.G. Williams, C.N.R. Rao, J.M. Thomas: L_2/L_3 white line intensity ratios in the EEL spectra of 3d transition metal compounds. Chem. Phys. Lett. **108**, 547–550 (1984)

4.141 J. Barth, F. Gerken, C. Kunz: Atomic nature of the LII,III white lines in Ca, Sc, and Ti metals by resonant photoemission. Phys. Rev. B **28**, 3608–3611 (1983)

4.142 L.M. Brown, C. Colliex, M. Gasgnier: Fine structure in EELS from rare earth sesquioxide thin films. J. Physique **45**, colloque C2, 433–436 (1984)

4.143 D.H. Pearson, B. Fultz, C.C. Ahn: Measurements of 3d state occupancy in transition metals using electron energy loss spectrometry. Appl. Phys. Lett. **53**, 1405–1407 (1988)

4.144 D.H. Pearson: Measurements of white line intensities in 4d transition metals using EELS. In *Proc. 47th Ann. Meeting EMSA*, ed. by G.W. Bailey (San Francisco Press, San Francisco 1989) pp.386–387

4.145 W.G. Waddington, P. Rez, I.P. Grant, C.J. Humphreys: White lines in the L23 electron energy loss and x-ray absorption spectra of 3d transition metals. Phys. Rev. B **34**, 1467–1473 (1986)

4.146 B.T. Thole, G. van der Laan: Branching ratio in x-ray absorption spectroscopy. Phys. Rev. B **38**, 3158–3171 (1988)

4.147 H. Kurata, N. Tanaka. Iron L23 white line ratio in nm-sized γ-iron crystallites embedded in MgO. Microsc. Microanal. Microstruct. **2**, 183–190 (1991)

4.148 T.I. Morrison, M.B. Brodsky, N.J. Zaluzec, L.R. Sill. Iron d-band occupancy in amorphous Fe_xGe_{1-x}. Phys. Rev. B **32**, 3107–3111 (1985)

4.149 T.I. Morrison, C.L. Foiles, D.M. Pease, N.J. Zaluzec. Relationships between local order and magnetic behavior in amorphous $Fe_{0.30}Y_{0.70}$: Extended x-ray absorption fine structure and susceptibility. Phys. Rev. B **32**, 3739–3744 (1987)

4.150 F.M.F. de Groot, J.C. Fuggle, B.T. Thole, G.A. Sawatzky: 2p x-ray absorption of 3d transition metal compounds: An atomic description including the crystal field. Phys. Rev. B **42**, 5459–5468 (1990)

4.151 R. Brydson, H. Sauer, W. Engel, E. Zeitler: Coordination fingerprinting and valency determination using electron energy loss spectroscopy. In *Electron Microscopy and Analysis 1991*, ed. by F.J. Humphreys, Inst. Phys. Conf. Ser. 119 (Inst. of Physics, Bristol 1991) pp.101–104

4.152 W. Engel, H. Sauer: EELS near-edge structures as a probe of solids. *Proc. 12th Int'l Congr. on Electron Microscopy*, ed. by L.D. Peachey and D.B. Williams (San Fransisco Press, San Francisco 1990) Vol.2, pp.71–72

4.153 R. Brydson, H. Sauer, W. Engel, E. Zeitler, J.M. Thomas, N. Kosugi, H. Kuroda: Electron energy loss and x-ray absorption spectroscopy of rutile and anatase: a test of structural sensitivity. J. Phys. Condens. Matter **1**, 797–803 (1989)

4.154 A. Bianconi, M. Dell'Ariccia, A. Gargano, C.R. Natoli: Bond length determination using XANES. In *EXAFS and Near Edge Structures*, ed. by A. Bianconi, L. Incoccia, L. Stipcich, Springer Ser. in Chemical Physics, Vol. 27 (Springer, Berlin, Heidelberg 1983) pp.57–61

4.155 J.H. Rask, B.A. Miner, P.R. Buseck. Determination of manganese oxidation states in solids by electron energy-loss spectroscopy. Ultramicroscopy **21**, 321–326 (1987)

4.156 E.A. Stern: Theory of EXAFS. In *X-ray-Absorption*, ed. by D.C. Koningsberger, R. Prins (Wiley, New York 1988) pp.3–52

4.157 R.D. Leapman, L.A. Grunes, P.L. Fejes, J. Silcox: Extended core-edge fine structure in electron energy-loss spectra. In *EXAFS Spectrosocpy Techniques and Applications*, ed. by B.K. Teo, D.C. Joy (Plenum, New York 1981) pp.217–239

4.158 D.E. Johnson, S. Csillag , E.A. Stern: Analytical electron microscopy using extended energy-loss fine structure (EXELFS). *Scanning Electron Microscopy 1981/I* (SEM Inc., AMF O'Hare 1981) pp.105–115

4.159 P.E. Batson, A.J. Craven: Extended fine structure on the carbon core ionization edge obtained from nanometer-sized areas with electron energy-loss spectroscopy. Phys. Rev. Lett. **42**, 893–897 (1979)

4.160 A.J. Bourdillon, S.M. El Mashri, A.J. Forty: Application of extended energy loss fine structure to the study of aluminium oxide films. Phil. Mag. **49**, 341–352 (1984)

4.161 A.E. Kaloyeros, W.S. Williams, R.B. Rizk, F.C. Brown, A.E. Greene: Study by extended x-ray absorption fine structure technique and microscopy of the amorphous state of titanium diboride thin films. J. Am. Ceram. Soc. **71**, 948–955 (1988)

4.162 J.M. Martin, J.L. Mansot: EXELFS analysis of amorphous and crystalline silicon carbide. J. Micr. **162**, 171–178 (1991)

4.163 R.D. Leapman: EXELFS Spectroscopy of amorphous materials. *Microbeam Analysis 1982*, ed. by K.F.J. Heinrich (San Francisco Press, San Francisco 1982) pp.111–117

4.164 M.M. Disko, O.L. Krivanek, P. Rez: Orientation dependent extended fine structure in electron energy-loss spectra. Phys. Rev. B **25**, 4252–4255 (1982)

4.165 M.M. Disko, C.C. Ahn, G. Meitzner, O.L. Krivanek: Temperature dependence of aluminium EXELFS. *Microbeam Analysis 1988*, ed. by D.E. Newbury (San Francisco Press, San Francisco 1988) pp.47–49

4.166 J.K. Okamoto, C.C. Ahn, B. Fultz: Temperature-dependent EXELFS of Al K and Fe L_{23} edges in chemically disordered and DO_3-ordered Fe_3Al. *Microbeam Analysis 1991*, ed. by D.G. Howitt (San Francisco Press, San Francisco 1991) pp.56–58

4.167 V. Serin, G. Zanchi, J. Sevely: EXELFS as a structural tool for studies of low Z-elements. Microsc. Microanal. Microstruct. **3**, 201–212 (1992)

4.168 P.H. Citrin: Transferability of phase shifts in extended x-ray absorption fine structure. Phys. Rev. Lett. **36**, 1346–1349 (1976)

4.169 P.A. Lee, G. Beni: New method for the calculation of atomic phase shifts: application to extended x-ray absorption fine structure (EXAFS) in molecules and crystals. *Phys. Rev. B* **15**, 2862–2883 (1977)

4.170 P. Rez: Energy loss fine structure. In *Transmission Electron Energy Loss Spectrometry in Materials Science*, ed by M.M. Disko, C.C. Ahn, B. Fultz (The Minerals, Metals and Materials Society, Warrendale 1992) pp. 107–129

4.171 P.S. Sklad, P. Angelini, J. Sevely: Extended electron energy loss fine structure analysis of amorphous Al_2O_3. Phil. Mag. A **65**, 1445–1461 (1992)

5. Quantitative Electron Energy-Loss Spectroscopy

Ray F. Egerton and *Richard D. Leapman*

Electron energy-loss spectroscopy in the transmission electron microscope provides the interpretative foundation for energy-filtered TEM imaging. Typically, we record the energy-dependence of intensity from a known area of a specimen, defined by the incident beam or by an image-plane (area-selecting) aperture. Recently it has also become feasible to record and process a complete energy-loss spectrum for each point in an image, although the information generated (a so-called spectrum-image) requires massive amounts of data storage [5.1,2]. Recording the entire spectrum (or large parts of it) is necessary for some types of spectrum processing, such as Fourier–log deconvolution, which removes the effects of plural scattering. In this chapter, we concentrate on quantitative aspects of EELS and on the instrumentation used in spectroscopy.

5.1 Instrumentation for Energy-Loss Spectroscopy

The imaging requirements of a spectrometer are considerably relaxed when the data come from a very small area of the specimen: the spectrometer can be mounted as an attachment below the column of the CTEM, or at the top in the case of a field-emission STEM with its electron source at the bottom. In almost all cases, a single-sector magnetic-prism spectrometer is used but there are two alternative methods of reading out the spectral data.

5.1.1 Serial Recording

To be useful, a spectrometer should give a digital output which can be fed into a multichannel analyser (MCA) or computer of some kind, for display and subsequent processing (Fig. 5.1). The simplest scheme for achieving this is serial recording. A narrow slit is placed at the image plane of the spectrometer and the current in the electromagnet is varied slightly to sweep the spectrum vertically across the slit. The ramp signal is supplied by the MCA system. The speed of the ramp (or dwell time per MCA channel) can be varied; a few milliseconds per channel is sufficient for recording the intense peaks at low energy loss, whereas 50 ms per channel or more gives lower noise when recording higher energy losses. A sensitive electron detector, usually a

Springer Series in Optical Sciences, Vol. 71
Energy-Filtering Transmission Electron Microscopy
Editor: Ludwig Reimer   ©Springer-Verlag Berlin Heidelberg 1995

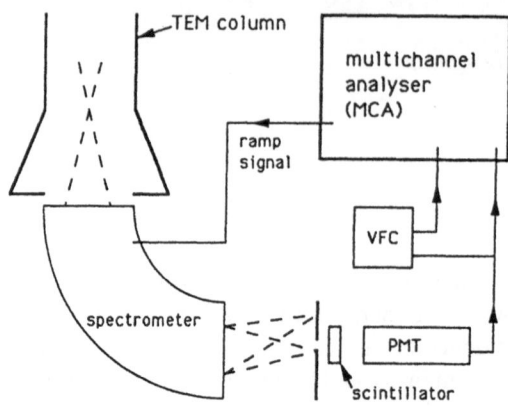

Fig. 5.1. A serial-recording EELS system below a CTEM column.

scintillator followed by a photomultiplier tube (PMT), generates a pulse for each electron that passes through the slit, provided the electron flux is sufficiently low. At high intensities, the pulses overlap and the resulting current output of the PMT is digitized via a voltage-to-frequency converter (VFC). A mechanical control varies the width of the detection slit. Large width gives a stronger signal which is relatively free of shot noise, but the energy resolution is poor. To increase resolution, the slit width is reduced while observing the zero-loss peak; the optimum setting is reached when the height of the peak starts to fall. Slit adjustment, focusing and aberration correction (to minimize the width of the zero-loss peak) are made more convenient by scanning the electron beam rapidly across the slit (by applying a ramp signal to the entrance dipole coils) and observing the output of the PMT directly on an oscilloscope screen. The large inductance of the main spectrometer scan coils limit the usual mode of scanning to no more than 1 spectrum per second.

A significant advantage of serial-recording detectors is that only one detection channel is involved, so the sensitivity (gain) of the system is the same (neglecting transient effects in the PMT) across the entire spectrum. The gain can be varied by changing the PMT supply voltage or switching to pulse counting during the acquisition, producing a "gain change" which is often found between the low-loss and core-loss regions of the spectrum. If necessary, the sensitivity can further altered by changing the dwell time per channel, and the sensitivity factor will be simply the ratio of the two dwell times.

The linearity of response is good in the analogue mode of recording, provided the anode current drawn from the PMT is small compared to the dynode current [5.3]. Overloading causes a transient increase in PMT dark current, giving a "tail" on intense peaks such as the zero-loss peak. This tail is less noticeable if the spectrum is scanned downwards in energy loss [5.4,5] or if the spectrum is recorded (at low incident intensity) in pulse-counting mode

[5.3]. In pulse-counting mode, linearity is within 15% for observed count rates f_{ob} up to 1 MHz. At higher count rates, the true count rate f_t is given by:

$$f_t = f_{ob}/(1 + f_{ob}T) \qquad (5.1)$$

where T is the deadtime of the detector, typically in the range 150-200 ns.

In the pulse-counting mode, the scintillator-PMT combination provides an almost noiseless detector which, despite the low signal intensity, is capable of recording ionization edges at least up to 2000 eV loss. But although the detective quantum efficiency (DQE) of the detector itself is high, the DQE of the whole system is low because, during most of the acquisition period, a large fraction of the energy-loss electrons are absorbed by the energy-selecting slit.

5.1.2 Parallel Recording

To obtain optimum noise performance, it is necessary to remove the slit and replace it with a position-sensitive detector. All energy losses are then recorded simultaneously, giving a stronger signal with lower shot noise.

Convenient forms of position-sensitive detector include the photodiode and charge-coupled-diode (CCD) arrays. Although these arrays respond directly to electrons, they are designed to respond to photons; there is less risk of damage if a thin scintillator is placed in front of the array and the light-optical image is transferred to the array by a lens or fibre optics. In the spectrometer marketed by the Gatan company, a thin layer of yttrium aluminium garnet (YAG) scintillator is directly mounted on a fibre-optic plate (Fig. 5.2). A set of quadrupole lenses precedes the detector, in order to magnify the dispersion in the vertical direction and achieve an energy resolution of 1 eV, or 0.5 eV in the case of a field-emission source.

Each photodiode behaves like a leaky capacitor. Before recording a spectrum, every diode is charged to the same potential. During spectrum acquisition, electron–hole pairs created by light emission in the scintillator cause the diodes to discharge by an amount proportional to the local photon intensity. At the end of the acquisition period, the charge remaining on each diode is sampled in turn and the spectrum is read into the MCA as a series of pulses, as in the case of serial recording.

The diodes also discharge by thermal generation of electron–hole pairs, giving rise to a background component in the energy-loss spectrum. This background is not constant; the leakage current is slightly different for each diode in the array, giving rise to an additional source of "noise" in the data. These effects can be corrected for by taking a readout with no electrons incident on the scintillator, then subtracting this "bias spectrum" from the EELS data by suitably programming the data-acquisition computer.

In addition, individual diodes in the array (and different parts of the scintillator) have slightly different sensitivities to irradiation. To partially remove this detector artifact, a "gain spectrum" is recorded by uniformly

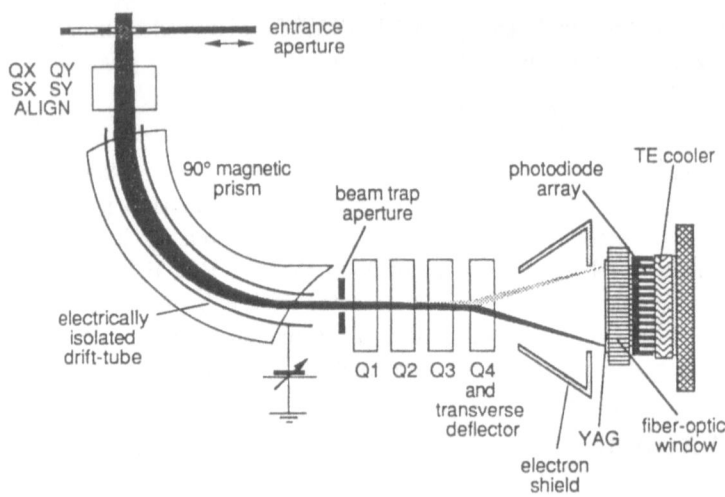

Fig. 5.2. Gatan model 666 parallel-recording spectrometer [5.6]. QX and QY are quadrupole coils for fine focussing of the energy-loss spectrum; SX and SY are sextupoles for minimizing second-order aberrations. The quadrupole lenses which increase the dispersion are labelled Q1–Q4.

illuminating the array; each background-subtracted energy-loss spectrum is divided by the gain spectrum within the data-analysing computer.

Currently available parallel-recording detectors suffer from memory effects; incomplete readout in the photodiode array causes a faint replica of previous spectra to appear in subsequent ones. This effect can be eliminated by making a sufficient number of readouts with the electron beam excluded, for example by lowering the viewing screen in the CTEM. Being an electrical insulator, a YAG scintillator accumulates internal charge in strongly irradiated regions, which is believed to increase locally its conversion efficiency. This artifact can be reduced by "annealing" the scintillator with a broad beam for several minutes.

Recently introduced cooled CCD detectors offer several advantages over photodiode arrays [5.7] in terms of linearity of intensity response, dynamic range, and detective quantum efficiency. In fact, the low noise properties of CCD detectors enable acquisition of spectra at nearly single electron sensitivity. Another advantage of CCD detectors is that, because their array elements are square, channel-to-channel gain variations can be removed with a simple division by a uniform illumination spectrum [5.7]. Photodiode arrays, on the other hand, have highly elongated elements whose sensitivity varies across their length, and their channel-to-channel gain variations cannot be exactly corrected because they depend on which part of the array is illuminated.

Because of lateral spreading of the electron beam and of the light generated within the scintillator, the recorded spectrum is broadened by the point-spread function (PSF) of the detector, which has a full width at half maximum (FWHM) of typically 3 channels and extended tails, which comprise roughly half the total intensity. These tails distort the shape of ionization edges (for example), but fortunately their effect can be removed by deconvolving the spectrum with the point-spread function (Fig. 5.3). Besides degrading the energy resolution on low-dispersion (high eV/channel) settings, the PSF reduces the detective quantum efficiency and the range of electron intensities over which high DQE is achievable [5.8].

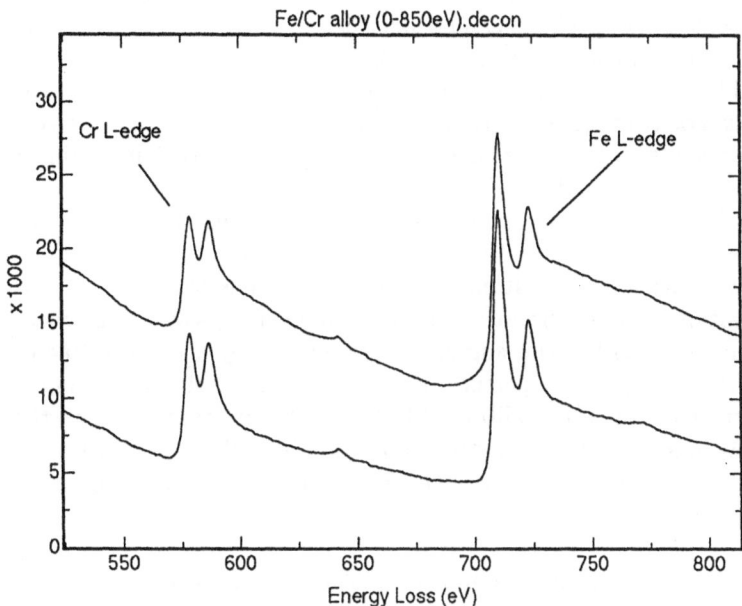

Fig. 5.3. Parallel-recorded energy-loss spectrum of nichrome alloy. The upper spectrum represents the raw data, the lower spectrum is after deconvolution with the point-spread function of the photodiode array.

Backscattering of the zero-loss beam is a serious problem in parallel-recording detectors, due to the absence of an energy-selecting slit, which would also discriminate against electrons travelling far from the optic axis. The effect has been reduced in the Gatan spectrometer by means of a beam-trap aperture and an automatic deflection system, which deflects the zero-loss beam when it falls outside the array. Schemes are available for removing the stray-scattering background, in cases where this is desirable [5.3,9].

Despite some shortcomings, parallel-recording detectors are generally preferable for the recording of ionization edges, especially at high energy loss where shot noise must be minimized to achieve an acceptable signal. Their advantage is greatest in the case of analysis of very small volumes of material and low concentrations. The reduced recording time needed to get adequate statistics results in less drift and contamination (or beam damage) of the specimen.

5.2 Elemental Analysis by EELS

As discussed in earlier chapters, analysis for specific elements is accomplished by making use of the ionization edges which occur at higher energy loss. Provided the energy-loss scale has been calibrated to within a few percent, elements present in the specimen can be identified by comparing the observed edge energies with tabulated values [5.11,24].

The height of an edge can give a rough indication of the amount of the element but is not truly quantitative on account of the near-edge fine structure, which depends on the spectrometer resolution and on the environment (chemical and crystallographic) of the atoms being detected. By integrating the core-loss intensity over an energy region that exceeds that of the fine structure, we obtain a quantity $I_{\mathrm{I}}(\alpha_{\mathrm{o}}, \Delta)$, which relates more directly to elemental concentration. I_{I} is written as a function of collection angle α_{o}, as well as integration width Δ, because we detect only core-loss electrons whose scattering angles lie within the angular range defined by the spectrometer entrance (SEA) or objective aperture. For quantitative analysis, the angle-limiting apertures need to be calibrated in terms of scattering angle. In the case of CTEM objective apertures, this calibration is easily done by comparing the "shadow" of the aperture with the diameters of diffraction rings from a crystalline test specimen.

5.2.1 Estimation of the Non-Characteristic Background

To ensure that $I_{\mathrm{I}}(\alpha_{\mathrm{o}}, \Delta)$ represents only the inner-shell component of scattering, we must make allowance for the background underlying an ionization edge (dashed line in Fig. 5.4). This background contains the tails of any preceding ionization edges, the tail of the valence-electron intensity (low-loss region) and plural inelastic scattering involving any large energy loss and a limited number of low-loss events, together with any instrumental background (due to stray scattering in the spectrometer and detector dark current) if this has not already been subtracted.

Atomic and solid-state physics suggests that the tails of lower-energy components will have a power-law dependence on energy loss E, and this relation is observed experimentally if the plural and instrumental scattering are negligible. We can therefore model the background intensity as:

$$J_{\mathrm{b}} = AE^{-r}. \tag{5.2}$$

The adjustable parameters A and r cannot be measured directly within the core-loss integration region, but they can be estimated by examining the energy dependence of the intensity over a region of width Γ directly preceding

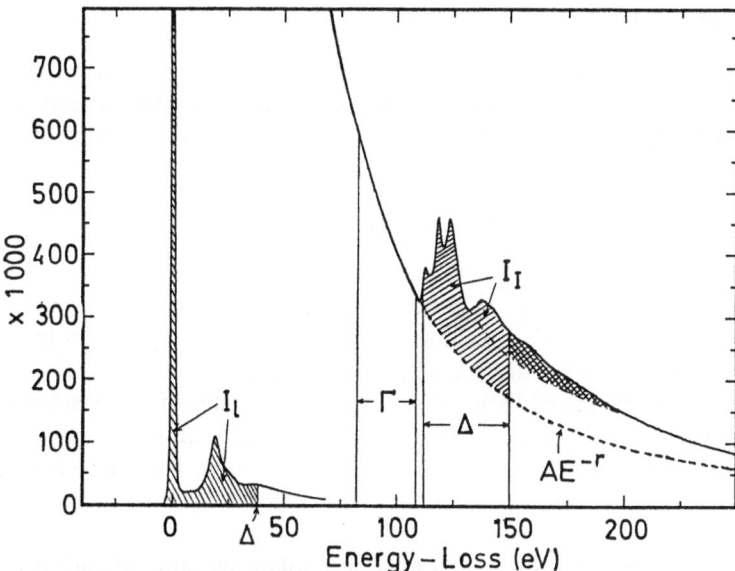

Fig. 5.4. Elemental analysis by integration of the core-loss intensity quadrupole lenses which increase the dispersion are labelled Q1–Q4.

the edge, on the assumption that their values change only slowly with energy loss. Bearing in mind this assumption, and the fact that the pre-edge background contains a noise component, the range Δ of extrapolation should not be too large. Values of the order of 50 eV are suitable for lower-energy edges but larger values (one or two hundred eV) may give better accuracy for higher-edges (where the data is comparatively noisy) provided other ionization edges do not occur within this region.

Least-squares fitting is normally used to deduce A and r, although more elaborate techniques may give improved accuracy [5.10]. Alternatively, the fitting region can be divided into two halves (of equal width) and the value of r found from the corresponding integrals I_1 and I_2 by applying the approximate formulae [5.11]:

$$r = \frac{2\log(I_1/I_2)}{\log(E_2/E_1)} \tag{5.3}$$

$$A = \frac{(1-r)(I_2+I_1)}{E_2^{1-r} - E_1^{1-r}}. \tag{5.4}$$

Note the factor of 2 in (5.3). Without it, the equation is correct only if I_1 and I_2 are integrals over narrow energy windows about the energies E_1 and E_2, a situation which is unfavourable because of the limited sampling of the background, which would cause the value of r to be more sensitive to spectral noise.

At lower energy loss, the power-law formula is less appropriate and more complicated functions may be required [5.12].

5.2.2 Quantification from Core-Loss Integrals

In any transmission measurement, the core-loss signal will depend on the number N of atoms per unit area of the specimen, sometimes called the areal density. The concept of a cross-section $\sigma(\alpha_o, \Delta)$ for scattering within our defined range of angle and energy loss would enable us to write:

$$I_1(\alpha_o, \Delta) = N I \sigma(\alpha_o, \Delta) \qquad (5.5)$$

where I is the incident-beam current, in the same units as I_1.

For routine microanalysis, (5.5) is inconvenient because it requires us to measure I (via a Faraday cage) and the sensitivity of the system (MCA counts per incident electron, which will vary with beam energy, for example). More importantly, the equation is realistic only for vanishingly thin specimens; for a typical TEM specimen, elastic scattering outside the angle-limiting aperture will cause I_1 to be reduced, while plural inelastic scattering will cause some of the core-loss intensity to appear outside the integration window (cross-hatched area in Fig. 5.4).

The situation is improved if we can make the following approximations:
1) that the probability of elastic scattering (outside the collection aperture) is the same for core-loss electrons and those in the low-loss region;
2) that the probability of inelastic scattering out of the core-loss integration window is the same as the probability of low-loss scattering beyond an energy loss Δ. We can then replace I in (5.5) by the intensity $I_1(\alpha_o, \Delta)$ within the low-loss region, giving:

$$I_1(\alpha_o, \Delta) = N I_1(\alpha_o, \Delta) \sigma(\alpha_o, \Delta). \qquad (5.6)$$

Although (5.6) remains accurate up to larger specimen thickness than (5.5), it can give misleading results when the thickness t exceeds the mean free path Λ for all inelastic scattering, which is typically 100 nm for 100 keV primary electrons. Above 1000 eV, edges are observable (with reasonable signal-to-background ratio) for t/Λ as large as 4 and are useful for qualitative or semi-quantitative analysis [5.13]. But if the specimen is crystalline, the angular distribution of inelastic scattering becomes quite complicated for thicker specimens [5.14] and there is currently no satisfactory quantification procedure for this case.

Approximation (2) can be avoided by removing plural scattering from the spectrum by deconvolution procedures, but the results are similar provided $\Delta \geq 50$ eV [5.11]. In the case of a very small probe of large convergence angle, (5.6) requires correction by a factor whose value depends on the edge energy as well as on the convergence and collection angles [5.11].

In most situations, we require only the ratios of elements (A and B), in which case (5.6) can be written:

$$\frac{N_A}{N_B} = \frac{I_A(\alpha_o, \Delta)}{I_B(\alpha_o, \Delta)} \frac{\sigma_B(\alpha_o, \Delta)}{\sigma_A(\alpha_o, \Delta)}. \qquad (5.7)$$

The low-loss region need not be measured, unless it is required for deconvolution. Equation (5.7) is analogous to the determination of elemental ratios in thin specimens by EDX analysis, except that in EELS we deal with atomic rather than weight ratios. The cross-section ratios are equivalent to k-factors [5.15] and can be determined experimentally by EELS of binary standards of known composition [5.15,16].

However, the individual cross-sections can also be calculated directly, on the basis of atomic physics. From a particular model of the atomic wavefunctions, one can evaluate the generalised oscillator strength (GOS) for a particular shell and element, as a function of momentum and energy transfer to the atom. The GOS is then integrated over the required range of energy loss and scattering angle, for a particular incident-electron energy. Commercial EELS software packages use hydrogenic theory, since the GOS is known in analytic form and can be evaluated rapidly on-line by the data-analysis computer. More accurate results, particularly for L- and M-shells, are obtainable from Hartree–Slater calculations, which are time-consuming and require more expertise.

A third option is to tabulate oscillator strengths for different elements, on the basis of experimental data and/or calculations. A great simplification is possible if the GOS can be set equal to the dipole oscillator strength, as used in optical theory. Dipole conditions apply provided we measure energy losses not too far from an ionization edge and over a restricted angular range, such that the angular distribution of core-loss scattering is Lorentzian. These conditions apply well to EELS analysis, provided we use an angle-restricting collection aperture. The energy-loss cross-section can then be evaluated from:

$$\sigma(\alpha_o, \Delta) = A \frac{\gamma}{\gamma + 1/\gamma} \frac{1}{E_0\langle E \rangle} [\ln(1 + \alpha_o^2/\theta_{\rm E}^2) + G] f(\Delta) \qquad (5.8)$$

where

$$
\begin{aligned}
A &= 1.30 \times 10^{-13} {\rm cm}^2 {\rm eV}^{-2}, \\
\langle E \rangle &= [E_{\rm I}(E_{\rm I} + \Delta)]^{1/2}, \\
\theta_{\rm E} &= (\langle E \rangle/E_0)\gamma/(1+\gamma), \\
\gamma &= 1 + E_0/511
\end{aligned}
$$

where E_0 is measured in keV and G is a correction for relativistic retardation, which can be neglected for $E_0 \leq 200$ keV; $f(\Delta)$ is the integrated optical oscillator strength, which is tabulated for useful values of the integration window Δ and can be obtained for intermediate values by interpolation [5.17].

5.2.3 Fitting to Spectral Standards

The above method of elemental quantification is difficult to apply to very noisy data, or if the edges are very weak or occur in close proximity to each other. If statistical noise in the spectrum is dominant, the uncertainty dI_I in the measurement of the core-loss integral I_I can be written [5.11] as:

$$dI_I = \sqrt{\mathrm{var}I_I} = \sqrt{I_I + hI_b}. \tag{5.9}$$

Here intensities are measured in terms of the number of energy-loss electrons; h is a number, which depends on the width of the fitting and integration regions and on the energy-dependences of the signal and background. If the

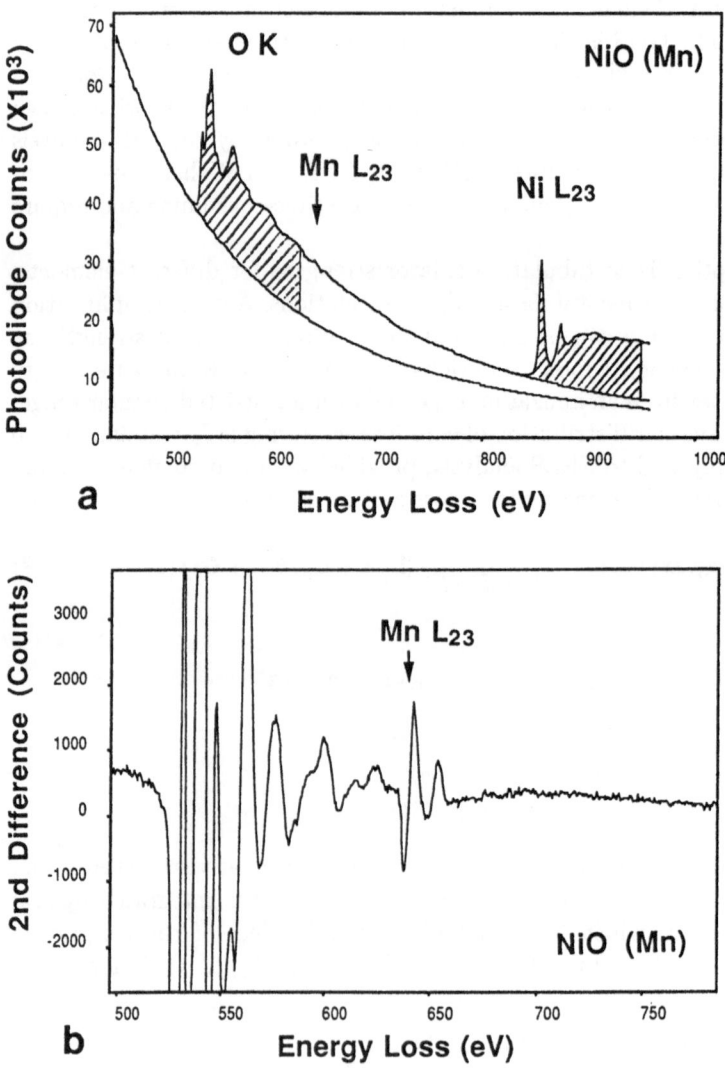

Fig.5.5. Caption see opposite page.

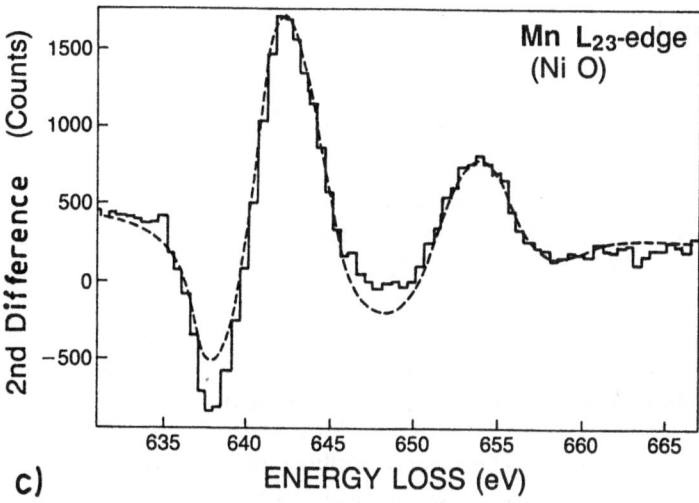

Fig. 5.5. (a) Energy-loss spectrum of NiO containing 0.5% manganese, recorded with 100 kV STEM with 20 mrad collection semi-angle, 0.8 nA probe current and 100 s' acquisition time. 100 eV integration windows are shown for the O K and Ni L_{23} edges. (b) Second-difference spectrum, showing manganese L_{23} edge. (c) Second-difference managanese edge fitted to the corresponding edge recorded from a potassium permanganate standard (dashed curve).

background were negligible, dI_1 would be simply the Poisson shot noise in the signal I_1 , but for small elemental concentrations the second term within the square root in (5.9) becomes dominant. For comparable fitting and integration widths, h is typically in the range 5 to 10, reflecting the fact that the pre-edge background noise becomes "amplified" in the process of extrapolation.

    The situation can be improved if the fitting procedure is extended across the edge; in other words, we use multiple least squares (MLS) techniques to fit the observed intensity $J(E)$ to an expression of the form:

$$F(E) = AE^{-r} + B_1 S_1(E) + ...$$ (5.10)

where $S_1(E)$ is a standard edge profile stored in the MCA. This procedure also has the advantage that further standards, $S_2(E)$ etc. (representing other ionization edges) can be added to the equation and values found for each fitting coefficient B. It does not matter if the edges are closely spaced, since we do not attempt to model the background for the higher-energy edges as a power law.

    If the specimen is very thin, such that plural scattering can be ignored, the standard spectra can be calculated from differential cross-sections [5.18] and the coefficients B are simply the product of the areal density of the appropriate element and the zero-loss intensity. Alternatively, each $S(E)$ can be measured from a test specimen containing the appropriate element [5.19], in which case the atomic ratio of two elements can be obtained from the fitting coefficients by:

$$\frac{N_1}{N_2} = \frac{B_1}{B_2} \frac{I_1(\alpha_o, \Delta)}{I_2(\alpha_o, \Delta)} \frac{\sigma_2(\alpha_o, \Delta)}{\sigma_1(\alpha_o, \Delta)} \tag{5.11}$$

where I_1 and I_2 are integrals of the standard spectra (over some convenient energy window Δ); σ_1 and σ_2 are the corresponding core-loss cross-sections.

Unfortunately, the region just above the edge (where the core-loss intensity is highest) contains prominent ELNES fine structure, which depends on chemical environment. Either the reference edges should be obtained from a chemically similar standard or this region should be excluded from the MLS fitting. Also, if the reference spectra are obtained from standards whose thickness is different from that of the specimen being analysed, it may be necessary to add terms to (5.10) to allow for plural scattering. Despite these complications, the procedure has been successfully applied to determine small concentrations of elements in biological matrices [5.19,20].

To remove the need for a background term in (5.10), it is possible to fit first or second differentials of the data, as in EDX analysis. In the case of a parallel-recording spectrometer, these differentials can be obtained directly from the detector, by recording the spectrum two or three times with a small (few eV) energy shift applied between acquisitions, then subtracting the spectra to form a first or second difference. This procedure has the advantage (over numerical differentiation in the computer) that, for a small edge superposed on a large background, gain variations between individual channels of the detector array largely cancel.

Application of this technique to determine low elemental concentrations is illustrated by considering the spectrum from a nickel oxide film containing manganese impurity (Fig. 5.5a). The Ni/O atomic ratio is obtained by measuring the integrated intensities for the oxygen K and nickel L_{23} edges and applying (5.7) to yield a result which is consistent with the nominal composition. The manganese L_{23} edge is barely discernable in the normal spectrum but is easily seen in the second-difference spectrum recorded with 4 eV offset (Fig. 5.5b). Furthermore, the manganese can be quantified by fitting a second-difference reference spectrum for the manganese L_{23} edge, recorded from a potassium permanganate specimen, as shown in Fig. 5.5c. Although edge shapes may differ considerably, depending on the chemical bonding or oxidation state, a reasonable estimate of the weak Mn signal is obtained from the fitting coefficient, considering also the corresponding standard deviation and the χ^2 parameter. Manganese is found to be present at a level of only 0.5% in this specimen.

Using a similar technique, concentrations below 10 ppm of calcium have been detected [5.20]. This capability is illustrated in Fig. 5.6, which shows clear detection of about 100 ppm of calcium in a freeze-dried cryosection of brain tissue [5.21]. In the raw spectrum, the calcium L_{23} edge is practically obscured by channel-to-channel gain variations but the signal (which is about a thousand times weaker than the background) is easily visible in the first-difference spectrum (inset of Fig. 5.6). It has also been demonstrated that distributions of such dilute elements can be obtained by STEM spectrum-imaging in the difference mode, where two or three parallel-recorded spectra, offset in energy, are accumulated at each point in a two-dimensional array of pixels. Such an approach requires long acquisition times but has the advantage of offering quantitative distribution images of minor elements, which give rise to essentially no visible features in the normal spectrum [5.22].

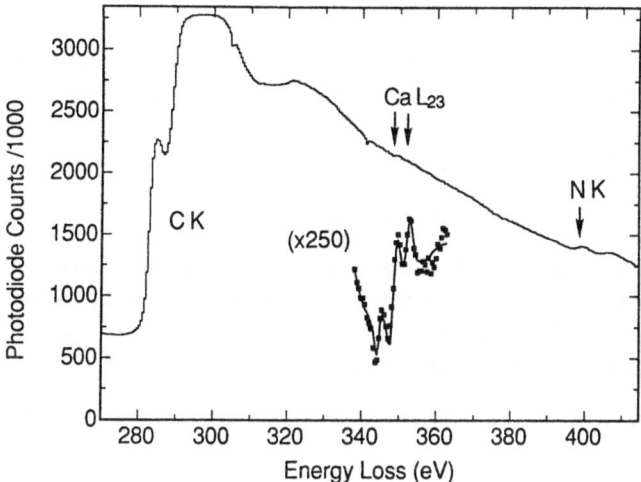

Fig. 5.6. Energy-loss spectrum of a freeze-dried cryosection of cerebellar cortex, in the region of endoplasmic reticulum contained in a postsynaptic terminal (specimen courtesy of S.B. Andrews). It was recorded in 200 s with 100 kV field-emission STEM at 5 nA probe current. The inset shows the first-difference spectrum (obtained with 6 eV shift) and a multiple-least-squares fit to a reference spectrum.

5.2.4 Sensitivity of EELS Elemental Analysis

The detection limits for microanalysis can be specified in two ways: as a minimum number of detectable atoms MDN (or minimum detectable mass: MDM) and as the minimum detectable atomic fraction f (or minimum mass fraction MMF). The best indication of these quantities is obtained experimentally, by pushing EELS to its limits with the best available technique and instrumentation. But it is also useful to calculate expected detection limits, to provide a guide to which experimental parameters are important and which limits arise from fundamental physics rather than instrumentation. A number of such calculations have been undertaken [5.11,23–25]. From (5.6), the core-loss signal I_I is:

$$I_I = I_l N_x \sigma_I(\alpha_o, \Delta) \qquad (5.12)$$

where N_x is the number of atoms (per unit area of specimen) to be detected. By analogy, the background I_b (due mainly to other "matrix" atoms, N_t per unit area) is:

$$I_b = I_1 N_t \sigma_b(\alpha_o, \Delta) \tag{5.13}$$

where σ_b is a cross section for inelastic scattering into the background. As a result of statistical noise in the signal and background and the limited detective quantum efficiency η of the spectrometer, the standard deviation of our measurement of I_I is, from (5.9):

$$dI_I = (\eta h I_b)^{1/2} \tag{5.14}$$

where we have assumed $I_b \gg I_I$, for a small detectable mass. We require that the signal I_I should be three times the standard deviation (Rose visibility criterion) to ensure that the signal I_I is genuine. Using a mean free path Λ_{el} to allow for elastic scattering (outside the collection aperture) by matrix atoms:

$$I_1 = IT \exp(-t/\Lambda_{el}) \tag{5.15}$$

where IT is the number of electrons (radiation dose D per unit area) received by the specimen during the acquisition period T. Combining these equations, we arrive at expressions for the minimum detectable atomic fraction f and the corresponding number of atoms MDN in a probe of diameter d:

$$f = (3/\sigma_I)[h\sigma_b/(\eta ITN_t)]^{1/2}\exp(t/2\Lambda_{el}), \tag{5.16}$$

$$MDN = (\pi/4)d^2 f N_t. \tag{5.17}$$

These quantities are given in Fig. 5.7 for the case of detection of Fe atoms on (or in) a 10 nm carbon film, assuming $h = 8$, $\eta = 1$ (ideal parallel detection), $E_0 = 100$ keV and hydrogenic cross sections. It is seen that detection of single atoms could be possible, but the radiation dose is very high ($D > 1 \times 10^{10}$ electrons/nm$^2 = 1.6 \times 10^5$ C/cm^2).

5.2.5 Comparison with EDX Spectroscopy

For elemental analysis, the technique most directly comparable to EELS is energy-dispersive x-ray spectroscopy carried out using a windowless, ultra-thin-window (UTW) or low-Z atmospheric-window detector. Both techniques take advantage of the ability to focus electrons into a very small probe, allowing analysis with a spatial resolution of the order of 10 nm in many cases.

The two techniques can even be performed simultaneously on the same TEM specimen. In CTEM, the objective aperture is normally removed to avoid errors in EDX quantification arising from electrons backscattered from the aperture, which generate characteristic x-rays from regions of the sample far outside the probe. Since EELS is best performed with an angle-selecting aperture, the CTEM would be operated in diffraction mode, with the spectrometer entrance aperture as the angle-selecting aperture.

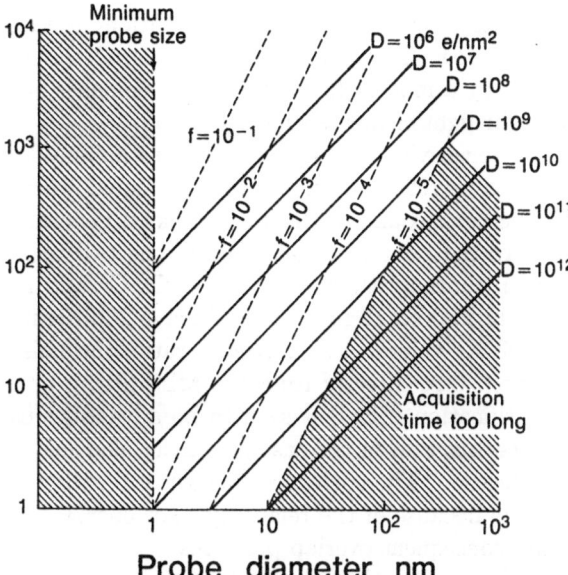

Fig. 5.7. Number of iron atoms detectable in a 10 nm carbon matrix by parallel-recording EELS [5.25]. Dashed lines give the corresponding atom fractions of iron; solid lines show the electron dose to the specimen, in electrons/nm². Shaded regions represent conditions which have been excluded on the basis of electron optics or excessive recording time.

The EDX signal (number of x-ray photons) is given by:

$$I_x = N\frac{IT}{e}\sigma_I(\pi, E_0)\omega\eta_x \tag{5.18}$$

where $\sigma_I(\pi, E_0)$ is the total ionization cross section for a particular atomic shell, ω is the corresponding x-ray fluorescence yield, and η_x is the collection efficiency of the x-ray detector. The ratio of the two signals may therefore be written as:

$$\frac{I_I}{I_x} = \frac{\sigma_I(\alpha_o, \Delta)}{\omega\sigma_x(\pi, E)}\frac{1}{\eta_x}\exp(-t/\Lambda_{el}) . \tag{5.19}$$

The exponential term, which is typically 0.3, represents loss of the EELS signal as a result of elastic scattering outside the collection aperture. Also, the electron spectrometer collects only part of the angular range of core-loss scattering and we analyse only a range Δ of the ionization edge, so the cross-section ratio in (5.19) is appreciably less than unity; a typical value is 0.1. However, the x-ray fluorescence yield ω, while close to unity for K-lines of heavy elements, is below 0.05 for photon energies below 2 keV and falls to 0.002 for carbon K-radiation. And whereas the DQE of a parallel-recording electron detector might exceed 0.5, measured values of collection efficiency for x-rays are typically no more than 0.003 (due largely to the limited solid angle of collection) and will be even lower for light-element radiation because of absorption in the detector window. Therefore, the EELS signal should exceed that available from EDX, by a modest factor for heavy elements but by a factor of several thousand for light elements such as carbon.

But as discussed earlier, the sensitivity of a spectroscopic technique depends on signal/noise considerations rather than just on the signal. Here, EDX has the advantage since the background to the x-ray peaks is generally lower than in EELS. To measure the relative sensitivities of the two techniques, test specimens with small concentrations of F, Na, P, Cl, Ca and Fe were deposited (from aqueous solution) on 10 nm carbon films [5.26]. Figure 5.8 shows the relative (EELS/EDX) values of signal/noise ratio, the noise component being obtained in each case directly from the MLS fitting procedure used for elemental quantification. EELS is seen to be advantageous in the case of very light elements, but also in the case of heavier elements if L-edges are used for the quantification. The corresponding x-ray L-emission peaks are in many cases unusable because of the relatively poor energy resolution of an EDX detector and consequent overlap problems.

The two techniques can also be compared on the basis of the accuracy of quantitative analysis. An accuracy of elemental ratios of 10% may be achievable from thin-film EDX spectroscopy, provided the system is calibrated (in terms of k-factor) for each element being analysed. However, this level of

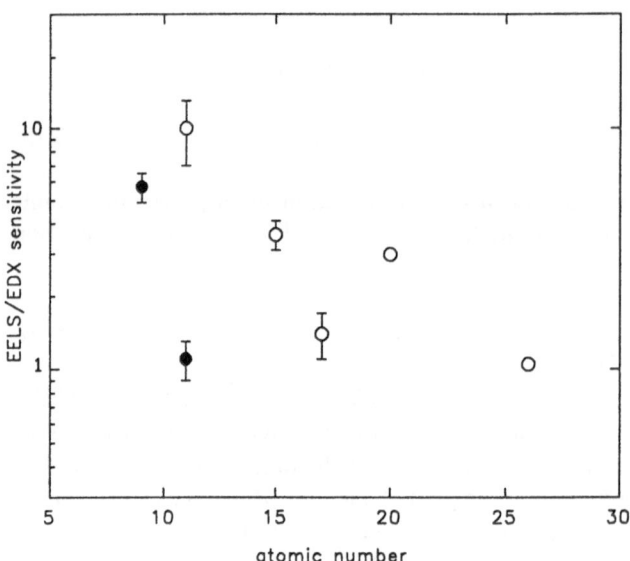

Fig. 5.8. Relative sensitivity of EELS (using parallel recording and first- or second-difference MLS processing of K and L edges), compared to EDX elemental analysis (using UTW detector and second-difference MLS processing of K emission peaks). Solid circles are for K shells, open circles for L shells. The increased sensitivity for Ca and Fe arises from the presence of white lines at the L edge threshold.

accuracy will be harder to achieve for light elements, where absorption of x-rays within the specimen (and at the detector) is not easy to quantify accurately. It is more difficult to specify the accuracy of EELS; the core-loss cross sections are known in some cases to within 10%, but plural scattering may introduce systematic errors in thicker specimens.

In terms of ease of use, EDX has a clear advantage. Software for elemental quantification is well established and, once set up, the spectrometer system requires little maintenance or adjustment. EELS requires more care and knowledge on the part of the operator, but in return the technique yields information beyond that of elemental composition, as we now illustrate.

5.3 Measurement of Specimen Thickness

A useful bonus to energy-loss spectroscopy is that it provides a convenient way of estimating the local thickness of a TEM specimen. This information is needed to convert the areal densities provided by EELS or EDX elemental analysis into true concentrations and for a variety of other procedures employed in electron microscopy.

The thickness information comes from measurement of the probability of inelastic scattering. The probability P_{0i} of no inelastic scattering is represented by the area I_0 under the zero-loss peak, relative to the total area I_t beneath the complete spectrum. This probability is also given by Poisson statistics, in terms of the total mean free path Λ for inelastic scattering (all energy-loss processes), namely $P_{0i} = I_0/I_t = \exp(-t/\Lambda)$, leading to:

$$t/\Lambda = \ln(I_t/I_0). \tag{5.20}$$

Although the value of Λ depends on the sample material, the incident-electron energy and the collection angle, it is of the order of 100 nm for 100 keV electrons, so a rough estimate of thickness is available immediately.

For thin specimens $(t/\Lambda < 2)$, most of the intensity lies below 200 eV energy loss and a single readout of the spectometer (without gain change) is sufficient. For thicker specimens, the spectrum may have to be recorded over a wider range.

To get a true measurement of thickness, we need to know Λ for the material being examined. Ideally, the value of Λ is obtained from a calibration sample, whose thickness is determined by other means. However, measurements on elements and simple inorganic compounds (with constituents of comparable atomic number) have shown that Λ can be parameterized in terms of mean atomic number:

$$
\begin{aligned}
\Lambda &= 106F(E_0/E_m)/\ln(2\beta E_0/E_m) \\
F &= (1 + E_0/1022)/(1 + E_0/511)^2 \\
E_m &= 7.6Z^{0.36}
\end{aligned}
\tag{5.21}
$$

where Λ is in nm, E_0 in keV and β in mrad. Because of the logarithmic dependence on β, the latter need not be known to high accuracy. The accuracy of (5.21) is believed to be 20% [5.27]. Values of E_m have been tabulated for some common materials, allowing greater accuracy in these cases [5.28].

For better accuracy, use can be made of a sum rule which relates the energy-loss intensity to optical properties via the Kramers–Kronig equations. It is necessary to know the quantity $\mathrm{Re}\{1/\epsilon(0)\}$, which in most materials can be taken as the reciprocal of the square of the optical refractive index, or as zero for a metal or semi-metal. Again, a low-loss spectrum is recorded, up to about 200 eV. The data processing can take the form of a short computer program, run on the data-analysis system, yielding thicknesses which are believed to be accurate to about 10% [5.29].

Another technique, based on the Bethe sum rule, provides an estimate of the mass-thickness of the specimen and is attractive for use with organic or biological specimens, since it requires no knowledge of the chemical composition [5.9]. However, it requires recording the spectrum over an extended range of energy loss, the removal of plural scattering by deconvolution and (in many cases) correction for the stray-scattering background of the spectrometer.

5.4 Spectrum Deconvolution

For detailed analysis of an energy-loss spectrum, it is useful to be able to remove the effects of plural scattering of the transmitted electrons. There are several alternative ways of doing this, of which the matrix procedure of Schattschneider and co-workers offers certain advantages [5.30,31]. But the most commonly employed procedures are based on fast Fourier transforms, and are known as the Fourier–log and Fourier–ratio methods [5.11].

The Fourier–log procedure is based on the fact that the recorded spectral intensity can be written in the form:

$$J(E) = \sum_n C_n [R(E) \otimes S_n(E)] \tag{5.22}$$

where $R(E)$ is a unit-area function representing the energy resolution of the spectrometer system, $\otimes$ represents a convolution with respect to energy loss, and $S_n(E)$ represents the n-fold self-convolution of the single-scattering intensity $S(E)$, also known as the single-scattering distribution (SSD). In order to satisfy Poisson statistics, the coefficient C_n must be given by:

$$C_n = I_0^{1-n}/n!. \tag{5.23}$$

If we take the Fourier transform of (5.24), each convolution becomes a simple product, giving:

$$j(f) = \sum_n C_n\, r(f)[s(f)]^n = I_0\, r(f) \exp[s(f)/I_0]. \tag{5.24}$$

Since $I_0 \cdot R(E)$ is just the zero-loss peak profile $Z(E)$, (5.26) can be written in the form [5.32]:

$$s(f) = I_0 \ln[j(f)/z(f)] \tag{5.25}$$

The SSD is therefore obtained by calculating the Fourier transform of the experimental spectrum, dividing by the transform of the zero-loss peak, taking the (complex) natural logarithm of the ratio and finally an inverse Fourier transform to obtain $S(E)$. However, $S(E)$ is the ideal SSD, unbroadened by the spectrometer resolution function $R(E)$. While it would be desirable to know this function, $S(E)$ can be successfully recovered (without excessive noise amplification) only if $J(E)$ is noise-free or coarsely sampled [5.32,33]. Usually we have to settle for a single-scattering distribution $J_1(E) = R(E) \otimes S(E)$ whose Fourier transform is given by:

$$j_1(f) = I_0\, r(f) \ln[j(f)/z(f)] = z(f) \ln[j(f)/z(f)]. \tag{5.26}$$

A common simplification [5.34] is to ignore instrumental broadening by replacing the zero-loss peak by a delta function of equal area I_0 before computing the transform $j(f)$. Then $z(f) = I_0$ in (5.28) and the deconvolution reduces to one Fourier transform and a simple numerical division, followed by an inverse transform. Although this approximate method can create spectral artifacts, these are noticeable only when the method is applied to plasmon-loss spectra of moderately thick specimens [5.33].

When applied to a region of the spectrum which contains ionization edges, the Fourier–log method removes plural scattering from both the core-

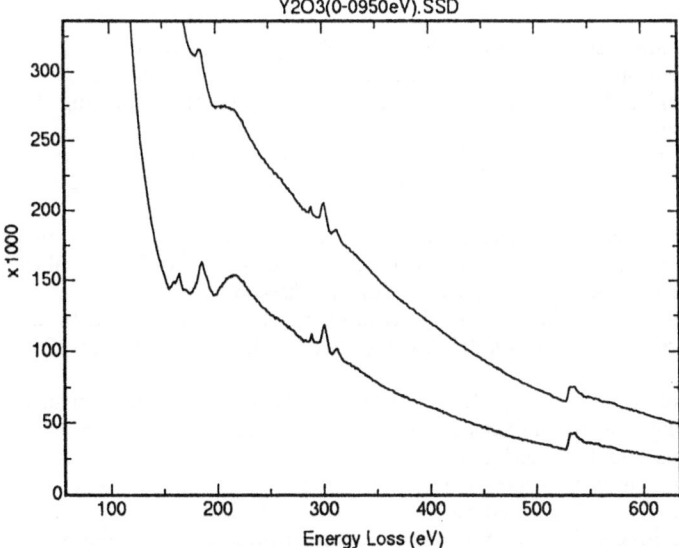

Fig. 5.9. Spectra of yttrium oxide on carbon, before (upper trace) and after (lower trace) Fourier-log deconvolution.

loss intensity and from the background. As a result, the visibility (signal/background ratio) of the edges is improved, particularly at lower energy loss (Fig. 5.9).

The Fourier–ratio method of removing plural scattering is applicable only to limited regions of a spectrum, such as an ionization edge or series of edges. The background preceding the edge of lowest energy loss must first be subtracted, using (5.2), and the intensity extrapolated to sufficiently large energy loss, such that the core-loss intensity $J_I(E)$ approximates to zero at both ends of the region. The Fourier transform $j_I^1(f)$ of the single-scattering distribution $J_I^1(E)$ is given by [5.11]:

$$j_I^1(f) = g(f)\, j_I(f)/j_v(f) \tag{5.27}$$

where $j_I(f)$ and $j_v(f)$ are Fourier transforms of the core-loss SSD and of the low-loss (valence-loss) spectrum (including the zero-loss peak), respectively; $g(f)$ is the Fourier transform of a "modification function" $G(E)$, required to prevent noise amplification. If $G(E) = Z(E)$, the resolution of $J_I^1(E)$ is the same as that of recorded data $J_I(E)$. Alternatively, $G(E)$ can be taken as a Gaussian distribution, representing the central portion of $Z(E)$ but without extended tails due to the point-spread function of the electron detector; deconvolution via (5.29) then removes not only plural scattering but also the effect of the tails, as in Fig. 5.3. A delta-function approximation is also available, which reduces the computing time [5.11,34].

References

5.1 C. Jeanguillaume and C. Colliex: Spectrum-image: the next step in EELS digital acquisition and processing. Ultramicroscopy **28**, 252–257 (1989)

5.2 J.A. Hunt and D.B. Williams: Electron energy-loss spectrum-imaging. Ultramicroscopy **38**, 47–73 (1991)

5.3 A.J. Craven and T.W. Buggy: Correcting electron energy loss spectra for artefacts introduced by a serial collection system. J. Micr. **136**, 227–239 (1984)

5.4 M.M. Disko: Practical methods for quantitative analysis of energy loss spectra. In *Microbeam Analysis 1986*, ed. by A.D. Romig and W.F. Chambers (San Francisco Press, San Francisco 1986) pp.429–433

5.5 D.C. Joy and D. M. Maher: The electron energy-loss spectrum: facts and artifacts. In *Scanning Electron Microscopy 1980/I*, (SEM Inc., Chicago 1980) pp.25–32

5.6 O.L. Krivanek, C.C. Ahn and R.B. Keeney: Parallel detection electron spectrometer using quadrupole lenses. Ultramicroscopy **22**, 103–106 (1987)

5.7 A.J. Gubbens and O.L. Krivanek: Applications of a post-column imaging filter in biology and materials science. Ultramicroscopy **51**, 146–159 (1993)

5.8 R.F. Egerton, Y.-Y. Yang and S.C. Cheng: Characterization and use of the Gatan parallel-recording electron energy-loss spectrometer. Ultramicroscopy **48**, 239–250 (1993)

5.9 P.A. Crozier and R.F. Egerton: Mass-thickness determination by Bethe-sum-rule normalization of the electron energy-loss spectrum. Ultramicroscopy **27**, 9–18 (1989)

5.10 J. Van Puymbroek, W. Jacob and P. Van Espen: Methodology for spectrum evaluation in quantitative electron energy-loss spectrometry using the Zeiss CEM902. J. Micr. **166**, 273–286 (1992)

5.11 R.F. Egerton: *Electron Energy-Loss Spectroscopy in the Electron Microscope* (Plenum, New York, 1986)

5.12 H. Tenailleau and J.M. Martin: A new background subtraction for low-energy EELS core edges. J. Micr. **166**, 297–306 (1992)

5.13 R.F. Egerton, Y.Y. Chen and Y.-Y Yang: EELS of "thick" specimens. Ultramicroscopy **38**, 349–352 (1991)

5.14 Z.L. Wang: Dynamical simulations of energy-filtered electron diffraction patterns. Acta. Cryst. A **48**, 674–688 (1992)

5.15 T. Malis and J.M. Titchmarsh: A k-factor approach to EELS analysis. In *Electron Microscopy and Analysis 1985* (Institute of Physics, Bristol 1985) pp.181–191

5.16 F. Hofer: Determination of inner-shell cross sections for EELS quantification. Microsc. Microanal. Microstruct. **2**, 215–230 (1991)

5.17 R.F. Egerton: Oscillator-strength parameterization of inner-shell cross sections. Ultramicroscopy **50**, 13–28 (1993)

5.18 J.D. Steele, J.M. Titchmarsh, J.N. Chapman and J.H. Paterson: A single stage process for quantifying electron energy-loss spectra. Ultramicroscopy **17**, 273–276 (1985)

5.19 R.D. Leapman and C.R. Swyt: Separation of overlapping core edges in electron energy loss spectra by multiple least squares fitting. Ultramicroscopy **26**, 393–403 (1988)

5.20 H. Shuman and A.P. Somlyo, Electron energy loss analysis of near-trace-element concentrations of calcium. Ultramicroscopy **21**, 23–32 (1987)

5.21 R.D. Leapman, J.A. Hunt, R.A. Buchanan and S.B. Andrews: Measurement of low calcium concentration in cryosectioned cells by parallel-EELS mapping. Ultramicroscopy **49**, 225–234 (1993)

5.22 R.D. Leapman and J.A. Hunt: Compositional imaging with electron energy loss spectroscopy. In *Microscopy, The Key Research Tool* (Electron Microscopy Society of America, Woods Hole, MA, USA 1992) pp.39–49

5.23 M. Isaacson and D. Johnson: The microanalysis of light elements using transmitted energy-loss electrons. Ultramicroscopy **1**, 33–52 (1975)

5.24 C. Colliex: Electron energy-loss spectroscopy in the electron microscope. In *Adv. in Optical and Electron Microscopy*, ed. by R. Barer and V.E. Coslett (Academic, London 1984) Vol. 9, pp.65–177

5.25 R.D. Leapman: EELS quantitative analysis. In *Transmission Electron Energy Loss Spectrometry in Materials Science*, ed. by M.M. Disko, C.C. Ahn and B. Fultz (TMS EMPMD Monograph Series No. 2, The Minerals, Metals and Materials Society, Warrendale 1992) pp.47-83

5.26 R.D. Leapman and J.A. Hunt: Comparison of detection limits for EELS and EDX. Microsc. Microanal. Microstruct. **2**, 231–244 (1991)

5.27 T. Malis, S.C. Cheng and R.F. Egerton: The EELS log-ratio technique for specimen-thickness measurement in the TEM. J. Electron Microsc. Techn. **8**, 193–200 (1988)

5.28 R.F. Egerton: A data base for energy-loss cross sections and mean free paths. *Proc. 50th Ann. Meeting of EMSA* (San Francisco Press, San Francisco 1992) pp.1264–1265

5.29 R.F. Egerton and S.C. Cheng: Measurement of local thickness by electron energy-loss spectroscopy. Ultramicroscopy **21**, 231–244 (1987)

5.30 D.S. Su and P. Schattschneider: Numerical aspects of the deconvolution of angle-integrated electron energy-loss spectra. J. Micr. **167**, 63–75 (1992)

5.31 D.S. Su and P. Schattschneider: Deconvolution of angle-resolved electron energy-loss spectra. Phil. Mag. A **65**, 1227–1140 (1992)

5.32 D.W. Johnson and J.C.H. Spence: Determination of the single-scattering probability distribution from plural-scattering data. J. Phys. D **7**, 771–780 (1974)

5.33 R.F. Egerton and P.A. Crozier: The use of Fourier techniques in electron energy-loss spectroscopy. *Scanning Microscopy, Suppl.2* (Scanning Microscopy International, AMF O'Hare, Chicago 1988) pp.245–254

5.34 C.R. Swyt and R.D. Leapman: Plural scattering in electron energy-loss (EELS) microanalysis. Scanning Microscopy **1**, 73–82 (1982)

6. Electron Spectroscopic Diffraction

Joachim Mayer, Christine Deininger, and *Ludwig Reimer*

6.1 Advantages of Electron Spectroscopic Diffraction

In principle, electron diffraction studies in a transmission electron microscope (TEM) provide much the same crystallographic information as does X-ray or neutron diffraction. Compared to the latter two, two important problems arise in electron diffraction owing to the much stronger interaction of electrons with matter:

1) The kinematical approximation is valid only for very thin films of 1–5 nm and

2) Inelastic scattering processes contribute to a diffuse background. For thicker specimens this background dominates the entire diffraction pattern.

The first problem can be solved by using the dynamical theory of electron diffraction, which describes the interactions between the primary beam and the diffracted beams. This theory allows the intensities in the diffraction patterns of crystals to be calculated with high accuracy when the Fourier coefficients V_g (structure amplitudes) of the crystal lattice potential are known. A summary of the dynamical theory is presented in Sect. 6.3.1.

For the second problem, removal of the inelastic scattering background improves the quality of any electron diffraction pattern, irrespective of the type of pattern and material. Energy filtering has no disadvantages. It is therefore desirable to use an EFTEM in the electron spectroscopic diffraction (ESD) mode, whenever possible, for all electron diffraction studies. In the following, we begin by discussing typical applications for different classes of materials (amorphous, polycrystalline and single-crystalline). For all of these studies, selected-area electron diffraction with parallel illumination of the sample in an area defined by a field-limiting aperture is used. The scattered intensities depend on scattering angle and scattering direction. For amorphous materials, information can be obtained on radial distribution functions and pair correlations (Sects. 6.2.1 and 6.2.2). The contrast of Debye–Scherrer rings from polycrystalline films can be increased and thicker films can be investigated when the elastically scattered electrons are selected by filtering (Sect. 6.2.3). For crystalline materials, the crystallographic parameters, such as the lattice type, dimensions and orientation, can be determined with higher accuracy and weak (e.g. superlattice) reflections can be detected for thicker specimen foils.

Springer Series in Optical Sciences, Vol. 71
Energy-Filtering Transmission Electron Microscopy
Editor: Ludwig Reimer   ©Springer-Verlag Berlin Heidelberg 1995

Whereas conventional electron diffraction needs illumination by electron beams or probes with apertures in the range 10^{-5}–10^{-3} rad for sharp Bragg spots, convergent-beam electron diffraction (CBED) requires electron probes with an illumination cone of $\simeq 10^{-2}$ rad, which can be produced only in modern TEMs. The electron probe is focused on the crystalline specimen and very small (nanometre-sized) areas or crystallites can then be investigated. Two major advantages of CBED compared to conventional electron diffraction are that

1) very small specimen areas can be analysed and

2) dynamical scattering information is generated for many different incident beam directions, which can be recorded simultaneously.

In many applications, this second advantage of CBED is much more important than the first, since it helps to overcome the ambiguities introduced in conventional diffraction patterns by the dependence of the Bragg diffracted intensities on the actual excitation error.

The presence of inelastically scattered electrons has prevented the development of quantitative electron crystallography comparable to X-ray crystallography. With the advent of energy filtering in TEM, it has become possible to acquire very accurate electron diffraction data, in particular with the CBED technique, which will be described in Sect. 6.4. The capability to measure structure factors for small areas, and hence the possibility of performing electron crystallography on a very local scale, is becoming an exciting new area and may prove to be one of the most important applications of electron spectroscopic diffraction (ESD). The methods and experimental techniques required are still under development. The quantitative comparison between calculated and experimental diffraction intensities is a numerically very demanding problem, requiring computing powers that have only recently become available.

By means of quantitative CBED, other important parameters, such as the specimen thickness and the crystal orientation with respect to the beam, can be determined accurately. CBED also allows the arrangement of the reflections in the high-order Laue zones (HOLZ) to be studied, by examining the deficiency lines, also referred to as HOLZ lines, in the (000) disc. From the geometry of the HOLZ lines, the lattice parameters, changes in the stoichiometry of the material or the accelerating voltage of the microscope can be measured (Sect. 6.4.3).

All the applications discussed above are based on elastic filtering of diffraction patterns. Inelastic scattering processes can be classified in four groups according to the different energy losses:

1) Thermal diffuse scattering (TDS) by electron-phonon interaction,

2) Plasmon excitation,

3) Inner-shell excitation, and

4) Compton scattering.

Only the latter three result in energy losses that can be separated by energy filtering. The background produced by thermal diffuse electron–phonon scattering shows energy losses only below 10 meV and cannot be removed from the diffraction patterns except in the case of large-angle CBED (Sect. 6.4.2), where TDS is also removed. In most cases, the TDS background can be neglected because its intensity is weak. However, longer exposures show that the intensity distribution of the TDS shows characteristic features (e.g. thermal diffuse streaks), which are associated with the phonon dispersion of the crystal. Elastic filtering improves the visibility of this contribution because the dominating background produced by plasmon losses and inner-shell ionizations can be removed by elastic zero-loss filtering.

In ESD, most applications require zero-loss filtering. In electron spectroscopic imaging (ESI) (Chap. 7), most of the analytical methods involve inelastically scattered electrons whereas their use is very limited in ESD. Examples of the use of inner-shell ionization losses will be given in Sect. 6.3.2. Finally, an imaging energy filter makes it possible to image and analyse the dispersion of plasmon and Compton scattering (Sect. 6.5).

6.2 Diffraction at Amorphous and Polycrystalline Specimens

6.2.1 Amorphous Diffraction Patterns

The diffuse diffraction maxima of amorphous films (Fig. 6.1) can be used to determine the radial density distribution $\rho(r)$ of atoms where $4\pi r^2 \rho(r) dr$ is the probability of finding neighbouring atoms inside a shell between the radial distances r and $r + dr$. In a single-scattering approximation, the radial intensity distribution satisfies the equation

$$\frac{1}{I_0}\frac{dI}{d\Omega} = N|f(q)|^2 \left[1 + \int_0^\infty 4\pi r^2 \rho(r)\frac{\sin(2\pi qr)}{2\pi qr}dr\right] + N\frac{dI_{in}}{d\Omega} \qquad (6.1)$$

(I_0 = incident electron current contributing to the diffraction pattern, dI = intensity recorded with a detector of solid angle $d\Omega$, $q = \theta/\lambda$ = spatial frequency, θ = scattering angle); it oscillates around the curve $N|f(q)|^2$, which would be observed for independent elastic scattering at the N atoms contributing to the diffraction pattern with a scattering amplitude $f(q)$ [6.1]. This distribution is superimposed on the angular distribution $dI_{in}/d\Omega$ of inelastically scattered electrons. After forming the normalized function

$$i(q) = \frac{1}{I_0 N|fq|^2}\left[\frac{dI}{d\Omega} - N|f(q)|^2 - N\frac{dI_{in}}{d\Omega}\right] \qquad (6.2)$$

the radial density distribution is obtained by an inverse Fourier transform [6.1,2] (Sect. 6.2.2).

Zero-loss filtering of an amorphous diffraction pattern can remove the inelastic contribution, whereupon the maxima and minima of the diffracted intensities become more pronounced and it is easier to fit the elastic contribution $N|f(q)|^2$, this will not, however, be exactly proportional to the differential elastic cross-section $d\sigma/d\Omega=|f(q)|^2$ because of multiple elastic scattering, which also occurs in thin films. The search for a good average curve around which the maxima and minima oscillate is hence the largest problem for a quantitative analysis.

The modes of electron spectroscopic diffraction (ESD) for generating energy-filtered diffraction patterns are described in Sect. 1.4.3. The intensity in amorphous electron diffraction patterns can be recorded sequentially with the aid of Grigson coils [6.3,4] situated between the last projector lens and the final screen; these are used to scan the pattern radially across a diaphragm or slit in front of a scintillator or semiconductor detector. The diffraction pattern can also be scanned across a slit by rocking the illuminating electron beam, which shifts the diffraction pattern on the final screen. Another possibility is densitometry of exposed photographic emulsions. For quantitative zero-loss filtering of amorphous and other diffraction patterns, it is necessary to record the intensity up to diffraction angles of a few tens of milliradians. Owing to the second-order aberration of filter lenses, only the central part of the diffraction pattern will be filtered and accurate zero-loss filtering is possible only when this aberration is corrected. When the incident beam is rocked, only a diffracted beam parallel to the optic axis is recorded and the shift of the selected energy window across the pattern has no influence.

Figure 6.1 shows the radial intensity distributions of a 15 nm germanium film, unfiltered, zero-loss filtered (elastically scattered electrons only) and filtered at an energy loss of $E = 18$ eV (plasmon loss). It demonstrates the increase of the relative intensities of the diffraction maxima and minima caused by zero-loss filtering, whereas the oscillation between maxima and minima is much lower for inelastically scattered electrons with $E = 18$ eV. This shows that the inelastically scattered electrons are also diffracted. However, instead of the primary beam of a small illumination aperture associated with elastic scattering, the intensity distribution has to be convolved with the angular distribution (1.19) of inelastically scattered electrons, proportional to $(\theta^2 + \theta_E^2)^{-1}$. The characteristic angle $\theta_E \simeq E/2E_0$ increases with increasing energy loss E. Though the angular distribution of this Lorentzian profile shows a strong peak in the range $0 < \theta < \theta_E$, it falls slowly with increasing θ and is overlapped by Compton scattering (Sect. 6.5.1), especially when multiplied with $d\Omega = 2\pi\theta d\theta$ to include all inelastically scattered electrons between θ and $\theta+d\theta$ irrespective of azimuth. With increasing film thickness, the multiple elastic scattering also contributes to the decrease of the amorphous diffraction maxima and minima.

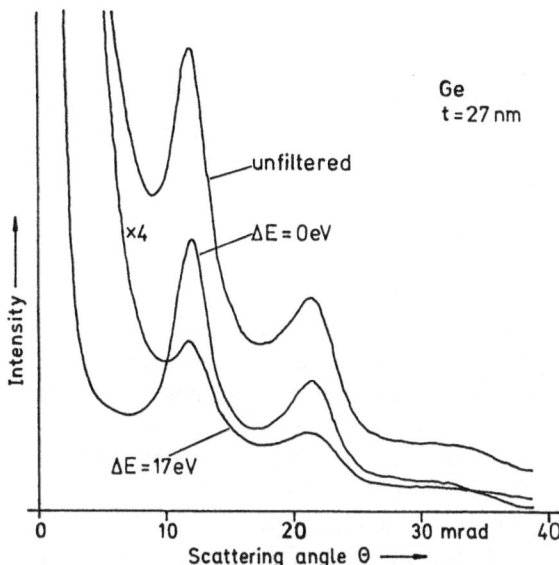

Fig. 6.1. Radial intensity distribution of the diffraction pattern of an evaporated a-morphous 27 nm germanium film: unfiltered, zero-loss and plasmon-loss filtered at $E=18$ eV.

6.2.2 Determination of Pair Correlation

The radial density distribution function $\rho(r)$ as introduced in Sect. 6.2.1 is a real space function, which shows maxima (minima) at specific radii r, at which the probability of finding a n-th neighbour of a given atom is high (low). For large r, the density $\rho(r)$ approaches the average density $\rho_0(r)$ and hence the radial density function asymptotically approaches a parabolic function, which tends to infinity for large r. A more useful quantity is obtained by forming the difference between $\rho(r)$ and the average density ρ_0:

$$4\pi r[\rho(r) - \rho_0] = G(r) = 8\pi \int_0^\infty i(q) \sin(2\pi q r) q \, dq. \qquad (6.3)$$

$G(r)$ is called the reduced density function (RDF) [6.2] and is also frequently referred to as the pair correlation function. It describes the deviation of the radial density $\rho(r)$ from the average density ρ_0 and is obtained by an inverse Fourier transform of the normalized intensity function $i(q)$ (6.2).

For amorphous specimens, $G(r)$ exhibits in general the following characteristic features:
1) A pronounced first maximum at a radius r_1, which is essentially determined by the radius of the first coordination shell around an atom. With increasing r, further maxima occur but these are much less pronounced. The positions of these maxima indicate the radii of the n-th neighbour coordination shells.
2) By definition, $\rho(r)$ is zero for $0 \le r \le r_1$. From equation (6.3), it follows that

$$G(r) = -4\pi r \rho_0 \qquad (6.4)$$

in this range. Thus the average density of the amorphous material can be calculated from the slope of experimental $G(r)$ curves at small r values.

3) In practice, the upper integration limit of (6.3) is limited by an upper experimental value q_{max}. This leads to the so called truncation effect, i.e., the occurrence of spurious maxima in $G(r)$ for small r, which have no physical meaning.

For amorphous materials consisting of several types of atoms, the above analysis is only valid when the approximation

$$f_i(\theta) = C_i f(\theta) \tag{6.5}$$

is valid, where C_i is a constant independent of the scattering angle θ. This approximation means that the angular dependence of the scattering amplitude $f(\theta)$ is assumed to be independent of the atomic number. For materials with different types of atoms, a maximum in $\rho(r)$ and hence $G(r)$ corresponds to a maximum in an individual $\rho_{ij}(r)$ where i and j may be atoms of different $(i \neq j)$ or the same $(i = j)$ type. For this situation, the positions of the maxima in $G(r)$ result in the most probable distances between atoms of type i and j. The identification of these distances in terms of the actual atom pairs can often be inferred from a knowledge of neighbour distances in relevant crystal structures. In practice, this procedure for determining most probable distances for different pairs of atoms can only be used if the individual $\rho_{ij}(r)$ do not overlap. In the case of severe overlap, reliable information can be obtained only by using more than one kind of radiation, by combining electron diffraction with x-ray or neutron diffraction for example.

If the peaks in $G(r)$ do not overlap, coordination numbers $C(r_1, r_2)$ can be calculated by evaluating the integral [6.2]:

$$C(r_1, r_2) = \int_{r_1}^{r_2} 4\pi r^2 \rho(r) dr, \tag{6.6}$$

where r_1 and r_2 are the distance limits between which the peak in $G(r)$, i.e. the corresponding interatomic distance, occurs. With (6.3) we obtain

$$C(r_1, r_2) = \frac{4}{3}\pi(r_2^3 - r_1^3)\rho_0 + \int_{r_1}^{r_2} r \, G(r) \, dr \tag{6.7}$$

for the case of an amorphous material consisting of only one type of atom. In the case of alloys (containing different types of atoms), a similar expression can be established, in which the contribution G_{ij} of atom pairs i, j has to be weighted with the atomic scattering amplitudes [6.2].

6.2.3 Debye–Scherrer Patterns of Polycrystalline Specimens

The Debye–Scherrer ring patterns from polycrystalline films show an increasing background of inelastically scattered electrons with increasing film thickness, though part of the background is also caused by thermal-diffuse scattering with negligible energy loss. Zero-loss filtering therefore results in an

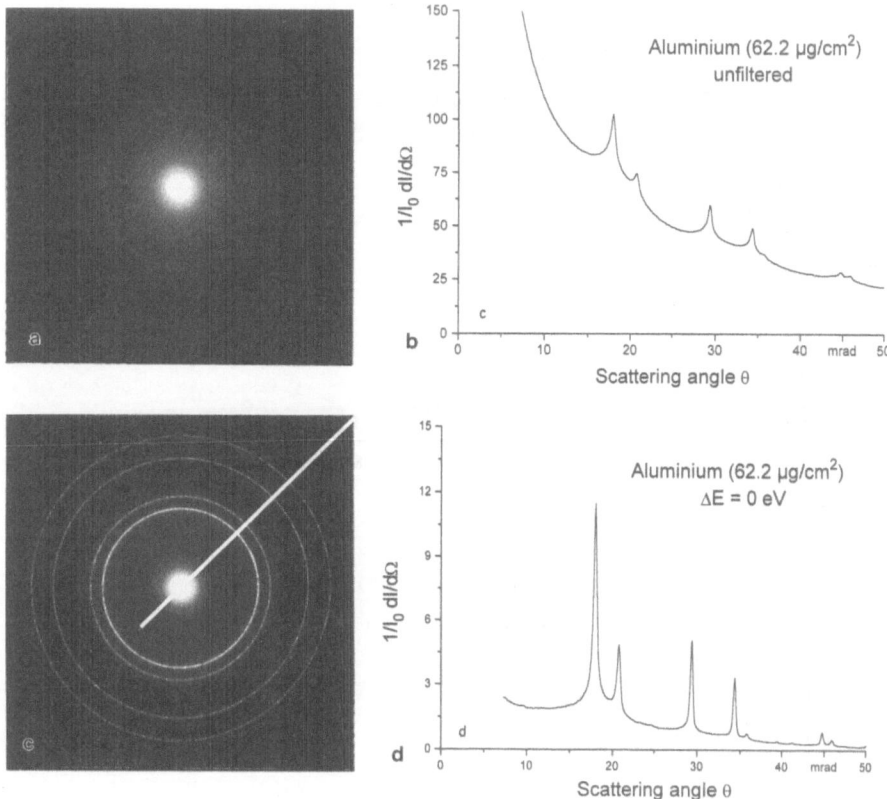

Fig. 6.2. Diffraction patterns of a 62 $\mu g/cm^2$ (230 nm) evaporated aluminium film (**a**) unfiltered and (**c**) zero-loss filtered with corresponding normalized radial intensity distributions (**b**) and (**d**) [note the different scales of $(1/I_0)$ dI/dΩ].

increase of the ring intensities relative to the background, as firstly demonstrated by Boersch by placing a retarding grid in front of a photographic emulsion (Fig. 2.43). In the past, zero-loss filtering was also used for quantitative work on the influence of temperature and dynamical theory on ring intensities [6.5–7] and Grigson coils [6.3] were employed to record zero-loss filtered amorphous and Debye–Scherrer patterns by scanning the pattern across a diaphragm of solid angle $\Delta\Omega$ in front of an electron detector [6.8,9].

Figure 6.2 demonstrates the strong influence of zero-loss filtering on the Debye-Scherrer diagram of a 230 nm evaporated aluminium film. For a quantitative comparison of intensities in electron diffraction patterns from different

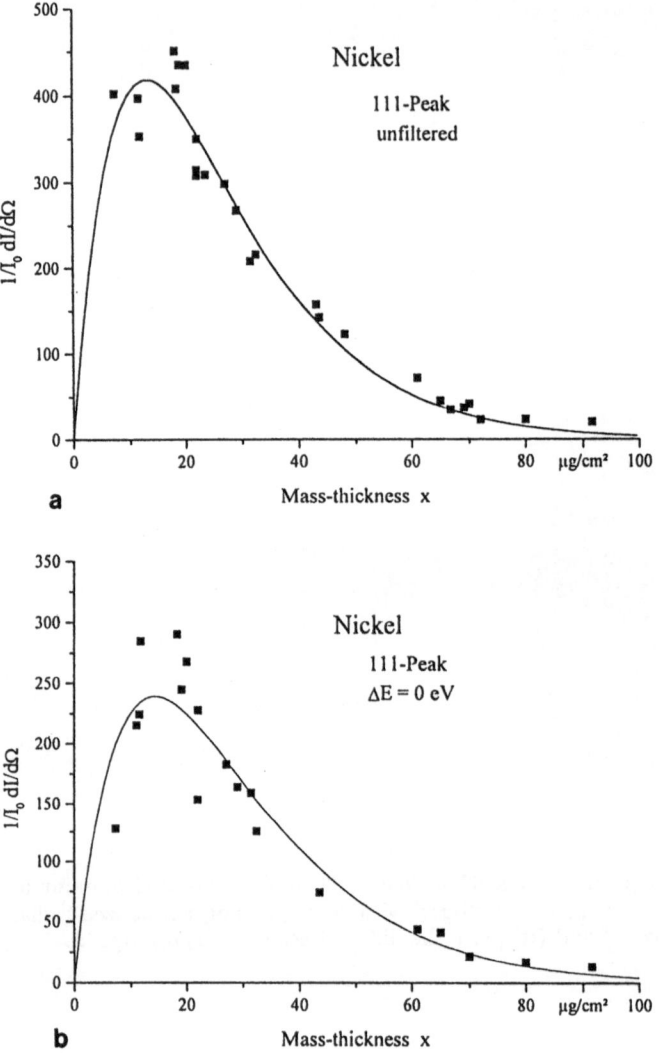

Fig. 6.3. Caption see opposite page.

film thicknesses and elements, it is necessary to record a normalized intensity distribution $(1/I_0)\ \Delta I/\Delta\Omega$. The current (intensity) ΔI passing through a solid angle element $\Delta\Omega$ at a scattering angle θ is recorded in arbitrary units by a scintillator–photomultiplier combination. Though I_0 will be the intensity of the primary beam spot in the absence of a specimen, direct measurement in the diffraction pattern overloads the photomultiplier. The transmissions

$$T(\alpha_i) = \frac{I'(\alpha_i)}{I_0'} = \frac{1}{I_0} \int_0^{\alpha_i} \frac{\Delta I}{\Delta\Omega} 2\pi\theta\, d\theta \qquad (6.8)$$

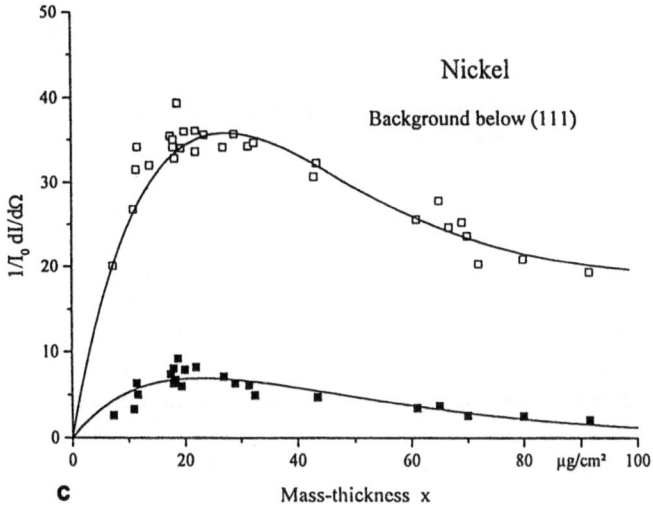

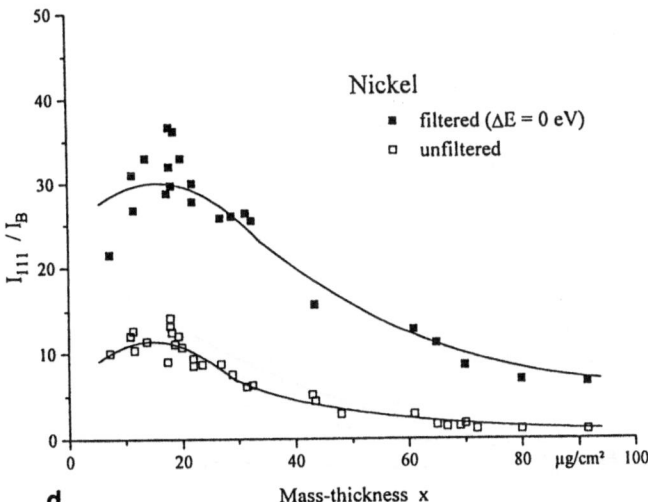

Fig. 6.3. Peak intensities I_{111} of 111 Debye–Scherrer rings of evaporated nickel films as a function of mass–thickness $x = \rho t$ in (a) unfiltered and (b) zero-loss filtered diffraction patterns. (c) Background intensity I_B below the 111 ring and (d) ratios I_{111}/I_B (unfiltered: open and zero-loss filtered: full squares).

$(i=1,2)$ are therefore measured for two apertures $\alpha_1 = 9.2$ and $\alpha_2 = 38.7$ mrad (objective diaphragms of 50 and 200 μm). (The dash on I' is introduced to distinguish intensities measured in the image mode from those in the diffraction pattern). The difference $T(\alpha_2) - T(\alpha_1)$ can be compared with the intensity ΔI between $\theta = \alpha_1$ and α_2, which results in

$$\frac{1}{I_0}\frac{dI}{d\Omega}[T(\alpha_2) - T(\alpha_1)] = \int_{\alpha_1}^{\alpha_2}\frac{\Delta I}{\Delta\Omega}2\pi\theta\, d\theta \quad . \tag{6.9}$$

Such normalized intensity records are shown in Fig. 6.2, demonstrating that the intensity of the zero-loss filtered background is about 20 times weaker in Fig. 6.2b than the background of the unfiltered pattern in Fig. 6.2a.

As a function of mass thickness $x = \rho t$, Fig. 6.3 shows (a) the unfiltered peak intensity of the 111-ring of evaporated nickel films, (b) the zero-loss filtered peak intensities, (c) the background intensities below the 111-ring and (d) the peak-to-background ratios. The benefit of zero-loss filtering can be quantified by calculating the ratio (gain) of peak-to-background intensities in Fig. 6.3d for zero-loss filtered and unfiltered diffraction patterns with increasing film thickness (Fig. 6.4). The gain is largest for low atomic number and decreases as $1/Z$ owing to the decrease of the ratio ν (1.25) of inelastic-to-elastic cross-sections. Zero-loss filtering is hence also of interest for diffraction experiments with carbon-rich organic sections, e.g. apatite crystals in calcified tissue sections where the resin or organic matrix causes a strong inelastic background [6.10].

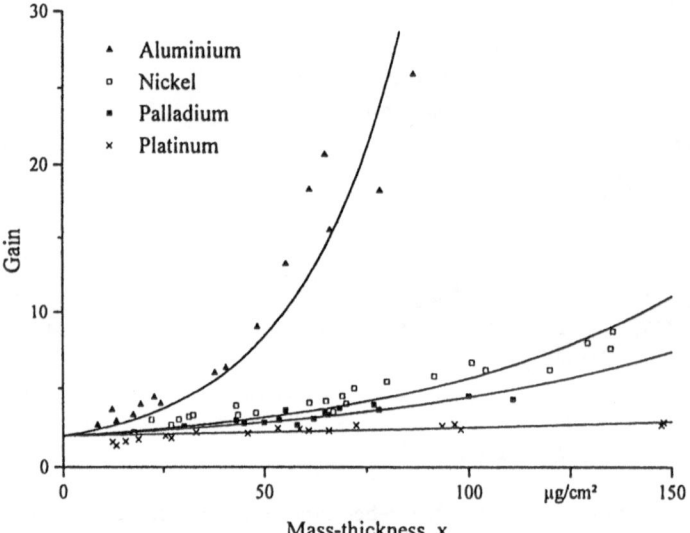

Fig. 6.4. Gain of contrast $G = (I_{111}/I_B)_{fil}/(I_{111}/I_B)_{unf}$ in Debye–Scherrer patterns of evaporated films of Al, Ni, Pd and Pt.

The results in Figs. 6.3 and 6.4 are obtained with fine-crystalline films that satisfy the kinematical diffraction theory. When the films contain coarse crystals, extending through the film from top to bottom, the intensity of the Debye–Scherrer rings is the result of averaging over dynamical reflection intensities, as demonstrated by the experiments of *Horstmann* and coworkers [6.5–7].

When recording the intensities in a Debye–Scherrer pattern with an energy window at higher energy losses E, the rings, including the primary beam, are blurred (convolved) by the angular distribution of multiple inelastic scattering, which increases with increasing E. Thus, unfiltered diffraction patterns contain the elastic diffraction rings with a width equal to the illumination aperture superposed on a broader ring consisting of rings belonging to the spectrum of energy losses. For narrow rings, the absence of zero-loss filtering can make it difficult to extrapolate the background below the rings.

Zero-loss filtering can also be of interest for diffraction from larger particles a few micrometres in size, to remove the inelastic background from thicker regions.

6.2.4 Small-Angle Electron Diffraction

Large specimen periodicities Λ result in small scattering angles $\theta = \lambda/\Lambda$. The largest periodicity detectable by electron diffraction depends on the illumination aperture α_i, which can be reduced to about $10^{-4} - 10^{-5}$ rad. This results in a maximum periodicity $\Lambda_{max} = \lambda/2\alpha_i = 20$–200 nm. Small-angle EDP can be recorded by strongly focusing the first condenser lens, switching off the second one and using the largest possible camera length for diffraction [6.11–13]. The method can be employed to study periodicities in collagen and catalase [6.11,14], the mean distance of crystallites in evaporated films with island structure [6.12], platinum aggregates in an annealed Pt/C film [6.15,16], conglomerates of latex spheres and virus particles [6.17,18] or polymers [6.19], for example.

Figure 6.5 shows the radial intensity distribution in a small-angle diffraction pattern of a thin Ag film on a carbon substrate. Zero-loss filtering (Fig. 6.5a) shows the central primary beam and a better resolved halo than in an unfiltered diffraction pattern. The most probable scattering angle of the halo appears at $\theta \simeq \lambda/d = 0.4$ mrad, which is related to the mean distance $d \simeq 10$ nm of the crystallites [6.20]. Filtering at $E = 25$ eV (Fig. 6.5b) shows that the halo disappears. This absence of the halo can be explained in terms of the excitation volume of plasmons, which is of the order of 5 nm and hence smaller than the mean distance between crystallites.

Periodicities Λ in the range 0.1–1000 nm can also be investigated quantitatively by light-optical Fraunhofer diffraction of micrographs on developed photographic films or by a digital Fourier transform; the periodicities are then $M\Lambda \simeq 0.05$–1 mm (M = magnification) [6.21]. This technique has also been proposed for determining transfer gaps and the envelope of the phase contrast transfer function by laser diffraction of the granularity of phase contrast [6.22] or the ripple structure of ferromagnetic films [6.21]. Zero-loss filtering will also increase the corresponding contrast and intensity modulation in these diffraction patterns. The intensity of directly recorded small-angle

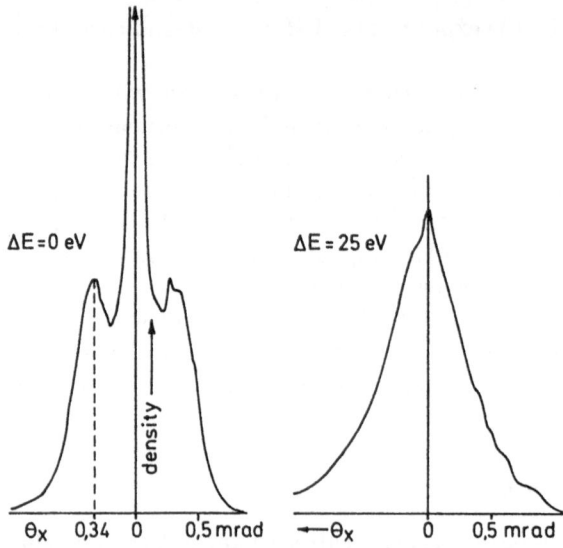

Fig. 6.5. Radial densitometer record of a small-angle electron diffraction pattern of a Ag film with island structure evaporated on a carbon film, (a) zero-loss and (b) plasmon-loss filtered with $E=25$ eV.

diffraction patterns is influenced by the phase shift of the electrons in the specimen, whereas the intensity of light-optical diffraction patterns at micrographs depends on the image intensity influenced by the imaging mode. The granularity caused by phase contrast and depending on defocus can be seen only in light-optical diffraction; the ripple structure of ferromagnetic domains can be seen in small-angle EDP in focus and in light-optical diffraction at strong defocus (Fresnel mode of Lorentz microscopy), for example.

6.3 Diffraction at Single-Crystalline Specimens

6.3.1 Dynamical Theory of Electron Diffraction

In the kinematical theory of electron diffraction, the amplitude $F(q)$ of an electron diffraction pattern (EDP) results from the superposition of partial waves scattered at the N atoms of the irradiated specimen area contributing to the EDP. The phase shift between two atoms a distance $\boldsymbol{r}$ apart becomes

$$\phi = 2\pi(\boldsymbol{k} - \boldsymbol{k}_0) \cdot \boldsymbol{r} = 2\pi\boldsymbol{q} \cdot \boldsymbol{r} \qquad (6.10)$$

where

$$|\mathbf{q}| = 2\sin(\theta/2)/\lambda \simeq \theta/\lambda \qquad (6.11)$$

and $\mathbf{k}_0$ and $\mathbf{k}$ with magnitude $1/\lambda$ are the wave vectors for the incident and scattered wave, respectively. This results in the diffraction intensity

$$I(q) \propto |F(q)|^2 = |\sum_{i=1}^{N} f_i \exp(-2\pi i \boldsymbol{q} \cdot \boldsymbol{r}_i)|^2$$

$$= |\sum_k f_k \exp(-2\pi i \boldsymbol{q} \cdot \boldsymbol{r}_k)|^2 \cdot |\sum_m \sum_n \sum_o \exp(-2\pi i \boldsymbol{q} \cdot \boldsymbol{r}_g)|^2 \qquad (6.12)$$

with f_k = scattering amplitude, $r_i = r_k + r_g$, r_k = position of the k-th atom in a unit cell, and

$$r_g = ma_1 + na_2 + oa_3 \quad (m, n, o \text{ integer}) \tag{6.13}$$

are the positions of the origin of unit cells with the fundamental translation vectors a_i. In the last triple sum of (6.12), the waves only interfere constructively when all $q \cdot r_g$ are integers or, using (6.13),

$$q \cdot a_i = h_i \quad (\text{integer}) \text{ for } i = 1, 2, 3 \quad . \tag{6.14}$$

A solution of this set of Laue equations is

$$q = k - k_0 = g = ha_1^* + ka_2^* + la_3^* \tag{6.15}$$

where g is a vector in the reciprocal lattice with the translation vectors a_i^*. Equation (6.15) is the Bragg condition in vector notation. Since $|q| = \theta/\lambda$ (6.11) and $|g| = 1/d_{hkl}$, this yields the Bragg law

$$2d_{hkl} \sin \theta_B = \lambda \tag{6.16}$$

The first sum in (6.12) gives the structure amplitude of the unit cell. Equation (6.15) can be used for the Ewald construction. A sphere of radius $|k| = 1/\lambda$ is drawn around the centre M, the starting point of the vector k_0, which ends at the origin of the reciprocal lattice; the Bragg condition is satisfied when this Ewald sphere passes through a reciprocal lattice point with the reciprocal lattice vector g (Fig. 6.6).

A detailed discussion of the intensity of the primary and the Bragg reflected beams, of convergent beam electron diffraction patterns (Sect. 6.4) and of the electron spectroscopic imaging (ESI) of crystals in Chap. 7 must be based on the dynamical theory of electron diffraction [6.23–26].

The potential $V(r)$ of a crystal, regarded as a superposition of all the atomic potentials, shows the lattice periodicity and can be expanded in a Fourier series

$$V(r) = -\sum_g (V_g + iV_g') \exp(-2\pi i g \cdot r) \quad . \tag{6.17}$$

A value V_g can be attributed to each reciprocal lattice point g. In the following, the imaginary part V_g' will describe the attenuation of the Bloch waves caused by scattering in angles between the Bragg spots due to thermal diffuse or inelastic scattering. In the Born approximation (1.12), $F(\theta)$ is also proportional to the Fourier transform of the scattering potential; the V_g in units eV are therefore related to the structure amplitude $F(\theta_g)$ in cm, ($\theta_g = 2\theta_B$):

$$V_g = \frac{\lambda^2 E^*}{2\pi V_e} F(\theta_g) = \frac{h^2}{2\pi m V_e} F(\theta_g) \tag{6.18}$$

where V_e is the volume of the unit cell.

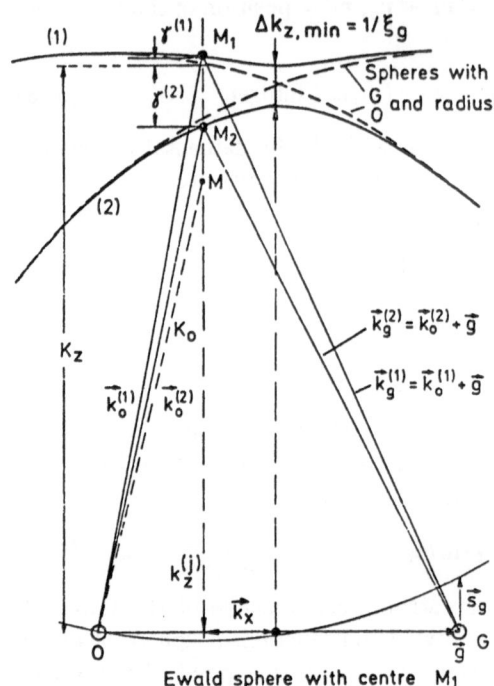

Fig. 6.6. Branches $j =$ (1) and (2) of the dispersion surface for the two-beam case with a least distance $\Delta k_{z,\min} = 1/\xi_g$ in the Bragg condition ($k_z=0$). Construction of the excitation points M_1 and M_2 on the dispersion surface for the tilt parameter k_x and the four wave vectors $k_0^{(j)}$ and $k_g^{(j)} = k_0^{(j)} + g$ ($j=1,2$) to the reciprocal lattice points O and G.

The propagation of electrons in crystals has to satisfy the Schrödinger equation

$$\left[\frac{\hbar^2}{2m}\nabla^2 + E_0^* - V(r)\right]\psi = 0 \quad , \tag{6.19}$$

where

$$E_0^* = E_0\frac{E_0 + 2m_0c^2}{2(E_0 + m_0c^2)} \quad . \tag{6.20}$$

The solutions of (6.19) will also reflect the lattice periodicity and are called Bloch waves

$$b^{(j)}(\boldsymbol{k},\boldsymbol{r}) = \sum_g C_g^{(j)} \exp[2\pi i(k_0^{(j)} + g)\cdot\boldsymbol{r}]\exp(-2\pi q^{(j)}z) \quad . \tag{6.21}$$

We have to sum over the n excited points $\boldsymbol{g} = \boldsymbol{g}_1, \boldsymbol{g}_2, \ldots, \boldsymbol{g}_n$ of the reciprocal lattice, including the incident beam direction ($\boldsymbol{g}_1 = 0$). A number $j =1,\ldots,n$ of different Bloch waves is needed to describe the propagation of electron waves in a crystal and to satisfy the boundary condition at the vacuum–crystal interface. This requires a superposition of n^2 different waves with wave vectors $k_g^{(j)} + \boldsymbol{g}$ and amplitude factors $C_g^{(j)}$.

We substitute (6.17) and (6.21) for ψ in (6.19) and introduce the abbreviation

$$K^2 = [2m_0 E_0(1 + E_0/2m_0 c^2) + 2m_0 V_0(1 + E_0/m_0 c^2)]/h^2 \qquad (6.22)$$

where $\boldsymbol{K}$ represents the wave vector inside the crystal, which depends on the sum of the kinetic energy and the coefficient $V_o = eU_i$ (inner potential) of the Fourier expansion (6.17). This gives

$$4\pi^2[K^2 - (k_0^{(j)} + g)^2]C_g^{(j)} + \sum_{h \neq 0} \frac{2m}{h^2} V_h \exp[2\pi i\, \boldsymbol{h} \cdot \boldsymbol{r}] C_g^{(j)} \exp[2\pi i\, (k_0^{(j)} + g) \cdot \boldsymbol{r}]$$

$$(6.23)$$

for all $\boldsymbol{g}$. This system of n equations can be satisfied if the coefficients of identical exponential terms are simultaneously zero. After collecting up terms containing the factor $\exp[2\pi i(k_0^{(j)} + g) \cdot \boldsymbol{r}]$, we obtain the fundamental equations of dynamical theory

$$[K^2 - (k_0^{(j)} + g)^2]C_g^{(j)} + \sum_{h \neq 0} \frac{2m}{h^2} V_h C_{g-h}^{(j)} = 0 \quad ; g = g_1, \cdots, g_n \quad . \quad (6.24)$$

The $k_0^{(j)} + g$ are the wave vectors of the Bloch waves, the magnitudes of which are not identical with K. As shown for the example of a two-beam case in Fig. 6.6, the excitation points M_i are the starting points of the vectors $k_0^{(j)}$, which end on the origin O of the reciprocal lattice. The M_i lie on Ewald spheres around the lattice points $\boldsymbol{g}$ but these spheres do not intersect.

For purposes of calculation, the position of M_j is described by the following vector (Fig. 6.6)

$$\boldsymbol{k}^{(j)} = k_z^{(j)} + \boldsymbol{k}_x = (K_z + \gamma^{(j)})\boldsymbol{u}_z + k_x \boldsymbol{u}_x \quad . \quad (6.25)$$

where $\boldsymbol{u}_x, \boldsymbol{u}_z$ are unit vectors. The component k_x depends on the tilt angle $\Delta\theta$ or the excitation error s_g as follows

$$\Delta\theta = k_x/K = s_g/g \quad . \quad (6.26)$$

Recalling that $K \gg g$, $K + k_z^{(j)} \simeq 2K$ and introducing the difference $(k_z^{(j)} - K_z) = \gamma^{(j)}$ from (6.25), the first factor of (6.24) becomes

$$[K^2 - (k_0^{(j)} + g)^2] = (K + |k_0^{(j)} + g|)(K - |k_0^{(j)} + g|) \simeq 2K(s_g - \gamma^{(j)}) \quad . \quad (6.27)$$

The excitation error s_g is negative when the reciprocal lattice point g lies outside the Ewald sphere, as in Fig. 6.6. With the aid of (6.27), the system of equations (6.24) can be written in matrix form, after dividing by $2K$. We obtain

$$\begin{bmatrix} A_{11} & A_{12} & \cdots & A_{1n} \\ A_{21} & A_{22} & \cdots & A_{2n} \\ \cdots & \cdots & \cdots & \cdot\mathord{\cdot}\cdot \\ A_{n1} & A_{n2} & \cdots & A_{nn} \end{bmatrix} \begin{bmatrix} C_1^{(j)} \\ C_2^{(j)} \\ \cdots \\ C_n^{(j)} \end{bmatrix} = \gamma^{(j)} \begin{bmatrix} C_1^{(j)} \\ C_2^{(j)} \\ \cdots \\ C_n^{(j)} \end{bmatrix} \qquad \text{for} \quad j = 1, ..., n$$

$$(6.28)$$

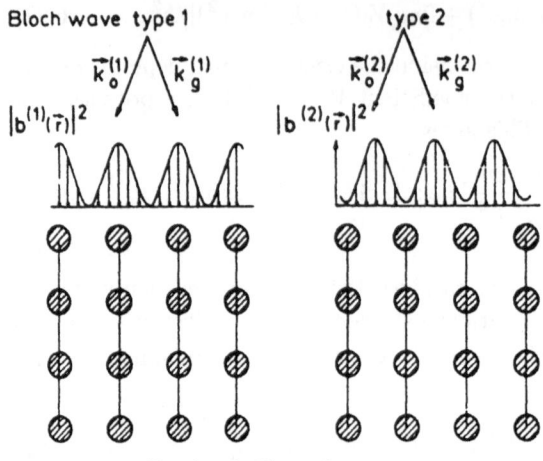

Bloch wave type 1   type 2

reflecting lattice planes

Fig. 6.7. Squared amplitude (probability density) $|\psi|^2$ of Bloch waves types 1 and 2 in the two-beam case with antinodes and nodes at the nuclei and lattice planes, respectively.

with the matrix elements

$$A_{11} = 0; \; A_{gg} = s_g; \; A_{hg} = A_{gh} = mV_{g-h}/Kh^2 = 1/2\xi_{g-h} \quad . \tag{6.29}$$

This is an eigenvalue problem for which computer subroutines exist. (The meaning of the excitation length ξ_g will be discussed below.) An orthogonal matrix $[A]$ has n different eigenvalues $\gamma^{(j)}$ ($j =1,...,n$) associated with the orthogonal eigenvectors $C_g^{(j)}$ ($g = g_1,\cdots,g_n$). The absorption parameters $q^{(j)}$ of the Bloch waves (6.21) become

$$q^{(j)} = \frac{m}{h^2K} \sum_g \sum_h C_h^{(j)} V'_{g-h} C_g^{(j)} . \tag{6.30}$$

The superposition (6.21) of waves with wave vectors $k_0^{(j)} + g$ sharing the same excitation point M_j results in Bloch waves propagating parallel to the lattice planes. Their amplitude is modulated perpendicular to their direction of propagation with a period equal to the lattice plane distance d_{hkl} (Fig. 6.7).

The total wave function, the solution of (6.19), will be a linear combination of the Bloch waves $b^{(j)}(\boldsymbol{k},\boldsymbol{r})$ (6.21), which are a superposition of all waves from the same point M_j to the different reciprocal lattice points $\boldsymbol{g}$:

$$\psi_{\text{tot}} = \sum_j \epsilon^{(j)} b^{(j)}(\boldsymbol{k},\boldsymbol{r}) = \sum_j \epsilon^{(j)} \sum_g C_g^{(j)} \exp[2\pi i(k_0^{(j)}+g)\cdot\boldsymbol{r}] \exp(-2\pi q^{(j)}z) .$$

$$\tag{6.31}$$

The $\epsilon^{(j)}$ are the Bloch-wave excitation amplitudes, which are also governed by the boundary condition and become $C_0^{(j)}$ for normal incidence on the crystal foil. The Bloch-wave field intensity $|\psi_{\text{tot}}|^2$ (probability density) shows either nodes or antinodes at the lattice planes (Fig. 6.7) depending on the crystal tilt. This explains the differences in the $q^{(j)}$ (see Fig. 6.8b, for example)

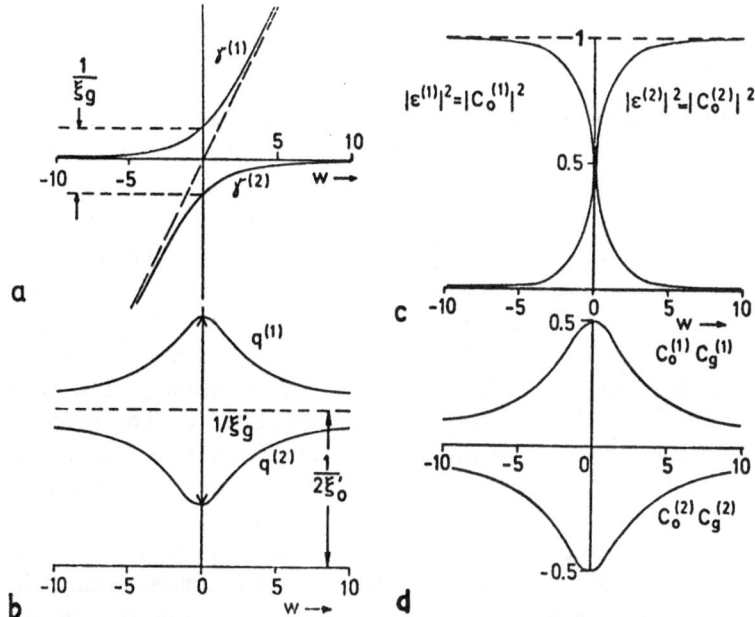

Fig. 6.8. Dependence of the Bloch wave parameters of the two-beam case on the tilt parameter $w = s\xi_g$ (Bragg condition: $w=0$): (a) $\gamma^{(j)}$ (where $\gamma^{(2)} - \gamma^{(1)}$ = distance between the two branches of the dispersion surface), (b) absorption paramaters $q^{(j)}$ and (c) and (d) wave amplitudes of the four excited Bloch waves.

because electron–specimen interactions that are concentrated at the nuclei, like thermal-diffuse and large-angle scattering or ionization of inner shells, show the highest probability of excitation when antinodes are at the nuclei and a decreased probability in the case of nodes at these sites.

The n eigenvalues $\gamma^{(j)}$ correspond to n Bloch waves (6.21) with wave vectors $k_o^{(j)} + g$. Their starting points do not lie on a sphere of radius K around O but at modified points M_j given by (6.25). For different tilts $\Delta\theta$ of the lattice planes relative to the Bragg position either the component k_x or the excitation error s_g (6.26) is used (Fig. 6.6). The starting points M_j of the wave vectors (dispersion surface) can be obtained by the following construction. The K_0 vector parallel to the incident electron wave determines the point M in Fig. 6.6. A straight line is drawn through M parallel to the surface normal. The intersections with the n-fold dispersion surface are the excitation points M_j. This construction satisfies the boundary condition that the tangential components of the waves have to be continuous at the vacuum–crystal interface.

The amplitude ψ_g of a particular reflected wave can be obtained by summing over all $j =1,...,n$ waves from the different excitation points M_j to one reciprocal lattice vector g,

$$\psi_g = \sum_j \epsilon^{(j)} C_g^{(j)} \exp(2\pi i \gamma^{(j)} z) \exp(-2\pi q^{(j)} z) \qquad (6.32)$$

if a constant phase factor is omitted and where $z = t$ is the foil thickness.

For the two-beam case, the eigenvalues $\gamma^{(j)}$, the absorption parameters $q^{(j)}$ and the combinations of eigenvector components appearing in (6.31) and (6.32) are plotted in Fig. 6.8 versus the tilt parameter $w = s\xi_g$. At $w=0$ (Bragg position), the wave vectors $k_0^{(j)}+g$ from different branches of the dispersion surface show a difference $\gamma^{(1)} - \gamma^{(2)} = 1/\xi_g$ (Fig. 6.8a) and the superpostion in (6.29) results in the pendellösung effect (thickness fringes or edge contours). With increasing foil thickness t, the transmitted (T) and Bragg reflected (R) intensities oscillate (Fig. 6.9) with a period ξ_g. The wave vector difference $\gamma^{(1)} - \gamma^{(2)}$ increases with increasing tilt w. The effective value of ξ_g and the amplitude of the thickness fringes decrease, the latter because the amplitudes of the two superposing waves become different (Fig. 6.8c). The absorption parameters $q^{(j)}$ result in an exponential attenuation.

The tilt dependence (rocking curve or bend contours) of T and R is shown in Fig. 6.10 for a thin and a thick foil. The transmission becomes asymmetric in w with anomalous transmission or absorption for positive or negative w, respectively. This can be explained by Fig. 6.7. At positive w, the Bloch-wave amplitude $\epsilon^{(2)} = C_0^{(2)}$ is largest (Fig. 6.8c) and the corresponding $q^{(2)}$ is low (Fig. 6.8b). At negative w, the Bloch wave with large $q^{(1)}$ is excited with the larger amplitude $\epsilon^{(1)} = C_0^{(2)}$. The rocking curve of a 40 nm Cu foil at $E_0=100$ keV for a three-beam case $(-g, 0, +g)$ with $g = 220$ is shown in Fig. 6.11a. The reflected amplitude for the $\overline{2}20$-reflection will be mirror-symmetric in the tilt parameter k_x/g.

When the sum

$$\sum_g I_g = \sum_i |C_0^{(i)}|^2 \exp(-4\pi q^{(i)} z) \qquad (6.33)$$

of the primary and Bragg reflected beams is formed, the pendellösung fringes of T and R cancel but the sum is not constant because of anomalous absorp-

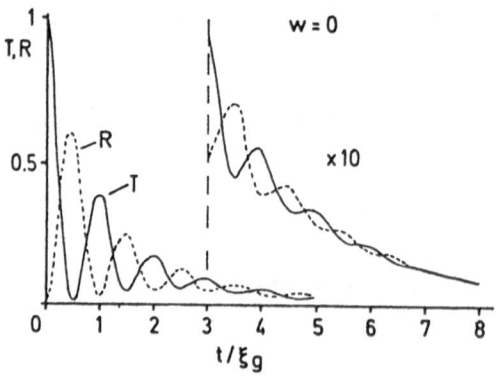

Fig. 6.9. Pendellösung fringes for the transmitted (T) and a Bragg-reflected beam (R) at the Bragg condition ($w=0$).

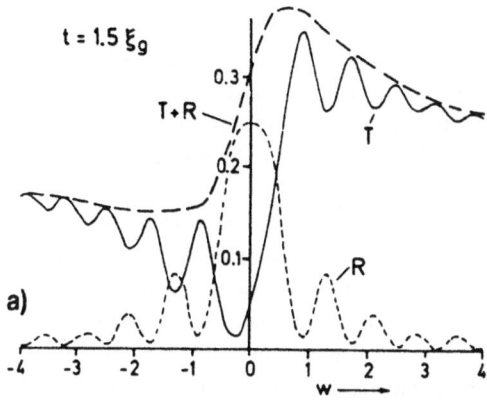

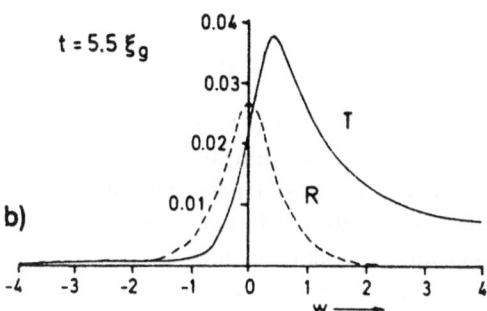

Fig. 6.10. Rocking curves of the transmitted (T) and Bragg-reflected (R) intensities for the dynamical two-beam case and foil thicknesses of (a) $t = 1.5\xi_g$ and (b) $t = 5.5\xi_g$.

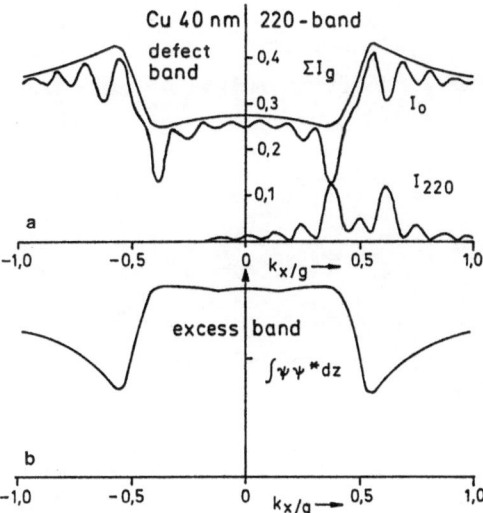

Fig. 6.11. (a) Rocking curve of the primary-beam intensity I_0 and Bragg-reflected intensity I_g and of ΣI_g (defect Kikuchi band) for the 3-beam case of Cu at $E_0 = 100$ keV; (b) large-angle scattering probability $\int \psi \psi^* dz$ (excess Kikuchi band).

tion (Figs. 6.9 and 6.10) and a defect band is formed (Fig. 6.11a). Conversely, excess Kikuchi bands are formed by scattering with wave vectors k between the Bragg spots as a result of thermal-diffuse and large-angle scattering (Fig. 6.11b). The theorem of reciprocity tells us that the intensity becomes proportional to the Bloch-wave field intensity $|\psi_{tot}|^2$ for a wave that is formally incident in the $-k$ direction [6.27].

This schematic presentation of dynamical electron diffraction only describes scattering between the reflections in the primary Bloch-wave field, whereas multiple elastic and inelastic scattering has to be taken into account for most TEM specimens. However, the inclusion of inelastic dynamic electron diffraction for thicknesses $t > \xi_g$ [6.28–33] is extremely time-consuming in computation and also needs more accurate values of the parameters of inelastic scattering.

6.3.2 Energy-Filtered Diffraction at Single Crystals

The contribution of inelastically scattered electrons increases with decreasing atomic number and increasing foil thickness in the diffraction patterns of single-crystalline specimens just as it does in those of polycrystalline films (Sect. 6.2.3). Zero-loss filtering of elastically scattered electrons can be used to enhance the Bragg spot intensity, especially of weak reflections, and the quasi-elastic thermal-diffuse streaks caused by electron–phonon scattering; energy filtering at high energy losses can separate the contributions from plasmon and inner-shell ionization losses to Kikuchi lines and bands [6.15,20,34–39].

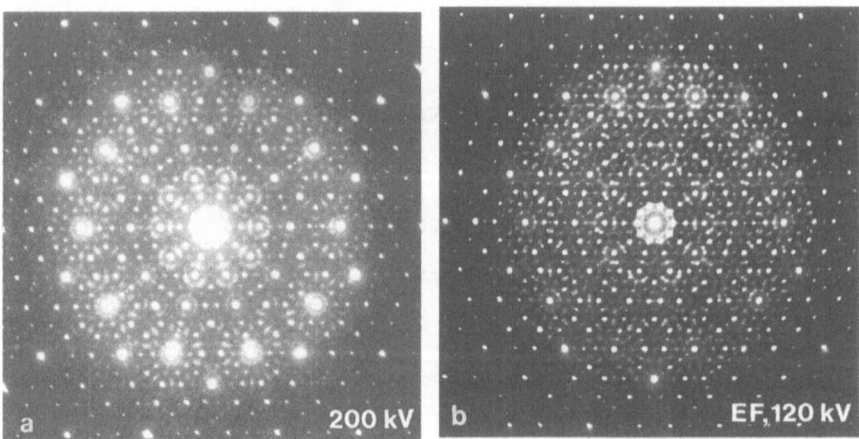

Fig. 6.12. Diffraction patterns of a decagonal quasicrystal in CoNiAl obtained with a conventional TEM at 200 kV and an EFTEM at 120 kV. Note the reduction of background and the much better visibility of the small reflections in the energy filtered pattern: (a) unfiltered, (b) zero-loss filtered.

An example of the enhancement of weak superlattice reflections from Al_3Li precipitates in an Al–7wt%Li alloy will be found in Fig. 7.19, together with electron spectroscopic images. Plasmon-loss scattering occurs at very small scattering angles and results in a broadening of the Bragg diffraction spots. This broadening can be particularly harmful if intensities and positions of the reflections are to be analysed quantitatively or if weak reflections surround very strong reflections, as is the case in large-unit-cell crystals or in crystals with a long-period super-cell. As an example, Fig. 6.12 shows the diffraction pattern of a decagonal quasicrystal along the tenfold axis. In the unfiltered pattern (recorded at 200 keV in a conventional TEM), the inelastic background due to plasmon losses broadens the reflections and in particular the primary beam. The details and weak reflections around the primary beam and the other strong reflections only become visible in the zero-loss filtered pattern recorded at 120 keV in the Zeiss EM912 Omega [6.40].

Totally destructive interference of elastically scattered electrons between the Bragg diffraction spots, which are caused by constructive interference, can only be expected for ideal undisturbed lattices. Diffuse scattering caused by thermal vibrations can be treated as electron–phonon scattering using a Debye phonon model, for example [6.33,41]. The scattering is inversely proportional to the square of phonon frequency $\nu(q)$. The diffuse streaks are therefore mainly generated by transverse acoustic phonons of low frequency with wave vectors k perpendicular to one of the atomic-chain directions and polarization vectors parallel to it [6.42–46]. Figure 6.13 shows as an example thermal-diffuse streaks in unfiltered and zero-loss filtered diffraction patterns of Si and Sn foils [6.38]. In thin foils, the streaks only appear in the zero-loss pattern, whereas in thicker foils they can also be observed in plasmon-loss filtered patterns due to elastic-inelastic scattering (Fig. 6.14b and c, for example). The streaks then appear more diffuse as a result of the convolution with the angular distribution of the plasmon loss. This separation of the streaks

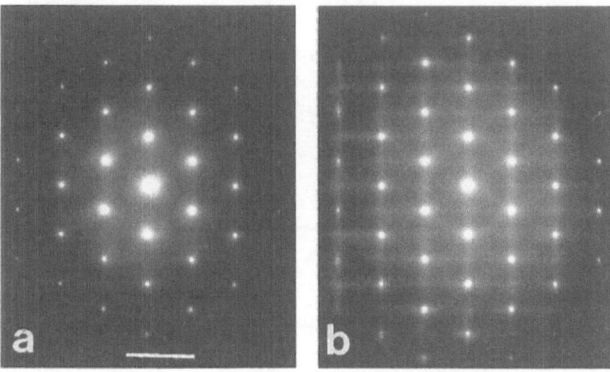

Fig. 6.13. Thermal diffuse streaks in single-crystal diffraction patterns of Sn: (a) unfiltered, (b) zero-loss filtered.

by zero-loss filtering shows that a quantitative investigation is almost impossible in a conventional diffraction pattern, which also contains the strong
background of Kikuchi lines and bands formed by inelastically scattered electrons. The qualitative and quantitative use of thermal-diffuse streaks will
therefore become of interest in the future.

Though Kikuchi lines are also generated by elastically scattered electrons,
as shown in Fig. 6.14b, the main contribution is concentrated at low energy
losses in the plasmon region. Because the excitation error s_g (6.26) of a Bragg
spot can be read from the distance between the spot and the corresponding
Kikuchi line (coincidence in case of exact Bragg condition), it can be an advantage to scan with an energy-selecting window over the low-loss part of the
EELS to obtain optimum contrast of Kikuchi lines. This is demonstrated in
Fig. 6.14b, where the orientation of the Si foil was adjusted by the goniometer
to exact 111 orientation, so that all the Kikuchi lines coincide with the six
{220} Bragg reflections.

The series of filtered diffraction patterns in Fig. 6.14 of a $\simeq$ 50 nm thick
111-oriented silicon foil shows the contributions of elastically and inelastically
scattered electrons of increasing energy loss E to the conventional unfiltered
pattern (a), which is typical of all single-crystal diffraction patterns. The zero-
loss filtered pattern (b) of elastically scattered electrons shows sharp Bragg
spots and Kikuchi lines as well as thermal diffuse streaks as discussed above.

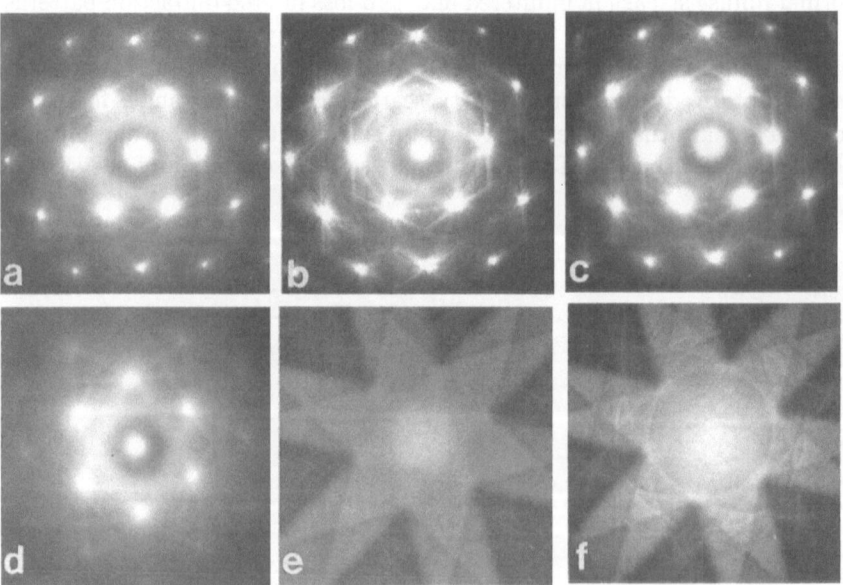

Fig. 6.14. Series of electron spectroscopic diffraction (ESD) patterns of a 111-
oriented thin Si foil ($t \simeq 50$ nm) with excess Kikuchi bands at high energy losses:
(a) unfiltered, (b) $E=0$ eV, (c) 16 eV, (d) 100 eV, (e) 1800 eV, and (f) 2000 eV.

Near the plasmon loss (c), the Bragg spots and also the weak thermal diffuse streaks are already blurred by the angular distribution of the inelastically scattered electrons. This blurring increases and the intensity of the Bragg spots decreases with increasing E as shown for 100 eV in (d). Above a few hundred electronvolts (e.g. just below the Si K edge in (e)), the pattern consists only of excess Kikuchi bands with a band profile shown in the example of Fig. 6.11b. These high energy losses are concentrated near the nuclei and, as discussed in Sect. 6.3.1, the intensity becomes proportional to the Bloch wave intensity $|\psi_{tot}|^2$ (6.31). Beyond the Si K edge at $E=1840$ eV, as shown for $E=2000$ eV in (f), the intensity and the contrast of the excess Kikuchi bands increase and the pattern of excess Kikuchi HOLZ lines from high-order Laue zones can be observed at the centre. These HOLZ lines are found in conventional patterns only in the CBED mode (Fig. 6.18, for example).

When the foil thickness is increased, all diffraction patterns at different energy losses consist of defect Kikuchi bands as demonstrated in Fig. 6.15, which shows ESD patterns of a $\simeq$ 800 nm Si foil. This can be explained [6.27] by the broadening of the angular distribution of the scattered electrons due to multiple scattering at the entrance part of the foil. A diffuse cone of scattered electrons then emerges from the foil. The numbers of electrons with k vectors on each of the Kossel cones are of equal magnitude and excess and defect Kikuchi lines are cancelled but the anisotropy of the anomalous absorption remains, resulting in defect Kikuchi bands as shown schematically in Fig. 6.11a. The same defect bands can be observed in Kossel patterns when the external aperture of the electron probe is increased, so that the CBED discs overlap and the intensity becomes proportional to $\sum I_g$ (6.33) [6.47,48] or when the incident electron beam becomes diffuse by traversing a thick amorphous carbon foil in front of the single-crystal specimen [6.49]. For semi-thin foils, diffraction patterns in front of the Si K edge can also change to defect band patterns, as shown with excess bands in Fig. 6.14e, whereas a pattern beyond the K edge still can show the excess band character

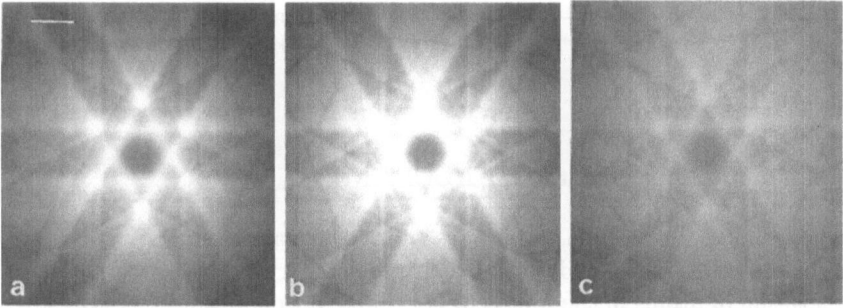

Fig. 6.15. Series of ESD patterns of a thick 111-oriented Si foil ($t \simeq$ 800 nm) with defect Kikuchi bands: (a) $E=100$ eV, (b) 500 eV and (c) 1300 eV.

(Fig. 6.14f). This is explained schematically in Fig. 6.16. In the case (a) of thin foils, the electrons are directly scattered from the unscattered incident beam or the low-energy plasmon losses to high losses near the edge, which results in dominant excess bands as shown in Figs. 6.14e and f. In the case (b) of semi-thin foils, the electrons scattered below the K edge contain a large fraction (D) of multiply and diffusely scattered electrons, resulting in defect bands, and only a small fraction (E) of electrons is directly scattered to high losses. This fraction (E) increases beyond the K edge and the excess character again dominates. At medium energy losses, the excess and defect types of band contrast can result in a cancellation of band contrast. In case (c) of thick foils, all the electrons are multiply and diffusely scattered. The edge profile is strongly blurred by multiple plasmon and higher energy losses and the pattern will only show defect bands, like those in Fig. 6.15.

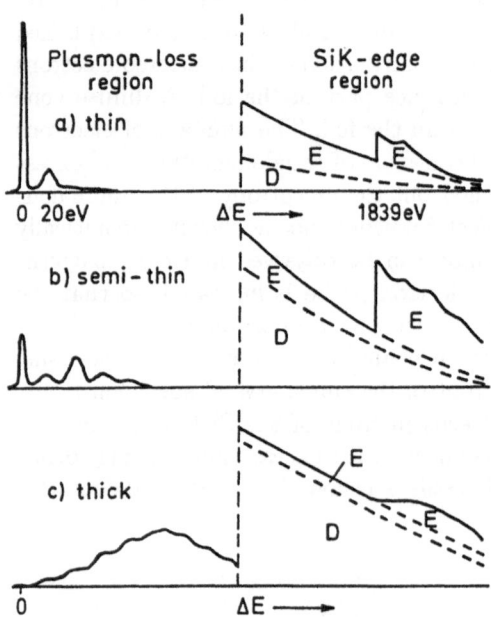

Fig. 6.16. Schematical EELS for (**a**) thin, (**b**) semi-thin and (**c**) thick foils with contributions to defect (D) and excess (E) Kikuchi bands.

6.3.3 Channelling Effects and ALCHEMI

As shown in Sect. 6.3.1, the amplitudes and the formation of nodes and antinodes of the Bloch-wave field depend sensitively on crystal orientation and on the position of atoms in the unit cell. For different orientations, across a low-indexed Kikuchi band for example, the amplitudes and hence the probability densities can vary at different nuclei. A high probability density results in an enhanced intensity of all interaction processes that are localized near the nuclei, such as backscattering or large-angle elastic scattering, observed as electron channelling patterns (ECP) in scanning electron microscopy [6.50,51], and inner-shell ionization, which renders the resulting EELS [6.52] and x-ray emission [6.53,54] orientation-dependent.

In the method of ALCHEMI (Atom Location by CHannelling Enhanced MIcroanalysis), the orientation dependence of characteristic x-rays excited by different intensities of antinodes at different lattice sites can be used to determine which sites are occupied by substitional impurity atoms [6.55,56],

The localization L of ionization processes in real space can be described approximately by [6.57]

$$L = \lambda/\theta_E \simeq \lambda E_0/\Delta E \qquad (6.34)$$

or by more quantitative calculations [6.58], where θ_E (1.21) is the characteristic angle of inelastic scattering. For a lattice plane spacing d, a location constant [6.59]

$$C_L = d/2L = \Delta E/4\theta_B E_0 \qquad (6.35)$$

can be defined with a value of about 0.5 for inner-shell excitations with an ionization energy of 1 keV [6.60]. Lower energy losses, which are frequently used in EELS analysis, do not exhibit this channelling effect.

ALCHEMI is also of interest in EELS analysis [6.52,55,61] because more inelastic scattering processes can be detected than in x-ray microanalysis owing to the better collection efficiency; the higher background in EELS is a disadvantage, however. Though no systematic studies of ALCHEMI by EELS exist, electron spectroscopic diffraction patterns beyond the ionization edges of the elements should show differences in intensity at different points of the ESD patterns, across a low-indexed Kikuchi band, for example. As in quantitative elemental analysis, an unspecific background subtraction will be possible by recording an ESD below and beyond an edge.

6.4 Convergent-Beam Electron Diffraction (CBED)

6.4.1 Basic Principles of CBED

CBED patterns are obtained by projecting a demagnified image of the electron source onto the specimen. A convergent cone-shaped beam is thus formed. The convergence angle is defined by the diameter of the condenser diaphragm but can be varied electron optically within certain limits. With this illumination geometry, the incident beam directions vary continuously within the cone. In the back focal plane of the objective lens, a pattern that consists of discs is formed rather than the spot pattern obtained for parallel illumination (Fig. 6.17). The discs have sharp edges and their diameter is

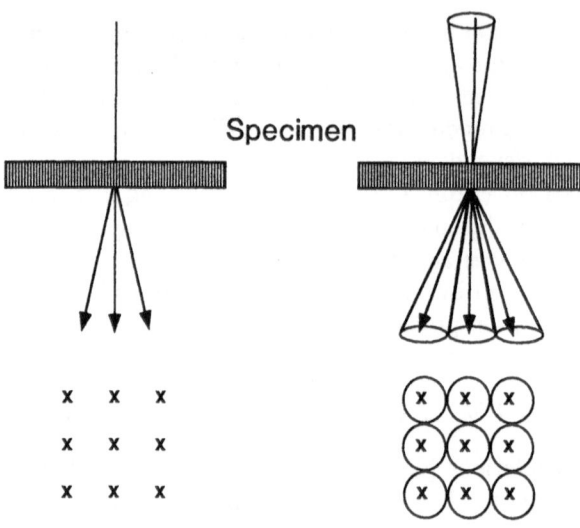

Specimen

x   x   x

x   x   x

x   x   x

Fig. 6.17. The geometry of electron diffraction by (a) parallel illumination and (b) convergent-beam electron diffraction (CBED).

proportional to the convergence angle. Each disc in the CBED pattern corresponds to one Bragg reflection g of the crystal. For an incoherent electron source (tungsten, LaB_6), the continuous variation of incident beam directions is equivalent to successive variation of the beam tilt for parallel illumination except that, for the latter, the illuminated area is larger. For each incident beam direction, a spot pattern is produced in the back focal plane, as illustrated in Fig. 6.17. The reflection intensity at any point inside a disc is independent of the intensities at all the other points obtained for different incident beam directions. Interference effects between the individual beams only have to be taken into account if a nanometre-sized specimen area is illuminated with a coherent probe, available only in instruments equipped with a field-emission gun. The consequences of interference between coherent beams will not be discussed here and the reader is referred to [6.62–64].

In thicker specimens, the pendellösung of dynamical diffraction creates minima and maxima of the individual reflections, which are influenced by the specimen thickness and the incident beam direction. In the CBED discs (Fig. 6.18), the minima can be seen as broad dark contours. In addition, sharp dark lines are visible in the central disc. These HOLZ lines can be attributed to reflections at reciprocal lattice points in a high-order Laue zone, which satisfy the Bragg condition for the corresponding incident beam direction.

The effect of zero-loss filtering on CBED patterns is shown in Fig. 6.19b for the (000) and (220) discs of an excited systematic row of reflections in an Si foil. Both sets of discs contain parallel and complementary dark and bright pendellösung fringes and line-scans through the discs correspond to rocking curves (Sect. 6.3.1). In CBED, complete rocking curves can be obtained in one single exposure. Without energy filtering, the inelastic scattering background in Fig. 6.19a obscures most of the information. Figure 6.19c shows

Fig. 6.18. Typical CBED pattern of Si along the $\langle 111 \rangle$ zone axis.

a quantitative comparison of rocking curves (line-scans) from the right-hand (220) discs in Figs. 6.19a and b. Energy filtering is therefore indispensable for quantitative evaluation and comparison with theory.

In a simulation of the intensity distribution in the CBED patterns, individual spot patterns are calculated for an array of incident beam directions and then superposed side by side to obtain a two-dimensional plot of the intensity distribution. The calculations are carried out using the plane wave formalism described in Sect. 6.3.1. It was shown there that the intensities in the diffraction patterns depend on specimen thickness, specimen tilt, accelerating voltage and the structure factors. In principle, all these parameters

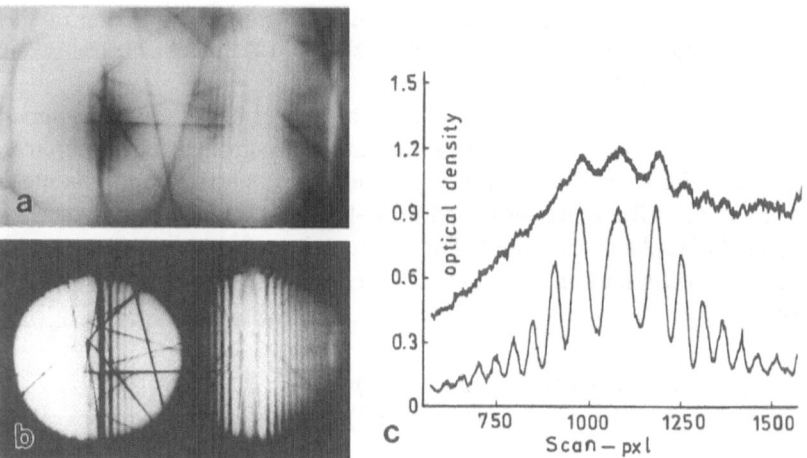

Fig. 6.19. (a) Unfiltered CBED pattern of Si in a two-beam orientation, (b) zero-loss filtered, (c) line profiles across the (220) discs of (a) and (b).

can be determined from a single CBED pattern. However, there is no way of inverting the experimentally observed intensity distribution directly to obtain values of these parameters. Instead, simulations have to be made with a range of values of the parameters until a best fit is obtained. Details of the technique will be described in Sects. 6.4.3 and 6.4.4.

In the evaluation of CBED patterns, we have to distinguish between low-index reflections belonging to the zero-order Laue zone (ZOLZ) and high-index reflections belonging to the high-order Laue zones (HOLZ). As discussed above, low-index reflections create the discs in the CBED pattern. The incident beam direction may vary as much as two Bragg angles within a single disc (Fig. 6.17) if the discs just touch each other. For a given incident beam direction, the intensity is determined by the specimen thickness and the structure factors; after removal of the inelastic scattering background by energy-filtering, the structure factors of the low-index reflections can be determined. Since CBED also allows the phases of the structure factors to be measured with an accuracy of better than one degree, one of the major problems of X-ray crystallography can be overcome. Furthermore, bonding effects in crystals can be studied by calculating charge density distributions from the low-index reflections (Sect. 6.4.5).

In contrast to the low-index reflections, scattering into the HOLZ-reflections involves large scattering angles and is only possible for a very narrow angular range. The scattering into the HOLZ reflections is mostly investigated by examining the defect lines (the so-called HOLZ-lines) within the (000) disc. The arrangement of the HOLZ lines is very sensitive to small changes in accelerating voltage, local lattice parameter or composition. Energy filtering allows thicker specimens to be inestigated, thereby increasing the contrast and sharpness of the HOLZ lines. Energy filtering also prevents the broadening and the shift of intensity minima of HOLZ lines by single or multiple plasmon losses. Furthermore, for an accurate quantification of the HOLZ line positions, the dynamical shift of the HOLZ lines has to be taken into account (Sect. 6.4.3).

Further important applications of CBED are the study of crystal symmetries and the determination of point groups and space groups. These applications are based on a qualitative interpretation of symmetries and dynamical extinctions in the CBED patterns and do not benefit as much from energy-filtering as the applications above. For a discussion of the point group and space group determination from CBED the reader is therefore referred to [6.65,66].

6.4.2 Large-Angle Patterns (LACBED) and Convergent-Beam Imaging (CBIM)

In the standard CBED method (Fig. 6.17), the maximum convergence angle of the incident beam corresponds to twice the Bragg angle θ_B of the reflection nearest to the (000) beam. Otherwise, the overlap of the discs obscures the information within them. This is a severe limitation in many applications, in particular for crystals with large unit cells and hence closely spaced reciprocal lattice points.

The first technique that was proposed to overcome this problem was the beam-rocking method [6.67,68]. In standard CBED, the whole pattern is obtained in parallel acquisition, whereas in the beam-rocking method it is acquired sequentially. The beam tilt coils are used to change the incident beam direction of a parallel beam in small steps across the angular range that is to be investigated. This can be achieved by either using an existing STEM unit or driving the beam tilt coils from an external computer control unit. Below the specimen, a spot diffraction pattern is formed, which shifts as the incident beam direction changes. In order to select a particular beam in the diffraction pattern, the shift has to be compensated by means of a second set of deflection coils below the specimen. The intensity in one beam can then be displayed on a monitor using the STEM unit or recorded digitally as a function of the incident beam direction. In this method, energy filtering and digital recording can be achieved very easily by passing the selected beam through a serial EELS spectrometer [6.69].

Ways of preventing overlap of the discs, while using large convergence angles in an incident cone of illumination, have been developed by *Tanaka* and co-workers [6.68,70] and are commonly referred to as "large angle CBED" or Tanaka patterns. The principle of this method is illustrated in Fig. 6.20. An electron probe with a large convergence angle is focused onto the specimen and an image of this spot can be observed in the image mode. The specimen is now lifted by an amount Δz, whereupon the single spot splits into a number of spots. The arrangement of these spots corresponds exactly to the reflections of a diffraction pattern, even though the spots are observed in the image mode! The magnification of this spot "diffraction" pattern increases as the specimen shift Δz is increased. In other words, this pattern observed in the image plane in LACBED is formed by the tips of all the cones produced by Bragg diffraction, which is not the case in the standard CBED technique where the tips of all the cones coincide (Fig. 6.17). In the LACBED technique, overlapping of the discs in the final pattern is prevented by selecting one of the cones with an aperture, as illustrated in Fig. 6.20. This can be done in any plane conjugate to the object plane at $\Delta z = 0$. An obvious choice is the first intermediate image plane, where the selected-area diaphragm is located. The centring of the diaphragm can be observed in the image mode. In a final step, the microscope is switched to the diffraction mode, where the LACBED disc formed by the selected beam is observed. As an example, Fig. 6.21 shows an energy-filtered LACBED pattern of the (000) beam of a TiAl crystal in $\langle 110 \rangle$ zone axis orientation.

A certain disadvantage of the LACBED technique is that by lifting the specimen within the cone of illumination, an increasingly large area of the specimen is illuminated (up to several hundred nm). One incident beam direc-

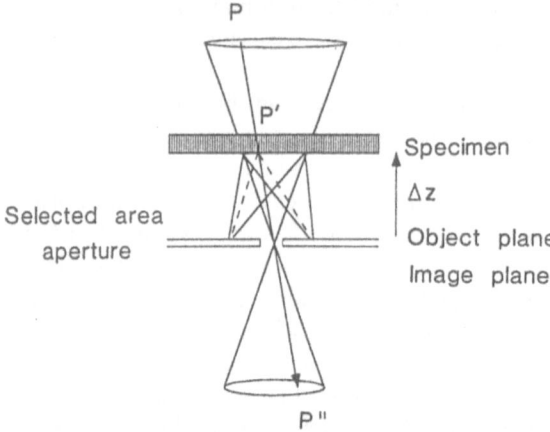

Fig. 6.20. The geometry of large-angle convergent-beam electron diffraction (LACBED). The specimen is lifted by Δz from the normal position. To prevent overlapping of the discs, the selected-area aperture can be used to select the (000) disc from the small diffraction pattern that is formed at the normal specimen and the conjugate planes.

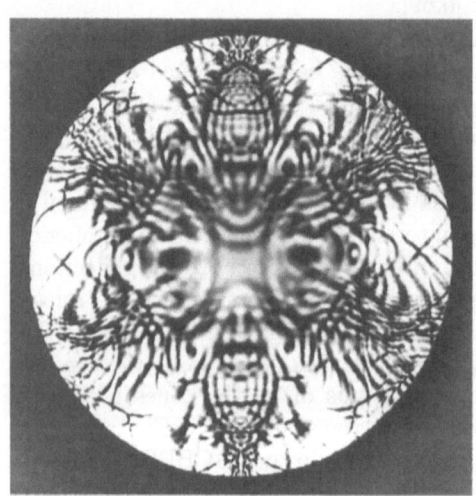

Fig. 6.21. A zero-loss filtered LACBED pattern from TiAl along the $\langle 110 \rangle$ zone axis.

tion (defined by point P in Fig. 6.20) corresponds to one point P′ in the specimen and produces one image point P″ in the LACBED disc. In the LACBED pattern, diffraction information is thus mixed with image information. This results in distortions and symmetry breaking in the pattern when the specimen is bent and/or inhomogeneous in thickness. On the other hand, this special feature of LACBED can be used to image distortions, such as strain distributions, or to analyse dislocations within the illuminated area. Figure 6.22 shows how the strain distribution in liquid-phase-expitaxy-(LPE)-grown SiGe films can be imaged with LACBED [6.71].

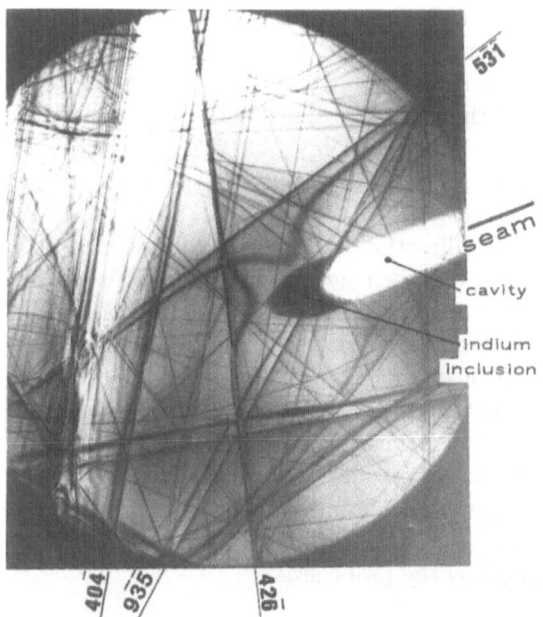

Fig. 6.22. LACBED pattern at the boundary of a LPE grown SiGe film. The HOLZ lines bend under the lattice strain caused by an indium inclusion.

In the technique for producing LACBED patterns described above, all the lenses and thus all the focal planes are left in their original settings while the specimen height is changed to separate the diffracted orders. In the original Tanaka method, the specimen is left at the same height and the focused spot as well as the focal plane are moved away from the specimen by changing the lens currents. The methods are equivalent and produce the same results. On a microscope with a digital z control, the method described above is easier to perform and to reproduce since only one parameter (the specimen height) has to be changed.

The required separation of the spots in the image plane depends on the diameter of the selected-area diaphragm that is used to select an individual spot. For smaller diaphragms the required shift Δz is smaller and hence the illuminated areas are also smaller. This reduces the effect of thickness variations and specimen bending. Ideally a 5 μm diaphragm is used.

If a very small selected-area diaphragm is used and the changes in specimen height are large, only a very narrow angular range of scattered electrons is allowed to pass through the aperture and to contribute to the LACBED pattern. Electrons inelastically scattered through angles larger than the selected aperture will also be intercepted by the diaphragm and can thus be removed from the pattern. The LACBED technique hence provides a kind of energy filtering by the aperture being used [6.72]. Owing to the small changes of scattering vector during plasmon losses, large changes in specimen height or lens current would be necessary to provide a substantial amount of energy filtering for plasmon loss electrons. The latter are therefore best removed by using an energy filter, even in LACBED. On the other hand, thermal diffuse scattering may lead to large scattering angles and the angular distribution has a minimum in the incident beam direction. It can be removed very efficiently in LACBED even for small changes in specimen height or lens current (while it cannot be removed by an imaging energy filter). Zero-loss filtered LACBED thus provides scattering information about the specimen almost free of inelastically scattered electrons.

In the LACBED technique, the diffraction pattern is in focus and HOLZ lines, for example, can be observed with their natural line width. The image information in the pattern is however blurred since the sample is no longer in the focal plane. The resolution in any image detail is limited by the probe size. Conversely, in the convergent-beam imaging (CBIM) method developed by *Humphreys* et al. [6.73], the image is in focus while a blurred diffraction pattern is superimposed on it. CBIM is realised by means of the geometrical condition shown in Fig. 6.20 but focusing (while in diffraction mode) on the bottom surface of the specimen. The image information is then transferred with the nominal resolution of the microscope, while the angular resolution in the diffraction pattern is limited by the probe size.

6.4.3 Quantitative Information from HOLZ Lines

Excitation of reflections in the higher order Laue zones involves much larger scattering angles than for the reflections of the ZOLZ because of the large radius of the Ewald sphere (Fig. 6.23). A clear distinction can always be made between ZOLZ and HOLZ reflections (although physically there is no difference and the distinction is made for purely geometrical reasons). In CBED, many HOLZ reflections can be excited simultaneously by the continuous variation of the incident beam direction, which is equivalent to rocking the Ewald sphere in the reciprocal lattice. For low-index reflections close to the origin, the rocking Ewald sphere intersects the needles (relrods) in the whole angular range given by the convergence angle (Fig. 6.24). The length of these needles is inversely proportional to the foil thickness. This explains in geometrical terms why complete discs are formed by the low-index reflections in CBED patterns. Conversely, the HOLZ reflections can be excited only if the incident beam direction (and thus the Ewald sphere) stays within a very narrow angular range. If the Bragg condition is satisfied for a HOLZ reflection, the incident beam direction can still be varied perpendicular to the plane of Fig. 6.24. The HOLZ reflections are therefore extended into narrow lines and corresponding defect lines appear in the (000) disc.

The angular width of a HOLZ line can be calculated in the two-beam approximation discussed in Sect. 6.3.1. The intensity of a reflection g in two-beam theory is given by [6.73,78]:

$$I_g = \left(\frac{\pi e^{-M}}{\xi_g}\right)^2 \frac{\sin^2(\pi t s)}{\pi s} \quad \text{with} \quad s'^2 = s^2 + \xi_g^2, \tag{6.36}$$

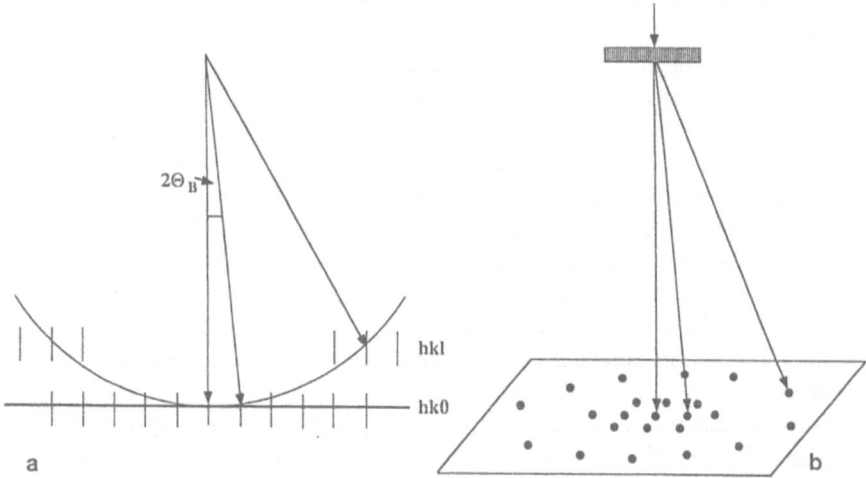

Fig. 6.23. (a) Ewald-sphere construction of the diffraction pattern of a thin single crystal, (b) Real space diagram of the scattering geometry.

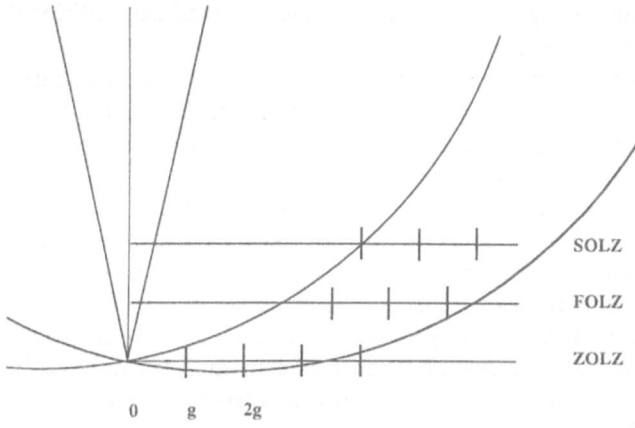

Fig. 6.24. Ewald-sphere construction of a convergent-beam electron diffraction pattern. Needles (relrods) of low-index reflections from the ZOLZ are excited in the whole angular region given by the convergence angle. HOLZ reflections are excited only in a small angular range.

where ξ_g is the extinction distance for the reflection g and e^{-M} is the Debye–Waller factor. From this equation, the angular range $\Delta\theta$ over which the diffracted intensity is appreciable can be derived:

$$\Delta\theta = 2e^{-M}/g\xi_g \text{ for } t \gg \xi_g/e^{-M} \text{ and } 2/gt \text{ for } t \ll \xi_g/e^{-M}. \tag{6.37}$$

For the low-index reflections of the ZOLZ, $\xi_g/e^{-M} \simeq 10$ nm (for 100 keV electrons), which yields $\Delta\Theta = 10^{-2}$ rad, showing that ZOLZ reflections can be excited over a wide angular range. In contrast, the extinction distances for HOLZ reflections are large, $\xi_g/e^{-M} \simeq 2000$ nm , and hence $\Delta\Theta \leq 10^{-4}$ rad. Variations in lattice plane distances d and crystallographic distortions can therefore be measured much more accurate by using HOLZ lines than by using low-index reflections of the ZOLZ. The angular width of the HOLZ lines is so small that, in principle, an accuracy of $\Delta d/d = 10^{-4}$ can be achieved.

Since distortions of the diffraction patterns cannot be avoided for large scattering angles, the HOLZ reflections themselves cannot be used for accurate lattice parameter measurement. Instead, the geometrical arrangement of the HOLZ lines in the (000) disc of CBED patterns is used. The arrangement of the HOLZ lines is completely independent of any imaging parameters of the microscope and depends only on the accelerating voltage and the crystal structure.

From equation (6.37) it can be seen that, by lowering the specimen temperature, sharper HOLZ lines are produced because reduction of the temperature increases e^{-M}, which improves the accuracy in quantification. Since a lower specimen temperature also reduces contamination, cooling is advantageous in most cases. However, if strain distributions at the interfaces of dissimilar materials are being studied, cooling might introduce additional strain due to differences in thermal expansion.

Energy filtering is very important since, according to (6.37), the sharpness of the HOLZ lines increases with increasing specimen thickness but the probability of single or multiple plasmon losses also increases. This leads to a broadening of the HOLZ lines (owing to the angular distribution of the plasmon loss electrons) and to a shift of the intensity minimum of the HOLZ lines. Exciting a plasmon loss of 20 eV at 100 keV is equivalent to a change of 2×10^{-4} in accelerating voltage with a corresponding change in Bragg angle and thus in HOLZ line position. This can be avoided, even in very thick specimens, by zero-loss energy filtering. The use of thick specimens is also very important to avoid the detrimental effect of thin film relaxations. In particular, in the investigation of heavily strained materials, energy filtering helps to provide much more reliable results.

For a quantitative evaluation of the HOLZ line positions, kinematical electron diffration can only be used as a rough first order approximation. Accurate results can be obtained only if the dynamical effects, which lead to a HOLZ line shift from the kinematical Bragg position, are taken into account. Theoretically, this and related effects are described in a three-beam theory and are well established [6.74–77]. The kinematical approach is, however, still used in many experimental studies even though it yields rather inaccurate results.

The shift of the HOLZ lines due to dynamical effects is explained schematically in Fig. 6.25 [6.78]. In the kinematical approach, the incident beam direction for which Bragg scattering into a reflection g occurs can be constructed by drawing spheres with radius $|K|$ around the origin and the point

g in the reciprocal lattice. ($|K|$ is the magnitude of the k-vector corrected for the mean inner potential of the crystal.) In kinematical theory the position of the HOLZ-line corresponding to a reflection h is given by the intersection of the spheres around the origin and point h (incident beam direction K_{kin} in Fig. 6.25). In dynamical theory, the interaction of the beam electron with the crystal leads to a splitting of the two spheres around the origin and g into two dispersion surfaces 1 and 2. As a consequence, the HOLZ reflection h can no longer be excited along K_{kin}, since the states associated with K_{kin} are no longer allowed in the crystal. If we assume that the HOLZ reflection h can still be treated kinematically, then the dynamical HOLZ line position can be found by following the sphere around h to the points of intersection with the dispersion surfaces. In the diffraction pattern, the HOLZ line shifts to the position given by the incident beam direction K_{dyn}. In most cases, the HOLZ line position will be given by the intersection with the top branch of the dispersion surface. In special cases, however, the intersections with the lower branches of the dispersion surface can be observed as well. Furthermore, for strong HOLZ reflections, a kinematical treatment is not accurate enough and the dynamical interactions between HOLZ and ZOLZ reflections have to be taken into account. The solutions of three- or many-beam theory have then to be used.

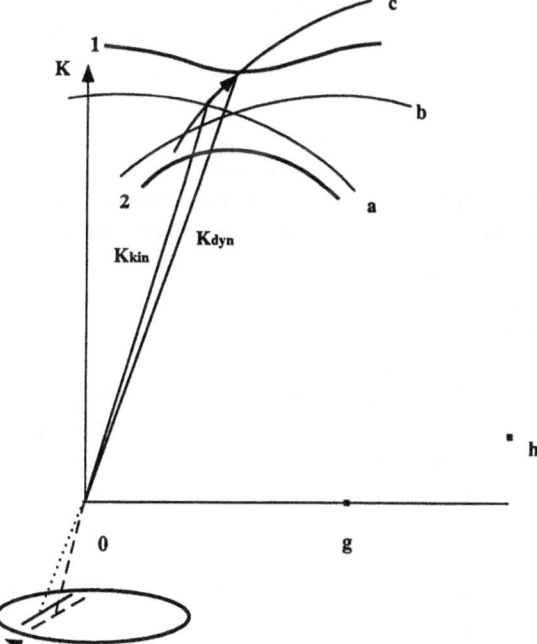

Fig. 6.25. The dynamical shift of HOLZ lines caused by the formation of the dispersion surfaces.

The errors introduced by not taking into account the dynamical HOLZ line shifts may be as high as several kV in the determination of the microscope accelerating voltage [6.79], or several percent if lattice parameters are being measured. The dynamical HOLZ line shift is highest for low-index zone axes and lowest for the incident beam direction, which avoids strong scattering into low-index reflections.

Various correction schemes for the dynamical HOLZ line shift have been discussed [6.78]. As a first approximation the simulations can be corrected by using an effective high voltage which is higher than the kinematical value [6.80,81]. This corresponds to an expansion of the spheres drawn in Fig. 6.25 and only produces useful results if the dispersion surfaces can be approximated by a sphere. By introducing an effective high voltage, the accuracy of measurement of lattice parameters or the accelerating voltage can be increased to 10^{-3}.

To obtain the highest accuracy, fully dynamical simulations are required, which yield maximum errors close to 10^{-4} [6.81,82]. With the computing power of modern workstations such dynamical simulations can be performed in reasonable times and are thus recommended even if the maximum possible accuracy is not needed. For the measurement of a three-dimensional strain state, a least-squares refinement program has been developed [6.82], which is capable of varying several lattice parameters independently until a best fit with the experimental pattern is obtained.

As an example, Fig. 6.26 shows the determination of the accelerating voltage U using a Si single crystal as a standard with an accuracy of 100 V. The experimental pattern (Fig. 6.26a) was obtained with the LACBED technique close to the $\langle 331 \rangle$ zone axis. Computer simulations in which the value of U was varied in the search for a best fit used only those 50 HOLZ lines which change their positions as U is varied. A complete simulation of all visible HOLZ lines would have required the inclusion of about 200 reflections.

HOLZ lines can also be used to measure lattice parameters or lattice strains very accurately. Figure 6.27 illustrates the determination of lattice strains at an interface between an aluminium film and a SiC substrate. The lattice strains are due to the difference in thermal expansion coefficients and build up as the material is cooled down [6.83]. The patterns were obtained along the $\langle 114 \rangle$ zone axis. Owing to the lattice distortions, the mirror symmetry along the vertical centre line in Fig. 6.27 is broken. The patterns were simulated for various tetragonal and trigonal distortions until a best fit could be obtained for a trigonally distorted lattice with $\Delta a/a = 2.5 \times 10^{-3}$ and $\Delta \gamma/\gamma = 2.2 \times 10^{-3}$, where γ is the angle between the a- and the b-axis. For the calculations, the structure factors of neutral spherical atoms were used because only the deviations from the mirror symmetry were used for quantification. The error introduced by not using the correct structure factors was compensated by using an effective high voltage, in this case 124.6 kV. From

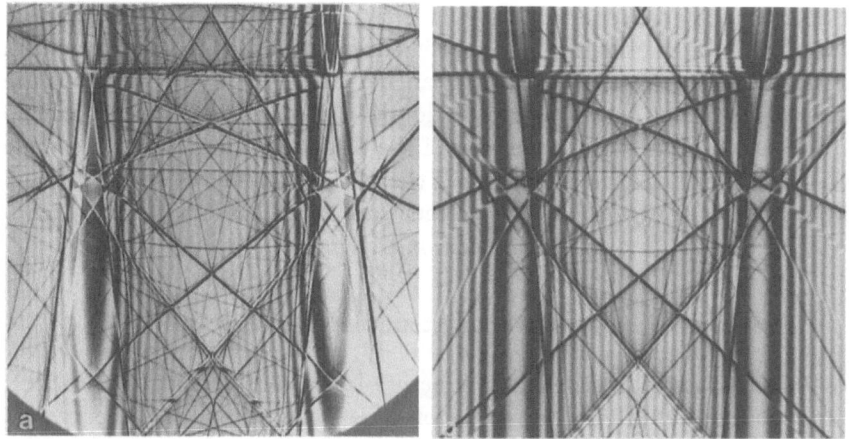

Fig. 6.26. Determination of the accelerating voltage. (a) Energy filtered LACBED pattern obtained from Si along the $\langle 331 \rangle$ zone axis, (b) computer simulation of the location of the HOLZ lines matched to the experimental pattern (a) for an accelerating voltage of 121.7 kV.

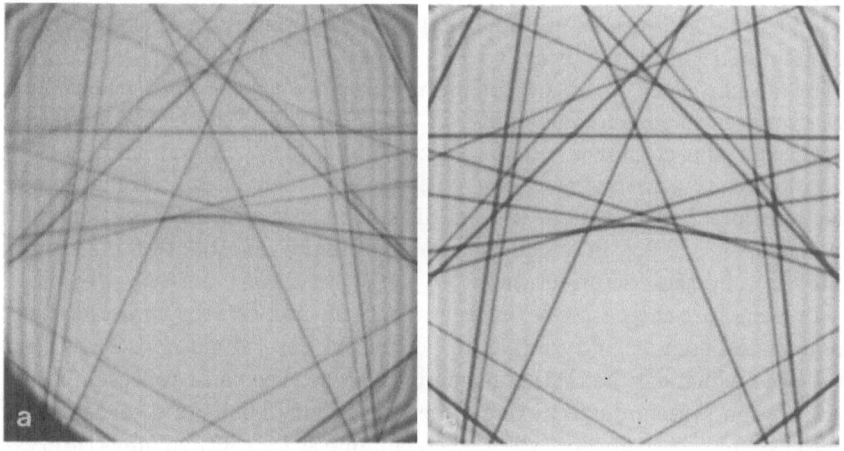

Fig. 6.27. Measurement of lattice strain. (a) CBED pattern of Al along the $\langle 114 \rangle$ zone axis. (b) Computer simulation for a trigonal strained lattice with $\Delta a/a = 2.5 \times 10^{-3}$ and $\Delta \gamma/\gamma = 10^{-3}$.

Fig. 6.27, it can be seen that much smaller distortions would still lead to a visible distortion of the HOLZ line arrangement and that lattice distortions of 10^{-4} could still be measured.

Special care has to be taken if local composition changes are to be studied with CBED. In principle, a change in composition leads to a proportional change in lattice parameter, which can be measured accurately from the change in the HOLZ line positions. However, the composition change also leads to a small shift in the position of the dispersion surface and thus to a corresponding HOLZ line shift, even if the lattice parameters do not change. Only with proper dynamical calculations can the HOLZ line shifts be analysed correctly in the case of compositional changes.

6.4.4 Refinement of Low-Order Structure Factors

Most of the classical methods for structure-factor measurement by electron diffraction are based on the results of three-beam dynamical theory and in particular on the occurrence of degeneracies. Examples are the analysis of three-phase structure invariants [6.84], the critical voltage method [6.85] and the analysis of degeneracies in three-beam cases for both centrosymmetric crystals [6.86] and non-centrosymmetric crystals [6.87,88]. Reviews of structure-factor measurement by electron diffraction can be found in [6.26,78,89–93]. None of the techniques mentioned above requires the use of energy-filtered diffraction patterns since they are all based on qualitative interpretation of the variations of the intensities. In the present account, we restrict the discussion to novel techniques, based on a quantitative analysis of the intensity distributions in CBED patterns by comparison of experimental and simulated CBED patterns.

For measurement of the low-order structure factors, various methods have been proposed [6.93–95]. They differ in the geometry of the CBED patterns, i.e. whether the crystal is tilted to a zone axis [6.94], a systematic orientation [6.93], or a many-beam case with several strongly excited high-index reflections [6.95]. We limit the present discussion to the method developed by *Zuo* and *Spence* [6.93], in which the reflection under investigation is strongly excited in a systematic orientation.

For structure-factor phase measurements in non-centrosymmetric crystals, a three-beam orientation has to be chosen. Several methods have been developed, which can be distinguished by whether the third reflection is excited in a non-systematic or a systematic orientation [6.96]. For measurements in the systematic orientation, the same technique as for centrosymmetric crystal can be used.

For the calculation of the intensities, the Bloch-wave formalism introduced in Sect. 6.3.1 is employed. For an accurate calculation of the intensities, a large number of reflections of the ZOLZ and HOLZ have to be included in the calculation. As explained in Sect. 1.4.1, each point in the CBED discs corresponds to a different incident beam direction, for which the eigenvalue equation (6.28) has to be solved by diagonalizing the matrix and calculating the reflection intensities for the given thickness. This numerical calculation has to be repeated for each new set of parameters and is hence very time-consuming. Even with very fast computers, the time required to simulate the whole pattern repeatedly for the whole parameter range would be much too high and calculations are restricted to one or two line profiles through the CBED discs.

The parameter space which appears in the calculation can be divided into "geometrical" and "physical" parameters. The separation into these two categories is somewhat arbitrary and is only made here for clarification. The geometrical parameters are: the coordinates of the starting and the end points of the line profiles, the incident beam direction and the radius of the CBED discs. In the calculation, all these geometrical parameters have to be treated as variables because none of them can be determined experimentally with sufficient accuracy to align the experimental and the calculated line profiles properly. The geometrical parameters have to be varied only within very narrow boundaries corresponding to a few pixels in the line profile.

The physical parameters include the structure amplitudes and the crystal thickness. For centrosymmetric crystals, the Fourier coefficients of the lattice potential (6.17) are composed of real parts V_g describing the elastic scattering and imaginary parts V_g' describing the absorption. For non-centrosymmetric crystals, both the elastic and absorptive part of the Fourier coefficients are complex quantities:

$$V(r) = \sum_g \left[|V_g| \exp(i\phi_g) + i|V_g'| \exp(i\phi_g') \right] \cdot \exp(-2\pi \, g \cdot r). \qquad (6.38)$$

Depending on the number of beams excited in the systematic row and the geometry of the line scan, the structure factors of one or two reflections can be determined.

The many-dimensional parameter space consisting of the geometrical and physical parameters cannot be separated into independent parts. Changes of the physical parameters always require adjustments of the geometrical parameters to achieve the best fit. As an example, it is not possible to determine the incident beam direction a priori and then refine the structure factors. A typical refinement problem thus involves a set of 12 to 16 non-separable parameters.

The determination of the incident beam direction is the most important problem in structure-factor determination. The incident beam direction is measured in a coordinate system that has its origin in the nearest low-index zone axis. The geometry is explained in detail in [6.93]. The experimental procedure is as follows: an approximate starting value can be obtained by locating the zone axis on diffraction patterns taken with a very low camera-length. Much more accurate values are then obtained by precisely locating the HOLZ line positions within the CBED discs. The technique is based on the use of maps obtained with the large angle CBED technique [6.97]. Such a map (Fig. 6.28) consists of a series of LACBED patterns, which are obtained for increasing tilt away from a high symmetry zone axis. In the map, the HOLZ line crossings visible in the CBED pattern, which has to be evaluated, can easily be located. The incident beam direction corresponding to these HOLZ line positions is then calculated using the program "HOLZ" [6.78]. The resulting values are used as starting values for the refinement. However, the incident beam direction always has to be refined simultaneously with the structure factors, since a change in the value of the structure factor being refined leads to a dynamical shift of the HOLZ lines (see Sect. 6.4.3) and thus to a slightly different value of the incident beam direction. The correct incident beam direction is only obtained together with the structure factors after completing the refinement.

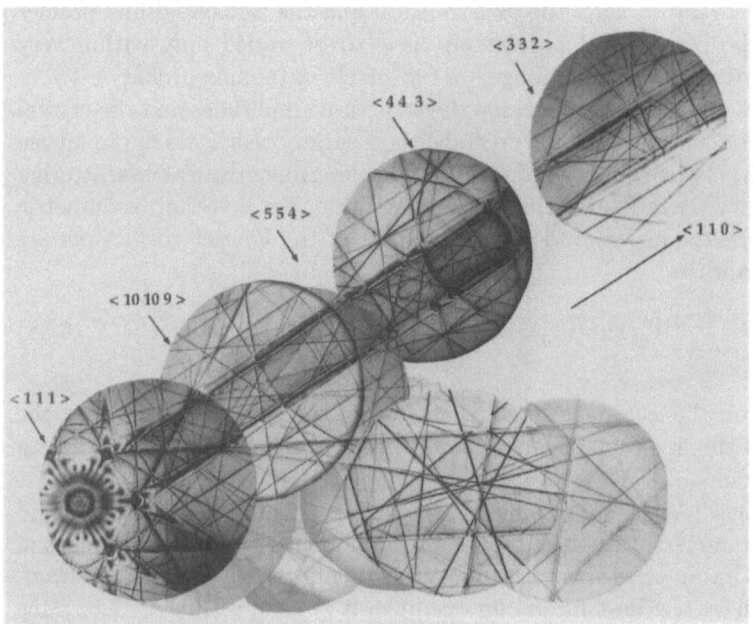

Fig. 6.28. Map of LACBED from Si near the ⟨111⟩ zone axis (left side). HOLZ line crossings from the central discs of experimental CBED patterns (obtained in a systematic orientation) can be identified on this map and used to determine the incident beam direction.

The program HOLZ also calculates the indices of the reflections that are excited in the chosen incident beam direction and hence have to be included in the Bloch wave calculation. More than a hundred reflections can be at least weakly excited. We have found that about 30 reflections of the ZOLZ plus a few relatively strong reflections of the HOLZ are sufficient to obtain accurate results. Strongly excited reflections are included in the matrix to be diagonalized; weakly excited reflections are treated by perturbation theory, which reduces the calculation time [6.93].

The refinement of the calculated line profiles with respect to the experimental data is performed with the aid of the χ^2 test; in statistical theory, χ^2 is defined as [6.93]:

$$\chi^2 = \sum_{i=1}^{n} \frac{w_i(cf_i^{\text{theo}} - f_i^{\text{exp}})^2}{\sigma_i^2}. \tag{6.39}$$

The parameters f_i^{exp} and f_i^{theo} are the experimental and theoretically calculated intensity values of the point i within the linescan, c is a normalization constant. The sum runs over all points i within the line-scan. The difference between the experimental and calculated value can be weighted by a factor w_i. The quantity σ_i is the standard deviation if each of the experimental values f_i^{exp} is obtained by averaging over several measurements. If only one measurement is made and Poisson statistics assumed for the experimental values, then σ_i^2 is set equal to f_i^{exp}.

The experimental noise of the intensity data acquired with the CCD camera frequently cannot be described by Poisson statistics; rather, it is independent of the intensity value f_i^{exp}. Setting σ_i^2 equal to f_i^{exp} thus overestimates the significance of the low-intensity values and we therefore set $\sigma_i = const$ [6.97], yielding:

$$\chi^2 = \sum_{i=1}^{n} w_i(cf_i^{\text{theo}} - f_i^{\text{exp}})^2. \tag{6.40}$$

For the minimization of the function χ^2, a refinement algorithm is used. In the refinement the whole set or a subset of the parameters described above is varied until a minimum of the function χ^2 is found. The refinement algorithms can be distinguished by whether they are able to escape from local minima and to find the global minimum (global refinement) or whether they are liable to get trapped in local minima (local refinement). In the original method developed by *Zuo* and *Spence* [6.93], a local refinement algorithm, SIMPLEX, is used. An iterative procedure then has to be used to find the global minimum. Details of the technique are described in [6.78] and experimental results have been published on a number of low-index reflections in GaAs [6.98], MgO [6.93], and BeO [6.99]. In general, a quadratic refinement problem with n parameters may exhibit up to 2^n local minima [6.100]. In the structure-factor determination, the density of local minima increases with increasing specimen thickness since the number of oscillations in the intensity data increases. On the other hand, the accuracy in the structure-factor determination also increases with the number of intensity oscillations. For the given experimental ranges, several local minima frequently exist. As an example, Fig. 6.29 shows the χ^2 surface as a function of the crystal thickness and the structure factor amplitude for a Si $\langle 111 \rangle$ reflection. The occurrence of several minima is obvious. *Deininger* et al. [6.97] have developed a refinement program, which is based on a modified Simulated Annealing (SA) algorithm [6.101,102], and can vary all the geometrical and physical parameters simultaneously. Local minima will be escaped from with a certain probability, which

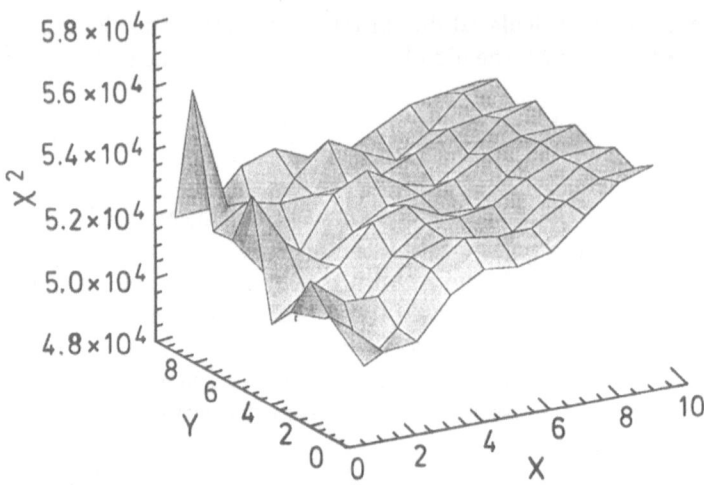

Fig. 6.29. χ^2 surface of a two-dimensional parameter subspace. x-axis: variation of thickness in steps of 10 nm starting at 95 nm. y-axis: variation of $|U_{111}|$ in steps of 10^{-3} nm starting from 4.258×10^{-3} nm. The surface shows deep local minima.

depends on the cooling rate with which the system is cooled down. Whether the refinement is successful depends on the correct choice of the starting "temperature" and the "rate of cooling".

As an example, we discuss the determination of the structure factors of the (220) reflection of silicon [6.97]. In the experiment, the sample was tilted to a systematic orientation, only two reflections being strongly excited. The tilt angle from the $\langle 110 \rangle$ zone was $\simeq 5°$. The specimen was cooled to 110 K, which corresponds to an estimated 140 K in the irradiated area. The CBED pattern was recorded by a CCD camera. Two line profiles along the lines indicated in Fig. 6.30 have been extracted. The first line crosses the discs perpendicular to the pendellösung fringes, the second line is placed in the (000) disc so that it crosses HOLZ lines and dark fringes that are due to non-systematic reflections. These positions ensure maximum sensitivity to the unknown structure factor and the incident beam direction, respectively.

The two lines intersect in one point to reduce the number of geometrical parameters to 12. The final result is shown in Fig. 6.30. No readjustments were necessary during the refinement, which is a major advantage of the SA algorithm. Several local minima were found and quitted during the refinement process. The following values were obtained for the structure amplitude and the thickness t:

$$U_{220} = 2mV_{220}/h^2 = (3.698 \pm 0.029)\,\text{nm}^{-2} \quad ; \quad t = (232.04 \pm 9.0)\,\text{nm}.$$

U_{220} is the normalized structure amplitude commonly in use. A Debye–Waller factor $\exp(-Bs^2)$ with $B = 2.7 \times 10^{-3}\,\text{nm}^2$ was used, corresponding to a specimen temperature of 140 K.

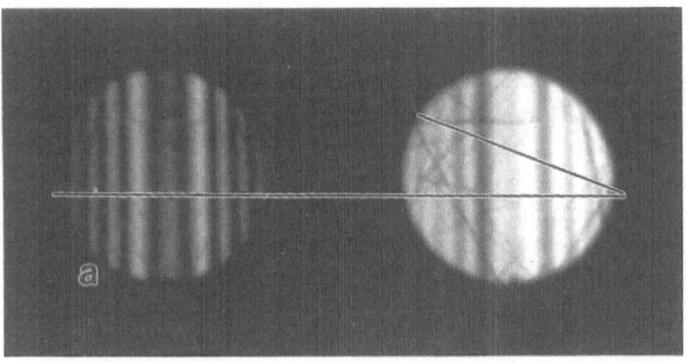

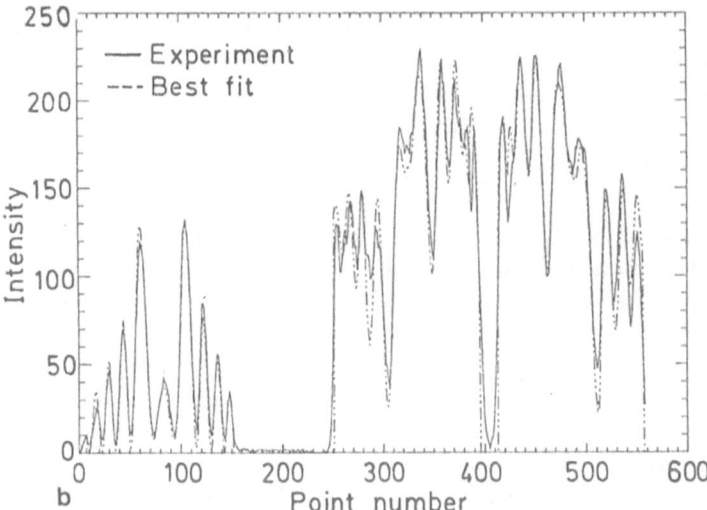

Fig. 6.30. Refinement of the Si (220) structure amplitude. (a) Systematic row of CBED discs recorded digitally by a CCD camera. (b) Experimental line profile and best computer fit.

In this example, the Debye–Waller factor has been extrapolated from tabulated values. While this is feasible for Si, tabulated values of the Debye–Waller factors do not exist for many other materials. In such cases, it is possible to determine the Debye–Waller factor from high-index reflections using the same CBED method as discussed above. It is assumed that, for increasing scattering vectors, the bonding effects have less influence on the structure factors, which then approximate those of neutral spherical atoms. Furthermore, the influence of the Debye–Waller factor on the total structure factor increases as a function of g^2 and is thus stronger for high-index reflections. From the difference between measured and tabulated structure factors [6.103], the Debye–Waller factors can be determined. One problem is that in the corresponding refinement the low-order structure factors have to be known but their values cannot be determined accurately without a knowledge of the Debye–Waller factor. Further research is necessary before the method can be generally used for a large variety of materials.

In the above discussion, it has also been assumed that the Debye–Waller factor is isotropic. While this is a good approximation for many metals, it may not be valid in general for covalently or ionically bonded materials. It is expected that measurements of direction-dependent Debye–Waller factors will in future be very important and reveal further information on atomic bonding.

6.4.5 Determination of Charge Density Distributions

The low-index reflections carry information on the bonding within the crystal. This information can be extracted from the structure factors by calculating the charge density distribution. The charge density distribution is related to the electrostatic potential via Poisson's equation:

$$\nabla^2 V(r) = -\frac{e}{\varepsilon_0}(\rho_n - \rho_e). \qquad (6.41)$$

The charge density can be separated into the contributions from the nucleus, ρ_n, and the contribution from the electrons, ρ_e. The former, ρ_n, can be expressed as a point charge of magnitude Z_i at the lattice position r_i of the atom i. The potential $V(r)$ can be calculated by summation over all Fourier coefficients V_g as in (6.18). Of the total electronic charge density, the bonding part is of interest. The contribution from the inner shells can be assumed to be spherical and does not directly influence any of the properties of the crystal. The charges forming, for example, the covalent bonds in a crystal can be composed of charges as small as 10^{-4} of the total electronic charge density. This contribution will thus not be visible in plots of the total charge density within a unit cell. The problem can be solved by plotting the difference charge density

$$\Delta\rho(r) = \rho_{\exp}(r) - \rho_{at}(r), \qquad (6.42)$$

where $\rho_{\exp}(r)$ is the charge density of the real crystal, which has been measured experimentally, and $\rho_{at}(r)$ is the charge density of a hypothetical crystal with neutral spherical atoms at the lattice sites. The difference charge density map shows areas of positive and negative charge density, which indicate accumulation and depletion of electronic charges due to the formation of covalent or ionic bonds. Covalent bonds can be identified in this map by an accumulation of charge density between two atoms. The strength of the bond can be estimated by integrating over the three-dimensional area with a surplus of electronic charges. The presence of ionic bonds can be inferred from a positive or negative charge balance at the atom positions.

For the calculation of the charge density $\rho_{at}(r)$, the X-ray scattering amplitudes tabulated in the *International Tables for X-ray Crystallography*

[6.103] are used. In the first Born approximation the atomic scattering amplitude for electrons is related to the atomic scattering amplitude for X-rays via the Mott formula:

$$V_g = \frac{e}{16\pi^3\varepsilon_0\Omega} \sum_i \frac{Z_i - f_i^x(s)}{s^2} \exp(-B_i s^2) \exp(-2\pi i g \cdot r). \qquad (6.43)$$

The first term of the sum describes the Rutherford scattering of the electrons at the nucleus, which depends on the charge Z_i of the nucleus i. This term does not occur in the atomic scattering amplitude for X-rays because the X-rays are scattered only by the electrons. Starting from the X-ray scattering amplitudes, the V_g are the potential coefficients for an arrangement of neutral spherical atoms on the lattice sites.

The X-ray structure amplitude F_g^x is obtained by summation over all the scattering amplitudes f^x of the atoms in the unit cell:

$$F_g^x = \sum_i f_i^x(s) \exp(-B_i s^2) \exp(-2\pi i g \cdot r). \qquad (6.44)$$

with $s = \sin\theta_B/\lambda = |g|/2$. The influence of the thermal vibrations of the atoms is taken into account by including the Debye–Waller factor $\exp(-B_i s^2)$, in which B_i is the temperature factor of atom i. The influence of the Debye–Waller factor increases quadratically with the magnitude of the scattering vector s. The Debye–Waller factors can be calculated from the tabulated values in the *International Tables for X-ray Crystallography* or, if available, taken from X-ray or neutron diffraction. As discussed above, it is also possible to measure the Debye–Waller factors by applying the CBED technique described in Sect. 6.4.4 to high-index reflections.

The number of reflections that has to be taken into account in the calculation of the charge density difference map depends strongly on the temperature. At higher temperatures, the decreasing Debye–Waller factor reduces the influence of reflections with large scattering vectors. Furthermore, the difference maps also depend strongly on the structure factors calculated for the neutral reference atoms. It has therefore been suggested [6.78] that all the research groups should use the same standards given in the *International Tables*. The structure factors, however, should not be calculated using the classical method of *Doyle* and *Turner* [6.104] but rather with the more accurate methods given in [6.105] or [6.106], especially for the high-index reflections and their absorption.

The calculation of charge density maps requires that the crystallography of the phases being studied be known. Changes of the composition by doping, for example, are only possible if they do not lead to a change of the atom positions or to a distortion of the unit cell. If this is the case, high-index reflections have to be measured as well and in a first step accurate atom positions have to be found. Since the determination of atom positions and of the charge density distribution are expected to become standard techniques in the near future, electron crystallography will be an important tool in many fields of materials science.

6.5 Imaging of the Dispersion of Plasmon and Compton Scattering

6.5.1 Compton Scattering

As shown in Chapts. 1 and 4, scattering of electrons at quasi-free atomic electrons results in the Bethe ridge or Compton peak (Figs. 1.23 and 4.3) with a maximum at the Compton angle

$$\sin^2 \theta_C = \frac{E}{E_0}[1 + (E_0 - E)/2m_0c^2]^{-1} \simeq E/E_0 \ . \tag{6.45}$$

This Compton peak can be seen in an ESD as a diffuse ring, the diameter of which increases as the square root of the energy loss E (Fig. 6.31). A quantitative record of the intensity across the ridge can be obtained (Fig. 1.23) either by a radial scan of the diffraction pattern at a fixed energy loss E (Fig. 6.32a) or by recording an EELS at a fixed scattering angle of about 5° [6.20,39,108], which also contains the carbon K edge in Fig. 6.32b. Another possibility of observation and quantitative analysis is the method of angularly resolved EELS (Fig. 1.22) [6.107].

The analogous Compton scattering of X-rays [6.109,110] is in common use for testing calculations of atomic orbitals in solids and has also been applied to line-scans by electron diffraction [6.108]. The intensity profile of the Bethe ridge is proportional to the projection of the momentum distribution of atomic electrons on the scattering direction (z):

$$J(p_z) = \int \int n(p_x, p_y, p_z) \mathrm{d}p_x \mathrm{d}p_y \tag{6.46}$$

where $n(\boldsymbol{p})$ is the momentum probability density. The Fourier transform

$$B(z) = \int J(p_z) \exp(-\mathrm{i}p_z z/\hbar) \mathrm{d}p_z \tag{6.47}$$

of a recorded intensity profile is the autocorrelation function of the ground-state wave function. An advantage of electron Compton scattering is the shorter exposure time and the size of the selected area, a disadvantage is the need for thin specimens to avoid multiple scattering effects and the influence of Bragg reflections on the intensity profile [6.111].

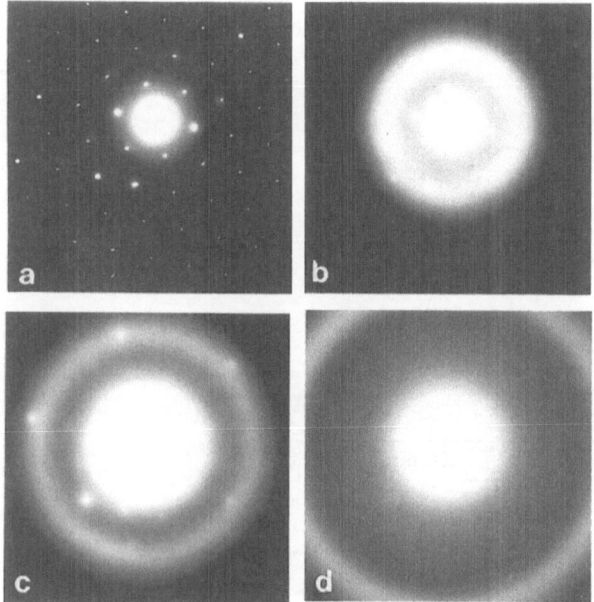

Fig. 6.31. Series of electron spectroscopic diffraction patterns of a graphite foil with energy losses of (a) 0, (b) 200, (c) 400 and (d) 800 eV showing the diameter of the Compton maximum increasing as $E^{1/2}$.

6.5.2 Plasmon-Loss Scattering

Plasmon losses show a dispersion (Sect.3.2.1)

$$E_{\mathrm{pl}}(\theta) = E_{\mathrm{pl}}(0) + 2E_0\alpha\theta^2 \qquad (6.48)$$

for angles smaller than the cutoff angle

$$\theta_{\mathrm{c}} = \lambda E/h v_{\mathrm{F}} \qquad (6.49)$$

where the intensity of plasmon losses drops rapidly (v_{F} = Fermi velocity). The agreement of the theoretical value of the constant

$$\alpha = \frac{3}{5} E_{\mathrm{F}}/E \qquad (6.50)$$

in (6.48) with experimental values is variable [6.112].

The dispersion and cutoff of plasmon losses can be imaged by angularly resolved EELS (Fig.1.22). The dispersion can also be imaged in a series of ESD patterns with a narrow width $\delta E \simeq 1$ eV of the energy window, as demonstrated in Fig. 6.33 for a single-crystalline Sn film with a plasmon loss at $E_{\mathrm{pl}}(0) = 17$ eV [6.113]. A disc of plasmon-loss scattered electrons can be observed around the primary beam and the Bragg reflections for losses near

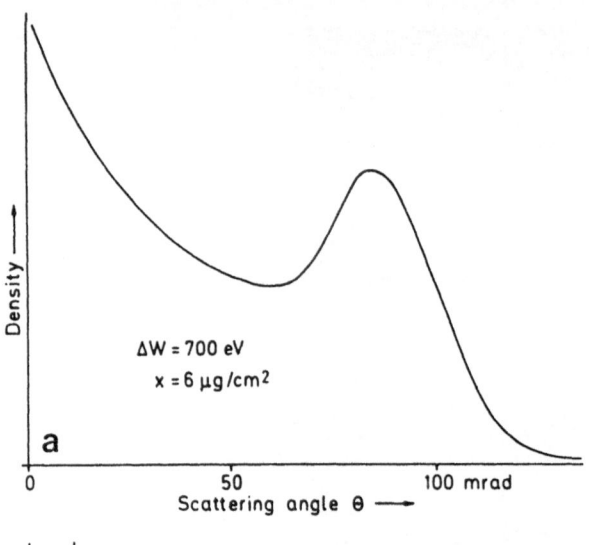

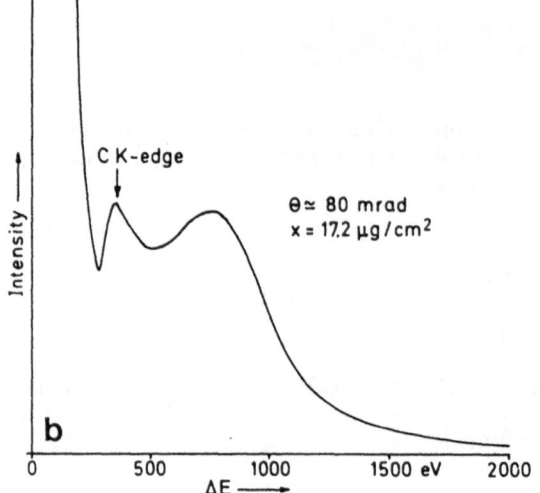

Fig. 6.32. Intensity scans through a diffraction pattern of an amorphous carbon film (see schematical diagram in Fig. 1.23) showing (**a**) the Compton maximum (Bethe ridge) as a radial scan with a constant energy loss $E=700$ eV, and (**b**) as an EELS at a fixed angle $\theta = 5°$.

$E_{pl}(0)$ (Fig. 6.33a). The parabolic dispersion $\propto \theta^2$ in (6.48) results in a diffuse ring increasing in diameter as the selected energy is raised by a few eV (Figs. 6.33b–d). On solving (6.48) for θ, the maximum of the ring is seen to be at a scattering angle

$$\theta^2 = [E_{pl}(\theta) - E_{pl}(0)]/2\alpha E_0 \quad . \tag{6.51}$$

Beyond 28 eV the ring disappears due to cutoff. The measured value $q_c = \theta_c/\lambda = 18$ nm^{-1} agrees well with $q_c = 15$ nm measured at 50 keV [6.114]. In

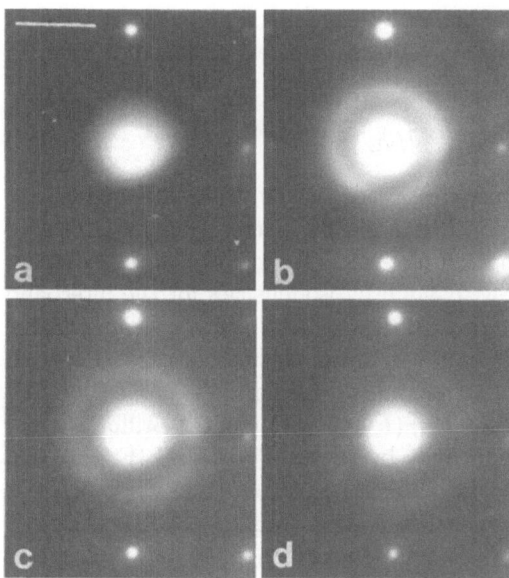

Fig. 6.33. Series of electron spectroscopic diffraction patterns of a single-crystal evaporated Sn film recorded with an energy window $\delta E=1$ eV at (**a**) $E=17$, (**b**) 21, (**c**) 23, and (**d**) 28 eV showing a section through the dispersion of plasmon losses (scale bar = 10 mrad).

diffraction patterns of a [111]-oriented Si foil, for example, the plasmon rings can also be observed around the {220}-type Bragg diffraction spots. This demonstrates that plasmon losses near the cutoff angle θ_c are also intraband-scattered, conserving the Bloch-wave field. Plasmon losses and intraband transitions can show an anisotropy in anisotropic crystals when the dielectric tensor $\epsilon(\omega,q)$ depends on the direction **q** of momentum transfer. When this tensor is transformed to its orthogonal main axes with diagonal components ϵ_{ii} the intensity of the plasmon losses becomes proportional to

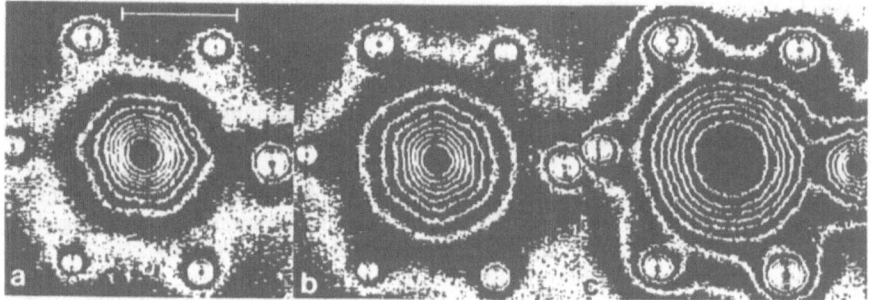

Fig. 6.34. Intensity contours (digital isodensities) in diffraction patterns of a graphite foil showing the azimuthal anisotropies for the interband transition at (**a**) $E=7$ and (**b**) 13 eV and no anisotropy for (**c**) the plasmon loss at $E=31$ eV (scale bar = 10 mrad).

$\mathrm{Im}\{-1/\sum_i \epsilon_{ii}q_i^2\}$, where the q_i are the components of $\mathbf{q}$ along the main axes [6.115]. The angle η between the momentum transfer $\hbar\mathbf{q}$ and the primary electron direction is given by

$$\tan\eta = \theta/\theta_{\mathrm{E}} \tag{6.52}$$

where $\theta_{\mathrm{E}} = E/mv^2$ (1.21). Thus, η approaches 90° for scattering angles $\theta \gg \theta_{\mathrm{E}}$. This anisotropy can be observed in ESD patterns as an azimuthal anisotropy of the angular distribution for scattering angles $\theta \le \theta_{\mathrm{c}}$. Evidence for this anisotropy is shown in Fig. 6.34a,b by isodensities for the interband transition of π-electrons in graphite with the incident electron beam parallel to the c-axis [6.113]. The intensity contours of the plasmon scattering around the primary beam show hexagons with corners directed towards the six $\{220\}$ Bragg reflections at $E = 7\,\mathrm{eV}$ (Fig. 6.34a) ($\mathit{\Gamma}$Q-direction of the Brillouin zone) but with the corners rotated azimuthally by 30° at $E = 13\,\mathrm{eV}$ (Fig. 6.34b). In contrast to this, the intensity contours of the plasmon loss of graphite at $E = 31\,\mathrm{eV}$ show an isotropic angular distribution (Fig. 6.34c). These results agree with direct measurement of the anisotropy of energy losses in graphite by sequential recording of EELS at different scattering angles and azimuths [6.116,117], whereas the contours in Fig. 6.34 form a parallel record of a two-dimensional map of anisotropy effects. Weaker anisotropies have also been observed with Si foils [6.113].

References

6.1 R. Leonhardt, H. Richter, W. Rossteutscher: Elektronenbeugungsuntersuchungen zur Struktur dünner nichtkristalliner Schichten. Z. Physik **165**, 121–150 (1961)

6.2 D.J.H. Cockayne, D.R. McKenzie: Electron diffraction analysis of polycrystalline and amorphous thin films. Acta Cryst. A **44**, 870–878 (1988)

6.3 C.W. Grigson: Improved scanning electron diffraction system. Rev. Sci. Instr. **36**, 1587–1593 (1965)

6.4 P.N. Denbigh, C.W.B. Grigson: Scanning electron diffraction with energy analysis. J. Sci. Instr. **42**, 305–311 (1965)

6.5 M. Horstmann, G. Meyer: Eine Gegenfeldanordnung zur Messung von Energie– und Winkelverteilungen gestreuter Elektronen. Z. Physik **159**, 563–583 (1960)

6.6 M. Horstmann, G. Meyer: Messung der Elektronenbeugungsintensitäten polykristalliner Aluminiumschichten bei tiefen Temperaturen und Vergleich mit der dynamischen Theorie. Z. Physik **182**, 380–397 (1965)

6.7 M. Horstmann, G. Meyer: Einfluß der Kristalltemperatur auf die Intensitäten dynamischer Elektroneninterferenzen. Z. Physik **183**, 375–393 (1965)

6.8 M.F. Tompsett: Review: Scanning high-energy electron diffraction in materials science. J. Mat. Sci. **7**, 1069–1079 (1972)

6.9 S. Kuwabara, J.M. Cowley: The effect of energy losses in aluminium on electron diffraction intensities. J. Phys. Soc. Jpn. **34**, 1575–1582 (1973)

6.10 U. Plate, H.J. Höhling, L. Reimer, R.H. Barckhaus, R. Wienecke, H.P. Wiesmann, A. Boyde: Analysis of the calcium distribution in predentine by EELS and of the early crystal formation in dentine by ESI and ESD. J. Micr. **166**, 329–341 (1992)

6.11 H. Mahl, W. Weitsch: Kleinwinkelbeugung mit Elektronenstrahlen. Z. Naturforschg. **15a**, 1051–1055 (1960)

6.12 R.H. Wade, J. Silcox: Small angle electron scattering from vacuum condensed metallic films. Phys. stat. sol. **19**, 57–76 (1967)

6.13 R.P. Ferrier: Small angle electron diffraction in the electron microscope. Adv. Opt. Electron Microscopy, ed. by R. Barer, V.E. Cosslett (Academic, London 1969) Vol.3, pp.155–218

6.14 R.T. Murray, R.P. Ferrier: Biological applications of electron diffraction. J. Ultrastruct. Res **21**, 361–377 (1967)

6.15 R. Castaing: Quelques applications du filtrage magnétique des vitesses en microscopie électronique. Z. angew. Phys. **27**, 171–178 (1989)

6.16 P. Duval, L. Henry: Intéret du filtrage des énergies en diffraction électronique. J. Appl. Cryst. **6**, 113–116 (1973)

6.17 J. Smart, R.E. Burge: Small-angle electron diffraction patterns of assemblies of spheres and viruses. Nature **205**, 1296–1297 (1965)

6.18 V. Drahoš, A. Delong: Low-angle electron diffraction from defined specimen area. *Microscopie Electronique 1970*, ed. by P. Favard (Société Francaise de Microscopie Electronique, Paris 1970) Vol.2, pp.147–148

6.19 G.A. Bassett, A. Keller: Low-angle scattering in an electron microscope: application to polymers. Phil. Mag. **9**, 817–828 (1964)

6.20 L. Reimer, I. Fromm, I. Naundorf: Electron spectroscopic diffraction. Ultramicroscopy **32**, 80–91 (1990)

6.21 L. Reimer, H.G. Badde, E. Drewes, H. Gilde, H. Kappert, H.J. Höhling, D.B. von Bassewitz, A. Rössner: Laserbeugung an elektronenmikroskopischen Aufnahmen. Forschungsber. des Landes Nordrhein–Westfalen Nr. 2314 (Westdeutscher Verlag, Opladen 1973)

6.22 F. Thon: Zur Defokussierungsabhängigkeit des Phasenkontrastes bei der elektronenmikroskopischen Abbildung. Z. Naturforschg. **21a**, 476-478 (1966)

6.23 H.A. Bethe: Theorie der Beugung von Elektronen an Kristallen. Ann. Physik (Leipzig) **87**, 55–129 (1928)

6.24 M. von Laue: *Materiewellen und ihre Interferenzen* (Akad. Verlagsgesellschaft, Leipzig 1944)

6.25 P.B. Hirsch, A. Howie, R.B. Nicholson, D.W. Pashley, M.J. Whelan: *Electron Microscopy of Thin Crystals* (Butterworths, London 1965)

6.26 L. Reimer: *Transmission Electron Microscopy, Physics of Image Formation and Microanalysis*, 3rd ed. (Springer, Berlin, Heidelberg 1993)

6.27 L. Reimer: Electron diffraction methods in TEM, STEM and SEM. Scanning **2**, 3–19 (1979)

6.28 I. Gjønnes, J. Taftø: Bloch wave treatment of electron channeling. Nucl. Instrum. Metho. **132**, 141–148 (1976)

6.29 C.J. Humphreys: The scattering of fast electrons by crystals. Rep. Progr. Phys. **42**, 1825–1887 (1979)

6.30 Y. Kainuma, M. Kogiso: Many-beam dynamical theory of the line in the middle of a Kickuchi band. Acta Cryst. A **24**, 81–84 (1986)

6.31 D. Lynch, C. Rossouw: Non zeroth-order Laue zone effects and ALCHEMI. Ultramicroscopy **21**, 69–76 (1987)

6.32 Y.H. Ohtsuki: *Charged Beam Interactions with Solids* (Taylor and Francis, London, 1983)

6.33 P. Rez, C.J. Humphreys, M.J. Whelan: The distribution of intensity in electron diffraction patterns due to phonon scattering. Phil. Mag. **35**, 81–86
(1977)

6.34 M. Creuzberg, H. Dimigen: Energieanalyse im Elektroneninterferenzbild von
Si-Einkristallen. Z. Physik **174**, 24–34 (1963)

6.35 G. Meyer–Ehmsen, A. Siems: Contribution of plasmon and quasi-elastic scattering to the Kikuchi structure of Si. Phys. Stat. Solidi (b) **63**, 577–586
(1974)

6.36 J.G. Philip, M.J. Whelan, R.F. Egerton: The contribution of inelastically
scattered electrons to the diffraction pattern and images of a crystalline
specimen. *Electron Microscopy 1974*, ed. by J.V. Sanders, D.J. Goodchild
(Australian Acad. of Science, Canberra 1974) Vol. I, pp.276–277

6.37 R.F. Egerton, J.G. Philip, P.S. Turner, M.J. Whelan: Modification of a TEM
to give energy-filtered images and diffraction patterns, and energy loss spectra. J. Phys. E **8**, 1033–1037 (1975)

6.38 L. Reimer, I. Fromm: Electron spectroscopic diffraction at (111) silicon foils.
Proc. 47th Ann. Meeting EMSA (San Francisco Press, San Francisco 1989)
pp.382–383

6.39 L. Reimer, I. Fromm, R. Rennekamp: Operation modes of electron spectroscopic imaging and EELS in a TEM. Ultramicroscopy **24**, 339–354 (1988)

6.40 J. Mayer, C. Deininger: Omega energy filtered convergent beam electron
diffraction. *Electron Microscopy 1992*, ed. by A. Ríos, J.M. Arias, L. Megias–
Megias, A. López-Gallindo (Secr. Publ. Universidad de Granada, Granada
1992) Vol. 1, pp.181–182

6.41 C.R. Hall, P.B. Hirsch: Effect of thermal diffuse scattering on propagation
of high energy electrons through crystals. Proc. Roy. Soc. (London) A **286**,
158–177 (1965)

6.42 G. Honjo, S. Kodera, N. Kitamura: Diffuse streak diffraction patterns from
single crystals. J. Phys. Soc. Jpn. **19**, 351–367 (1964)

6.43 K. Komatsu, K. Teramoto: Diffuse streak patterns from various crystals in
x-ray and electron diffraction. J. Phys. Soc. Jpn. **21**, 1152–1159 (1966)

6.44 N. Kitamura: Temperature dependence of diffuse streaks in single-crystal Si
electron diffraction patterns. J. Appl. Phys. **37**, 2187–2188 (1966)

6.45 H.P. Herbst, G. Jeschke: Diffuse streak-patterns from PbJ_2- and Bi-single
crystals and their temperature dependence. *Electron Microscopy 1968*, ed. by
D.S. Bocciarelli (Tipografia Poliglotta Vaticana, Rome 1968) Vol.1, pp.293–
294

6.46 E.M. Hörl: Thermisch-diffuse Elektronenstreuung in As-, Sb- und Bi-Kristal-
len. Optik **27**, 99–105 (1968)

6.47 L.E. Thomas, C.J. Humphreys: Kikuchi patterns in a high voltage electron
microscope. Phys. Stat. Solidi (a) **3**, 599–615 (1970)

6.48 J.M. Cowley, D.J. Smith, G.A. Sussex: Application of a high voltage STEM.
Scanning Electron Microscopy 1970 (ITTRI, Chicago 1970) p.11–16

6.49 Y. Nakai: Excess and defect Kikuchi bands in electron diffraction patterns.
Acta Cryst A **26**, 459–460 (1970)

6.50 D.G. Coates: Kikuchi-like reflection patterns obtained with the SEM. Phil.
Mag. **16**, 1179–1184 (1967)

6.51 H. Drescher, E.R. Krefting, L. Reimer: The orientation dependence of the
electron backscattering coefficient of gold single crystal films. Z. Naturforschg. **29a**, 833–837 (1974)

6.52 J. Taftø, O.L. Krivanek: Site-specific valence determination by EELS. Phys.
Rev. Lett. **48**, 560–563 (1982)

6.53 D. Cherns, A. Howie, M.H. Jacobs: Characteristic x-ray production in thin crystals. Z. Naturforschg. **28a**, 565–571 (1973)

6.54 B. Neumann, L. Reimer: Anisotropic x-ray generation in thin and bulk single crystals. J. Phys. D **13**, 1737–1745 (1980)

6.55 J. Taftø, Z. Liliental: Studies of the cation atom distribution in $ZnCr_xFe_{2-x}O_4$ spinels using the channeling effect in electron induced x-ray emission. J. Appl. Cryst. **15**, 260–265 (1992)

6.56 J.C.H. Spence, J. Taftø: ALCHEMIE: a new technique for locating atoms in small crystals. J. Micr. **130**, 147–154 (1983)

6.57 J.C.H. Spence: *High Resolution Transmission Electron Microscopy*, ed. by. P.R. Buseck, J.M. Cowley, and L. Eyring (Oxford University Press, Oxford 1988)

6.58 A. Berger, H. Kohl: Optimum imaging parameters for elemental mapping in an EFTEM. Optik **92**, 175–193 (1993)

6.59 A.J. Craven, J.M. Gibson, A.Howi, D.R. Spalding: Study of single electron excitations by electron microscopy, I. Image contrast from delocalized excitations. Phil. Mag. A **38**, 519–527 (1978)

6.60 J. Taftø, G. Lehmpfuhl: Direction dependence in electron energy loss spectroscopy from single crystals. Ultramicroscopy **7**, 287–294 (1982)

6.61 A. Weickenmeier, E. Quandt, H. Kohl, H. Rose, H. Niedrig: Computation and measurement of characteristic energy-loss and large-angle convergent-beam patterns of molybdenum selenide. Ultramicroscopy **49**, 210–219 (1993)

6.62 W.J. Vine, R. Vincent, P. Spellward, J.W. Steeds: Observation of phase contrast in convergent-beam electron diffraction patterns. Ultramicroscopy **41**, 423–428 (1992)

6.63 J.M. Cowley: Coherent interference in CBED and shadow imaging. Ultramicroscopy **4**, 435–450 (1979)

6.64 J.C.H. Spence: Practical phase determination of inner dynamical reflections in STEM. Scanning Electron Microscopy 1978/I (SEM Inc., AMF O'Hare 1978) pp.61–68

6.65 D.J. Eaglesham: Applications of convergent beam electron diffraction in materials science. J. Electr. Micr. Techn. **13**, 66–75 (1989)

6.66 J.W. Steeds, E. Carlino: Electron crystallography. In *Electron Microscopy in Materials Science*, ed. by. P.E. Merli and V. Antisari (World Scientific, Singapore 1992) pp.279–313

6.67 J.A. Eades: Another way to form zone axis patterns. Inst. Phys. Conf. Ser. No. 52 (Inst. of Physics, London 1980) pp.9–13

6.68 M. Tanaka, R. Saito, X. Ueno, Y. Harada: Large-angle convergent-beam electron diffraction. J. Electron Microsc. **29**, 408–439 (1980)

6.69 A. Higgs, O.L. Krivanek: Energy-filtered, double-rocked zone axis patterns. Proc. 39th Ann. Meeting EMSA (San Fransisco Press, San Francosco 1981) pp.346–347

6.70 M. Tanaka, M. Terauchi: Whole pattern in convergent-beam electron diffraction using the hollow-cone method. J. Electron Micr. **34**, 52–55 (1985)

6.71 F. Banhart, N. Nagel: Convergent beam electron diffraction studies of epitaxial Si/SiO_2 systems. Phil. Mag. A **70**, 341–357 (1994)

6.72 I.K. Jordan, C.J. Rossouw, R. Vincent: Effects of energy filtering in LACBED patterns. Ultramicroscopy **35**, 237–243 (1991)

6.73 C.J. Humphreys, D.M. Maher, H.L. Fraser, D.J. Eaglesham: Convergent-beam imaging – a transmission electron microscopy technique for investigating small localized distortions in crystals. Phil. Mag. A **5**, 787–798 (1988)

6.74 H.K. Shinohara: Diffraction of cathode rays by single crystals, Part III. – Simultaneous reflection. Sci. Papers Inst. Phys. Chem. Res. Tokyo **20**, 39–51 (1932)

6.75 J. Gjønnes, D. Watanabe: Dynamical diffuse scattering from magnesium oxide single crystals. Acta Cryst. **21**, 297–302 (1966)

6.76 J. Gjønnes, R. Høier: Multiple beam-dynamic effects in Kikuchi patterns from natural spinels. Acta Cryst. A **25**, 595–602 (1969)

6.77 B.F. Buxton: Bloch waves and higher order Laue zone effects in high energy electron diffraction. Proc. Roy. Soc. (London) A **350**, 335–361 (1976)

6.78 J.C.H. Spence, J.M. Zuo: *Electron Microdiffraction* (Plenum Press, New York 1992)

6.79 Y.P. Lin, A.R. Preston, R. Vincent: Lattice parameter measurement by convergent-beam electron diffraction. *EMAG 1987*, Inst. Phys. Conf. Ser. 90 (Inst. of Phys., London 1987) pp.115–116

6.80 Y.P. Lin, D.M. Bird, R. Vincent: Errors and correction terms for HOLZ line simulations. Ultramicroscopy **27**, 233–240 (1989)

6.81 J.M. Zuo: Perturbation theory in high-energy transmission electron diffraction. Acta Cryst. A **47**, 87–95 (1991)

6.82 J.M. Zuo: Automated lattice parameter measurement from HOLZ lines and their use for the measurement of oxygen content in $YBa_2Cu_3O_{7-\delta}$. Ultramicroscopy **41**, 211–223 (1992)

6.83 S.J. Rozeveld, J.M. Howe, S. Schmauder: Measurement of residual strain in an Al-SiC_w composite using convergent-beam electron diffraction. Acta met. mater. **40**, Suppl., 173–183 (1992)

6.84 K. Kambe: Study of simultanious reflexion in electron diffraction by crystals. J. Phys. Soc. Jpn. **12**, 13–25 (1957)

6.85 F. Nagata, A. Fukuhara: 222 electron reflection from aluminium and systematic interaction. Jpn. J. Appl. Phys. **6**, 1233–1235 (1967)

6.86 J. Gjønnes, R. Høier: The application of non-systematic many-beam dynamic effects to structure-factor determination. Acta Cryst. A **27**, 313–316 (1971)

6.87 K. Marthinsen, R. Høier: Many beam-dynamic effects and phase information in electron channeling patterns. Acta Cryst. A **42**, 484–492 (1986)

6.88 D.M. Bird, R. James, A.R. Preston: Direct measurement of crystallographic phase by electron diffraction. Phys. Rev. Lett. **59**, 1216–1219 (1989)

6.89 J.M. Cowley: Crystal structure determination by electron diffraction. In *Progress in Materials Science* (Pergamon, Oxford 1967)

6.90 J.M. Cowley: The determination of structure factors from dynamical effects in electron diffraction. Acta Cryst. A **25**, 129–134 (1969)

6.91 P. Goodman: Accurate structure factor and symmetry determination. *EMAG 1978*, Inst. Phys. Conf. Ser. No. 41 (Inst. of Physics, London 1978) pp.116–117

6.92 J. Gjønnes, H. Matsuhata, J. Taftø: Structure factor determination from critical voltages in electron diffraction. Proc.47th Ann. Meeting EMSA (San Fransisco Press, San Francisco 1989) pp.490–491

6.93 J.M. Zuo, J.C.H. Spence: Automated structure factor measurement by convergent-beam electron diffraction. Ultramicroscopy **35**, 185–196 (1991)

6.94 D.M. Bird, M. Saunders: Sensitivity and accuracy of CBED pattern matching. Ultramicroscopy **45**, 241–251 (1992)

6.95 R. Høier, L.N. Bakken, K. Marthinsen, R. Holmestad: Structure factor determination in non-centrosymmetric crystals by a 2-dimensional CBED-based multi-parameter refinement method. Ultramicroscopy **49**, 159–170 (1993)

6.96 J.M. Zuo, K. Gjønnes, J.C.H. Spence: FORTRAN source listing for simulating three-dimensional convergent beam patterns with absorption by the Bloch wave method. J. Electr. Micr. Techn. **12**, 29–55 (1989)

6.97 C. Deininger, G. Necker, J. Mayer: Determination of structure factors, lattice strains and accelerating voltage by energy-filtered CBED. Ultramicroscopy **54**, 15–30 (1994)

6.98 J.M. Zuo, J.C.H. Spence, M. O'Keefe: Bonding in GaAs. Phys. Rev. Lett. **61**, 353–356 (1988)

6.99 J.M. Zuo, J.C.H. Spence, J. Downs, J. Mayer: Measurement of individual structure-factor phases with tenth-degree accuracy: The 00.2 reflection in BeO studied by electron and x-ray diffraction. Acta Cryst. A **49**, 422–429 (1993)

6.100 H.P. Schwefel: *Numerical Optimization of Computer Models* (Wiley, London 1981)

6.101 L. Ingber: Very fast simulated re-annealing. Math. Comput. Modelling **12**, 967–973 (1989)

6.102 L. Ingber: Simulated Annealing: Practice versus theory. Statistics and Computing (1993)

6.103 C.H. Macgillavry. G.D. Riek (eds.): *International Tables for X-ray Crystallography* (Kynoch, Birmingham 1968) Vol.III

6.104 P.A. Doyle, P.S. Turner: Relativistic Hartree–Fock x-ray and electron scattering factors. Acta Cryst. A **24**, 390–397 (1968)

6.105 A. Weickenmeier, H. Kohl: Computation of absorptive form factors for high-energy electron diffraction. Acta Cryst.A **47**, 590–597 (1991)

6.106 D.M. Bird, Q.A. King: Absorptive form factors for high-energy electron diffraction. Acta Cryst. A **46**, 202–208 (1990)

6.107 L. Reimer, R. Rennekamp: Imaging and recording of multiple scattering effects by angular-resolved EELS. Ultramicroscopy **28**, 258–265 (1989)

6.108 B.G. Williams, T.G Sparrow, R.F. Egerton: Electron Compton scattering from solids. Proc. Roy. Soc. (London) A **393**, 409–422 (1984)

6.109 B.G. Williams: *Compton Scattering. The Investigation of Electron Momentum Distribution* (McGraw–Hill, London 1977)

6.110 M.J. Cooper: Compton scattering and electron momentum determination. Rep. Progr. Phys. **48**, 415–481 (1985)

6.111 P. Jonas, P. Schattschneider, P. Pongratz: Removal of Bragg–Compton channel coupling in electron Compton scattering. *Electron Microscopy 1990*, ed. by L.D. Peachey, D.B. Williams (San Francisco Press, San Francisco 1990) Vol.2, pp.24–25

6.112 H. Raether: *Excitations of Plasmons and Interband Transitions by Electrons* (Springer, Berlin, Heidelberg 1980)

6.113 I. Fromm, L. Reimer, R. Rennekamp: Investigation and use of plasmon losses in energy-filtering TEM. J. Micr. **166**, 257–271 (1992)

6.114 P. Schmüser: Anregung von Volumen- und Oberflächenplasmaschwingungen in Al und Mg durch mittelschnelle Elektronen. Z.Physik **180**, 105–126 (1964)

6.115 J. Daniels, C. von Festenberg, H. Raether: Optical constants of solids by electron spectroscopy. In *Springer Tracts in Modern Physics*, ed. by G. Höhler (Springer, Berlin, Heidelberg 1970) Vol.54, pp.77–135

6.116 K. Zeppenfeld: Anisotropie der Plasmaschwingungen in Graphit. Z. Physik **211**, 391–399 (1968)

6.117 K. Zeppenfeld: Nichtsenkrechte Interbandübergänge in Graphit durch unelastische Elektronenstreuung. Z. Physik **243**, 229–243 (1971)

7. Electron Spectroscopic Imaging

Ludwig Reimer

Whereas an image in a conventional transmission electron microscope (CTEM) contains contributions from elastically and inelastically scattered electrons that pass the objective diaphragm, the electron spectroscopic imaging (ESI) mode of an EFTEM allows us to separate the contributions by inserting an energy-selecting slit in the energy-dispersive plane of a filter lens or a prism spectrometer; energy-loss windows $E \pm \Delta/2$ with a width $\Delta = 1$–20 eV can be selected (see Sect. 1.4.2). The choice between zero-loss filtering and the use of an energy-loss window in the loss-spectrum of inelastically scattered electrons depends on the specimen and the information wanted. A schematical energy-loss spectrum in Fig. 7.1 shall demonstrate the ESI modes [7.1–5] which are discussed in detail in this Chapter.

7.1 Zero-Loss Imaging

The energy-selecting slit is centred at the zero-loss peak of energy width δE formed by the unscattered and elastically scattered electrons that pass the objective aperture α with scattering angles $\theta \geq \alpha$. This width is limited by the energy spread of the electron gun, $\delta E \simeq 0.5$–2 eV for thermionic tungsten or LaB_6 cathodes due to the Boersch effect and $\simeq 0.2$–0.5 eV for W/ZrO-Schottky or field-emission cathodes. For most specimens, the EELS below 5 eV shows no plasmon losses, and the contribution of all inelastically scattered electrons can be absorbed by a slit of width $\Delta \simeq 5$–10 eV.

Because the theory of phase and Bragg contrast and of crystal-lattice imaging with a large number of Bragg diffracted beams assumes that the incident wave is plane and is only phase-shifted by the specimen, zero-loss filtering not only increases the contrast but also allows a better comparison with simulated images to be made. In CTEM, the image intensity is overlapped by the contribution of inelastically scattered electrons, which may or may not show preservation of Bragg contrast and the scattering process can be delocalized resulting in a reduced resolution. Owing to the increasing probability of inelastic scattering [ratio of inelastic-to-elastic total cross-sections ν (1.25)] with decreasing atomic number, the increase in contrast is largest for material of low Z, polymers and biological material for example.

Springer Series in Optical Sciences, Vol. 71
Energy-Filtering Transmission Electron Microscopy
Editor: Ludwig Reimer ©Springer-Verlag Berlin Heidelberg 1995

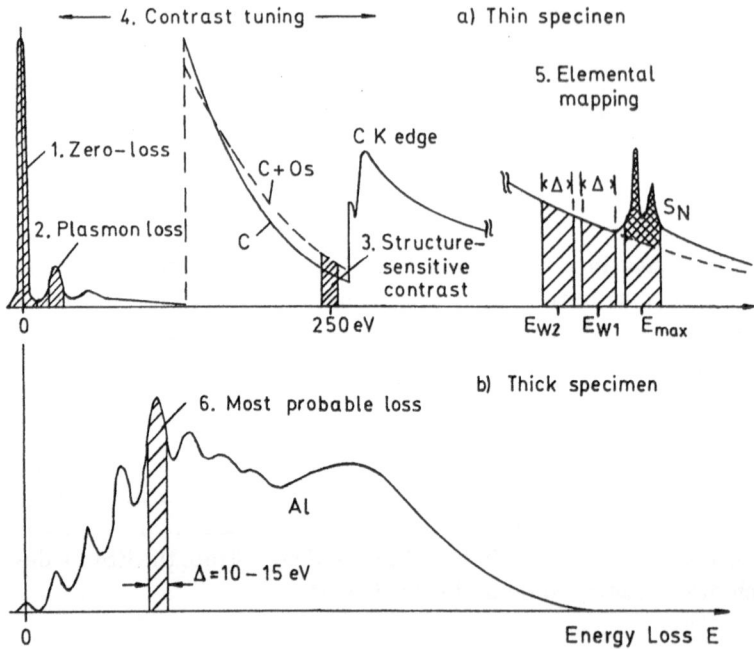

Fig. 7.1. Imaging modes of electron spectroscopic imaging (ESI) with selected energy windows at different parts of the electron energy loss spectrum (EELS).

Elastic scattering provides the largest contribution to scattering contrast caused by electrons scattered through angles θ larger than the objective aperture α, because a large fraction of inelastically scattered electrons is concentrated at scattering angles $\theta \leq \alpha$ and pass the objective diaphragm. All the contrast modes discussed below will show an overlap of scattering contrast due to multiple elastic–inelastic scattering processes. Therefore, not all contrast effects observed with an energy window of inelastically scattered electrons are specific for the selected energy loss but can be caused by additional elastic scattering. Inelastic scattering can also influence the zero-loss image intensity (see also Sect. 2.1).

7.1.1 Types of Contrast in TEM

The discussion of contrast needs a treatment of differences and mutual relations between scattering and phase contrast. Absorption in the thin specimens used in TEM can be neglected. With the exception of backscattered electrons with scattering angles $\theta > 90°$, all electrons are transmitted. Scattering contrast is caused by the interception of electrons at the objective diaphragm, about 20 μm to 200 μm in diameter, when the scattering angle is larger than the objective aperture α of about 4 to 40 mrad. The fraction of electrons (transmission) that passes through the objective aperture ($\theta < \alpha$) decreases with increasing specimen thickness and increasing atomic number. Therefore, scattering contrast in biological specimens is generated by introducing heavy atoms during the fixation with osmium tetroxide or potassium permanganate solutions, by staining sections with lead or uranium salt solutions, or by use of the negative staining technique in which viruses or macromolecules are embedded in phosphotungstic acid, for example. Polymers also can be stained by osmium or ruthenium tetroxide. Surface replicas or organic macromolecules on a supporting film shadowed with heavy metal coatings can be treated by the scattering contrast algorithm presented below.

The electrons passing the aperture create phase contrast, which depends on defocusing. Before embarking on the full discussion of scattering contrast, we want to mention that scattering and phase contrast can be explained by a unique theory when complex scattering amplitudes (1.11) are introduced [7.6–9]. The total cross-section σ_{el} of elastic scattering is related to the imaginary part of $f(\theta)$ for $\theta = 0$ via the optical theorem of scattering theory:

$$\sigma_{el} = 2\lambda \text{Im} f(0) = 2\lambda \mid f(0) \mid \sin \eta(0), \tag{7.1}$$

and the partial cross-section for scattering through the interval $\alpha \leq \theta \leq \pi$ becomes

$$\sigma_{el}(\alpha) = 2\lambda \mid f(0) \mid \sin \eta(0) - \int_0^\alpha \mid f(\theta) \mid^2 2\pi \sin \theta \, d\theta. \tag{7.2}$$

The last term is proportional to the number of scattered electrons passing the aperture. In a layer of mass–thickness $dx = \rho dz$ with $N dx$ atoms per unit area ($N = N_A/A =$ Avogadro number / atomic number = number of atoms per gram) the incident intensity I is decreased by

$$dI(\alpha) = -IN\left[\sigma_{el}(\alpha) + \sigma_{in}(\alpha)\right]dx = -I dx/x_{k,unf}(\alpha) \tag{7.3}$$

due to scattering through angles $\theta > \alpha$; note that the corresponding inelastic cross-section $\sigma_{in}(\alpha)$ is included. Integration of (7.3) over the mass–thickness $x = \rho t$ of the film results in the exponential decrease of transmission in the unfiltered mode

$$T_{unf} = I(\alpha)/I_0 = \exp[-x/x_{k,unf}(\alpha)] \tag{7.4}$$

where I_0 is the intensity without specimen. Mass–thickness x in $\mu g/cm^2$ and thickness t in nm are related by

$$t = 10\, x/\rho \tag{7.5}$$

with the density ρ in g/cm^3 . The contrast thickness x_k (see Table 7.1. for 80 keV in the unfiltered and zero-loss filtered mode) is the value of x for which the transmission falls to $1/e = 37\%$.

Phase contrast is caused by the interference of the incident wave of amplitude ψ_0 (intensity $= \mid \psi_0 \mid^2$) and the scattered wave of amplitude $\psi_0 f(\theta)$, which is phase-shifted by the wave aberration [7.10]

Table 7.1. Contrast thicknesses $x_{k,unf}$ and $x_{k,fil}$ in units of $\mu g/cm^2$ of 80 keV electrons for the unfiltered and zero-loss filtered mode and different objective apertures α in units of mrad.

Mode	unfiltered			zero-loss filtered		
Aperture	4 mrad	10 mrad	20 mrad	4 mrad	10 mrad	20 mrad
Carbon	17.5	25.0	38.0	9.7	10.0	10.4
Germanium	14.2	16.9	21.6	11.1	11.6	13.2
Platinum	15.0	16.4	19.6	13.2	14.0	15.7

$$W(\theta) = \frac{\pi}{2\lambda}(C_s\theta^4 - 2\Delta z\theta^2) \qquad (7.6)$$

of the objective lens where C_s = spherical aberration constant, Δz = defocusing distance. The image amplitude $\psi(r)$ is obtained by taking the inverse Fourier transform of $f(\theta)$ multiplied by the phase factor $\exp[-iW(q)]$ and the factor $\exp(i\pi/2)=i$, which represents the general phase shift of $\pi/2$ between unscattered and scattered waves:

$$\psi(r) = \psi_0 \left\{ \delta(0) + \frac{i}{\lambda} \int_{q\leq\alpha/\lambda} f(q)\exp[-iW(q)]\exp(2\pi i q \cdot r)\,d^2q \right\} \qquad (7.7)$$

where $q = 1/\Lambda = \theta/\lambda$ is the spatial frequency of an object periode Λ, and the delta function $\delta(0)$ indicates that the direction of the incident wave is parallel to the optic axis $q = 0$. The image intensity is given by $I(r) = \psi\psi^*$. In scattering contrast (7.4) as well as phase contrast (7.7), the number of electrons passing through the aperture will be the same because the integrations only extend from $\theta=0$ to α. In phase contrast a small particle on a supporting film becomes darker in underfocus (Δz positive) with a bright ring and bright in overfocus with a dark ring. Even in focus ($\Delta z=0$), phase contrast is observed because the wave aberration $W(\theta)$ never becomes simultaneously zero for all scattering angles θ. The granulation of carbon films changes with defocusing and this is a typical phase contrast effect [7.11] caused by the varying phase shifts of spatial frequencies associated with the wave aberration (7.6). At low magnifications when the phase contrast effects are not resolved or when the illumination aperture and/or energy spread of the electron gun are greater (decrease of spatial and temporal coherence, respectively), an averaged intensity is observed, which is described by the scattering contrast.

When investigating crystalline specimens, the Bragg reflections at lattice planes are absorbed by the objective diaphragm when the Bragg diffraction angle $2\theta_B$ is larger than the objective aperture α. The resulting Bragg contrast is seen as thickness fringes and bend contours and allows strain fields around dislocations and other lattice defects to be imaged. The intensity of Bragg reflections can be calculated by the dynamical theory of electron diffraction. When several Bragg reflected waves including the primary wave can pass a large objective diaphragm ($2\theta_B < \alpha$) their interference results in a crystal-lattice image, which is a typical phase contrast image because the contrast is strongly affected by changes in the wave aberration (7.6) and by phase shifts caused by the excitation errors (deviations from the exact Bragg condition).

7.1.2 Scattering Contrast of Amorphous Specimens

Scattering Contrast Without Filtering. The contrast thickness $x_k(\alpha)$ in the exponential law (7.4) of transmission can be calculated from the differential elastic (1.13) and inelastic (1.23) cross-sections [7.12]:

$$\frac{1}{x_{k,\text{unf}}(\alpha)} = \frac{1}{x_{\text{el}}(\alpha)} + \frac{1}{x_{\text{in}}(\alpha)} = \frac{N_A}{A} \int_\alpha^\infty \left(\frac{d\sigma_{\text{el}}}{d\Omega} + \frac{d\sigma_{\text{in}}}{d\Omega}\right) 2\pi\theta d\theta$$

$$= \frac{1}{x_{\text{el}}[1 + (\alpha/\theta_o)^2]} + \frac{1}{Zx_{\text{el}}} \left\{ -\frac{1}{[1 + (\alpha/\theta_o)^2]} + 2\ln[1 + (\theta_o/\alpha)^2] \right\} \quad (7.8)$$

with the total mean free path

$$x_{\text{el}} = \frac{A}{N_A\sigma_{\text{el}}} = \frac{\pi A a_H^2}{N_A Z^2 R^2 \lambda^2 (1 + E_0/m_0c^2)^2} \quad (7.9)$$

of elastic scattering. The disagreement between calculated and experimental values of x_k increases with increasing atomic number and WKB calculations show a better but not complete agreement [7.12]. However, the values x_{el} and θ_o depending on electron energy and atomic number can be fitted to experimental values of $x_{k,\text{unf}}(\alpha)$ (see Tables 1.1 and 7.1), after which the dependence on aperture α can be described by using these values in (7.8). The dependence of $x_{k,\text{unf}}$ on atomic number Z for a given aperture α and electron energy E can be approximated by a power law [7.13,14]

$$x_{k,\text{unf}} = \frac{A}{bZ^a} \quad (7.10)$$

with $b = 1.15 \exp[-(2.213 + 86.13 \, \alpha - 696.0 \, \alpha^2)]$ and the exponent $a = 1.136 + 16.19 \, \alpha - 186.7 \, \alpha^2$ (x_k in μg/cm^2 and α in rad). This relation is used in Sect. 7.4.6 in a method of determining the mass–thickness by measuring the transmission.

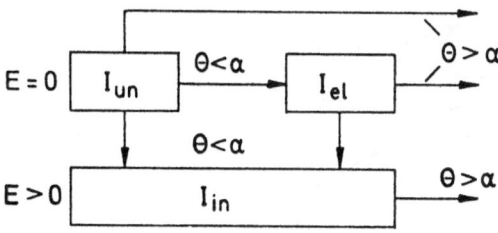

Fig. 7.2. The flow of intensities I_{el}, I_{in} and I_{un} of elastically, inelastically and unscattered electrons, respectively, to angles θ smaller and larger than the objective aperture α.

Scattering with Zero-Loss Filtering. For the discussion of the influence of energy-filtering on transmission, we consider that contributions from the unscattered (primary), elastically and inelastically scattered electrons are superimposed on the unfiltered transmission (7.4) ($T_{\mathrm{unf}} = T_{\mathrm{un}} + T_{\mathrm{el}} + T_{\mathrm{in}}$). The simple flow diagram of scattering processes with scattering angles $\theta < \alpha$ and $\theta > \alpha$ in Fig. 7.2 results in the following system of coupled differential equations [7.15]:

$$dT_{\mathrm{un}} = -T_{\mathrm{un}}\frac{dx}{x_{\mathrm{t}}}, \tag{7.11}$$

$$dT_{\mathrm{el}} = T_{\mathrm{un}}dx\Big(\frac{1}{x_{\mathrm{el}}} - \frac{1}{x_{\mathrm{el}}(\alpha)}\Big) - T_{\mathrm{el}}\frac{dx}{x_{\mathrm{in}}} - T_{\mathrm{el}}\frac{dx}{x_{\mathrm{el}}(\alpha)}, \tag{7.12}$$

$$d(T_{\mathrm{un}} + T_{\mathrm{el}} + T_{\mathrm{in}}) = -(T_{\mathrm{un}} + T_{\mathrm{el}} + T_{\mathrm{in}})\frac{dx}{x_{\mathrm{k,unf}}(\alpha)} \tag{7.13}$$

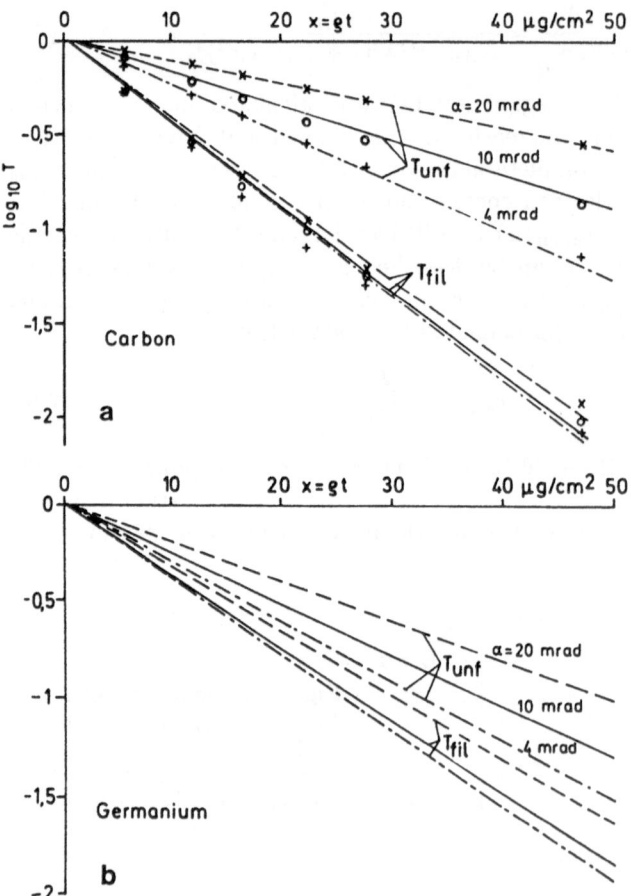

Fig. 7.3. Caption see opposite page.

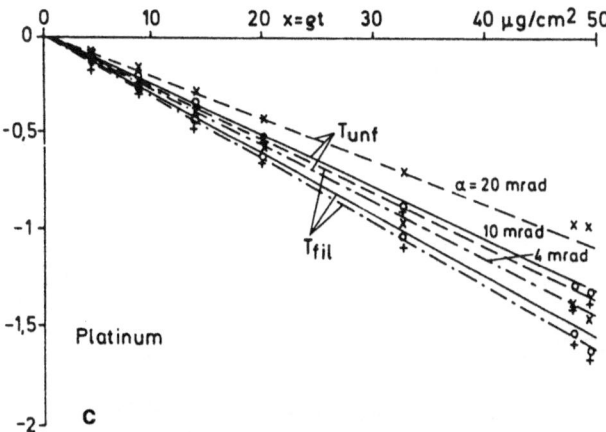

Fig. 7.3. Transmission T_{unf} and T_{fil} of (a) carbon, (b) germanium and (c) platinum films in unfiltered and zero-loss filtered images as a function of mass–thickness $x = \rho t$ and objective aperture α for 80 keV electrons.

with $1/x_t = 1/x_{el} + 1/x_{in}$. The solution making use of the ratio ν from (1.25) is

$$T_{un} = \exp[-x/x_t] = \exp[-x(1+\nu)/x_{el}], \quad (7.14)$$

$$T_{fil}(\alpha) = T_{un} + T_{el} = \exp\left[-\frac{x}{x_{el}}\left(\frac{1}{[1+(\alpha/\theta_o)^2]}+\right)\right] \quad (7.15)$$

$$T_{unf} = T_{un} + T_{el} + T_{in} = \exp[-x/x_{k,unf}(\alpha)]. \quad (7.16)$$

Equation (7.16) is identical with the unfiltered transmission T_{unf} in (7.4). Figure 7.3 shows the measured transmissions T_{unf} and T_{fil} of carbon, germanium and platinum films in a semi-logarithmic plot; the resulting straight line is a proof of the exponential laws (7.15,16). Such a law is obeyed up to mass–thicknesses x of $\simeq 50$ μg/cm^2 though these film thicknesses are larger than the total mean free path $x_{el}/(1+\nu)$ and multiple elastic/inelastic scattering will occur, whereas the formulae (7.4) and (7.14–16) assume single scattering. Deviations resulting in a lower decrease of $\log_{10}T_{unf}$ with increasing mass–thickness beyond $x \simeq 50$ μg/cm^2 [7.12,13,16] (see Fig. 7.27 below) are caused by rescattering from $\theta > \alpha$ to $\theta < \alpha$.

This theory of scattering contrast also allows us to calculate the fractions T_{el} of elastically and T_{in} of inelastically scattered electrons passing through an objective aperture α ($\alpha = 20$ mrad in Fig. 7.4). Carbon shows the lowest fraction T_{el} and a high T_{in} due to the large value of $\nu = 3.3$, whereas T_{el} is larger than T_{in} for platinum ($\nu = 0.25$, Fig. 1.9).

Contrast Enhancement in Zero-Loss Filtered Images. The scattering contrast depends on the local distribution of mass–thickness x. In the case of biological sections or surface replicas, the carbon supporting film of mass–thickness x_C results in an approximately uniform decrease of transmission.

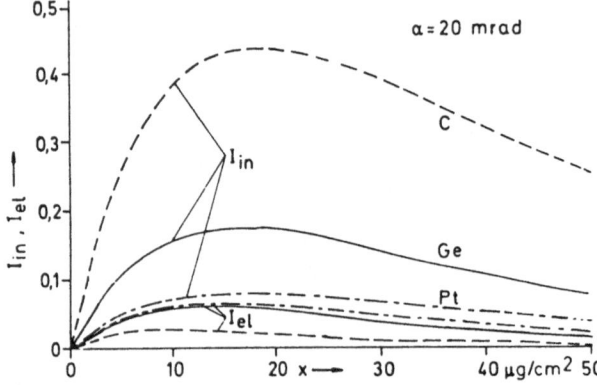

Fig. 7.4. The fractions T_{el} and T_{in} of elastically and inelastically scattered 80 keV electrons passing through an objective aperture $\alpha = 20$ mrad versus the mass–thickness x.

A transmission of $T = 10^{-3} = 0.1\%$ is the practical limit and the maximum mass–thicknesses of about 70–90 $\mu g/cm^2$ for 80 keV electrons can be read from Figs. 7.3a-c (see also Fig. 7.11). This maximum mass thickness increases with increasing electron energy and published x_{el} and θ_o values – [7.13] and Table 1.1 – can be used to calculate the transmission in the unfiltered and zero-loss filtered mode using (7.15) and (7.16), respectively.

An additional mass–thickness Δx of carbon or another element (E) on a carbon substrate (C) of mass–thickness x_C results in a lower transmission

$$T_{unf,C+E} = \exp\left[-\left(\frac{x_C}{x_{k,C}} + \frac{\Delta x_E}{x_{k,E}}\right)\right], \qquad (7.17)$$

for example. The bright-field contrast for unfiltered and zero-loss filtered images becomes

$$C_{unf} = \frac{T_{unf,C+E} - T_{unf,C}}{T_{unf,C}} = \exp(-\Delta x_E/x_{k,E}) - 1 \simeq -\Delta x_E/x_{k,E}, \qquad (7.18)$$

$$C_{fil} = \frac{T_{fil,C+E} - T_{fil,C}}{T_{fil,C}} \simeq -\Delta x_E\left[\frac{\nu_E}{x_{el,E}} + \frac{1}{x_{el,E}(\alpha)}\right]. \qquad (7.19)$$

This results in the following gain of contrast when switching from the unfiltered to the zero-loss filtered mode [7.15]:

$$G_0 = \frac{C_{fil}}{C_{unf}} = \frac{x_{k,E}}{x_{el,E}}\left[\nu_E + \frac{1}{1 + (\alpha/\theta_o)^2}\right] \qquad (7.20)$$

which is plotted in Fig. 7.5 as a function of objective aperture α. The strongly decreasing gain with increasing atomic number means that carbon mass-thickness variations Δx_E cause a large gain of contrast with zero-loss filtering, and unfortunately of the chatter in microtome sections, whereas the gain of contrast by staining with heavy atoms or of shadowed surface replicas is much lower. As an example, we measured a gain of about 1.3 on and between myelin lamellae stained with OsO_4 and embedded in epon. A gain of 1.6 was found

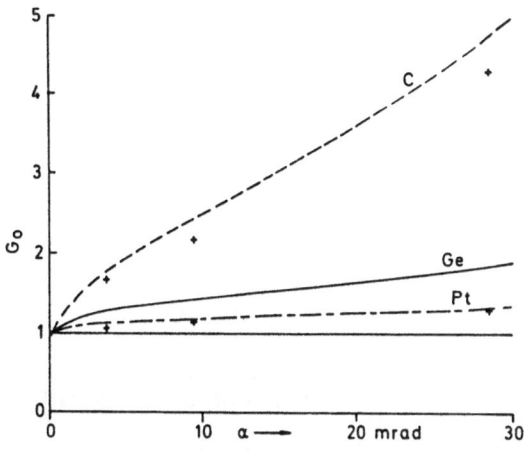

Fig. 7.5. Increase of the gain $G_0 = C_{\mathrm{fil}}/C_{\mathrm{unf}}$ in contrast by zero-loss filtering with increasing objective aperture for thin films of C, Ge and Pt on a supporting film. Experimental points: carbon and platinum films on a carbon substrate.

for tobacco mosaic virus [7.17]. An increase of contrast by zero-loss filtering has also been shown for copolymers, owing to differences in density [7.18].

Influence of Chromatic Aberration on Contrast. The energy loss spectrum of inelastically scattered electrons with a spread δE of energy losses results in a chromatic error disc of diameter

$$d_{\mathrm{c}} = C_{\mathrm{c}}(\delta E/E_0)\alpha. \tag{7.21}$$

For example, with a chromatic aberration constant $C_{\mathrm{c}} = 1.7$ mm, a width $\Delta = 50$ eV of the energy window for a 0.2 μm section, and $\alpha = 10$ mrad, we get $d_{\mathrm{c}} \simeq 10$ nm. In practice d_{c} increases less than linearly with increasing α as shown for the resolution of edges [7.19], where a diameter $d_{\mathrm{c}} = 7$ nm has been measured for $C_{\mathrm{c}} = 2$ mm, $\alpha = 10$ mrad after the passage of 100 keV electrons through 0.23 μm of polystyrene.

As a consequence of chromatic aberration, an image blurred by a convolution with the chromatic error disc of the fraction of inelastically scattered electrons, which increases with increasing thickness and aperture, is superimposed on the well-resolved image generated by the fraction of unscattered and elastically scattered electrons. The contrast for details smaller than d_{c} relative to their blurred background therefore becomes

$$C_{\mathrm{unf}} = \frac{T_{\mathrm{fil,C+E}} - T_{\mathrm{fil,C}}}{T_{\mathrm{unf,C}}} \tag{7.22}$$

instead of (7.18) and the gain will now be [7.15]

$$G_{\mathrm{c}} = \frac{C_{\mathrm{fil}}}{C_{\mathrm{unf}}} = \frac{T_{\mathrm{unf,C}}}{T_{\mathrm{fil,C}}} = \exp\left[x_{\mathrm{C}}\left(\frac{1+\nu_{\mathrm{C}}}{x_{\mathrm{el,C}}} - \frac{1}{x_{\mathrm{k,C}}}\right)\right]. \tag{7.23}$$

The gain G_{c} achieved by avoiding chromatic aberration is plotted in Fig. 7.6 and does not depend on the elemental composition. This gain of contrast

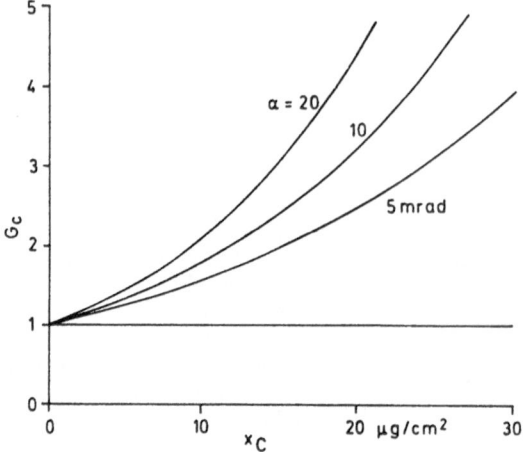

Fig. 7.6. The increase of the gain G_c in the case of dominant chromatic aberration with increasing carbon mass–thickness x_C for different objective apertures α.

is demonstrated in Fig. 7.7 for a 0.2 μm epon–embedded liver section ($x \simeq 20\mu$g/cm^2). For $\alpha = 10$ mrad the gain from Fig. 7.6 is $G_c = 3.3$. The unfiltered image (Fig. 7.7a) shows the blurring caused by chromatic aberration. The zero-loss filtered image (Fig. 7.7b) shows sharper details (smaller than d_c) and an increase in the contrast of the nucleoplasm and the endoplasmic reticulum. The increase in contrast of the microtome chatter with a period larger than d_c is described by G_0 (7.20). Figure 7.8 shows a corresponding comparison of unfiltered and zero-loss filtered images of a section of a copolymer of polyethylene and polypropylene stained with ruthenium oxide [7.5].

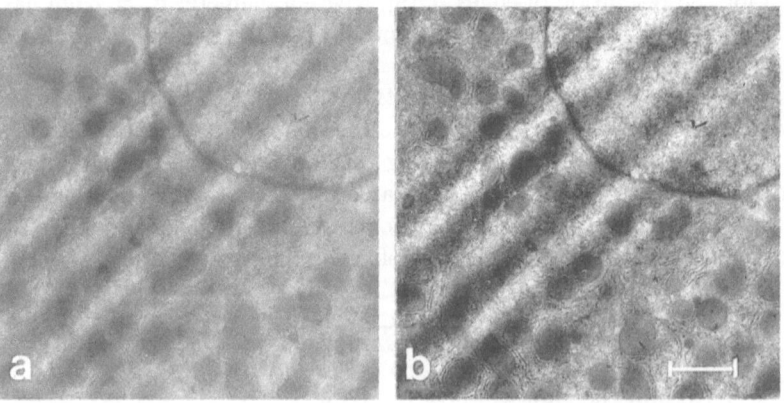

Fig. 7.7. Comparison of (**a**) an unfiltered and (**b**) zero-loss filtered image of an epon embedded liver section (OsO$_4$ fixation, uranyl-acetate stained, $t = 0.2$ μm, bar = 1 μm).

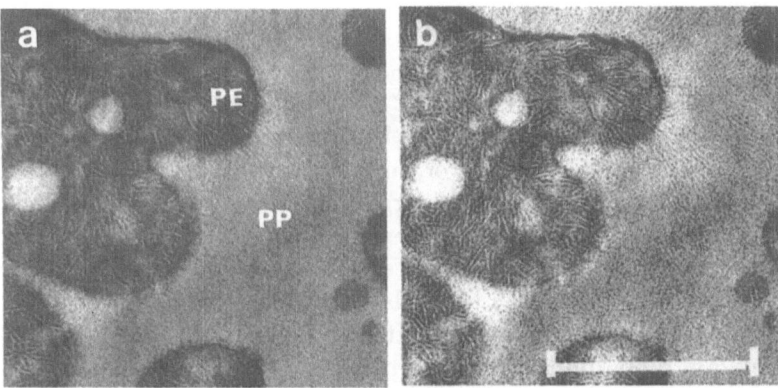

Fig. 7.8. Comparison of (a) an unfiltered and(b) zero-loss filtered image of a thin section of a copolymer of polyethylene (PE) and polypropylene (PP) stained with ruthenium oxide (bar = 0.5 μm).

Top–Bottom Effect. Though zero-loss filtering avoids chromatic aberration and allows us to observe thicker sections, the resolution will be limited by the increasing width of the angular distribution of scattered electrons caused by multiple scattering. This causes a lateral distribution increasing in width with increasing specimen thickness. In dedicated STEM and the STEM mode of a CTEM, this results in the top–bottom effect [7.20]. Details at the top (entrance side) are scanned by the unbroadened electron probe of diameter (0.2–2 nm) and those at the bottom (exit side) by the broadened probe of about $\Delta x \simeq 15$ nm in diameter for 80 keV electrons passing through 1 μm polystyrene. The same lateral spread also limits the lateral resolution in x-ray microanalysis and can be estimated from [7.9,21]

$$\Delta x = 1.05 \times 10^5 (\frac{\rho}{A})^{1/2} \frac{Z}{E_0} t^{3/2} \qquad (7.24)$$

with Δx and the film thickness t in cm, and E_0 in eV.

The theorem of reciprocity between CTEM and STEM suggests that an inverse effect should be observable in the TEM mode with a sharper imaging of structures at the bottom (exit side). Calculations for different electron energies [7.19] showed that this effect will be totally blurred by the larger chromatic aberration. When avoiding the chromatic aberration in EFTEM by selection of inelastically scattered electrons with an energy width at the most probable energy loss (Sect. 7.3.2), a top–bottom effect can be observed, which results in a blurring of 8 nm of structures at the top of 1 μm polystyrene spheres [7.22]. The fact that the broadening is less serious than that observed in the STEM mode can be explained considering that the instrument is focused on a virtual plane of least confusion inside the thick film.

7.1.3 Z-Ratio Contrast

The idea of Z-ratio contrast arises from the fact that the ratio $\sigma_{el}/\sigma_{in} = \nu^{-1}$ in (1.25) and Fig. 1.9 is proportional to Z. In a dedicated STEM, a fraction of the elastically scattered electrons can be collected either by an annular detector for backscattered electrons or, in greater numbers, by an annular detector for transmitted electrons with large scattering angles in front of the spectrometer. The recorded signal S_{el} is predominately caused by elastically scattered electrons. The magnetic prism spectrometer, for example, allows us to separate a large fraction of the inelastically scattered electrons (S_{in}), predominately concentrated at small scattering angles. For thin specimens, both signals are proportional to the corresponding cross-sections and the mass–thickness, and the ratio S_{in}/S_{el} becomes proportional to the mean local atomic number Z. This mode of Z-ratio contrast using a dedicated STEM has been used in biology [7.23–27] and materials science [7.28].

Ottensmeyer and Arsenault [7.29] proposed that the idea of Z-ratio contrast should be transfered to the ESI mode of an EFTEM. The difference signal

$$T_{el}(\alpha_1, \alpha_2) = T_{fil}(\alpha_2) - T_{fil}(\alpha_1) \qquad (7.25)$$

of zero-loss filtered images with a large aperture $\alpha_2 \geq 20$ mrad and a low aperture $\alpha_1 \leq 5$ mrad can be calculated as an "elastic" image, and

$$T_{in}(\alpha_2) = T_{unf}(\alpha_2) - T_{fil}(\alpha_2) \qquad (7.26)$$

as an "inelastic image". However, our measurements and calculations of $T_{el}(\alpha_1, \alpha_2)/T_{in}(\alpha_2)$ show that this ratio is not independent of the mass–thickness x. The expected values of $1/\nu = Z/20$ (1.25) can be obtained only for unreasonably large apertures and thin films [7.3,30]. This means that the Z-ratio contrast needs large scattering angles, which only can be realized with an annular detector in the STEM mode. However, the signal

$$K = \ln(T_{unf})/\ln(T_{fil}) \qquad (7.27)$$

in EFTEM is independent of x and increases with increasing Z though not linearly [7.30].

7.1.4 Lorentz Microscopy and Electron Holography

In ferromagnetic films of thickness t with a spontaneous magnetic induction $\boldsymbol{B_s}$ the Lorentz force $\boldsymbol{F} = -e\,\boldsymbol{v} \times \boldsymbol{B_s}$ results in an angular deflection

$$\epsilon = eB_s t/m. \qquad (7.28)$$

For example, an iron film of thickness $t = 50$ nm and $B_s = 2.1$ T results in $\epsilon = 0.1$ mrad for $E_0 = 100$ keV. The illumination aperture α_i has to be smaller than ϵ to observe ferromagnetic domains either in the Foucault or the Fresnel mode [7.31–33].

In the Foucault mode, the deflected beam from one domain is absorbed at the edge of a thin-film diaphragm exactly at the focal plane of the objective lens and the corresponding domains appear dark whereas those with opposite magnetization appear bright. The contrast is decreased by inelastically scattered electrons that are scattered through angles larger than ϵ.

In the Fresnel mode, defocusing results in divergent (dark) and convergent (bright) images of domain walls. An analysis of the intensity profile allows the wall-width to be measured [7.34]. Inelastically scattered electrons form a background [7.35] and zero-loss filtering can reduce this background in divergent wall images [7.36].

For convergent wall images the specimen works like a biprism, with the result that interference fringes are formed where the electron waves from both sides of the wall overlap. Biprism interference fringes created by an external electron-optical biprism are used in electron holography [7.37,38], where the transmitted and phase-shifted waves overlap with a reference wave in vacuum to form fringe distances down to 0.03 nm. In both cases, inelastic scattering generates a diffuse background and additional zero-loss filtering holds out the promise of better contrast and reconstruction of the phase and amplitude of the transmitted wave.

7.1.5 Influence of Zero-Loss Filtering on Bragg Contrast

Bragg contrast is caused by the interception of Bragg reflected electrons at the objective diaphragm when the aperture α is smaller than $\theta = 2\theta_B$, where θ_B fulfils the Bragg condition (6.16) for lattice planes with Miller indices hkl. The contrast is given by the intensity of the primary beam, which can be calculated by the dynamical theory of electron diffraction (Sect. 6.3.1). This results in the formation of pendellösung fringes with increasing thickness, shown for a two-beam case in Fig. 6.9, and of rocking curves with increasing tilt (excitation errors or tilt parameter k_x/g) out of the Bragg condition, shown in Fig. 6.10 and for a three-beam case in Fig. 6.11a.

Owing to the absorption of the Bloch-wave field, the intensity of the primary beam decreases and the number of inelastically scattered electrons that pass the aperture in the bright-field mode increases with increasing foil thickness. Zero-loss filtering can therefore remove the fraction of inelastically scattered electrons, which can be dominant for thicker foils and also causes an image blurring by chromatic aberration. Plasmon and higher energy losses can also show a preservation of Bragg contrast, which will be discussed in Sect. 7.2.4.

The influence of zero-loss filtering on contrast will be demonstrated by comparisons of unfiltered and zero-loss filtered images. The moiré fringes of graphite (Fig. 7.9a,b) and the thickness fringes in the grains of an aluminium film (Fig. 7.9c,d) are imaged with a stronger contrast. The microtwins in epitaxially grown Ag films (Fig. 7.10a) also show a stronger contrast with zero-loss filtering (Fig. 7.10b). When imaging the microtwins with their extra

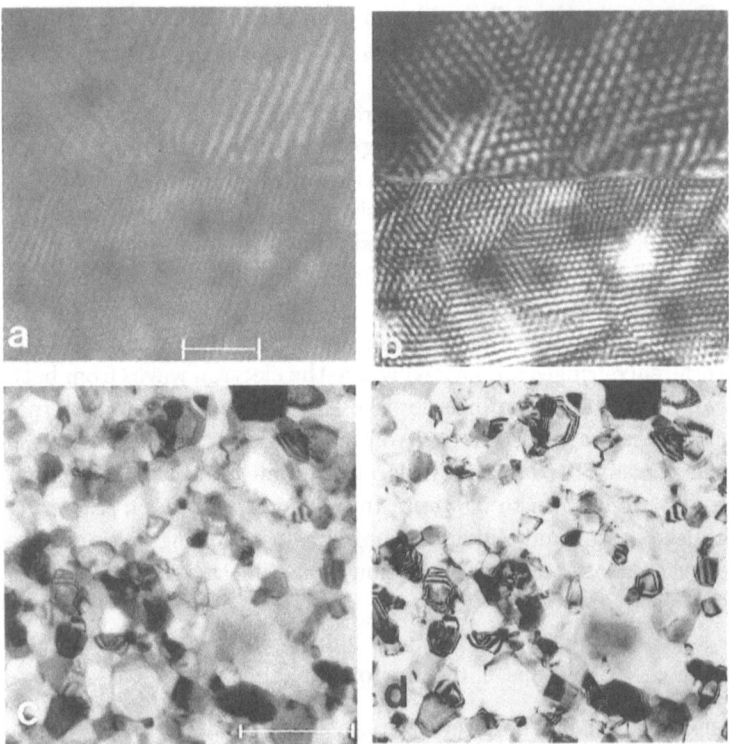

Fig. 7.9. Comparison of unfiltered (a,c) and zero-loss filtered (b,d) images of a graphite foil (a,b) (bar = 1 μm) and an evaporated 270 nm film of aluminium (c,d) (bar = 0.5 μm).

spots [7.39] by shifting the objective diaphragm (dark-field mode), the twins are blurred by strong chromatic-error streaks (Fig. 7.10c) which, disappear after zero-loss filtering (Fig. 7.10d). However, Bragg reflections at larger angles to the optic axis also result in additional image errors. The dark-field mode obtained by tilting the primary beam and adjusting the diffracted beam on axis will therefore still be the optimum method. The gain of contrast decreases with increasing atomic number owing to the decrease of the ratio ν of inelastic-to-elastic cross-sections (1.25), just like to the decrease of the gain G_0 (7.20) for amorphous specimens.

Figure 7.11 shows as solid lines the decrease of zero-loss transmission T_{fil} of amorphous films with increasing mass–thickness and measured values of the averaged transmission of polycrystalline films [7.40]. The latter show a higher transmission than amorphous films of comparable mass–thickness [7.41] because of the destructive interference of elastically scattered electrons between the Bragg spots, which is only destroyed by thermal-diffuse scattering. Since

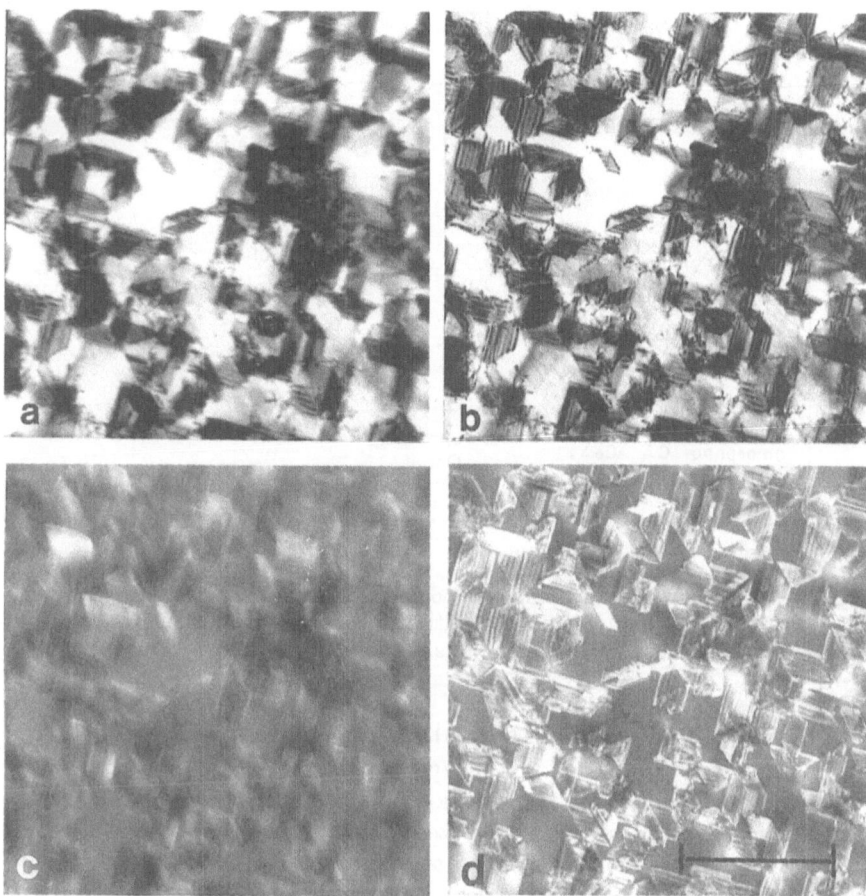

Fig. 7.10. Microtwins in an epitaxially grown silver film on rocksalt (**a**) unfiltered and (**b**) zero-loss filtered bright-field image; (**c**) unfiltered and (**d**) zero-loss filtered dark-field image obtained by shifting the objective diaphragm and selecting an extra spot in the diffraction pattern (bar = 0.5 μm).

a transmission $T = 10^{-3}$ is a practical limit, we can read from Fig. 7.11 that at 80 keV crystalline specimens can be investigated by zero-loss filtering up to mass–thicknesses of about 150 μg/cm^2 in comparison to about 70–90 μg/cm^2 for amorphous specimens (Fig. 7.3). (Thicker specimens can be investigated by most-probable-loss imaging (Sect. 7.3.3) up to $\simeq$ 300 μg/cm^2.) These mass–thicknesses are averaged values over all crystal orientations. The mean transmission can then be described by

$$T_\mathrm{m} = \exp(-2\pi\overline{q}t) \tag{7.29}$$

with the mean absorption length ξ_0' as the reciprocal absorption parameter q (6.21). When calculating the thickness showing a transmission $T = 10^{-3}$

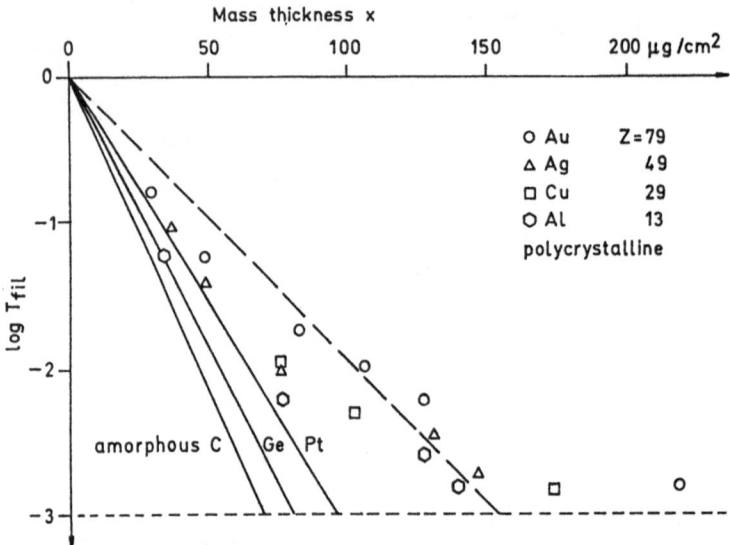

Fig. 7.11. Measured transmission of polycrystalline Al, Cu and Au films in a semilogarithmic plot and comparison with the transmission of amorphous C and Ge and fine-crystalline Pt films from Figs. 7.3a-c.

with the aid of (7.29), the values for Al, Cu and Au closely correspond to the experimental value of $x = \rho t \simeq 150 \mu g/cm^2$ [7.40].

In the case of anomalous transmission (low $q^{(j)}$), still larger foil thicknesses can be successfully investigated by zero-loss filtering [7.42]. The zero-loss filtered images at 80 keV correspond to unfiltered images at 200 keV [7.43], where the chromatic error being reduced owing to the lower probability of inelastic scattering and to the presence of the electron energy E_0 in the denominator of (7.21).

The method of weak-beam imaging of dislocations and other lattice defects results in sharper images of dislocations because the extinction distance ξ_g is considerably reduced. This allows the distance between dissociated dislocations to be resolved and the stacking fault energy to be determined [7.44]. However, the contrast decreases with increasing foil thickness because of the increasing background of inelastically scattered electrons. Figure 7.12 demonstrates the increase in contrast and sharpness of weak-beam images of dislocations in aluminium by zero-loss filtering. The moiré fringes produced by a small-angle grain boundary also appear with stronger contrast and it can be observed that one moiré fringe ends where the dislocation terminates at the interface [7.45].

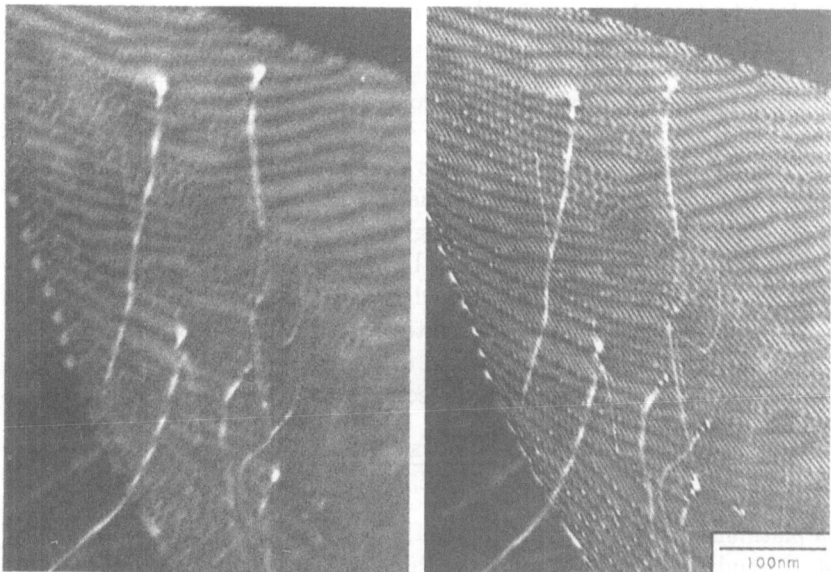

Fig. 7.12. Weak-beam dark-field image (a) unfiltered, (b) zero-loss filtered of dislocations in aluminium. A small-angle boundary is included and produces moiré fringes (by courtesy of *J. Mayer* [7.45]).

7.2 Plasmon-Loss Imaging

Shifting of the energy-selecting slit to one of the plasmon losses allows us to image the local variation of their excitation. This is of special interest for materials with a sharp plasmon loss that has been shifted by a local variation in concentration, or inside precipitates, for example. In a parallel recorded loss spectrum (PEELS) with a few pixels per eV, it is possible to detect shifts of the plasmon loss down to 0.1 eV. Separation of image contributions from plasmon losses separated by about 1 eV – limited by the energy spread of the primary beam – is possible with narrow energy-selecting slits. For homogeneous specimen foils, plasmon-loss imaging will show no interesting details. Preservation of the Bragg contrast is observed in crystalline specimens because inelastically scattered electrons are also Bragg diffracted and travel through the crystal as Bloch waves. Therefore nearly the same Bragg contrast as in zero-loss imaging is observed and thickness fringes, bend contours and images of lattice defects also appear in the plasmon-loss image. Only with increasing energy loss will the Bragg contrast effects be blurred, at losses of a few 100 eV, but differences in anomalous transmission can still be observed.

7.2.1 Scattering Contrast in the Presence
of Inelastically Scattered Electrons

The following discussion of scattering contrast by plasmon- or low-loss electrons assumes that we use a broad energy window Δ covering the main part of inelastically scattered electrons. This is approximately true for the investigation of stained sections or macromolecules on a supporting film. (In Sect. 7.2.5 we discuss the selective ESI using narrow energy windows and sharp plasmon losses.) Stained structures in biological sections may appear dark and differ little from elastic bright-field images or the contrast may be reversed and the structures appear brighter than the resin matrix [7.46]. The reasoning developed for the calculation of the transmission and resulting in (7.15–17) can be used to calculate the contrast

$$C_{\text{in}} = \frac{T_{\text{in,C+E}} - T_{\text{in,C}}}{T_{\text{in,C}}}. \tag{7.30}$$

With platinum for (E) as an example of a heavy element, calculations using (7.14–16) and the tabulated values in Table 1.1 show that stained structures can appear bright (positive C_{in}) in thin sections and dark (negative C_{in}) in thicker sections.

7.2.2 Phase Contrast by Inelastically Scattered Electrons

We showed in Sect.7.1.1 that phase contrast is caused by interference between the unscattered and scattered waves and depends on defocusing. The amplitude after passing a weak phase structure with a spatial frequency q can be written as [7.9]

$$\psi_{\text{s}} = 1 - i\phi_q \cos(2\pi qx). \tag{7.31}$$

Normally, a weak phase structure does not result in a contrast because of the phase shift of $\pi/2$ between scattered and unscattered wave [factor i in (7.31)]. The Fourier transform of ψ_{s} consists of two diffraction maxima of amplitude $F(\pm q) = i\phi_q/2$. Introducing the wave aberration $W(q)$, the wave amplitude at the image

$$\psi_{\text{m}}(x') = 1 + \sum_{\pm q} \frac{1}{2} i\phi_q e^{-iW(q)} e^{2\pi qx'} = 1 + i\phi_q e^{-iW(q)} \cos(2\pi qx') \tag{7.32}$$

is given by the inverse Fourier transform and the image intensity becomes

$$I_{\text{m}}(x') = \psi_{\text{m}}\psi_{\text{m}}^*/M^2 = [1 + 2\sin W(q) \cdot \phi_q \cos(2\pi qx') + ...]/M^2, \tag{7.33}$$

where $x' = Mx$ and $q' = q/M$. The factor of the term $\phi_q\cos(2\pi qx')$ is the contrast transfer function (CTF)

$$B(q) = -2\sin W(q) \tag{7.34}$$

of the phase structure [7.47]. The sign of $B(q)$ is chosen so that $B(q) > 0$ for positive phase contrast (decrease of I_m, $\Delta z > 0$, underfocus).

Only when the incident wave is parallel and monochromatic is the illumination coherent. In reality, we have partial coherence: limited spatial coherence, associated with the finite value of the illumination aperture α_i, and limited temporal coherence caused by the energy spread $\delta E = 1\text{–}2$ eV of the electron gun or by the width Δ of the selected energy window. Partial coherence results in an incoherent superposition of image intensities obtained from the superposition of the amplitudes of single-electron waves. The effect of partial spatial and temporal coherence can be represented by envelope functions, which modulate the CTF [7.48,49], causing a decrease of $B(q)$ at the resolution limit (high q). The incident and elastically scattered waves are coherent in spite of very small recoil energies, whereas waves of inelastically scattered electrons are incoherent relative to the incident and elastically scattered waves and also to inelastic waves that differ in the final object states [7.50]. These are, for example, the excitation of plasmons or single electrons with different transferred momenta $hq' = h(k_n - k_0)$ where k_0 and k_n are the wave vectors of the incident and inelastically scattered waves, respectively. Plasmon excitation also shows a dependence of energy loss on the scattering angle (dispersion), and inelastic electron waves with different k_n will also be incoherent.

The CTF can thus be calculated, including partial spatial coherence, by convolution with the angular distribution of inelastic scattering [7.30]. In elastic scattering, the partial spatial coherence is represented by a Gaussian distribution of incidence with a half-width about α_i. Though many inelastically scattered electrons are concentrated at low scattering angles $\theta = \lambda q$, the number $N(q)dq$ of inelastic electrons with spatial frequencies between q and $q+dq$, shows a long tail because this angular distribution $\propto (\theta^2 + \theta_E^2)^{-1}$ is Lorentzian. The fraction concentrated at high q totally blurs the CTF just like incoherent illumination and only the fraction with an angular width of about θ_E shows a damped oscillation of $\mathrm{CTF_{in}}$ with a decreased amplitude. Qualitatively this expected decrease of phase contrast can be observed for the granulation of carbon films on comparing the zero-loss and plasmon-loss filtered images [7.30]. Phase contrast structures with the carbon plasmon loss at lower defocusing and $\Delta = 4$ eV have also been observed in the ESI mode [7.51] and in a dedicated STEM [7.52]. The quantitative measurement of the phase contrast of colloidal gold particles on carbon films [7.53] shows that phase contrast of the colloids can be observed with zero-loss filtering up to carbon films of mass–thickness 40 μg/cm^2 whereas the phase contrast is invisible in unfiltered images because of the large fraction of inelastically scattered electrons.

7.2.3 Crystal-Lattice Imaging
and Inelastically Scattered Electrons

The interference of several Bragg reflected waves passing the objective diaphragm results in an image of the crystal structure. Because the Bragg reflected waves are phase-shifted by the wave aberration (7.6) and their amplitude is governed by the dynamical theory of electron diffraction, the intensity distribution depends sensitively on defocusing, foil thickness and orientation. A defocus series must therefore be compared with a series of computer simulations. Quantitative comparison is made difficult by the contribution of inelastically and thermal-diffusely scattered electrons that pass the objective diaphragm too. A knowledge of the contribution of the inelastically scattered electrons to lattice-fringe and crystal-structure images is hence important for the discussion of high-resolution images [7.54] because the necessary comparison with multislice calculations gives only the contribution of elastically scattered electrons. Zero-loss filtering will allow us to get good contrast in thicker areas where the contrast will be decreased by chromatic aberration. However, the contrast of crystal-structure images shows periodic contrast reversals with increasing foil thickness. Only very thin foil areas with t < 10 nm are used in practice therefore.

Lattice-fringe contrast in graphitized carbon [7.52] and crystal-structure images of silicon [7.55–57] have also been observed with reduced contrast in the plasmon-loss image because the inelastically scattered electrons preserve the Bragg contrast but show a spectrum of excitation errors s_g (see next Section). When calculating the plasmon-filtered images of crystal lattices, the incoherent superposition of inelastic images also has to be taken into account [7.58].

7.2.4 Bragg Contrast of Inelastically Scattered Electrons

The discussion of Bragg contrast in Sect. 7.1.5 under the aspect of zero-loss filtering is now continued for plasmon-loss and high-energy-loss filtering. We know from theory [7.59,60] and experiment [7.40,61–66] that the Bragg contrast is preserved in inelastic scattering processes that excite plasmons or inner-shell ionizations of low ionization energy, whereas the thermal-diffuse scattering does not preserve Bragg contrast. The inelastically scattered electrons around the primary beam and the Bragg spots belong to Bloch-wave fields that differ from the elastic Bloch-wave field only in their wave vectors $k_g^{(j)}$, the amplitude of which is slightly modified by the energy loss and the angle by the angular distribution of inelastic scattering. The dark-field images with Bragg reflections R_1 and R_2 from an aluminium foil are the same in the zero-loss filtered images in Figs. 7.13a,c and the plasmon-loss filtered images in Figs. 7.13d,f. With the aperture shifted half the distance between R_1 and R_2, the intensity in the plasmon-loss filtered image results from inelastically scattered electrons around R_1 and R_2 and Fig. 7.13e shows both

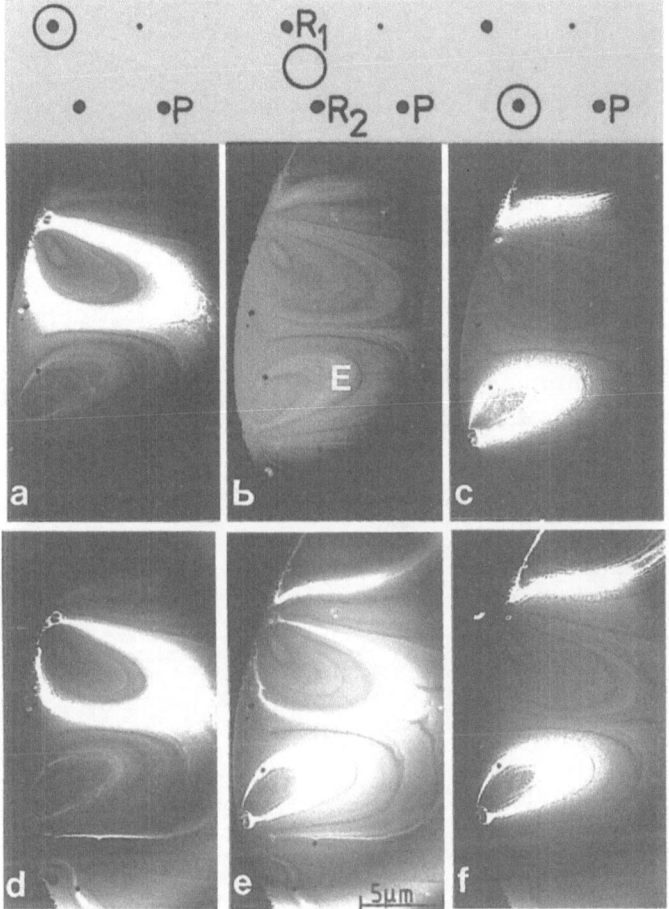

Fig. 7.13. Bend contours in an aluminium foil imaged (a–c) with zero-loss and (d–f) with plasmon-loss filtering (15 eV) for three positions (top) of the objective diaphragm on and between the Bragg spots R_1 and R_2 (E = bend contours of thermal diffusely scattered electrons).

sets of bend contours from the reflections R_1 and R_2 . The zero-loss filtered image in Fig. 7.13b with the thermal-diffusely scattered electrons shows no bend contours but only a structure (E) due to anomalous absorption because the intensity is proportional to $\sum I_g$ in (6.33) and Figs. 6.9–11.

However, the angular distribution of inelastically scattered electrons creates a spectrum of excitation errors and finally a blurring of edge and bend contours with increasing energy loss [7.40,67–69], analogous to incoherent illumination with an increased illumination aperture as in the STEM mode [7.70]. The image of crystalline specimens also shows only differences in anomalous transmission. Figure 7.14 shows a typical influence of energy loss on the

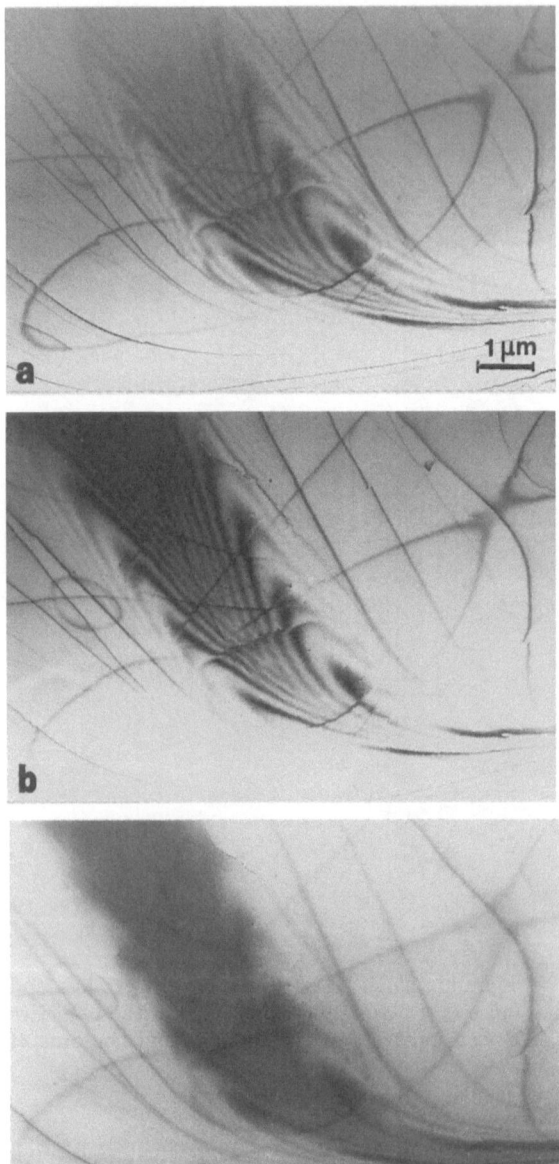

Fig. 7.14. ESI image of superposed edge and bend contours in an Al foil imaged by (a) zero-loss, (b) plasmon-loss filtering (E = 15 eV) and (c) energy filtering at E = 300 eV (thickness increases from bottom to top).

superposition of bend and thickness contours in a wedge-shaped and bent aluminium foil. Plasmon-loss filtering in Fig. 7.14b shows approximately the same contrast as zero-loss filtering (Fig. 7.14a). With increasing energy loss, the pendellösung fringes are blurred, as demonstrated by placing an energy window at E = 300 eV in Fig. 7.14c [7.40]. A two-beam calculation of thickness fringes in aluminium is shown in Figs. 7.15a and b for zero-loss imaging and excitation errors $w = s\xi_g = 0$ (Bragg condition) and 1.3, respectively. For the three-fold plasmon loss with $E = 45$ eV, the maximum of the fringes in Figs. 7.15c,d is shifted to higher thicknesses due to the Poisson distribution

$$P(t, E) = \left(\frac{t}{\Lambda_{\mathrm{pl}}} \right)^n \frac{\exp(-t/\Lambda_{\mathrm{pl}})}{n!} \qquad (7.35)$$

which gives the probability for a multiple plasmon loss $E = nE_{\mathrm{pl}}$. The dashed curves in Figs. 7.15c,d are obtained by superpositing fringe patterns corresponding to a spectrum of excitation errors associated with angular distribution of the inelastically scattered electrons. The blurring of the fringe contrast is small for $s = 0$ (Fig. 7.15c) and increases with increasing s (Fig. 7.15d).

This blurring has less influence on stacking-fault fringes, which can be observed up to energy losses of 300 eV [7.40,66], for example.

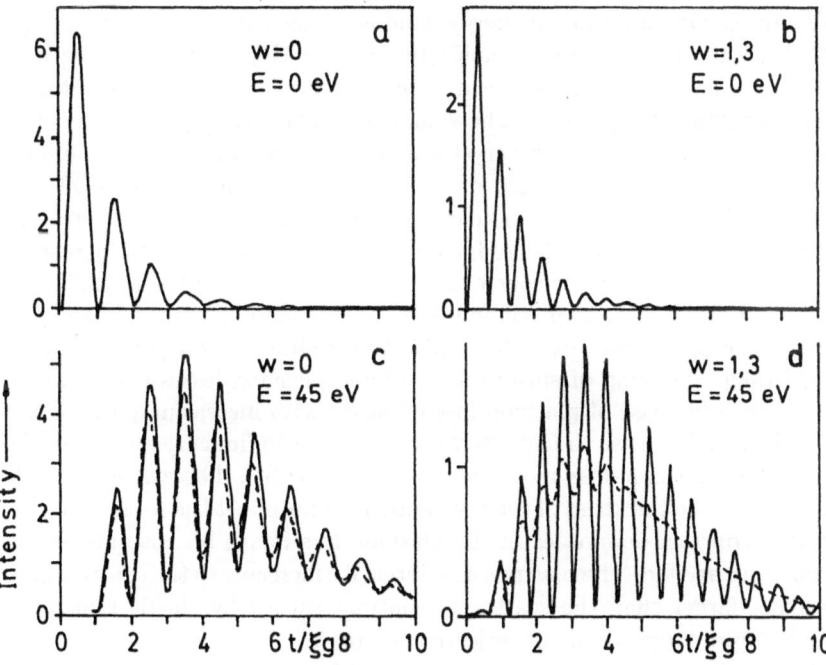

Fig. 7.15. Edge contours calculated for the dynamical two-beam case (**a**) at the Bragg condition and (**b**) $w = s\xi_g = 0.3$. Modifications for an energy loss $E = 45$ eV at (**c**) $w = 0$ and (**d**) $w = 1.3$ considering the Poisson distribution (———) and an additional blurring by the spectrum of excitation errors ($---$).

7.2.5 Selective Plasmon-Loss Imaging

The discussion of plasmon losses in Chapt. 3 shows that surface and volume plasmon losses can be excited, each whith a dispersion and a cut-off angle. The size and surface coating of small particles influence the excitation of surface and plasmon losses [7.71–76]. The investigation of the EELS plasmon-loss region and ESI with selected energy windows can be used not only for the confirmation of theoretical models but also for characterization of the particle. Selection of surface plasmon losses can result in a bright rim around particles because such losses can even be excited at distances of a few nanometres away from the particle; the electron does not strike the particle directly but causes a polarization inside it. The excitation probability decreases exponentially with distance and fringes can even be distinguished at distances of about 50 nm [7.71]. They have been observed in a dedicated STEM [7.71,75,76] and by ESI in an EFTEM [7.77], as demonstrated for an MgO particle in Fig. 7.16. The energy-loss spectrum at the indicated positions in Fig. 7.16d differ in the relative intensities of the volume (22 eV) and surface plasmon (16 and 12 eV) losses. On situating an energy window at the surface plasmon loss ($E = 16$ eV), a bright rim is seen outside the particle (Fig. 7.16b); placing it at the plasmon loss ($E = 22$ eV) creates an approximately uniform excitation inside the particle (Fig. 7.16c). The volume plasmon losses are more strongly damped in small than in larger crystals. This has been observed in the form of an increase of the half-width of the Al 15 eV plasmon loss of 50 keV electrons at $\theta = 0°$ from 1 eV for coarse-grained films to 3 eV for fine-grained crystalline films; the half-width of fine-grained films decreases with increasing θ, reaching the value for the coarse-grained films at $\theta = 2$ mrad [7.78,79]. At crystal boundaries, surface plasmon losses are excited predominantly and this decreases the probability for volume plasmon excitation because of the sum rule [7.71,80]. Coupling of surface and volume plasmon losses can result in an oscillatory increase of plasmon loss intensity with increasing particle size [7.80]. Figure 7.17 shows the intensity profiles across indium crystallites in the zero-loss and plasmon-loss mode with $E=11$ eV [7.81]. The zero-loss mode indicates that the thickness of the crystallite is uniform. At the periphery, this scattering contrast dominates in the plasmon-loss image because almost no plasmons are excited. The plasmon-loss intensity increases to the centre where it becomes larger than the scattering contrast caused by elastic scattering through angles larger than the objective aperture.

Plasmon-loss imaging can also be employed for selective imaging when different phases or precipitates and the matrix show separated sharp plasmon-losses. Because the plasmon energy (3.51) is proportional to the number of conduction electrons per unit volume, it can change as a function of concentration and phase, in intermetallic compounds and binary or ternary alloys [7.82–89] for example. Shifts of the plasmon-loss maxima can be measured with an accuracy of 0.1 eV, even when the width of the plasmon loss exceeds 1 eV or is convolved with the energy spread of the electron gun. In the case of

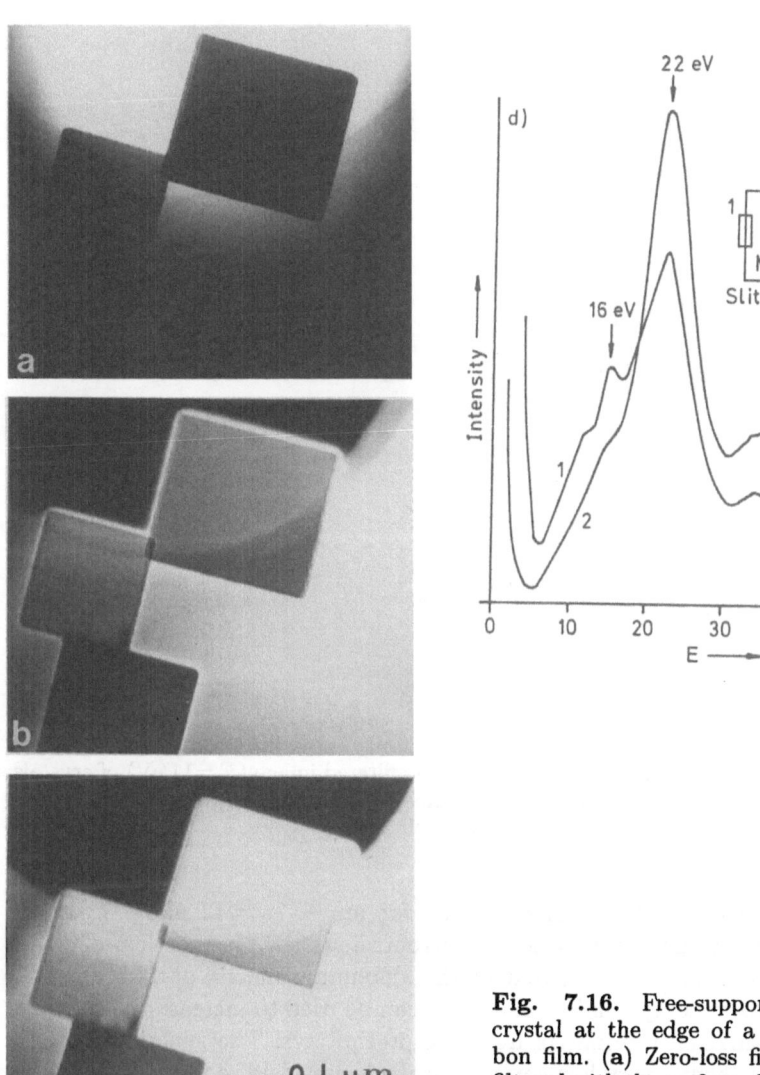

Fig. 7.16. Free-supported MgO crystal at the edge of a holey carbon film. (a) Zero-loss filtered, (b) filtered with the surface plasmon loss at $E = 16$ eV and (c) the volume plasmon loss at $E = 22$ eV; (d) EELS at the positions 1 and 2.

electron-spectroscopic imaging, the plasmon losses of different phases must differ by more than 1 eV. For example, bright images of helium bubbles in aluminium have been selected at $E = 11$ eV, which disappeared when the 15 eV loss of aluminium was selected [7.90], Be precipitates in Al appeared brighter at the plasmon loss (19 eV) of Be [7.91], In and Sn crystallites on opposite sides of a carbon film can be imaged separately by selecting their very sharp plasmon losses at $E = 12$ and 13 eV, respectively [7.3].

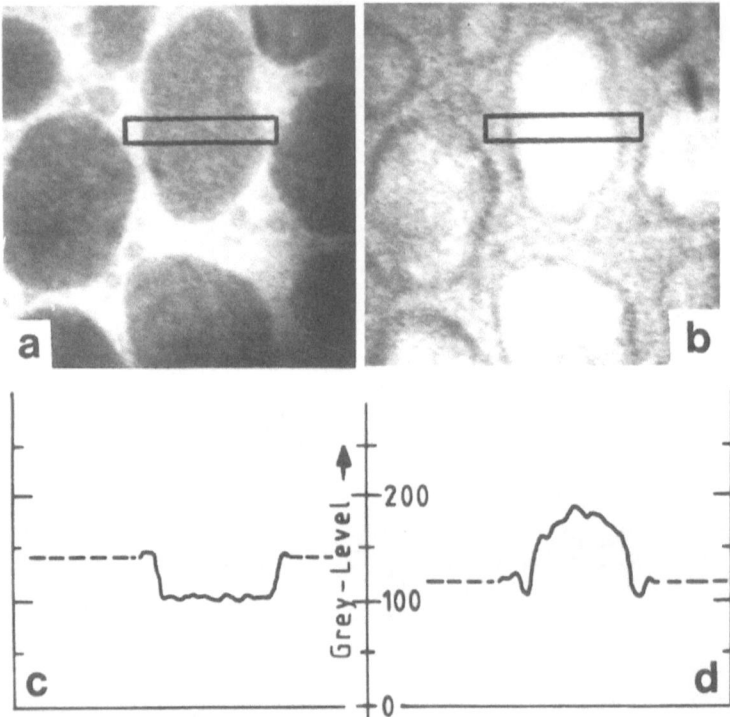

Fig. 7.17. (a) Zero-loss and (b) plasmon-loss filtered image ($E = 11$ eV) of crystals in an evaporated indium film. The linescans (c) and (d) are recorded across the rectangles indicated in (a) and (b) (bar = 50 nm).

This method will be demonstrated for an Al-7wt%Li alloy [7.81]. In Fig. 7.18c a parallel-recorded loss spectrum shows the plasmon loss of Al_3Li precipitates at 13.5 eV and that of the aluminium matrix at 14.5 eV. ESI with a narrow energy window $\Delta = 1$ eV can be used to increase the contrast and to distinguish the matrix from the precipitates. The spherical δ-Al_3Li precipitates show a very low contrast in the bright-field mode as seen in the unfiltered image in Fig. 7.18a. Filtering with the 13.5 eV plasmon loss of the precipitates creates bright images in a dark background (Fig. 7.18b) as has also been shown with a dedicated STEM [7.92]. At the 14.5 eV loss of the matrix (Fig. 7.18d), the contrast is reversed and, with an energy window at 14 eV between the plasmon peaks, the contrast vanishes totally. Another way of increasing the contrast of these precipitates is to form the dark-field image with a superlattice reflection from the precipitates [7.93]. The weak superlattice reflections can be more easily seen and selected in a zero-loss filtered diffraction pattern (Fig. 7.19). Figure 7.20 shows a comparison of ESI (a) with $E = 13.5$ eV and (b) with a zero-loss filtered dark-field image with a superlattice reflection. The latter shows the largest contrast for the

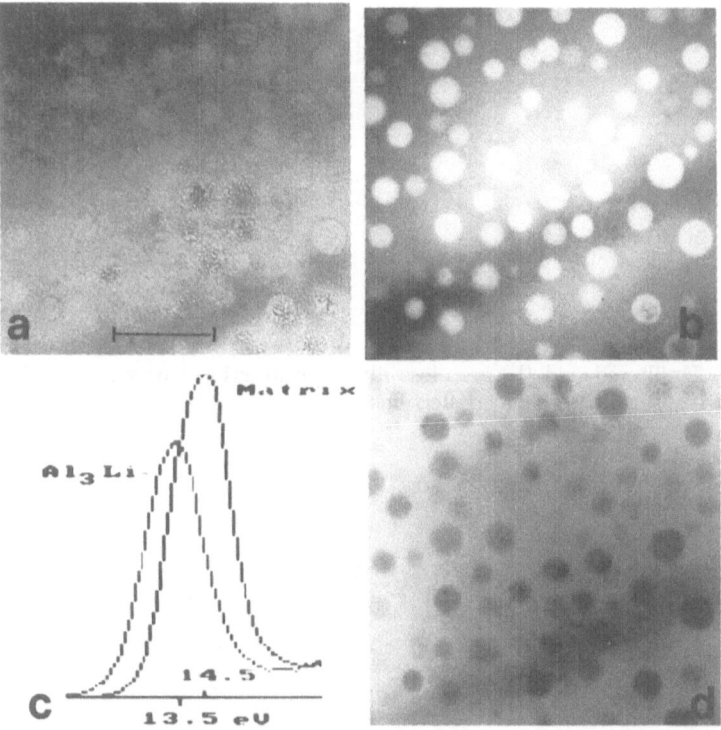

Fig. 7.18. (c) Parallel-recorded EELS of the plasmon losses of Al_3Li precipitates and the Al matrix in an Al-7wt%Li alloy. (a) unfiltered image, (b) ESI at 13.5 eV and (d) 14.5 eV (bar = 200 μm).

precipitates but the contrast is more influenced by diffraction effects than the 13.5 eV image; the precipitates of equal intensities indicated by arrows in (a) have different intensities in (b).

7.3 High Energy-Loss Imaging

7.3.1 Structure-Sensitive Imaging

Imaging of biological sections at an energy loss $E \simeq 250$ eV just below the carbon K edge at $E = 285$ eV offers greater "negative contrast" (bright) than the increase of positive contrast (dark) caused by zero-loss filtering [7.1,30,94]. This is demonstrated by the comparison of a zero-loss image (Fig. 7.21a) and a structure-sensitive image at $E = 250$ eV (Fig. 7.21b) of a biological liver section 60 nm thick (OsO_4-glutaraldehyde fixed, uranyl-acetate stained and epon-embedded). This allows an image of biological sections to be formed

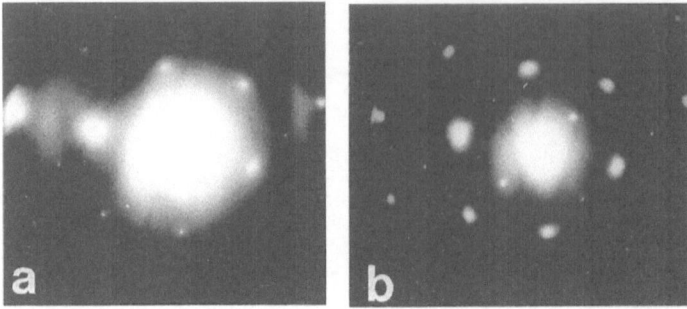

Fig. 7.19. (a) Unfiltered and (b) zero-loss filtered diffraction pattern of an Al-7wt%Li alloy with weak superlattice reflections.

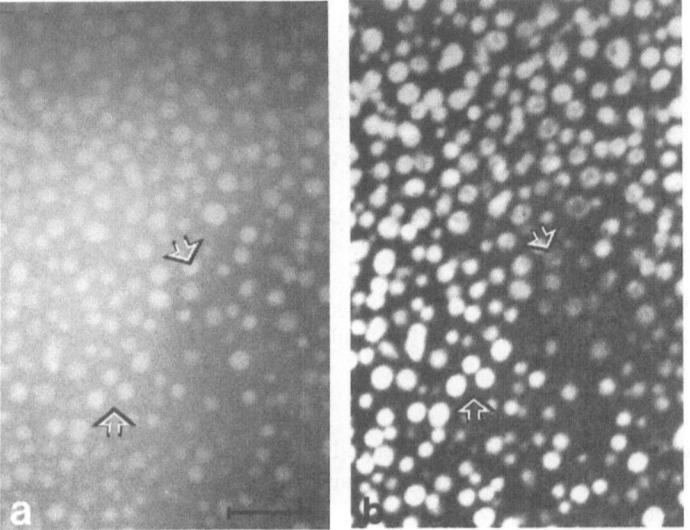

Fig. 7.20. Comparison of (a) plasmon-loss filtered (13.5 eV) bright-field image and (b) zero-loss filtered dark-field image of the precipitates with the superlattice reflection of an Al-7wt%Li alloy (bar = 100 nm).

with a minimum contribution from carbon and a relatively strong contribution from the noncarbon atoms (P, S and staining elements, for example) inside the biological structures though with unknown cross-sections. Beyond the carbon K edge, however, the contribution of the carbon atoms to the image intensity increases relatively rapidly and the contrast from the noncarbon atoms becomes very low. At medium energy losses ($E \simeq 50$ eV) the contrast changes from positive to negative contrast.

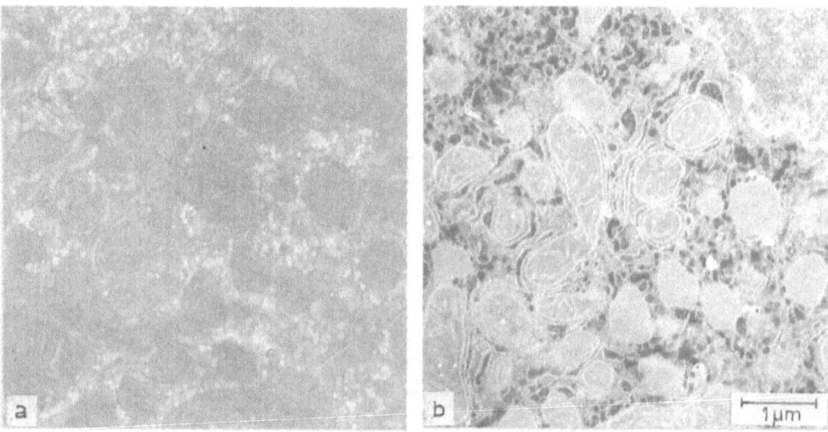

Fig. 7.21. ESI of a liver section 60 nm thick (OsO_4–glutaraldehyde fixed, uranyl-acetate stained and epon embedded) (a) zero-loss filtered and (b) ESI at $E=250$ eV with "structure-sensitive" contrast.

These contrast reversals have their parallel in intersections of the energy-loss spectra. Figures 7.22a,b show loss spectra near the carbon K edge of stained (nucleus) and unstained (epon) parts of a liver section. The first intersection is at $E = 50$–70 eV. Below the carbon K edge, the spectrum intensity of the strongly stained nucleus is nearly twice that of the pure resin, when using an objective aperture of $\alpha = 6$ mrad. This may be due to lower values of the exponent r in the decrease $\propto E^{-r}$ (4.8) and/or to the presence of L, M or N edges of the noncarbon elements (X) below $E = 250$ eV. Analogous contrast reversals can be observed in a model specimen containg the metal as an evaporated film of mass–thickness x_X on a carbon substrate of mass–thickness x_C [7.30]. The structure-sensitive contrast can be described by the following theoretical approach.

The inelastic scattering of the majority of the primary and plasmon-scattered electrons to an energy loss $E = 250$ eV is proportional to the mass thickness, and the number of electrons in the interval $E, E + \Delta$ becomes

$$\int_E^{E+\Delta} N(E)\mathrm{d}E = (N_C + N_X)\Delta = N_o(c_C x_C + c_X x_X)\Delta, \qquad (7.36)$$

where $c = [\mathrm{d}\sigma/\mathrm{d}E]_{250\mathrm{eV}} \cdot N_A/A$. This number of electrons below the carbon K edge increases linearly with increasing mass–thickness and reaches a maximum at about 100 μg/cm^2. We can therefore assume that the linear approximation of (7.36) will be valid for $x \leq 20\mu$g/cm^2 and we find

$$(N_X/N_C)_{250\mathrm{eV}} = c_r\, x_X/x_C. \qquad (7.37)$$

with $c_r = c_X/c_C$. A plot of measured values of

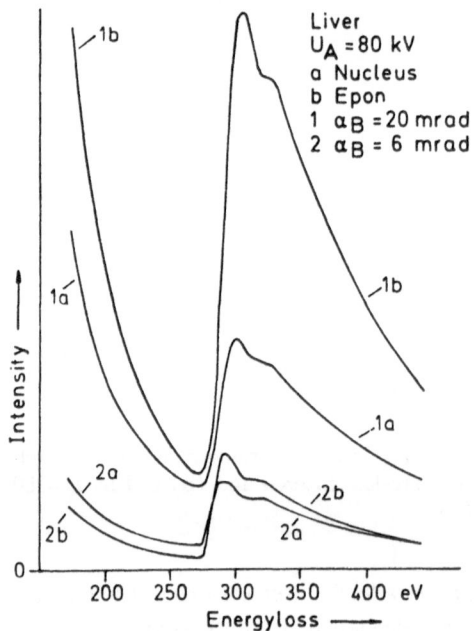

Liver
$U_A = 80$ kV
a Nucleus
b Epon
1 $\alpha_B = 20$ mrad
2 $\alpha_B = 6$ mrad

Intensity

0

200   250   300   350   400 eV
Energyloss ⟶

Fig. 7.22. EELS of section shown in Fig. 7.21 from areas at (**a**) the nucleus and (**b**) pure epon for two apertures α.

$$C = N_{Pt}/N_C = (I_{Pt+C} - I_C)/I_C \qquad (7.38)$$

versus the mass thickness ratio x_{Pt}/x_C is shown in Fig. 7.23 for Pt films on carbon. The intensities I_{Pt+C} and I_C are measured by a scintillator–photomultiplier combination for the carbon–metal layer and the pure carbon film, respectively, and result in $c_r = 2.3$. Measurements on layers of other elements also show an increase of structure-sensitive contrast with values of c_r between 1 and 4 but the increase shows large variations that can be attributed to the different distance of the edges below 250 eV, their different decrease with increasing E and to the unkown contribution of the tail of single electron excitation at $E = 250$ eV [7.30].

In a comparison between conventional techniques, the dark-field mode also shows bright stained structures in contrast to dark structures in the bright field mode. The dark field intensity also increases with increasing thickness up to a maximum for 10–20 μg/cm^2. The dark field contrast can be increased by zero-loss filtering because of the more frequent inelastic scattering in carbon but the factor c_d (equivalent to c_r) only lies between 1 and 2 [7.30].

This type of contrast is not restricted to biological sections. In doubly evaporated Ag–Au films on rocksalt, the Ag crystals appear bright just beyond the Ag M_{45} edge at $E = 440$ eV, below the edge where the contribution of Ag is a minimum – as it is for C below the C K edge – the Au crystals appear bright [7.95].

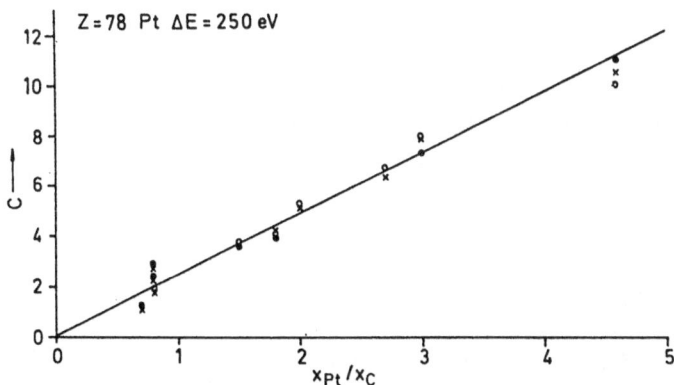

Fig. 7.23. Ratio $C = N_{Pt}/N_C = (I_{Pt+C} - I_C)/I_C$ of the number of electrons scattered at $E = 250$ eV by platinum and by the carbon film versus the ratio x_{Pt}/x_C of their mass–thicknesses.

7.3.2 Contrast Tuning

The energy-loss spectra from different parts of a specimen can intersect several times owing due to differences in the decrease of the background intensity with increasing energy loss and overlapping of the ionisation edges of different elements. This causes contrast reversals when the selected energy is tuned over a larger range of selected energy losses. This technique of contrast tuning [7.96,97] has been applied to thicker biological sections when the stained areas become very dark and cannot be recorded together with much brighter areas because of the limited range of grey levels of photographic emulsions. Tuning the energy loss can reveal an optimum condition, in which both parts are recorded with comparable contrast and intensity within the range of available grey levels. This is demonstrated in Figs. 7.24b-e with a series of energy-filtered ESI at the energy losses indicated in the loss spectrum (Fig. 7.24a) of the 0.5 μm epon section of liver. The spectrum shows that there is still enough intensity at the zero-loss peak to get a zero-loss filtered image but with very high contrast beyond the dynamical range of the photographic emulsion. Optimum contrast of all cell compartments can be obtained with $E = 200$ eV (Fig. 7.24d).

As another example of contrast tuning, Fig. 7.25 shows the same copolymer as in Fig. 7.8 but with less ruthenium oxide and a thicker section [7.5]. The unfiltered image (Fig. 7.25a) and the ESI at $E = 50$ eV (Fig. 7.25b) show no strong difference in contrast and the boundaries between polyethylene and polypropylene cannot be clearly distinguished. Maximum contrast and good separation of the phases can be obtained with $E = 200$ eV (Fig. 7.25c) and quantitative determination of the volume fractions by stereologic methods is possible. This contrast decreases again at $E = 300$ eV (Fig. 7.25d).

7.3.3 Most-Probable-Loss Imaging

The loss spectrum of thick specimens of mass–thickness $x = \rho t$ (Fig. 7.24a, for example) consists of a Landau distribution [7.98] with a most probable energy loss at

$$
\begin{aligned}
E_{\mathrm{p}} &= \xi[\ln(\xi/Q_{\min}) - \beta^2 + 0.198] \\
\text{with} \quad \xi &= \frac{2\pi e^4}{(4\pi\epsilon_o)^2 mv^2}\frac{N_{\mathrm{A}}Z}{A}\rho t \\
\text{and} \quad Q_{\min} &= J^2(1 - \beta^2)/2mv^2,
\end{aligned}
\tag{7.39}
$$

in which J = mean ionisation potential and $\beta = v/c$. Measurements in loss spectra on thick films with large apertures agree with this formula [7.99].

Values of E_{p} obtained from EELS in an EFTEM with small aperture are compared in Fig. 7.26 with (7.39). Better agreement, including the width of the loss maximum, is obtained when the loss spectra of thick specimens are calculated by a Fourier method [7.100,101] using single-scattering functions $S_1(E, \theta)$ for inelastic scattering. For an estimate, we can assume $E_{\mathrm{p}} = pE_{\mathrm{pl}}$, where $p = t/\Lambda_{\mathrm{pl}}$. With a mean-free-path $\Lambda_{\mathrm{pl}} = 90$ nm and a plasmon loss $E_{\mathrm{pl}} = 25$ eV in organic material, this gives $E_{\mathrm{p}} = 270$ eV for a 1 μm section whereas the zero-loss transmission after passing a 1 μm (100 μg/cm^2) layer is about 10^{-4}. The intensity at the most probable loss is large enough to record an image either by a dedicated STEM [7.102] or by EFTEM with a filter lens [7.96,103] as shown by measurements (MPL) in Fig. 7.27. This technique can increase the limit of zero-loss imaging of amorphous specimens ($x \leq 70\mu$g/cm^2) to $x \leq 150\mu$g/cm^2. Most-probable loss imaging also decreases the very strong chromatic aberration of unfiltered images which will, however, be limited by the width of the selected energy window ($\Delta = 10$–20 eV) and the large aperture necessary to obtain sufficient intensity.

In STEM modes, chromatic aberration can be completely avoided because there is no imaging lens behind the specimen. However, the top-bottom effect discussed in Sect. 7.1.2 has to be taken into account for both modes. High-voltage electron microscopy (HVEM) provides another way of observing thick sections. The chromatic aberration decreases with increasing energy due to the presence of the electron energy E_0 in the denominator of (7.21) and the width δE decreases because of the increase of the mean-free-path for plasmon losses. The product $C_{\mathrm{c}} \cdot \alpha$ can be assumed to be constant. ESI images of 0.7 μm sections at 80 keV are comparable with micrographs in a conventional TEM at 200 keV [7.96], though there are differences in contrast: the ESI image shows more details.

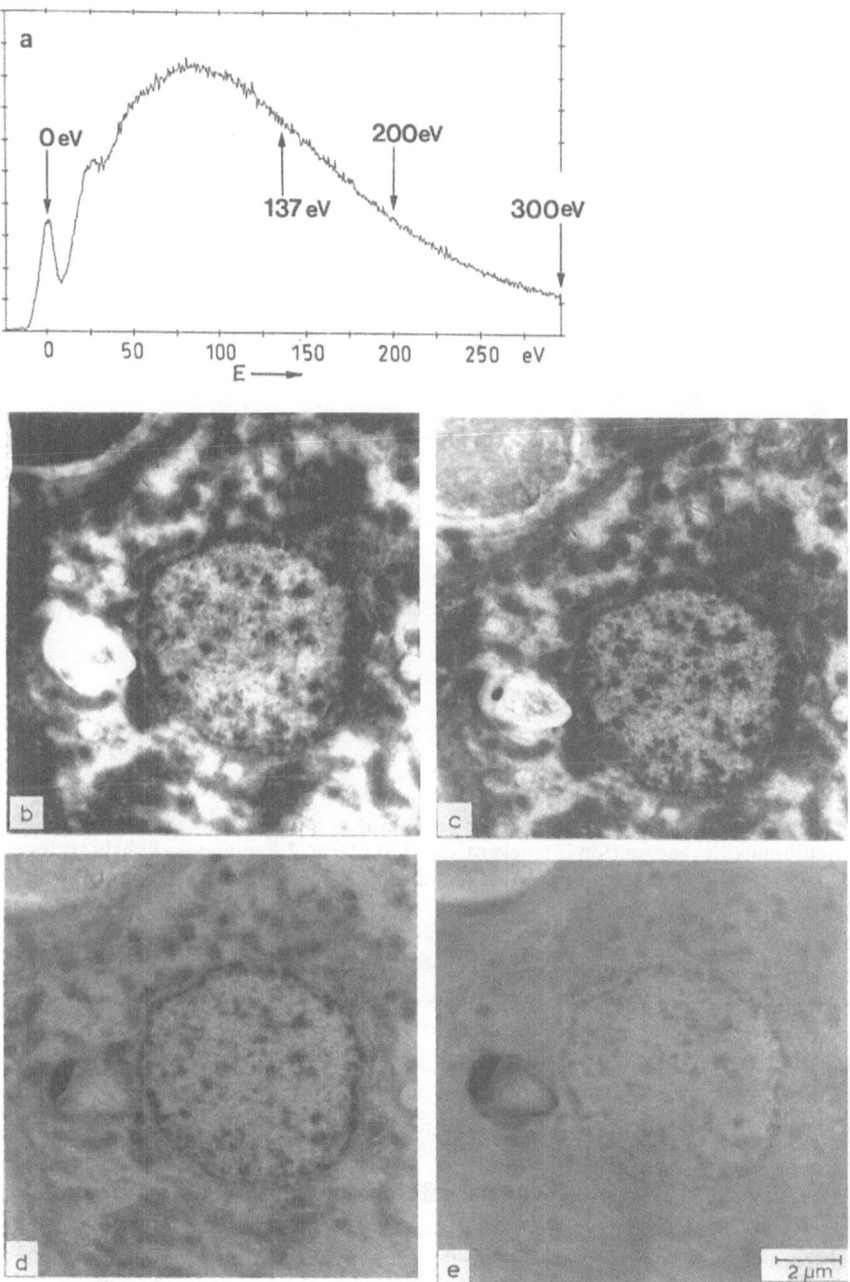

Fig. 7.24. (a) EELS of a thick epon-embedded liver section and contrast tuning with ESI images at (**b**) $E=0$ eV, (**c**) 137 eV, (**d**) 200 eV and (**e**) 300 eV indicated by arrows in (**a**).

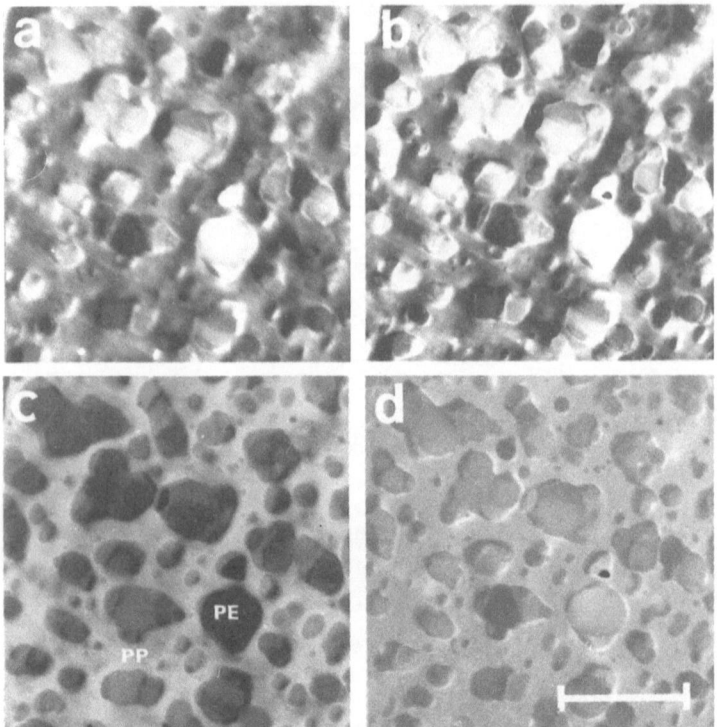

Fig. 7.25. Copolymer of polyethylene (PE) and polypropylene (PP) stained with ruthenium oxide; the section is thicker than that of Fig. 7.8 and more lightly stained. (a) Unfiltered, (b) E=50 eV, (c) 200 eV and (d) 350 eV (bar = 5 μm).

Most-probable-loss imaging of crystalline specimens increases the useful thickness from $\simeq 150\mu$g/cm^2 for zero-loss filtering (Fig. 7.11) to $\simeq 300\mu$g/cm^2 [7.40]. The greatest thicknesses for zero-loss and most-probable-loss imaging are about twice those mentioned above for amorphous specimens because the mean absorption coefficient μ_{cryst} for orientations without strongly excited low-order Bragg reflections is about half of $\mu_{\text{amorph}} = 1/x_k$ for amorphous specimens [$T = \exp(-\mu t)$] [7.40,41].

We discussed in Sect. 7.2.4 the influence of the angular distribution of inelastically scattered electrons on Bragg contrast, which is equivalent to a spectrum of excitation errors. The diffraction pattern of thick crystalline specimens shows no Bragg spots but only defect Kikuchi bands [7.104]. These are formed in the lower part of the thick specimen whereas the upper (entrance) parts acts more as a diffuser by multiple scattering. In thin specimens, the Bragg contrast is generated by the primary beam only and interception of the Bragg diffracted spots at the objective diaphragm. In thick specimens,

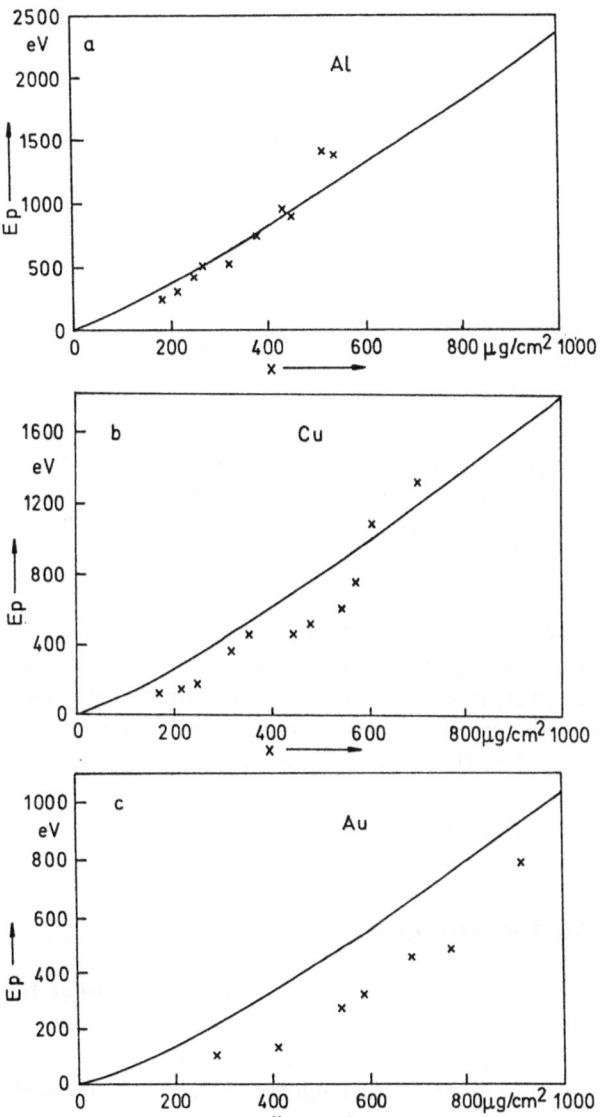

Fig. 7.26. Measured most probable energy losses E_p of (**a**) aluminium, (**b**) copper and (**c**) gold film versus mass–thickness $x = \rho t$ and comparison with the Landau formula (7.39).

directions exist in the diffuse cone of the scattered electrons that are Bragg diffracted to the primary beam direction and pass through the aperture. The image intensity becomes dependent on $\sum I_g$ (6.33), which cancels the pendellösung fringes of the dynamical theory and only variations in anomalous transmission (channelling) are observed (see the rocking curve of a three-beam case shown in Fig. 6.11a). The intensity variation caused by this channelling contrast is lower than the pure Bragg contrast and a bent foil appears more uniform. However, lattice defects can still be observed though with a

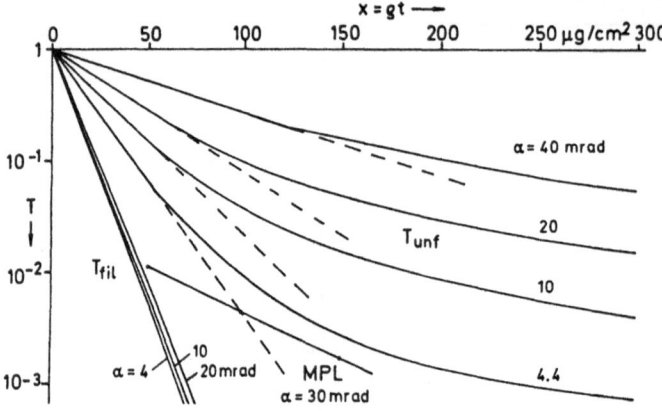

Fig. 7.27. Semi-logarithmic plot of the transmission T of carbon films for 80 keV electrons for films of large mass–thickness x and different objective apertures. Full curves: T_{unf} [8.8] and T_{fil} [8.19]; MPL: calculated and measured relative intensities at the most probable energy loss with $\Delta = 10$ eV [8.115].

reduced contrast. This contrast is analogous to the "multi-beam imaging" mode in HVEM [7.105], where the primary and diffracted beams can pass through a large aperture diaphragm and can overlap in the final image with negligible shift by spherical aberration. The same effect can be observed at 100 keV in the STEM mode when the large electron probe aperture that is necessary to get a small electron probe is employed [7.70]. Here the large aperture is externally generated by electron optics.

7.4 Element Distribution Images

The distribution of specific elements can be recorded in a TEM either by elemental mapping with the signal from an energy-dispersive x-ray analyser [7.106] or by digital calculation of an element distribution image from a series of ESI images below and beyond the ionization edge of the element of interest [7.107–109].

7.4.1 X-Ray Elemental Mapping

In the case of X-ray elemental mapping, the specimen is scanned by an electron probe in the STEM mode of a TEM or in a dedicated STEM. The X-ray quanta of energy $E_x = h\nu$ generate pulses of $N = E_x/E_i$ electron–hole pairs in a lithium-drifted silicon Si(Li) or high-purity germanium HPGe [7.110] crystal in which the mean energies for generating a pair are $E_i = 3.8$ and 3 eV, respectively. The pulse height behind a charge-sensitive amplifier is proportional to the X-ray energy E_x and is fed to a multichannel analyser. The statistics of electron–hole pair formation results in an energy resolution δE_x of about 100–200 eV. Pulses within a selected X-ray energy window at the characteristic line of the element are recorded as dots in an X-ray map. The density of dots is a measure of the concentration of the element. The X-ray continuum in the selected X-ray window of width δE_x also contributes to the background in the map. There are not enough recorded X-ray quanta per pixel in a reasonable frame-time for us to be able to subtract a background extrapolated from windows below and beyond the characteristic line unless scan-times of several hours are acceptable. The absorption of X-rays in the beryllium window of a Si(Li) detector only allows K lines to be detected for $Z \geq 11$ (Na) whereas K lines for $Z \geq 4$ (Be) can be seen with HPGe or thin-window Si(Li) detectors.

The characteristic X-rays are emitted isotropically and only a small solid angle of about 10^{-3}–10^{-2} sr is collected. When the vacancy in an ionized inner shell is filled by an electron from a higher state, the difference energy can either be emitted as an X-ray quantum with a fluorescence yield ω or transferred to another electron in the upper shell, which then leaves the atom as an Auger electron with a yield a $(\omega + a = 1)$. For low elements, the Auger yield a dominates and the X-ray fluorescence yield ω can become less than 10^{-2}. In this case it is better to analyse the Auger electrons and through-the-lens extraction methods have been proposed for the STEM mode [7.111]. Unfortunately, the escape depth of the Auger electrons with energies of 50–2000 eV is of the order of a few tens of nanometres. Conversely, the signal in X-ray microanalysis increases with increasing specimen thickness and up to 1 μm sections can be investigated. In most cases, the absorption correction can be neglected. However, the resolution decreases for thick sections because of the electron beam broadening by multiple scattering as described by (7.24). Though X-ray microanalysis is possible with nanometre probes, scattered electrons and X-ray fluorescence can produce spurious characteristic lines within an area of about one micrometre in radius around the probe. Therefore, errors can occur for low concentrations close to a structure containing a high concentration, whereas in EELS and ESI only the inner-shell ionization losses of the transmitted electrons are recorded regardless of the neighbouring concentration.

7.4.2 Element Distribution Images by ESI

In element distribution images (EDI) formed by ESI, structures containing a specific element become brighter as the selected energy loss is raised and passes an ionization edge of this element, and a series of images with 10^5–10^6 pixels can be sequentially recorded below and beyond the ionisation edge. In a dedicated STEM, the complete energy-loss spectrum of sequentially recorded pixels can be parallel-recorded with a few hundred channels. These records contain a great deal of information but we have the problem of digital storage capacity. From the view of radiation damage (Sect. 7.4.4) the STEM mode has the advantage that the minimum of charge density $q = jt$ in units of C/cm^2 strikes the specimen because the pre-irradiation during focusing can be directed onto a nearby or smaller area.

ESI images can be recorded and digitized either by scanning photodensitometry of exposed photographic emulsions or by direct recording with a CCD (charge-coupled device) camera coupled to a fluorescent layer by a fibre plate [7.112,113]. The fluorescent layer consists of a powder layer of phosphor, the thickness of which is adapted to the primary electron energy, or a thin slice of a YAG single crystal. Peltier-cooled CCD arrays can work at temperatures of about −40°C which allows weak image intensities to be integrated over longer exposure times of about 100 s. Another possibility is to record the intensity on an intermediate screen, which is imaged by a photo-objective on a CCD or SIT (semiconductor intensifying target) camera.

All recorded images should be offset-corrected, which means that the output behind the amplifier and ADC must become zero when the electron beam is switched off. It will be better to adjust the zero level to a small positive value, which can be subtracted digitally, than to work with this level below the lower threshold of the recording system.

The image recording systems can introduce an unwanted fixed-pattern image corresponding to the structure of the fluoresent screen and to differences in the light guide from the scintillator to the CCD and in the detection efficiency of the CCD pixels. Additionally, shading effects can appear, caused by a decrease of recorded intensity with increasing distance from the centre of a SIT camera, for example. This fixed pattern image can be recorded by illumination of the fluorescent screen with uniform electron current density either without any specimen or by blurring the image by introducing a very strong defocus. The recorded local variations in intensity are superposed multiplicatively in the recorded intensity and a corrected image is obtained by dividing the recorded image by the fixed-pattern image pixel per pixel and multiplying the result with a constant factor so that the full scale of the bit range (0–255, for example) is occupied.

As an example of an element distribution image, Fig. 7.28 shows the microstructure of a ceramic material derived from polymer precursors [7.114]; the bright-field image and the elemental distribution images obtained with C K, N K and O K edges, reveal the distribution of SiC, Si_3N_4 and the amorphous oxide. The background in the carbon map is due to the carbon coating of the sample and in the oxygen map, to surface oxydation of the Si_3N_4 grains.

7.4.3 Methods of Background Subtraction

For the ESI mode, it is essential to reduce the number of images to a minimum. The standard procedure is to record with an energy window Δ two images below the edge and a third beyond the edge of ionisation energy E_I (Fig. 7.29). The former are used to calculate the background intensity I_B

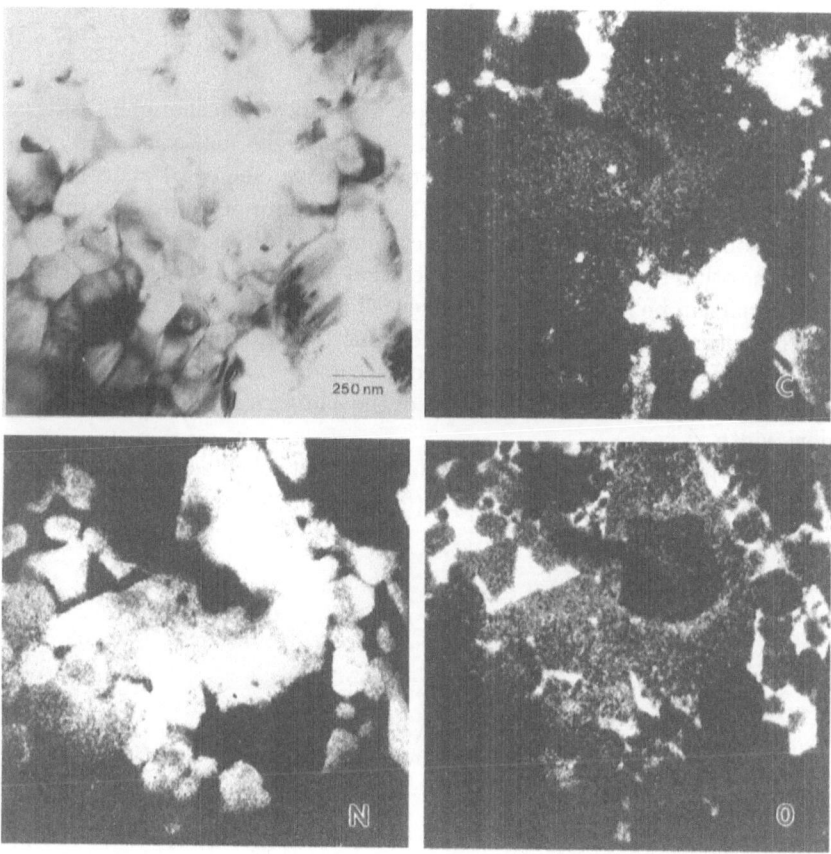

Fig. 7.28. Elemental distribution images using C, N and O K edges showing the distribution of SiC, Si$_3$N$_4$ and the amorphous oxide (by courtesy of *J. Mayer*).

below the third image pixel per pixel. The net signal difference $S_N(\alpha, \Delta)$ for EDI depends on the aperture α and the window width Δ :

$$S_N(\alpha, \Delta) = \int_{E_I}^{E_I+\Delta} \left[\frac{dI(\alpha)}{dE} - \frac{dI_B(\alpha)}{dE} \right] dE = S(\alpha, \Delta) - S_B(\alpha, \Delta) \quad (7.40)$$

where $dI(\alpha)/dE$ is the EELS intensity recorded with a narrow energy window dE. The number of atoms per unit area of an element of interest is given by

$$N = \frac{1}{\sigma(\alpha, \Delta)} \frac{S_N(\alpha, \Delta)}{S_0(\alpha, \Delta)} \quad (7.41)$$

whereby $S_0(\alpha, \Delta)$ is the signal from a window of equal width containing the zero-loss and plasmon loss. The partial ionisation cross-section

$$\sigma(\alpha, \Delta) = \int_0^\alpha \int_{E_I}^{E_I+\Delta} \frac{d^2\sigma}{dEd\Omega} 2\pi\theta \, d\Omega \, dE \qquad (7.42)$$

depends on the aperture α and the window width Δ, which should be as large as possible to cover a large fraction of electrons in the ionisation edge, and values of $\Delta = 50$–100 eV are used for quantitative elemental analysis. For recording EDI, only widths $\Delta = 10$–20 eV (maximum 50 eV, depending on magnification) are viable because of the chromatic aberration of the objective lens. On the other hand, it should be noted that, with small widths $\Delta = 10$–20 eV, the signal can be more strongly influenced by the ELNES structure of the edge. The optimal energy loss E_{max}, the value that produces the maximum net signal difference $S_N(\alpha, \Delta)$ at an edge, can be obtained from an ESI-simulation [7.115]. This is a convolution of the energy-loss spectrum with the window width Δ.

The following methods can be used to extrapolate the background and obtain the signals $S_B(\alpha, \Delta)$ and $S_N(\alpha, \Delta)$ (for a specification of sets of windows and signals see Figs. 7.1 and 7.29).

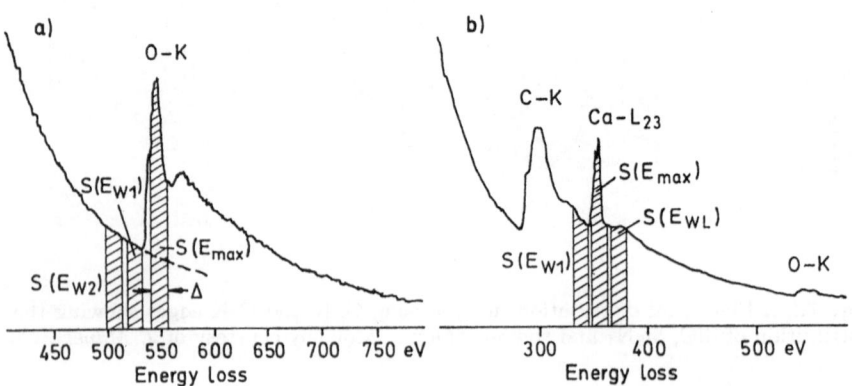

Fig. 7.29. Signals and energy windows used for (**a**) the two- and three-window methods and (**b**) the white-line method to calculate an element distribution image (EDI) by (7.43) (7.44) and (7.47), respectively.

Two-Window Method. Recording with an energy window beyond the edge at E_{max} gives the signals $S(E_{max})$ per pixel and at E_{W1} below the edge the signals $S(E_{W1})$. The net signal is given pixel by pixel by

$$S_N = S(E_{max}) - c \, S(E_{W1}) \qquad (7.43)$$

in which the empirical correction factor $c < 1$ takes into account the decrease of the background. This method can however only be applied in exceptional cases where there is a large jump ratio at the edge.

Three-Window Method. For two energy windows E_{W1} and E_{W2} below the edge (Fig. 7.29a), the corresponding pixel signals are recorded. To eliminate the background signal the power law $S_B \propto AE^{-r}$ is applied, where the factor A and exponent r are calculated pixel by pixel from the signals $S(E_{W1})$ and $S(E_{W2})$

$$S_N = S(E_{max}) - AE_{max}^{-r} \tag{7.44}$$

with

$$r = 2 \frac{\log[S(E_{W1}/S(E_{W2})]}{\log(E_{W2}/E_{W1})} \tag{7.45}$$

$$A = \frac{(1-r)[S(E_{W1}) + S(E_{W2})]}{E_{W1}^{1-r} - E_{W2}^{1-r}}. \tag{7.46}$$

Optimal windows E_{W1} and E_{W2} are given in Ref. [7.115].

White-Line Method. This method was developed to calculate EDI, for which the earlier methods cannot be applied because of the convolving effects of low-loss electrons or edge superpositions (e.g. a Ca L spectrum superposed on the ELNES of C K). This method works with windows on the white line at E_{max} and below and beyond the white lines at E_{W1} and E_{WL}, respectively (Fig. 7.29b) and interpolates the background linearly [7.116]:

$$S_N = S(E_{max}) - \left\{ S(E_{WL}) + \frac{E_{max} - E_{W1}}{E_{WL} - E_{W1}}[S(E_{WL}) - S(E_{W1})] \right\}. \tag{7.47}$$

As an example, Fig. 7.30a shows an EDI of Ca in apatite crystals inside epon sections of the epiphyseal growth plate at an early stage of bone formation [7.117]. The apatite crystals appear dark in the bright-field mode because many electrons have been scattered through angles larger than the objective aperture. This also decreases the intensities of the Ca L peak and of the Ca EDI. As a consequence, the contrast of the apatite crystals becomes small relative to the spurious signals in the surrounding tissue, which are not specific for Ca due to inaccuracies in background stripping.

Correction of Scattering Contrast. It should be noted that in structures that already appear darker in the unfiltered and zero-loss filtered bright-field image, electrons are elastically scattered at scattering angles θ larger than the objective aperture α. In a first approximation, this applies to the same degree for EDI on account of double elastic-inelastic scattering. When, for example, epon-embedded calcium apatite crystals in the early stages of bone formation appear dark in the bright-field image, because of the stronger elastic scattering, the white lines of the Ca L edge are also weaker as a result of the elastic scattering; the EDI, obtained from the net signal S_N, hence shows a lower concentration than it should. This can be avoided if the signal S_l of a bright-field image containing the low-loss electrons (zero-loss and plasmon losses) with the same window width Δ of the selected energy window is divided pixel by pixel as in (7.41) [7.117–119]. Care should be taken that the

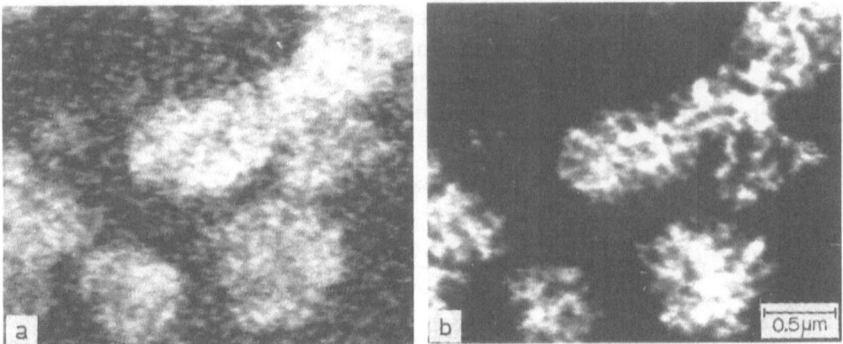

Fig. 7.30. Element distribution image of Ca in apatite crystals embedded in epon. (a) Net image with the signal S_N of the white-line method (7.47) and (b) the ratio signal S_N/S_0 for correction of elastic large-angle scattering (S_0 = signal from the zero- and plasmon-loss region with the same energy window of width Δ).

large intensity discrepancies between the signals S_1 and S_N are taken into consideration. Figure 7.30b shows that such a ratio signal S_N/S_1 results in an increase of intensity in the Ca containing apatite crystals and a stronger relative decrease of unspecific structures outside the crystals. (In both Figs. 7.30a and b, the largest intensity is expanded to full scale.) When the partial cross-section $\sigma(\alpha, \Delta)$ is known, the number N of atoms per unit area can be estimated by the relation (7.41).

7.4.4 Limitations for Element Distribution Images

Extraction of Elements During Preparation. A major problem with biological material is the extraction of the element of interest during chemical fixation, dehydration, embedding and sectioning (contact with water during floating). New embedding resins such as Nanoplast can decrease the extraction [7.120]. An excellent method of preservation is cryo-fixation, cryo-sectioning and cryo-transfer to the microscope. We can therefore expect that cryo-sectioning will become of increasing interest because mass loss by radiation damage is also reduced (see below). The zero-loss mode can be used to increase the contrast of unstained frozen-hydrated sections [7.46,121] because the ratio of inelastic-to-elastic total cross-sections (1.25) increases to $\nu \simeq 4$ for ice.

Most of the staining, cytochemical and antigen-labelling methods developed for TEM use heavy metals to obtain sufficient contrast. EDI also allows us to use staining and precipitation with low-Z atoms that show a pronounced edge, like fluorine or boron [7.122], for example, which can be less toxic for the cell.

Specimen Thickness. An important condition for EDI is that specimens thicknesses must be smaller or of the order of the mean free path Λ_{pl} of plasmon losses, which is about 80 nm in biological sections and about 40 nm in aluminium foils for 80 keV electrons. Deconvolution methods as used for EELS cannot be applied because they need the whole loss spectrum. In the case of inorganic material, EDI is often possible only at the edges of electrolytically thinned foils, which are not representative of the bulk material and can have another composition. More uniform layers can be prepared by ion-beam etching. This restriction to thin specimens is a handicap of EELS and EDI not shared by x-ray analysis and elemental mapping. An increase of the mean free path by using higher electron energies will therefore become very important for the EDI and EELS of thicker specimens, though the mean free path of plasmon losses shows a saturation at electron energies of a few hundred keV [7.123].

Radiation Damage. The inelastic scattering processes result in electron excitations that can destroy chemical bonds in organic and inorganic material. Free radicals are formed and these can generate secondary damage processes and provoke cross-linking of molecules. As a consequence, a mass-loss of about the fraction of non-carbon atoms and fading of electron diffraction patterns occur. The damage depends on the energy dissipated per unit volume, which is proportional to the number of incident electrons $n = j\tau/e$ per unit area, where j is the current density and τ is the irradiation time, or to the charge density (dose) $q = j\tau = en$. A dose of $q = 1$ C/cm^2 corresponds to $n = 6 \times 10^4$ nm^{-2} and for 60 keV electrons to a transferred energy density of 1.8×10^4 eV/nm^3. Current densities j of about 10^{-2} A/cm^2 are necessary for bright-field imaging at a magnification $M = 10000$; for a high-resolution micrograph a charge density $q = 0.5$ C/cm^2 is needed for the exposure of a photographic emulsion to a density $S=1$.

Amino acids and hydrocarbon molecules are irretrievably damaged by $q = 10^{-3}$–10^{-2} C/cm^2, aromatic molecules and nucleotides are destroyed at $\simeq 10^{-1}$ C/cm^2 [7.9,124]. The cross-section for the damage of aromatic compounds corresponds to the ionization cross-section of the carbon K shell and indeed measurement of the dose needed to destroy molecules shows that the probability of damage decreases rapidly for electron energies below 2 keV as the carbon K edge is approached [7.125].

These processes result in a mass-loss of most of the non-carbon atoms and after about one second the specimen consists of a cross-linked carbon-rich polymer. Phosphorus, sulphur and sodium, for example, can also partly leave the specimen though their mass-loss may saturate at charge densities of a few tens or hundreds of C/cm^2. The loss of non-carbon atoms can also be recognized in a decrease of their ionization edges [7.126–129].

Radiation damage can be decreased by lowering the specimen temperature and about 4–10 times higher doses can be applied at 4K (liquid helium cooling) [7.130–132]. However, the primary damage to the molecules is not suppressed but only the mobility of the fragments, so that the dose for a given loss of mass is increased. When a specimen irradiated at 4K is subsequently observed at room temperature, the mass loss is about the same at that seen after irradiation with the same dose at room temperature [7.133].

Inorganic specimens are normally more resistant because most of the damage processes are reversible. For example, irradiation of alkali halides causes additional maxima to appear in the plasmon-loss region, which can be attributed to the formation of alkali colloids [7.134] while NiO_2 shows a loss of oxygen at very high doses larger than 10^4–10^5 C/cm^2 [7.135]. EELS offer an excellent method for the in-situ study of radiation damage processes in inorganic material as well.

Detection Sensitivity. When discussing analytical sensitivities, we have to distinguish two cases. When the element being investigated is distributed in mass fractions (wt%) of about 10^{-5}–10^{-2}, the minimum detectable mass fraction (MMF) is of interest. In cases where the atoms are concentrated in macromolecules and smaller clusters or precipitates, we consider the minimum detectable number of atoms (MDN) or minimum detectable mass (MDM). In the former case, PEELS will be the best technique because a large number of channels is needed below and beyond the edge for background subtraction or for detection of the edge by first and second derivatives of the loss spectrum. For EDI by the three-window method, the small number of windows will often be insufficient for accurate suppression of the background. Otherwise, the last case will be the domain for EDI, especially when the specific atoms are concentrated in clusters so that they show a sufficient high intensity relative to the background and when the distribution of clusters is of interest.

Estimates of the MMF and MDN have been discussed for EELS [7.136–138] on the assumption that the core signal of the investigated element should be a factor $k=3$–5 times the noise of the background. The formulae can also be used for EDI though we have to take into account that the background in EELS can be averaged over a larger number of pixels and that we have only two to three images in EDI and normally use an energy window of only $\Delta =$ 10–20 eV. The background in EDI is still modulated by the inhomogeneous structure of the matrix and the increase in intensity from an element of interest should be significantly larger than the background, which will not necessarily be generated by atoms of the same element.

The most frequent application in biology concerns the element distribution images of P, Ca and S [7.139–147]. Analytical sensitivities of clusters of 150 phosphorus atoms have been observed in EFTEM [7.147]. By EELS under favourable conditions, 200 P atoms in a 10 nm-size region at the centre of tobacco mosaic virus have been detected, 320 copper atoms in a single haemocyanin molecule in EELS, even four iron atoms in a single molecule of haemoglobin though with an extremely high dose [7.148], and only a few uranium atoms in STEM [7.149], for example. However, embedding in sections increases the background and its noise and such high sensitivity needs an incident charge density q much higher than 1 C/cm^2 $\simeq 6 \times 10^4$ electrons per nm^2 and the doses mentioned for damage above.

7.4.5 Spatial Resolution

High resolution of element distribution images needs a sharp image disc of single atoms and sufficient intensity to increase the signal-to-noise ratio. The latter can be achieved by both increasing the width Δ of the selected energy window and increasing the illumination aperture α_i to increase the current density j_s at the specimen due to the conservation of gun brightness $\beta = j_s/\pi\alpha_i^2$. Increasing Δ increases the effect of chromatic aberration and increasing α_i reduces the spatial coherence. Therefore, an optimum compromise of the imaging parameters has to be found [7.150]. As an example, Fig. 7.31 shows (a) the radial intensity distribution for an elastic dark-field image and (b) for an image with the oxygen K loss at $E = 540$ eV. The dashed curves are for an ideal lens ($C_s = C_c = 0$) and the solid curves for $C_s = C_c = 2.7$ mm with Δ=20 eV and an optimum focus $\Delta z = 0.82\sqrt{C_s\lambda} = 78$ nm in (a) and and $\Delta z = 9.5$ nm in (b). The dotted curve in (b) results from a calculation with $\alpha_i = \alpha = 10$ mrad. These results agree qualitatively with the good resolution of 2.4 nm oxygen lattice planes in an AlON ceramic (Fig. 2.48).

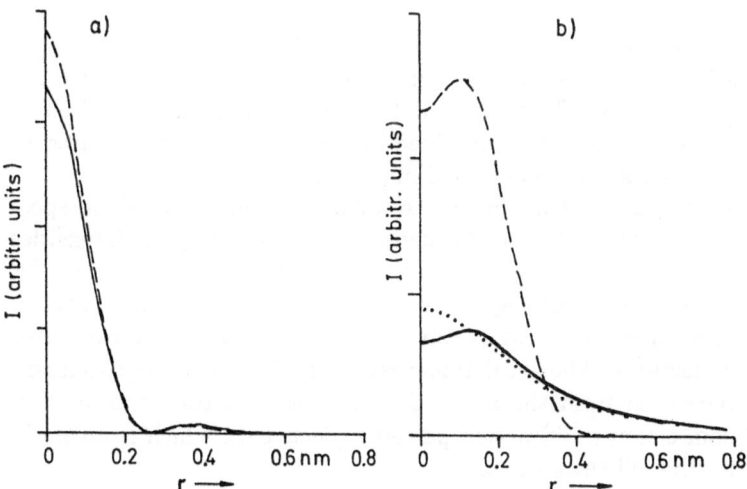

Fig. 7.31. Radial intensity distribution of the image of an oxygen atom (a) by elastic dark-field and (b) with the O-K edge. The dashed curves are calculated for an ideal lens ($C_s = C_c = 0$), the solid curves for $C_s = C_c$ =2.7 mm and (a) $\Delta z =$ 78 nm, (b) $\Delta z = 9.5$ nm and $\Delta = 20$ eV. The dotted curve in (b) is for $\alpha_i = \alpha =$ 10 mrad [7.147].

7.4.6 Methods of Measuring the Specimen Thickness

A knowledge of the local specimen thickness is necessary for quantitative interpretation of energy loss spectra and element distribution images.

Estimation of the thickness of biological sections by interference colour can only be used to control the sectioning process. The best in-situ method of measuring the mass–thickness $x = \rho t$ will be by electron transmission using Eq.(7.4) [7.151] or the calibration curves for unfiltered (T_{unf}) or zero-loss transmission (T_{fil}) in Fig. 7.3. This method works only for amorphous and polycrystalline specimens. For a compound with p_i atoms of atomic number Z_i , and atomic weight A_i per molecule, the molecular weight is $M = \sum_i p_i A_i$. The number n of molecules per unit area and the number n_i of atoms of type i per unit area are

$$n = x N_A / M \quad \text{and} \quad n_i = p_i n. \tag{7.48}$$

The transmission of a compound becomes

$$T = \exp[-\sum_i x_i / x_{ki}(\alpha)], \tag{7.49}$$

or using (7.10) with $x_i = n_i A_i / N_A$:

$$T = \exp\left(-\frac{x}{M} \sum_i p_i b Z_i^a\right). \tag{7.50}$$

which can be solved for x. The mass–thickness of evaporated films can be monitored during evaporation by the quartz-oscillator method, in which the thickness is obtained from the frequency shift caused by the deposited mass. The geometrical thickness can be obtained by Tolansky multi-beam interferometry with an accuracy of about ± 0.5 nm [7.115].

Another in-situ method applicable to amorphous and crystalline specimens makes use of the ratio of plasmon-loss to zero-loss peak intensities (Sect. 5.5.4). This needs a knowledge of the mean-free path of plasmons for the selected values of objective aperture α and energy window Δ. However, in crystalline specimens, the results can depend also on the excitation error s_g of Bragg reflections. The local thickness of single-crystalline specimens can also be measured from the fringe distance $\Delta\theta$ in a two-beam case of convergent-beam electron diffraction patterns (Sect. 6.4), which is inversely proportional to the foil thickness t:

$$\Delta\theta = d_{hkl}/t \tag{7.51}$$

where d_{hkl} is the lattice-plane distance.

References

7.1 Bauer: Electron spectroscopic imaging: an advanced technique for imaging and analysis in TEM. In *Methods in Microbiology*, ed. by F. Mayer (Academic, London 1988) Vol.20, pp.113–146

7.2 L. Reimer, A. Bakenfelder, I. Fromm, R. Rennekamp, M. Ross-Messemer: Electron spectroscopic imaging and diffraction. EMSA Bull. **20**, 73–80 (1990)

7.3 L. Reimer: Energy-filtering transmission electron microscopy. Adv. Electr. Electron Phys. **81**, 43–126 (1991)

7.4 L. Reimer, I. Fromm, P. Hirsch, U. Plate, R. Rennekamp: Combination of EELS modes and electron spectroscopic imaging and diffraction in an energy-filtering electron microscope. Ultramicroscopy **46**, 335–347 (1992)

7.5 L. Reimer, I. Fromm, Ch. Hülk, R. Rennekamp: Energy-filtering transmission electron microscopy in materials science. Microsc. Microanal. Microstruct. **3**, 141–157 (1992)

7.6 H. Niehrs: Optimale Abbildungsbedingungen und Bildintensitätsverlauf bei einer Elektronenmikroskopie von Atomen. Optik **30**, 273–293; **31**, 51–71 (1969)

7.7 L. Reimer, H. Gilde: Scattering theory and image formation in the electron microscope. In *Image Processing and Computer-aided Design in Electron Optics*, ed. by P.W. Hawkes (Academic, London 1973) pp.138–167

7.8 L. Reimer: Elektronenoptischer Phasenkontrast. II. Berechnung mit komplexen Atomstreuamplituden für Atome und Atomgruppen. Z. Naturforschg. A **24**, 377–389 (1969)

7.9 L. Reimer: *Transmission Electron Microscopy. Physics of Image Formation and Microanalysis.* Springer Ser. in Optical Sciences Vol. 36, 3rd ed. (Springer, Berlin, Heidelberg 1993)

7.10 O. Scherzer: The theoretical resolution limit of the electron microscope. J. Appl. Phys. **20**, 20–29 (1949)

7.11 F. Thon: Zur Defokussierungsabhängigkeit des Phasenkontrastes bei der elektronenmikroskopischen Abbildung. Z. Naturforschg. A **21**, 476–478 (1966)

7.12 F. Lenz: Zur Streuung mittelschneller Elektronen in kleinste Winkel. Z. Naturforschg. A **9**, 185–204 (1954)

7.13 L. Reimer, K.H. Sommer: Messungen und Berechnungen zum elektronenmikroskopischen Streukontrast für 17 bis 1200 keV Elektronen. Z. Naturforschg. A **23**, 1569–1582 (1968)

7.14 L. Reimer: Messung der Abhängigkeit des elektronenmikroskopischen Bildkontrastes von Ordnungszahl, Strahlspannung und Aperturblende. Z. angew. Phys. **13**, 432–434 (1961)

7.15 L. Reimer and M. Ross-Messemer: Contrast in the electron spectroscopic imaging mode of a TEM. I. Influence of zero-loss filtering on scattering contrast. J. Micr. **155**, 169–182 (1989)

7.16 E. Zeitler, G.F. Bahr: Contributions to the quantitative interpretation of electron microscope pictures. Exp. Cell Res. **12**, 44–50 (1957)

7.17 J.P. Langmore, B.D. Athey: Removal of inelastically scattered electrons substantially increases phase contrast on frozen-hydrated molecules. *Proc. 45th Ann. Meeting of EMSA* (San Francisco Press, San Francisco 1987) pp.652–653

7.18 M. Kunz, M. Möller, H.J. Cantow: The net distribution of elements by element specific electron microscopy – ESI. Makromol. Chemie Rapid Commun. **68**, 401–410 (1987)

7.19 L. Reimer, P. Gentsch: Superposition of chromatic error and beam broadening in TEM of thick carbon and organic specimens. Ultramicroscopy **1**, 1–5 (1975)

7.20 P. Gentsch, H. Gilde, L. Reimer: Measurement of the top bottom effect in scanning transmission electron microscopy of thick amorphous specimens. J. Micr. **100**, 81–92 (1974)

7.21 J.I. Goldstein, J.L. Costley, G.W. Lorimer, S.J.B. Reed: Quantitative X-ray analysis in the electron microscope. In *Scanning Electron Microscopy 1977/I*, ed. by O. Johari (IIT Research Institute, Chicago 1977) pp.315–324

7.22 L. Reimer, M. Ross-Messemer: Top-bottom effect in energy-selecting TEM. Ultramicroscopy **21**, 385–388 (1987)

7.23 A.V. Crewe, J. Wall,. J. Langmore: Visibility of single atoms. Science **168**, 1338–1340 (1970)

7.24 E. Carlemalm, C. Colliex, E. Kellenberger: Contrast formation in electron microscopy of biological material. Adv. Electr. Electron Phys. **63**, 269–334 (1985)

7.25 R. Reichelt, E. Carlemalm, A. Engel: Quantitative contrast evaluation for different STEM imaging modes. In *Scanning Electron Microscopy 1984/II*, ed. by O. Johari (SEM Inc., AMF O'Hare 1984) pp.1011–1021

7.26 C. Jeanguillaume, M. Tence: How to become a thickness independent image in a STEM. Ultramicroscopy **23**, 67–76 (1987)

7.27 M. Haider: Filtered dark-field and pure Z-contrast: two novel imaging modes in a STEM. Ultramicroscopy **28**, 240–247 (1989)

7.28 S.J. Pennycook, D.E. Jesson: High-resolution incoherent imaging of crystals. Phys. Rev. Lett. **64**, 938–941 (1990)

7.29 F.P. Ottensmeyer, A.L. Arsenault: Electron spectroscopic imaging and Z-contrast in tissue sections. In *Scanning Electron Microscopy 1983/IV*, ed. by O. Johari (SEM Inc., AMF O'Hare 1983) pp.1867–1875

7.30 L. Reimer, M. Ross-Messemer: Contrast in the electron spectroscopic imaging mode of a TEM. II. Z-Ratio, structure-sensitive and phase contrast. J. Micr. **159**, 143–160 (1990)

7.31 J.N. Chapman: The investigation of magnetic domain structures in thin foils by electron microscopy. J. Phys. D **17**, 623–647 (1984)

7.32 P.J. Grundy, R.S. Tebble: Lorentz electron microscopy. Adv. Phys. **17**, 153–242 (1968)

7.33 R.H. Wade: Lorentz microscopy or electron phase microscopy of magnetic objects. In *Adv. in Optical and Electron Microscopy*, ed. by R. Barer and V.E. Coslett (Academic, London 1973) Vol. 5, pp.239–296

7.34 R.H. Wade: The determination of domain wall thickness in ferromagnetic films by electron microscopy. Proc. Phys. Soc. **79**, 1237–1244 (1962)

7.35 L. Reimer, H. Kappert: Elektronen-Kleinwinkelstreuung und Bildkontrast in defokussierten Aufnahmen magnetischer Bereichsgrenzen. Z. angew. Phys. **27**, 165–170 (1969)

7.36 C. Mory, C. Colliex: Inelastic effects in Lorentz microscopy. Phil. Mag. **33**, 97–103 (1976)

7.37 H. Lichte: Electron holography approaching atomic resolution. Ultramicroscopy **20**, 293–304 (1986)

7.38 A. Tonomura: Applications of electron holography. Rev. Mod. Phys. **59**, 639–669 (1987)

7.39 L. Reimer: Untersuchungen zur Zwillingsbildung in Silberaufdampfschichten. Optik **16**, 30–34 (1959)

7.40 A. Bakenfelder, I. Fromm, L. Reimer, R. Rennekamp: Contrast in the electron spectroscopic imaging mode of a TEM. III. Bragg contrast of crystalline
specimens. J. Micr. **159**, 161–177 (1990)

7.41 L. Reimer: Deutung der Kontrastunterschiede von amorphen und kristalli-
nen Objekten in der Elektronenmikroskopie. Z. angew. Phys. **22**, 287–296
(1967)

7.42 G. Lehmpfuhl, D. Krahl, M. Swoboda: Electron microscopic channelling
imaging of thick specimens with medium-energy electrons in an energy-filter
microscope. Ultramicroscopy **31**, 161–168 (1989)

7.43 A. Bakenfelder, L. Reimer, R. Rennekamp: Comparison of images of crys-
talline specimens by energy-filtering TEM at 80 keV and CTEM at 200 keV.
Proc. XIIth Int'l Congr. for Electron Microscopy, ed. by L.D. Peachy and
D.B. Williams (San Francisco Press, San Francisco 1990) Vol.2, pp.62–63

7.44 D.J.H. Cockayne: The principles and practice of the weak-beam method of
electron microscopy. J.Micr. **98**, 116–134 (1973)

7.45 J. Mayer: Electron spectroscopic imaging and diffraction application in materials science. *Proc. 50th Ann. Meeting EMSA* (San Francisco Press, San
Francisco 1991) pp.616–617

7.46 W. Probst, E. Zellmann, R. Bauer: Electron spectroscopic imaging of frozen-
hydrated sections. Ultramicroscopy **28**, 312–314 (1989)

7.47 K.J. Hanszen: The optical transfer theory of the electron microscope: fundamental principles and applications. In *Adv. in Optical and Electron Microscopy*, ed. by R. Barer and V.E. Cosslett (Academic, London 1971) Vol.
4, pp.1–84

7.48 K.J. Hanszen, L. Trepte: Der Einfluß von Strom- und Spannungsschwankun-
gen sowie der Energiebreite der Strahlelektronen auf Kontrastübertragung
und Auflösung des Elektronenmikroskopes. Optik **32**, 519–538 (1971)

7.49 K.J. Hanszen , L. Trepte: Die Kontrastübertragung im Elektronenmikroskop
bei partiell kohärenter Beleuchtung. Optik **33**, 166–181 (1971)

7.50 H. Kohl, H. Rose: Theory of image formation by inelastically scattered electrons in the electron microscope. Adv. Electr. Electron Physics **65**, 173–200
(1985)

7.51 J.M. Martin, J.L. Mansot, M. Hallouis: Energy filtered electron microscopy
(EFEM) of overbased reverse micelles. Ultramicroscopy **30**, 321–328 (1989)

7.52 A.J. Craven, C. Colliex:. The effect of energy loss on phase contrast. *Inst.
of Physics Conf.Ser.* 36, ed. by D.L. Misell (Inst. of Physics, Bristol 1977)
pp.271–274

7.53 P. Hirsch, L. Reimer: Increase of zero-loss filtering on electron optical phase
contrast. J. Micr. **174**, 143–148 (1994)

7.54 C.B. Boothroyd, W.M. Stobbs: The contribution of inelastically scattered
electrons to high resolution [110] images of AlAs/GaAs hetereostructures.
Ultramicrocopy **31**, 259–274 (1989)

7.55 N. Ajika, H. Hashimoto, K. Yamaguchi, H. Endoh: Atomic structure image
formed by plasmon-loss electrons. Jpn. J. Appl. Phys. **24**, L41–L44 (1985)

7.56 O.C. Krivanek, C. Ahn: Energy-filtered imaging with quadrupole lenses.
XIth Int'l Congr. on Electron Microscopy, ed. by T. Imura, S. Maruse and
T. Suzuki (Jpn. Soc. of Electr. Micr., Tokyo 1986) Vol.I, pp.519–520

7.57 O.L. Krivanek: Developments in electron detectors and recording systems. In
Electron Microscopy 1992, ed. by A. Rios, J.M. Arias L. Megias-Megias, A.
López-Galindo (Secr. Publ. Univ. of Granada, Granada 1992) Vol.I., pp.83–
87

7.58 Z.L. Wang, J. Bentley: Theory of phase correlations in localized inelastic
electron diffraction and imaging. Ultramicroscopy **38**, 181–213 (1991)

7.59 A. Howie: Inelastic scattering of electrons by crystals. Proc. Roy. Soc. (London) A **271**, 268–287 (1963)

7.60 C.J. Humphreys, M.J. Whelan: Inelastic scattering of fast electrons by crystals. Phil. Mag. **20**, 165–172 (1969)

7.61 H. Watanabe: Energy selecting microscope. Jpn. J. Appl. Phys. **3**, 480–485 (1964)

7.62 R. Castaing, P. Henoc, L. Henry, M. Natta: Degre de coherence de la diffusion électronique par interaction électron-phonon. C.R. Acad. Sci. (Paris) **265**, 1293–1296 (1967)

7.63 S.L. Cundy, A.J.F. Metherell, M.J. Whelan: Contrast preserved by elastic and quasi-elastic scattering of fast electrons near Bragg beams. Phil. Mag. **15**, 623–630 (1967)

7.64 S.L. Cundy, A. Howie, U. Valdre: Preservation of electron microscope image contrast after inelastic scattering. Phil. Mag. **20**, 147–163 (1969)

7.65 S. Kuwubara, T. Uefuji: Variation of electron microscopic thickness fringes of Al single crystals with energy loss. J. Phys. Soc. Jpn. **38**, 1090–1097 (1975)

7.66 A.J. Craven, J.M. Gibson, A. Howie, D.R. Spalding: Study of single-electron excitations by electron microscopy. I. Image contrast from delocalized excitations. Phil. Mag. A **38**, 519–527 (1978)

7.67 P.H. Duval, L. Henry: Calcul de l'influence de la diffusion inélastique des électrons sur les images de monocristaux. Phil. Mag. **35**, 1381–1385 (1977)

7.68 S. Doniach, C. Sommers: Coherence of inelastically scattered fast electrons in crystals of finite thickness. Phil. Mag. A **51**, 419–427 (1985)

7.69 W.M. Stobbs, A.J. Bourdillon: Current applications of electron energy loss spectroscopy. Ultramicroscopy **9**, 303–306 (1982)

7.70 L. Reimer, P. Hagemann. STEM of crystalline specimens. In *Scanning Electron Microscopy 1976/I*, ed. by O. Johari (IITRI, Chicago 1976) pp.321–328

7.71 P.E. Batson: Inelastic scattering of fast electrons in clusters of small spheres. Surf. Sci. **156**, 720–734 (1985)

7.72 P.E. Batson: Surface plasmon coupling in clusters of small spheres. Phys. Rev. Lett. **49**, 936–940 (1982)

7.73 Z.L. Wang, J.M. Cowley: Surface plasmon loss excitation for supported metal particles. Ultramicroscopy **21**, 77–94 (1987)

7.74 Z.L. Wang, J.M. Cowley: Size and shape dependence of the surface plasmon frequencies for supported metal particle systems. Ultramicroscopy **23**, 97–108 (1987)

7.75 L.D. Marks: Observation of the image force for fast electrons near an MgO surface. Solid Stat Commun. **43**, 727–729 (1982)

7.76 A. Howie, R.H. Milne: Electron energy loss spectra and reflection images from surfaces. J. Micr. **136**, 279–285 (1984)

7.77 L. Reimer, I. Fromm, R. Rennekamp: Operation modes of electron spectroscopic imaging and electron energy-loss spectroscopy in a TEM. Ultramicroscopy **24**, 339–354 (1988)

7.78 C. von Festenberg: Zur Dämpfung des Al 15-eV Plasmaverlustes in Abhängigkeit vom Streuwinkel und der Kristallitgröße. Z. Phys. **207**, 47–55 (1967)

7.79 V. Krishan, R.H. Ritchie: Anomalous damping of volume plasmons in polycrystalline metals. Phys. Rev. Lett. **24**, 1117–1119 (1970)

7.80 D.B. Tran Thoai, E. Zeitler: Inelastic scattering of fast electrons by thin metal slabs. Phys. Stat. Solidi (a) **120**, 467–474 (1990)

7.81 I. Fromm, L. Reimer, R. Rennekamp: Investigation and use of plasmon losses in energy-filtering transmission electron microscopy. J. Micr. **166**, 257–271 (1992)

7.82 S.L. Cundy, A.J.F. Metherell, M.J. Whelan, P.W.T. Unwin, R.B. Nicholson: Studies of segregation and the initial stages of precipitation at grain boundaries in an Al-7wt%Mg alloy with an energy analysing electron microscope. Proc. Roy. Soc. (London) **307**, 267–275 (1968)

7.83 D.R. Spalding, A.J.F. Metherell: Plasmon losses in Al-Mg alloys. Phil. Mag. **18**, 41–48 (1968)

7.84 C. von Festenberg: Energieverlustmessungen an III-V Verbindungen. Z. Phys. **227**, 453–481 (1969)

7.85 A.J.F. Metherell: Energy analysing and energy selecting electron microscopes. Adv. in Optical and Electron Microscopy, ed. by R. Barer and V.E. Cosslett (Academic, London 1971) Vol. 4, pp.263–361

7.86 B. Bernert, P. Zacharias: Die optischen Konstanten von Silber-Gallium-Legierungen in der Nähe der Plasmafrequenz. Z. Phys. **241**, 205–216 (1972)

7.87 M. Schlüter: Die optischen Eigenschaften von Gold, Silber und Gold-Silber-Legierungen zwischen 2 und 40 eV aus Energieverlustmesungen. Z. Phys. **250**, 87–98 (1972)

7.88 H. Möller, A. Otto: Plasmon dispersion in aluminium-magnesium alloys. Phys. Rev. Lett. **45**, 2140–2143 (1980)

7.89 R. Grundler: Volume plasmon dispersion of polycrystalline films of the ternary semiconductors $ZnSnAs_2$ and $ZnSiAs_2$. Phys. Stat. Sol. (b) **140**, K19–K22 (1987)

7.90 P. Henoc, M. Natta, L. Henry: Pertes caractéristiques associées à la dimension et à la nature de petits précipités dans une matrice cristalline. In Microscopie Electronique 1970, ed. by P. Favard (Soc. Francaise de Micr. Electr., Paris 1970) Vol.II, pp.123–124

7.91 R. Castaing: Energy filtering in electron microscopy and electron diffraction. In Physical Aspects of Electron Microscopy and Microbeam Analysis, ed. by B.J. Siegel and D.R. Beaman (Wiley, New York 1975) pp.287–301

7.92 P. Sainfort, P. Guyot: High-spatial-resolution STEM analysis of transition micro-phases in Al-Li and Al-Li-Cu alloys. Phil. Mag. A **51**, 575–588 (1985)

7.93 D.B. Williams, H.W. Edington: The precipitation of $\delta'(Al_3Li)$ in dilute aluminium-lithium alloys. Met. Sci. **9**, 529–532 (1975)

7.94 W. Probst, R. Bauer: Technik und biologische Anwendung der elektronenspektroskopischen Abbildung (ESI) und Elektronen-Energieverlust-Spektroskopie (EELS). Abbildung (ESI) und Elektronen-Energieverlust-Spektroskopie (EELS). Verh. Dtsch. Zool. Ges. **80**, 119–128 (1987)

7.95 P. Keusch, J.R. Guenter, R. Bauer: Improvement of the epitaxial orientation of thin vapour deposited gold films on alkali halides by double evaporation. Proc. XIth Int'l Congr. on Electron Microscopy, ed. by T. Imura, S. Maruse and T. Suzuki (Jpn. Soc. of Electron Microscopy, Kyoto 1986) Vol.II, pp.1379–1380

7.96 R. Bauer, U. Hezel, D. Kurz: High-resolution imaging of thick biological specimens with an imaging electron energy loss spectrometer. Optik **77**, 171–174 (1987)

7.97 H.J. Wagner: Contrast tuning by electron spectroscopic imaging of half-micrometer-thick sections of nervous tissue. Ultramicroscopy **32**, 42–47 (1990)

7.98 L. Landau: On the energy loss of fast electrons by ionization. J. Phys. USSR **8**, 201 (1944)

7.99 L. Reimer, K. Brockmann, U. Rhein: Energy losses of 20-40 keV electrons in 150-600 μg cm^{-2} films. J. Phys. D **11**, 2151–2155 (1978)

7.100 L. Reimer: Calculation of the angular and and energy distribution of multiple scattered electrons using Fourier transforms. Ultramicroscopy **31**, 169–176 (1989)

7.101 L. Reimer, R. Senkel: Calculation of energy spectra of electrons transmitted through thin aluminium foils. J. Phys. D **25**, 1371–1376 (1992)

7.102 C. Colliex, C. Mory, A.L. Olins, D.E. Olins, M. Tence: Energy-filtered STEM imaging of thick biological sections. J. Micr. **153**, 1–21 (1989)

7.103 L. Reimer. R. Rennekamp, I. Fromm, M. Langenfeld: Contrast in the electron spectroscopic imaging mode of TEM: IV. Thick specimens imaged by the most-probable energy loss. J. Micr. **162**, 3–14 (1991)

7.104 L. Reimer: Electron diffraction methods in TEM, STEM and SEM. Scanning **2**, 3–19 (1979)

7.105 H. Hashimoto: HVEM contrast theory. In *High Voltage Electron Microscopy*, ed. by P.R. Swann et al. (Academic, London 1974) pp.9–21

7.106 A.P. Somlyo: Compositional mapping in biology: x-rays and electrons. J. Ultrastructure Res. **88**, 135–142 (1984)

7.107 K.M. Adamson-Sharpe, F.P. Ottensmeyer: Spatial resolution and detection sensitivity in microanalysis by electron energy loss selected imaging. J. Micr. **122**, 309–314 (1981)

7.108 F.P. Ottensmeyer: Elemental mapping by energy filtration: advantages, limitations, and compromises. Ann. New York Acad. Sci. **483**, 339–351 (1986)

7.109 H. Shuman, C.F. Chang, A.P. Somlyo: Elemental mapping and resolution in energy-filtered conventional electron microscopy. Ultramicroscopy **19**, 121–134 (1986)

7.110 T.J. White, D.R.Cousens, G.J. Auchterlonie: Preliminary characterization of an intrinsic germanium detector on a 400-keV microscope. J. Micr. **162**, 379–390 (1991)

7.111 P. Kruit: Auger spectroscopy in STEM, recent progress. In *Electron Microscopy 1992*, ed. by A. Rios, J.M. Arias, L. Megias-Megias, A. L'opez-Galindo (Secr. Publ. Universidad de Granada, Granada 1992) Vol.I, pp.215–218

7.112 P.T.E. Roberts, J.N. Chapman, A.M. MacLeod: CCD-based image recording system for the CTEM. Ultramicroscopy **8**, 385–398 (1984)

7.113 N.J. Zaluzec: Two-dimensional CCD arrays as parallel detectors in electron energy loss and x-ray wavelength dispersive spectroscopy. Ultramicroscopy **28**, 131–136 (1989)

7.114 J. Mayer: Energy filtered electron microscopy: applications in materials science. In *Electron Microscopy 1992*, ed. by A. Rios, J.M. Arias, L. Megias-Megias, A. López-Galindo (Secr. Publ. Universidad de Granada, Granada 1992) Vol.1, pp.269–270

7.115 L. Reimer, U. Zepke, J. Moesch, St. Schulze-Hillert, M. Ross-Messemer, W. Probst, E. Weimer: *EELSpectroscopy. A Reference Handbook of Standard Data for Identification and Interpretation of Electron Energy Loss Spectra and for Generation of Electron Spectroscopic Images.* (Zeiss, Oberkochen 1992)

7.116 C. Colliex: Electron energy loss spectroscopy analysis and imaging of biological specimens. Ann. New York Acad. Sci. **483**, 311–325 (1986)

7.117 R.H. Barckhaus, H.J. Höhling, I. Fromm, P. Hirsch, L. Reimer: Electron spectroscopic-diffraction and imaging of the early and mature stages of calcium phosphate formation in the epiphyseal growth plate. J. Micr. **162**, 155–169 (1991)

7.118 R.D. Leapman: Scanning transmission electron microscope (STEM) elemental mapping by electron energy-loss spectroscopy. Ann. New York Acad. Sci. **483**, 326-338 (1986)

7.119 H. Shuman, C.F. Chang, E.L. Bahe, A.P. Somlyo: Electron energy-loss spectroscopy: quantitation and imaging. Ann. New York Acad. Sci. **483**, 295-310 (1986)

7.120 H. Lehmann, U. Kunz, A. Jacob: A simplified preparation procedure of plant material for elemental analysis by ESI and EELS techniques. J. Micr. **162**, 77-82 (1991)

7.121 R.R. Schröder: Zero-loss energy-filtered imaging of frozen-hydrated proteins: model calculations and implications for future developments. J. Micr. **166**, 389-400 (1992)

7.122 J.L. Costa, D.C. Joy, D.M. Maher, K.L. Kirk, S.W. Hui: Fluorinated molecule as a tracer: difluoroserotonin in human platelets mapped by electron energy-loss spectroscopy. Science **200**, 537-539 (1978)

7.123 J. Sevely, J.P. Perez, B. Jouffrey: Energy losses of electrons through Al and C films from 300 keV up to 1200 keV. In *High Voltage Electron Microscopy*, ed. by P.R. Swann et al. (Academic, London 1974) pp. 38-47

7.124 L. Reimer: Methods of detection of radiation damage in electron microscopy. Ultramicroscopy **14**, 291-304 (1984)

7.125 A. Howie, F.J. Rocca, U. Valdre: Electron beam ionization damage processes in p-therphenyl. Phil. Mag. B **52**, 751-757 (1982)

7.126 R.F. Egerton: Chemical measurements of radiation damage in organic samples at and below room temperature. Ultramicroscopy **5**, 521-523 (1980)

7.127 R.F. Egerton: Measurement of radiation damage by electron energy-loss spectroscopy. J. Micr. **118**, 389-399 (1982)

7.128 R.F. Egerton: Organic mass loss at 100 K and 300 K. J. Micr. **126**, 95-100 (1982)

7.129 R.D. Leapman, R.L. Ornberg: Quantitative electron energy loss spectroscopy in biology. Ultramicrocopy **24**, 251-268 (1988)

7.130 L. Reimer, J. Spruth: Interpretation of the fading of diffraction patterns from organic substances irradiated with 100 keV electrons at 10-300 K. Ultramicroscopy **10**, 199-210 (1982)

7.131 M.K. Lamvik, D. Kopf, S.D. Davilla: Mass loss rate in collodion is greatly reduced at liquid helium temperature. J. Micr. **148**, 211-217 (1987)

7.132 R.H. Wade: The temperatur dependence of radiation damage in organic and biological samples. Ultramicroscopy **14**, 265-270 (1984)

7.133 G. Siegel: Der Einfluß tiefer Temperaturen auf die Strahlenschädigung von organischen Kristallen durch 100 keV-Elektronen. Z. Naturforschg. A **27**, 325-332 (1972)

7.134 M. Creuzburg: Entstehung von Alkalimetallen bei der Elektronenbestrahlung von Alkalihalogeniden. Z. Phys. **194**, 211-218 (1966)

7.135 P.A. Crozier, J.N. Chapman, A.J. Craven, J.M. Titchmarsh: Some factors affecting the accuracy of EELS in determining elemental concentrations in thin films. In *Analytical Electron Microscopy 1984*, ed. by D.B. Williams and D.C. Joy (San Francisco Press, San Francisco 1984) pp.79-82

7.136 M. Isaacson, D. Johnson: The microanalysis of light elements using transmitted energy-loss electrons. Ultramicroscopy **1**, 33-52 (1975)

7.137 D.C. Joy, D.M. Maher: Electron energy-loss spectroscopy: detectable limits for elemental analysis. Ultramicroscopy **5**, 333-342 (1980)

7.138 P. Rez : Detection limits and error analysis in energy-loss spectrometry. In *Microbeam Analysis 1983*, ed. by R. Gooley (San Francisco Press, San Francisco 1983) pp.153-155

7.139 A.L. Arsenault, F.P. Ottensmeyer: Stereoscopic representation of complex overlapping elemental maps in electron spectroscopic images. J. Micr. **133**, 69–72 (1984)

7.140 A.L. Arsenault, F.P. Ottensmeyer: Quantitative spatial distribution of calcium phosphorus and sulfur in calcifying epiphysis by high resolution spectroscopic imaging. Proc. Nat. Acad. Sci. USA **80**, 1322–1326 (1983)

7.141 A.L. Arsenault, F.P. Ottensmeyer: Visualization of early intramembranous ossification by electron microscopic and electron spectroscopic imaging. J. Cell Biol. **98**, 911–921 (1984)

7.142 D. Blottner, H.J. Wagner: Localization of calcium and phosphorus in early predentin-matrix matrix components by electron spectroscopic imaging (ESI)-analysis in rat molars. Cell and Tissue Res. **255**, 611–617 (1989)

7.143 M. Döpfner, C. Wiencke: Calcium compartmentation in ontarctic broade algae. Ultramicroscopy **32**, 7–11 (1990)

7.144 G. Harauz, F.P. Ottensmeyer: Nucleosome reconstruction via phosphorus mapping. Science **226**, 936–940 (1984)

7.145 U.R. Heinrich, M. Drechsler, W. Kreutz, W. Mann: Identification of precipitable Ca^{2+} by ESI and EELS in the organ of Corti of guinea pig. Ultramicroscopy **32**, 1–6 (1990)

7.146 H. Körtje, D. Körtje, H. Rahmann: The application of energy-filtering electron microscopy for the cytochemical localization of Ca^{2+}–ATPase activity in synaptic terminals. J. Micr. **162**, 105–114 (1991)

7.147 F.P. Ottensmeyer, D.W. Andrews, A.L. Arsenault, Y.M. Heng, G.T. Simon, G.C Weatherley: Elemental imaging by electron energy loss microscopy. Scanning **10**, 227–238 (1988)

7.148 R.D. Leapman, S. B. Andrews: Biological electron energy loss spectroscopy: the present and the future. Microsc. Microanal. Microstr. **2**, 387–394 (1991)

7.149 C. Mory, C. Colliex: Elemental analysis near the single-atom detection level by processing sequences of energy-filtered images. Ultramicroscopy **28**, 339–346 (1989)

7.150 A. Berger, H. Kohl: Optimum imaging parameters for elemental mapping in an energy filtering transmission electron microscope. Optik **92**, 175–193 (1993)

7.151 U. Plate, H.J. Höhling, L. Reimer, R.H. Barckhaus, R. Wienecke, H.P. Wiesmann, A. Boyde: Analysis of the calcium distribution in predentine by EELS and of the early crystal formation in dentine by ESI and ESD. J. Micr. **166**, 329–341 (1992)

8. Energy-Filtered Reflection Electron Microscopy

John C. H. Spence

8.1 Development of REM

The imaging of crystal surfaces at high resolution using electron beams has been a long-standing challenge for physicists. Techniques for this purpose promise wide application, from research into the kinetics and atomic mechanisms of crystal growth to studies of semiconductor interfaces and monolayer magnetism. Ideally, such a method should be capable of identifying atomic species, should provide real-time atomic resolution imaging over a wide range of temperatures, and should allow imaging of individual sub-surface layers. Currently, while no one technique can achieve all these goals, the following techniques come closest : secondary-electron scanning microscopy in ultra-high vacuum (UHV SEM) [8.1], scanning Auger microscopy (SAM) [8.1], scanning tunnelling microscopy (STM) [8.2], sub-surface backscattered channelling imaging [8.3,4], forbidden-reflection TEM lattice imaging in ultra-high vacuum (UHV HREM) [8.5] and the reflection electron microscopy (REM) technique reviewed in this chapter. (The field-ion microscope may also provide an even higher performance than these methods in certain specialised cases [8.2]). These electron beam techniques offer resolutions of about 1 nm for REM and SEM [8.1], 2 nm for SAM [8.2], and about 0.25 nm for STM and UHV HREM. The resolution of the low-energy electron microscopy (LEEM) technique [8.6] may also become competitive in the future if chromatic-aberration corrected designs can be implemented. The REM method provides crystallographic information rather than the topological information provided by SEM, and, unlike STM, allows high-speed recording of surface processes at high temperatures in real time. At high accelerating voltages, the technique suffers from the important disadvantage that a severely foreshortened image is produced, as discussed below. First developed by *Ruska* [8.7] using diffusely scattered electrons, it became powerful when Bragg diffracted beams were used instead to form the image in a conventional electron microscope [8.8]. But the real breakthrough and revival of interest in the method occurred when ultra-high vacuum conditions were obtained in a modified TEM instrument by *Yagi* [8.9,10], allowing remarkable images such as that shown in Fig. 8.1 to be produced from atomically clean surfaces [8.9]. The 2.3 nm fringes seen in the image are retained up to the surface structure transition temperature at 830° C. Using images similar to these, these workers

Springer Series in Optical Sciences, Vol. 71
Energy-Filtering Transmission Electron Microscopy
Editor: Ludwig Reimer   ©Springer-Verlag Berlin Heidelberg 1995

were able to observe directly the Si (111) (7×7) – (1×1) phase transition in real time, and to elucidate the atomic mechanisms involved and the order of the phase transition [8.10]. Equally impressive results have since been produced by the Novosibirsk group [8.11]. During the same period, the methods of energy-filtered scanning REM (SREM) were developed [8.12], from which the dramatic benefits of removing inelastically scattered electrons from the images could be seen. With the appearance of commercial imaging energy filters for TEM instruments, it has now become possible to record these elastically filtered REM images without scanning, as discussed below, thereby allowing much larger amounts of data to be recorded efficiently. The possibility of microanalysis using energy-loss spectroscopy in REM was demonstrated in 1988 [8.13]. In this chapter, however, we are concerned mainly with elastic energy filtering rather than surface channelling or microanalysis in the reflection geometry. In this brief review of REM, we summarize some high-

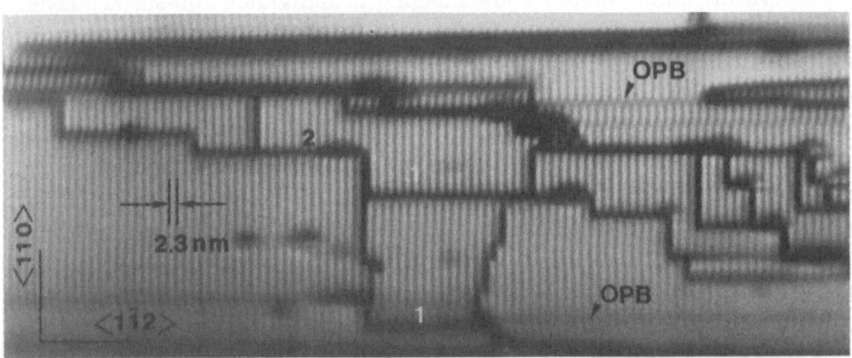

Fig. 8.1. REM lattice image of a Si(111) 7×7 reconstructed surface. Dark lines are steps. The fringe spacing is 2.3 nm. Out-of-phase boundaries (OPB) are also shown [8.8].

lights from the REM literature, outline the main concepts used to interpret REM images and, finally, suggest future developments, which might benefit from the more extensive use of energy filtering. Review articles on REM can be found elsewhere [8.14,15] and in *Ultramicroscopy* Vol.48 (1993). The field can be divided into work performed under surface science conditions using ultra-high vacuum machines and that undertaken in conventional TEM instruments. The later type of work has proven useful for the characterization of surface roughness and faceting on crystals, and for studies of the effect of heat-treatment and oxidation on oxide ceramics [8.16,17]. Thus, for example, oxidation at high temperature has been found by this method to produce large atomically flat regions on sapphire. The surfaces of diamond may also be characterised in this way, with implications for research into abrasion. The imaging of ferroelectric domains by REM has recently been demonstrated

[8.18], and imaging of magnetic domains by REM may also be possible, as described below. REM imaging under UHV conditions is concerned with the atomic mechanisms of crystal nucleation and growth, the kinetics of surface reactions, and the structure of interfaces. Recent applications of this method using an ultra-high vacuum transmission electron microscope operated in the reflection mode can be found, for example, in [8.19]. Here the movement of In and Cu atoms on Si(111) surfaces was studied under an applied electric field (electromigration). The atomic motion was found to follow the direction of the current. In addition, by observing the process in real time, the detailed mechanism of electromigration can be studied and its relationship to substrate features determined. In summary, the power of the REM method (by comparison with STM, for example) is that it allows direct observation of dynamic processes (using video recordings) on surfaces at high temperature. Its most important disadvantage is the presence of geometric distortion, or foreshortening, in the images, and the lack of chemical information. (This deficiency can be remedied to some extent in the scanning reflection (SREM) mode, which also offers energy filtering).

8.2 The Geometry of REM Imaging

Figure 8.2 shows a ray diagram for REM in a conventional TEM for the simple case where the specular beam is imaged. This beam makes an angle α with the surface. This angle may or may not be equal to the Bragg angle θ shown. The angular change β due to refraction by the mean inner potential is also shown. All angles are shown exaggerated for clarity - at 100 kV, $\alpha \simeq$ 20 mrad. All REM images are "dark field" images - the direct beam (the continuation of ray OA) is depleted by diffraction and inelastic scattering inside the sample. Beams other than the specular beam may also be used for imaging by selecting them with the objective aperture. A specular-Bragg diffraction condition gives high intensity and is therefore frequently used for REM, as discussed below. Since the distance from the detector screen to the lens is fixed, only one line at B (normal to the page) is in focus on the sample. This line can be identified by its characteristic appearance in the image. Object point A is in focus at A′ and out of focus on the screen at A″. The image is inverted left to right and along the optic axis z. The incident beam direction has been adjusted so that the reflected beam falls on the optic axis - this is an important requirement for high resolution REM imaging. A RHEED pattern is seen in the back-focal plane.

The most striking feature of REM images is their foreshortening. The geometrical distortion of the images is such that the magnification is different in two orthogonal directions. The line through B normal to the page is imaged at the normal magnification indicated on the microscope. The magnification for distances along the surface normal to this direction is reduced by a factor $(\sin\alpha)^{-1} \simeq 40$ for typical conditions at 100 kV. Here α is the angle made

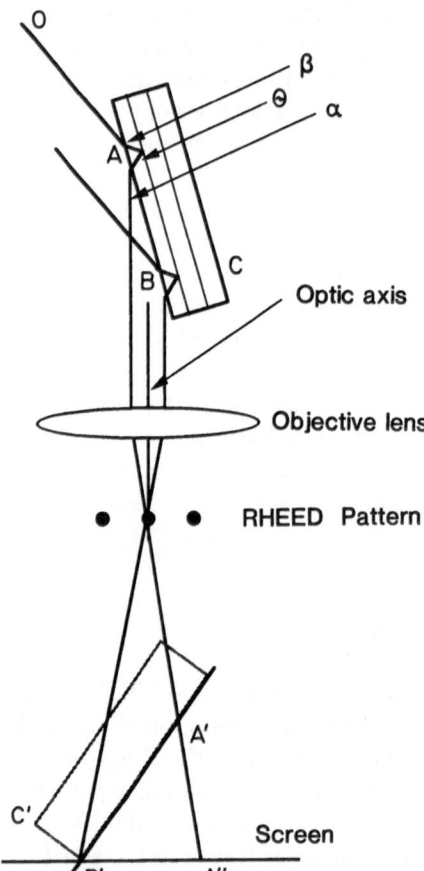

Fig. 8.2. The geometry of REM imaging. Note position of optic axis, Bragg angle θ, specular reflection angle α and refraction angle β. Conjugate points are labelled A,B,C and A',B',C'. Only B is in focus [8.13].

by the beam used for imaging with the crystal surface. The use of higher order reflections for imaging thus reduces foreshortening, but also the image intensity. The Ewald sphere construction, imposing energy and momentum conservation in RHEED, must be used to interpret the diffraction conditions in REM, and this involves reciprocal space rods rather than points. (For a review of RHEED theory, see for example [8.20,21]). Because of the foreshortening, a circle inscribed on the surface is imaged in REM as an ellipse. A line (e.g. a surface step) on the object which actually makes an angle ϕ with the normal to the in-focus line will appear in the image at an angle ψ to the in-focus line, where

$$\tan\psi = \tan\phi/\sin\theta. \qquad (8.1)$$

An analysis of the geometric distortions in REM images can be found in [8.15] and references therein.

8.3 Samples, Techniques and Instrumentation

A certain amount has been learnt about crystal surfaces and processes by using the REM technique to image samples prepared in air in conventional electron microscopes. The effect of the inevitable hydrocarbon adsorbate layer on the REM image will be small, since REM contrast depends on Bragg diffraction, local crystallography and lattice strain. But the delicate reconstructions characteristic of the equilibrium structure of clean surfaces cannot be expected to be seen. Work on samples prepared in air has thus been restricted to inert materials such as the noble metals, refractory oxides, semiconductors and the sulphides. Since crystallographic alignment is very difficult with small grain sizes, there is an additional requirement for single crystals. Thus the ideal REM sample for ex-situ preparation is a chemically inert, electrically conductive single crystal which cleaves readily and does not form a thick surface oxide layer. Examples include PbS, the noble metals, graphite and GaAs. Good results have also been obtained from MgO, sapphire and heavily doped III-IV semiconductors. For students learning the technique, dramatic images can easily be obtained of the surface facets of noble metal wires heated in air. These show growth steps and depressions due to vacancy nucleation [8.22]. Similarly, the effects of oxidation on the surfaces of sapphire and MgO have been studied extensively [8.16,17]. Work on graphite and PbS is described in [8.23].

For researchers attempting REM imaging for the first time, perhaps the simplest sample to start with is heavily doped GaAs. If freshly cleaved in air immediately before transferring the sample into the microscope, this material usually gives good REM images (unlike silicon). The sample may be mounted on a normal (or slightly modified) TEM grid, and no special facilities are required. In conventional TEM instruments, a cold-stage will be found invaluable for the reduction of contamination, which otherwise reduces the contrast of REM images rapidly. An [001] semiconductor wafer must be cleaved to dimensions of about 0.3 mm thick × 2.5mm × 1 mm , then placed on one side of a TEM grid, and the thin edge (whose normal is [110]) viewed in reflection. The sample height must be adjusted so that the midplane of the sample falls at the eucentric lens current setting – subsequent coarse focusing should be done with the sample height controls rather than the lens current. A sample that is too thick (along the beam direction) may show excessive surface roughness, so that the glancing angle beam always strikes protrusions and no RHEED pattern is found. If the sample is too thin, the REM image will appear as a thin line.

In practice a smooth edge region of the sample is sought, free of asperities, in the image mode, with the [110] surface inclined by about one Bragg angle from the plane containing the beam direction, as in Fig. 8.2. It is common to work at magnifications between 10000 and 40000. The RHEED pattern is then sought in the diffraction mode using the dark field beam tilts to tilt the beam into the surface. The specular and RHEED spots may be distinguished from transmission spots due to surface protrusions by the fact that the later do not move when the sample is tilted. To bring the specular spot onto the optic axis, it is then necessary to identify the sense of sample rotation – i.e. to distinguish the top and bottom shadow edges. The spot can be moved onto the optic axis using the sample tilts, since the surface acts as a mirror. The objective aperture must then be introduced to select one beam. Exposure times are those typical of dark-field microscopy – perhaps 5–30 s. It will be found that moving the sample into the beam moves the electron probe up against the surface, because of the sample tilt. More extensive practical details for setting up REM imaging can be found in [8.24], which also contains details of gold sphere sample preparation.

In order to study reproducible processes on atomically clean surfaces of interest to the surface science community, a vacuum of better than 10^{-9} mbar is required, so that the time for monolayer coverage of the surface becomes appreciable. The design and construction of suitable instruments for UHV REM work is a major undertaking, which has been completed in very few laboratories around the world. The design must take account of the fact that less than a monolayer of foreign atoms on a surface may inhibit a desired surface reconstruction. Differential pumping will be needed to isolate the camera chamber if film is used, and the entire sample chamber and specimen exchange mechanism must be rebuilt using UHV compatible materials. A double-tilt heating stage with near-atomic thermal and vibrational stability is required, capable of heating the sample to perhaps 1000°C [8.25]. If atomic resolution is desired, the space available in the polepiece gap will be reduced to a few millimetres by the requirement for small aberration coefficients. The double-tilt heating holder, objective aperture and and cold-finger must all fit within this small volume. If only the sample region of a TEM is made bakeable, the outgassing of molecules from the unbaked liner tube in the gun and projector lenses may then limit the ultimate vacuum. The number of O-rings must be minimized. (Baking temperatures are limited to about 120°C if Viton O-rings are present under compression, resulting in very long bake times.) The design of double tilt holders which do not use O-rings, however, requires complex and expensive mechanisms. In-situ sample preparation facilities will also be needed in an adjoining chamber, and the design is greatly complicated if facilities are also required for evaporation onto the sample while it is in the polepiece gap. Some impression of the complexity of these projects can be found in [8.25–28]. An important question for these instruments concerns the need to use an immersion lens for REM work. If such a lens is not used and the resulting resolution penalty of about 1 nm accepted, UHV design becomes greatly simplified, since the dimensions around the sample can then be increased [8.29]. Evaporation of material from the sample onto the surrounding cold-finger etc. may cause difficulty in these machines, in addition to problems caused by cross-contamination between different materials. The cold-finger surrounding the sample should act as a cryo-pump, but, unless it is carefully designed, it will also be heated by radiation from the sample, and may accumulate material during sample evaporation. Although more ambitious designs have been successful after many man-years of effort and refinement, the accumulation of experience to date suggests that a successful UHV machine for REM work should be designed for imaging a rather limited range of materials (e.g. silicon, noble metals, refractory metal oxides) at moderate resolution (e.g. 1 nm) using a non-immersion lens. Extensive in-situ sample preparation facilities should be provided in an adjoining chamber (e.g. evaporation, Auger detector, film thickness monitor, LEED, sample heating etc.). Fast video-rate recording is essential to combat the effects of thermal drift at the heated sample.

An alternative to rebuilding a conventional TEM for UHV work is the purchase of a commercial STEM, which may be supplied as a fully bakeable surface science instrument . This allows reflection imaging using scanning (SREM), as described below, and has the advantages of also providing excellent microanalytical and reflection microdiffraction facilities, a field-emission gun and serial energy filtering.

The severe foreshortening which occurs in REM at high voltages can be avoided by working at lower voltages. Several specialised ultra-high vacuum SREM instruments have been built for this purpose, operating at voltages of 5–30 kV [8.30–32]. The question of the optimum accelerating voltage for REM has been discussed in detail in the literature. This involves, for example, trade-offs between source brightness, penetration, foreshortening, inelastic scattering, polepiece dimensions and aberration coefficients, amongst other factors. The larger working space available at lower voltages, for example, allows the introduction of additional detectors and specimen treatment devices, but may result in poor resolution if the probe becomes too large. Experience has also shown that it is difficult for researchers working in a University environment to obtain the degree of vibrational isolation needed for sub-nanometre resolution together with adequate stray magnetic field shielding in a bakeable UHV machine. A field-emission gun is essential for SREM imaging and RHEED microdiffraction.

8.4 Image Interpretation for REM

Figure 8.3 shows a typical low resolution REM image. The gross features can be interpreted geometrically, as they can in SEM images, giving topological contrast. Small objects on the surface, however, appear with a twin (inverted) image, for the same reason that real and virtual images are seen of any object placed on a mirror. The obstruction of rays reflected from the surface by the object forms one shadow image, while the other is formed by incident rays which are prevented from reaching the reflecting surface by the object. Between the two images, an artifactual gap is seen, resulting from the Goos–Hanchen effect [8.33]. (In RHEED theory, this gap depends on the rate of change of dynamical scattering phase with diffraction angle). For such a small particle on a crystalline substrate, it has been suggested that the imaging conditions are similar to TEM, with the wave reflected from the surface providing the illumination [8.15]. Resolution is then limited by chromatic aberration due to the spread of energies in the reflected wave. Depth of focus is limited by the size of the illumination aperture.

The wave field excited within a sample used for REM work can be computed by standard RHEED methods if the surface is atomically flat. These computations show that the wave field penetrates only a few atomic layers. Thus stacking faults close to the surface and parallel to it have been seen in REM images of graphite [8.23]. The penetration depth may be measured from observations of dislocation lines emerging normal to the surface. At the surface resonance condition [8.20,34] at which a Kikuchi line crosses the specular reflection, a thin sheet of elastic intensity is excited at the crystal surface [8.15]. (The effects of defects such as surface steps interrupting the propaga-

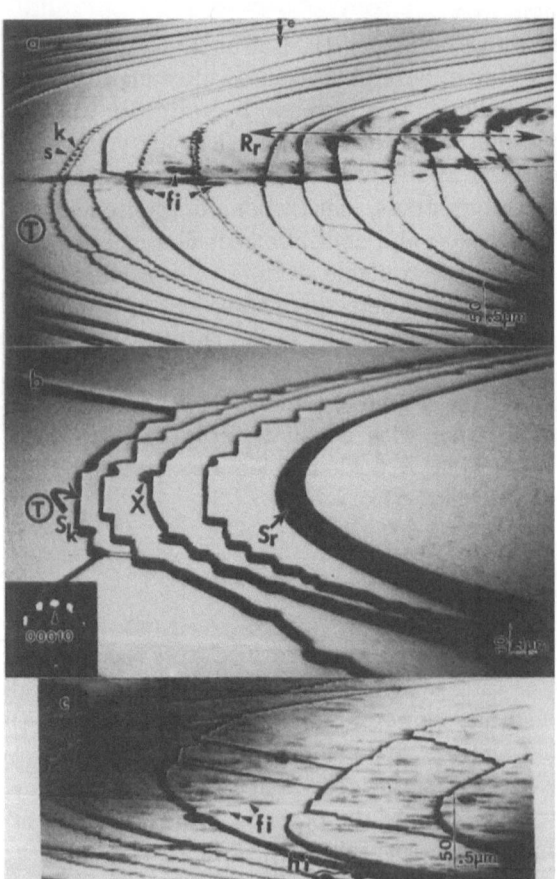

Fig. 8.3. The basal plane of bulk sapphire imaged by reflection electron microscopy after oxygen annealing (from [8.61]). Silver has been deposited on the surface in increasing amounts reading down the page. It is seen to grow out from the steps. A Philips EM400T TEM instrument was used.

tion of this sheet of intensity is described below). Both domain boundaries and surface steps are clearly seen in REM images of atomically clean surfaces. The image contrast at a step may be due to both phase contrast and strain contrast effects, and has been analysed in detail in the literature [8.35–37]. A phase-contrast mechanism, involving interference between waves reflected from the top and bottom of steps, was proposed by Cowley [8.38], and this produces the observed Fresnel-like fringe. The height of a step which terminates at an emerging dislocation of Burgers vector b can be taken to be $b \cdot n$, where n is the surface normal. Thus terminating (monatomic) steps are commonly seen in REM images of semiconductors prepared under UHV conditions or of noble metals heated in air. The sense of a step (up or down) can be determined from the bright or dark nature of the Fresnel fringes or from real-time recordings of step motion during sublimation. The important problem of determining the strain field at surface steps from REM images of clean surfaces has also been studied in detail [8.14,39]. The question of how much of the scattering phase-shift derives from strain may perhaps be resolved by reflection electron holography methods [8.40,41]. These holograms have shown that the phase change across certain steps on (111) Au and Pt is about π [8.14,40,41] but full dynamical computations are required for an accurate estimate.

Double images of steps are also predicted under the commonly used surface resonance condition [8.37]. Dynamical computations suggest that the surface resonance wave is largely unaffected when it encounters an up step, but emerges from the crystal to form an extra spot at a down step. These computational results have been tested using the scanning reflection (SREM) method. This allows reflection microdiffraction patterns to be obtained at individual steps together with their corresponding SREM images [8.42]. When the probe was situated at a single surface step, an extra spot was clearly seen. Images formed with this spot showed only the down steps, while those formed with the specular beam showed both types of step. This confirms that up steps do not produce extra spots in RHEED patterns, in accordance with dynamical computations. The interpretation of SREM images follows similar lines to that of REM imaging under reciprocal aperturing conditions, according to the theorem of reciprocity. Figure 8.4 shows a typical SREM image, obtained from two MgO cubes, using the Vacuum Generators HB5 STEM instrument operating at 100 kV [8.12].

Methods for computing dynamical RHEED patterns and REM images from defective surfaces have been developed recently. Most of the computational methods developed for RHEED (e.g. [8.43]) deal with atomically flat surfaces (or those for which only strains normal to the surface are permitted) and are therefore not useful for REM imaging of defects. Since an atomically flat surface establishes a steady state between incident and diffracted wave fields, these RHEED computations necessarily show no REM contrast. The earliest REM defect calculations were based on a column approxima-

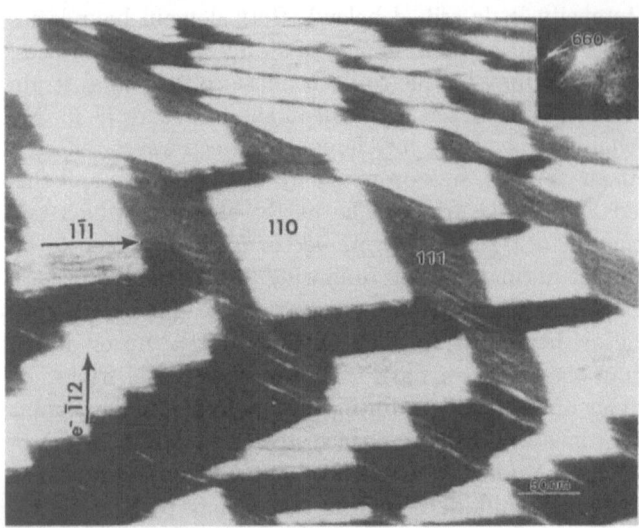

Fig. 8.4. Elastically filtered scanning reflection image of the terraced (110) vicinal surface of copper, showing (111) facets. The (660) specular beam was used at the surface resonance condition. This image was obtained on the Vacuum Generators HB5 STEM instrument [8.62]. By stopping the probe, microdiffraction patterns, X-ray spectra and energy loss spectra may be obtained.

tion of doubtful validity, with columns taken normal to the surface. Then, for a strained surface, the intensity diffracted from each local region can be computed using a calculation for a perfect crystal whose orientation is given by the local tilt [8.35,44,45]. Unfortunately, more accurate calculations show that any surface defect perturbs the RHEED wave field for a distance of many nanometres further "downstream", so that column widths of perhaps 50 nm may be required at 100 kV. These are then foreshortened in the image. More sophisticated computations for imperfect surfaces use either columns running parallel to the beam [8.46], or a variant of the multislice method, with slices taken normal to the direction of the electron beam, rather than parallel to the surface [8.15]. Then, for these slices, backscattering can be neglected and a small-angle approximation made. An artificial superlattice is constructed, containing a slab of crystal and adjacent vacuum [8.47]. For the region of perfect crystal preceding a defect, a computer library of steady-state solutions could conceivably be built up for given conditions of voltage, orientation and surface structure. Dynamical calculations for REM imaging at steps using the multislice method are reported in [8.42], and a comparison of Bloch-wave, real-space and Fourier methods has also been given [8.48]. Figure 8.5 shows the result of such a many-beam computation for the wavefield excited at the (110) surface of GaAs, covered by a monolayer of iron [8.49]. A 120 kV electron beam is travelling approximately along [001], normal to the page. The wavefield is seen to run close to the surface in Fig. 8.5a, where the (440) specular diffraction condition is satisfied for glancing incidence. In Fig. 8.5b, the incident angle has been increased to satisfy the (660) reflection, at the (001) azimuth. The z-coordinate is along the beam path, in the plane of the surface. Then these calculations give the total wavefield $\Psi(x, y, z)$ on planes normal to the beam at the sample, and so are useful for studying channelling effects on x-ray, Auger and energy-loss spectra (for a review, see [8.50]). To obtain the in-focus REM image as formed by a particular beam on the detector screen, it is necessary to Fourier transform $\Psi(y, z)$ near the in-focus z coordinate z_o of interest (e.g. just beyond a step). The beam used for imaging is then isolated, surrounded by diffuse elastic scattering from the step. The inverse transform of this distribution then gives the REM image over a range of distances running from z_o back to the step. Scaling for the foreshortening must also be included.

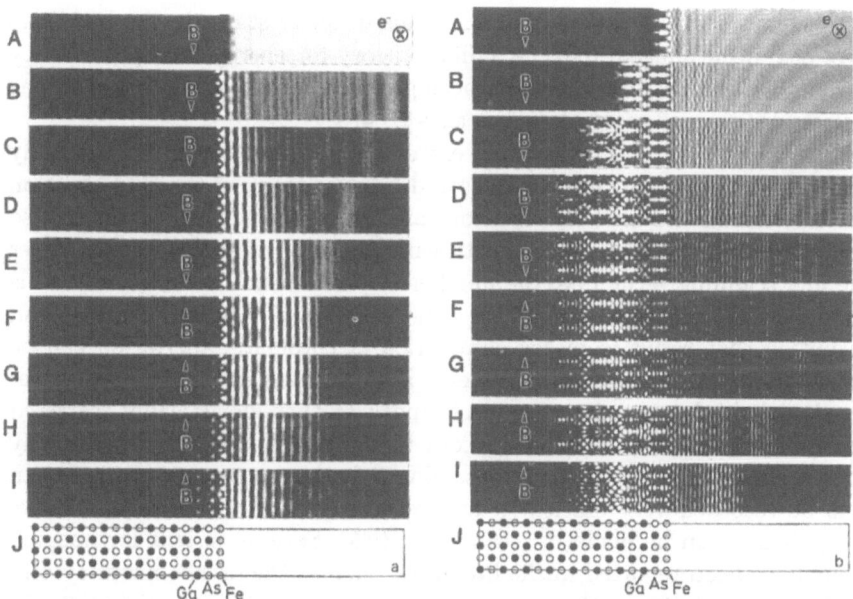

Fig. 8.5. (a) The two-dimensional real-space electron intensity distribution for 120 kV electrons incident on a monolayer of Fe on a GaAs (110) surface. Beam direction approximately normal to the page, satisfying the (440) specular condition. Each panel A–I shows the wave field at increments of 27.15 nm along the beam path, with the first panel A at 27.15 nm and the last (I) at 244.35 nm. The direction of the 2 T magnetic field inside the iron monolayer is indicated by the arrow. (b) Similar to Fig. 8.5a, but for the (660) specular reflection ([001] azimuth). Penetration is much greater. The structure is indicated below, with the position of the surface also shown. The dynamical calculations use hundreds of interacting beams. These figures indicate the depth of penetration of the electron wave field in REM.

8.5 Energy-Filtered REM

The resolution of an REM image is limited to about 1 nm in directions transverse to the beam by the effects of chromatic aberration and to perhaps 40 nm in the direction along the beam path by the foreshortening effect. Since a serial electron energy-loss spectrometer tuned to the elastic peak is normally used as the detector on STEM instruments (and because many STEM instruments are not fitted with any lenses following the sample), STEM scanning reflection images (SREM) are unaffected by chromatic aberration. The serial spectrometer often used on STEM instruments also has the advantage of very large dynamic range, since a photomultiplier detector is used, and may also be used for microanalysis and microdiffraction in the reflection geometry. The ability to stop the probe during SREM image formation on a region of interest and so collect the microdiffraction RHEED pattern is also extremely valuable. Inner-shell excitations have been observed in the reflection geometry at 100 kV [8.51] and localised defect states in the band-gap have also been imaged [8.52]. A unified theory for energy-loss spectroscopy in the reflection geometry has also been given by *Howie* [8.53] , based on the dielectric theory of energy loss (see also [8.54]).

The SREM method is therefore extremely powerful because it is readily combined with analytical techniques and because many of the field-emission STEM instruments used for this work are designed for UHV conditions, so that samples may be prepared in-situ with atomically flat surfaces. However like any scanning technique, the number of pixels which can be collected is limited by the recording time. The same benefits of improved contrast and resolution that result from the elimination of chromatic aberration can be obtained in REM on TEM instruments if an imaging energy-loss spectrometer is fitted, also tuned to the elastic peak so that inelasticallly scattered electrons are excluded. Then all pixels in the image field are differentially filtered simultaneously. Images containing a far larger number of pixels can thus be recorded very efficiently, and recordings that would require hours using SREM can be made in seconds by REM. Several designs for imaging filters have been reported in the literature, such as the prism–mirror system of *Castaing* and *Henry*, the Omega filter of *Rose* and *Plies*, or the alpha filter of *Pérez* (see [8.55] and Chap. 2 for a review and references). These may either be fitted integrally within the microscope column or added following the viewing screen. This second design allows image filtering capabilities to be added to existing TEM instruments [8.56]. We report here REM images obtained on the new Zeiss 912 Omega TEM/STEM instrument, fitted with an Omega filter [8.57] in the electron optical column. The machine is a fully digital side-entry 120 kV instrument, equipped with a turbopump and ion pump and fitted with an LaB_6 or W thermionic cathode. The energy-loss spectrum may also be displayed on the detector screen, so that if a CCD camera is fitted, the spectrum may be collected by parallel detection also. Although not an ultra-high vacuum instrument, the machine can provide much new information on the effects of energy losses on RHEED patterns and REM images within the limits set by contamination. This information, particularly for RHEED patterns, has implications for the interpretation of the RHEED patterns used to monitor crystal growth by molecular beam epitaxy [8.21]. In particular, it may be helpful for determining how much of the diffuse scattering in RHEED patterns arises from surface defects, and how much from inelastic scattering of various types.

Fig. 8.6. (a) Right: Unfiltered reflection image of InP cleaved on (110) with electron beam near [001] recorded at 100 kV. The maximum width of the object is 510 nm [8.63]. (b) Left: identical with Fig. 8.6a except for the introduction of an elastic imaging energy filter. The greatly improved sharpness of the image is obvious.

Figure 8.6 shows elastically filtered and unfiltered (all-electron) REM images obtained from a sample of cleaved InP at 100 kV from the (110) surface, with the beam near [001]. The specular beam has been aligned with the optic axis. Surface steps and an opaque object and its mirror image are seen. To obtain the filtered image, an 8 eV energy window was placed around the elastic peak in the energy-loss spectrum. Since the exposure time needed to be increased from 2 s to 10 s for the filtered micrograph to produce approximately the same optical density, we conclude that about a fifth of the scattering within the small objective aperture used is elastic. The elastic image is seen to be much sharper but the longer exposures place greater demands on the stability of the stage and on the reduction of contamination. The recording of RHEED patterns is much less demanding, since drift is unimportant and, since it is not necessary to align the beam with the optic axis, less contamination occurs. Figure 8.7 compares filtered and unfiltered RHEED patterns from the same sample. The filtering is seen to result in a large reduction in background between the Bragg spots, which is difficult to reproduce accurately in photographic prints.

The dominant inelastic processes affecting these patterns are phonon scattering, valence electron excitation and bulk and surface plasmon excita-

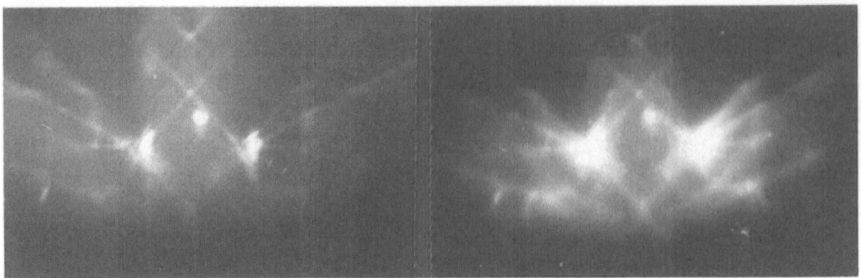

Fig. 8.7. (a) Left: Unfiltered RHEED pattern from the (110) surface of cleaved InP. Beam near [001], $E_0 = 100$ kV. Kikuchi parabolas and specular reflected beam can seen. (b) Identical experimental conditions except for the introduction of an 8 eV energy filtering window for elastic scattering. Mixed, multiple loss scattering has been removed – phonon scattering remains. The background is reduced.

tion. Combinations of these also occur. A calculation of the complete angle-dependent multiple-loss scattering pattern would be extremely complicated and has never been attempted. The filter does not remove the direct phonon losses but does remove all of the single and multiple surface and bulk plasmons (other contributions are small). The phonon-loss electrons may produce extra spots in RHEED patterns [8.58]. The effect of inelastic processes occurring outside the crystal [8.52] is negligible for REM. Provided that the dielectric response function is known (or is measureable from Kramers–Kronig analysis of transmission spectra), the angle-integrated reflection loss spectrum may be calculated [8.59]. The glancing angles used in high voltage REM imaging result in long optical paths for electrons skimming just below the surface [8.13] and hence a large probability for multiple inelastic scattering. This path length, which plays a role analogous to sample thickness in TEM, is approximately $L = 2d/\alpha$, where d is the penetration depth of the wave field shown in Fig. 8.5. We may thus understand the reduction in background in the filtered RHEED pattern as follows: most of the scattering between the Bragg spots is due to phonon scattering. But associated plasmon losses resulting from multiple scattering then take these loss electrons outside the filtering window, resulting in reduced background. A series of energy-filtered RHEED patterns taken with different energy loss peaks would further help to isolate the processes responsible for the features seen in the patterns. Thus it should be possible to determine in more detail how much of the Kikuchi lines, resonance parabolas, and streaks seen in RHEED are due to bulk and surface plasmons and how much to phonon and elastic diffuse scattering from surface steps and other defects. In this way it has been possible to test theoretical predictions [8.46,60]. One and two-dimensional channelling effects in RHEED patterns have recently been analysed in this way [8.61].

8.6 Future Trends

The use of REM imaging in ultra-high vacuum systems seems likely to be further developed in a few specialised research laboratories because, together with the UHV SEM technique, these methods provide the simplest method of obtaining a real-time image of a surface at high temperature, showing crystallographic and topological contrast. This information is vital to our understanding of the mechanisms involved in crystal growth and sublimation. The field has been well reviewed recently by Yagi [8.14]. For REM work in conventional TEM instruments, valuable contributions will continue to be made to our understanding of abrasion and of heat treatments to material surfaces, particularly for oxides and other inert systems. Special situations continue to be discovered for which the ability to study bulk samples in TEM is invaluable – for example, it has recently been shown that ferroelectric domains can be imaged by REM using the associated strains [8.18]. Similarly, magnetic domains may be visible – in this case using the Lorentz force to "detune" the resonance condition [8.49]. Semiconductor multilayers have been shown to give good contrast in REM, providing a quick characterisation method and allowing layer thicknesses to be measured from rapidly cleaved bulk material [8.62]. It has been suggested that the process of LPE growth for III-V semiconductors could be studied by REM, and the question of the step structure of dissociated dislocations emerging at a surface resolved in this way. The role of emerging dislocations which do not produce steps ($\mathbf{g} \cdot \mathbf{n} = 0$) in growth might also be studied. In all cases, images of higher contrast and improved resolution will be obtainable using an imaging filter tuned to the elastic energy loss peak [8.63,64]. For our understanding of the basic physics of RHEED, a series of RHEED patterns taken at different energy losses in the 0–30 eV range would resolve important questions regarding the electronic excitations responsible for the various features seen between Bragg reflections in RHEED [8.65]. Matching these with computations remains a considerable challenge for our theoreticians[8.61].

References

8.1 G.G. Hembree, J.S. Drucker, F.C. Luo, M. Krishnamurthy, J.A.V. Venables: Auger electron spectroscopy and microscopy with probe size limited resolution. Appl. Phys. Lett. **58**, 1–8 (1991)

8.2 D.P. Woodruff, T.A. Delchar: *Modern Techniques of Surface Science* (Cambridge University Press, Cambridge 1986)

8.3 P. Morin, M. Pitaval, D. Besnard, G. Fontaine: Backscattered channelling imaging. Phil. Mag. **40**, 511–520 (1979)

8.4 J.T. Czernuszka, N.J. Long, P.B. Hirsch: Electron channelling contrast imaging of dislocations. In *Proc. XIIth Int'l Congr. on Electron Microscopy*, ed. by R.M. Fisher (San Francisco Press, San Francisco 1990) Vol.2, pp.600–601

8.5 Y. Haga, K. Takayanagi: Single atom imaging in high resolution UHV electron microscopy: Bismuth on Si(111) surface. Ultramicroscopy **45**, 95–102 (1992)

8.6 E. Bauer, M. Maunschau, W. Swiech, W. Telieps: Surface studies by LEEM and PEEM. Ultramicroscopy **31**, 49–55 (1989)

8.7 E. Ruska: Die elektronenmikroskopische Abbildung elektronenbestrahlter Oberflächen. Z. Physik **83**, 492–503 (1933)

8.8 J. S. Halliday, R.C. Newman: Reflexion electron microscopy using diffracted electrons. Brit. J. Appl. Phys. **11**, 158–165 (1960)

8.9 K. Takayanagi, Y. Tanishiro, K. Kobayashi, K. Akiyama, K. Yagi: Surface structures observed by high resolution UHV electron microscopy at atomic level. Jpn. J. Appl. Phys. **26**, L957–L960 (1987)

8.10 N. Osakabe, Y. Tanishiro, K. Yagi, G. Honjo: Observation of Si(111) 7×7 – 1×1 transition by REM. Surf. Sci. **97**, 393–404 (1980)

8.11 A.V. Latyshev, A.L. Aseev, A.B. Krasilnikov: Reflection electron microscopy study of structural transformations on silicon. Surf. Sci. **227**, 24–34 (1990)

8.12 J.M. Cowley: Surface energies and surface structure of small crystals studied by use of a STEM instrument. Ultramicroscopy **114**, 587–606 (1982)

8.13 Z.L. Wang, J.M. Cowley: Reflection electron energy loss spectroscopy. Surf. Sci. **193**, 501–512 (1988)

8.14 K. Yagi: Reflection electron microscopy. Surface Science Reports **17**, 305–362 (1993)

8.15 J.M. Cowley: Reflection electron microscopy. In *Surface and Interface Characterisation by Electron Optical Methods*, ed by A. Howie and U. Valdrè (Plenum, New York 1988) pp.127–149

8.16 P.A. Crozier, M. Gajdardziska-Josifovska, J.M. Cowley: Observation of reconstruction on (111) MgO. In *Proc. XIIth Int'l Congr. on Electron Microscopy*, ed. by R.M. Fisher (San Francisco Press, San Francisco 1990) Vol.2, pp.280–281

8.17 N. Yao, Z.L. Wang, J.M. Cowley: REM and REELS identification of atomic terminations at β-alumina surfaces. Ultramicroscopy **28**, 533–549 (1989)

8.18 F. Tsai, J.M. Cowley: Observations of ferrolelectric domain boundaries in $BaTiO_3$ by REM. Ultramicroscopy **45**, 43–54 (1992)

8.19 A. Yamanaka, K. Yagi: Surface electromigration of In and Cu on Si(111) surfaces studied by REM. Surf. Sci. **242**, 181–190 (1991)

8.20 A. Ichimiya, K. Kambe, G. Lehmpfuhl: Observation of the surface state resonance by the CBED RHEED technique. J. Phys. Soc. Jpn. **49**, 684–691 (1980)

8.21 P.K. Larsen, P.J. Dobson: RHEED and Reflection Imaging of Surfaces. NATO ASI Series (Plenum, New York 1988) Vol. B188

8.22 T. Hsu, J.M. Cowley: Reflection electron microscopy of f.c.c. metals. Ultramicroscopy **11**, 239–250 (1983)

8.23 J. Spence, W. Lo, M. Kuwabara: Observation of the graphite surface by reflection electron microscopy during STM operation. Ultramicroscopy **33**, 69–82 (1990)

8.24 T. Hsu: A laboratory guide for reflection electron microscopy. Norelco Reporter **31**, 1–15 (1984)

8.25 P.R. Swann, J.S. Jones, O. Krivanek, D.J. Smith, J.A. Venables, J.M. Cowley: UHV conversion of a 300 kV HREM. *Proc. 45th Ann. Meeting of EMSA* (San Francisco Press, San Francisco 1987) pp.136–137.

8.26 Y. Kondo et al.: Design and development of UHV HRTEM. Ultramicroscopy **35**, 111–118 (1991)

8.27 K. Takayanagi, K. Yagi, K. Kobayashi, G. Honjo: Technique for routine UHV in-situ electron microscopy of growth process of epitaxial thin films. J. Phys. E **11**, 441–455 (1978)

8.28 K. Heinemann, H. Poppa: An ultrahigh vacuum multipurpose specimen chamber with sample introduction system for in-situ transmission electron microscopy investigations. J. Vac. Sci. Techn. A **4**, 127–138 (1986)

8.29 M.L. McDonald, J.M. Gibson, F.C. Unterwald: The design of an ultra-high vacuum specimen environment for a high resolution TEM. J. Sci. Instr. **86**, 13–21 (1988)

8.30 C. Elibol, H.J. Ou, G.G. Hembree, J.M. Cowley: Improved instrument for medium energy electron diffraction and microscopy of surfaces. Rev. Sci. Instr. **56**, 1215–1228 (1985)

8.31 J.A.V. Venables, A.P. Janssen, P. Akhter, J. Derrien, C.J. Harland: Surface studies in a UHV field-emission gun scanning electron microscope. J. Micr. **118**, 351–367 (1980)

8.32 M. Ichikawa, K. Hayakawa: Micro-probe RHEED. Jpn. J. Appl. Phys. **21**, 145–159 (1982)

8.33 K. Kambe: Anomalous reflected images of objects on crystal surfaces in REM. Ultramicroscopy **25**, 259–266 (1988)

8.34 S. Miyake, K. Hayakawa: Resonance effects in low and high energy electron diffraction by crystals. Acta Cryst A **26**, 60–77 (1970)

8.35 L.M. Peng, J.M. Cowley, T. Hsu: Diffraction contrast in REM - steps. Micron and Micr. Acta. **18**, 179–186 (1987)

8.36 T. Hsu, L-M. Peng: Experimental studies of steps in REM. Ultramicroscopy **22**, 217–224 (1987)

8.37 Y. Uchida, G. Lehmpfuhl: Observation of double contours of monatomic steps in REM. Ultramicroscopy **23**, 53–60 (1987)

8.38 J.M. Cowley, L. Peng: The image contrast of surface steps in REM. Ultramicroscopy **59**, 59–68 (1985)

8.39 N. Osakabe, Y. Tanishiro, K. Yagi, G. Honjo: Image contrast of dislocations and atomic steps on (111) silicon surface in reflection electron microscopy. Surf. Sci. **102**, 424–437 (1981)

8.40 H. Banzhof, K.H. Herrman, H. Lichte: Reflection electron microscopy and interferometry of atomic steps on gold and platinum. Microscopy Research and Technique **20**, 450–466 (1992)

8.41 N. Osakabe: Observation of surfaces by reflection electron holography. Microscopy Research and Technique **20**, 457–466 (1992)

8.42 Z.L. Wang, J. Liu, P. Lu, J. M. Cowley: Electron resonance reflections from perfect crystals and crystal surfaces with steps. Ultramicroscopy **27**, 101–112 (1989)

8.43 H.A. Bethe: Theorie der Beugung von Electronen an Kristallen. Ann. Physik (Leipzig) **87**, 55–76 (1928)

8.44 H. Shuman: Bragg diffraction imaging of defects at crystal surfaces. Ultramicroscopy **2**, 361–376 (1977)

8.45 L.M. Peng, J.M. Cowley: Diffraction contrast in REM. Screw dislocations. Micron and Micros. Acta **18**, 171–180 (1987)

8.46 A. Howie: Reflection high energy electron diffraction and reflection electron imaging of surfaces. In *RHEED and Reflection Imaging of Surfaces*, ed. by P. Larsen and P. Dobson (Plenum, New York 1988)

8.47 L.M. Peng, J.M. Cowley: Multislice for RHEED. Acta Cryst. A **42**, 545–552 (1986)

8.48 G.R. Anstis, X.S. Gan: Simulation for reflection electron microscopy. *Scanning Electron Microscopy 1992, Proc. Pfefferkorn Conf. in Signal Processing*, (SEM Inc., AMF O'Hare)

8.49 Z.L. Wang, J.C.H. Spence: Magnetic contrast in reflection electron microscopy. Surf. Sci. **234**, 98–108 (1990)

8.50 J. Spence, Y. Kim: Adatom site determination using channelling effects in RHEED on X-ray and Auger electron production. In *RHEED and Reflection Electron Imaging of Surfaces*, ed. by P.K. Larsen and P.J. Dobson (Plenum, New York 1988) pp.117–128

8.51 Z.L. Wang, J. Bentley: Optimum experimental conditions for quantitative surface microanalysis by REM. Micros. Microanal. Microstr. **2**, 301–314 (1991)

8.52 J. Bruley, L.M. Brown, S.D. Berger: A study of the electronic structure of diamond by STEM. *EMAG 1985*, ed. by G. Tatlock, Conf. Ser. 78 (Inst. of Phys., Bristol 1985) pp.561–564.

8.53 A. Howie: Surface reactions and excitations. Ultramicroscopy **11**, 141–148 (1983)

8.54 J. Schilling: Reflection ELS of Ga, In, Al, Si at 10 kV. Z. Physik B **25**, 61–67 (1976)

8.55 R.F. Egerton: *Electron Energy-Loss Spectroscopy in the Electron Microscope* (Plenum, New York 1986)

8.56 O.L. Krivanek, A.J. Gubbens, N. Dellby: Developments in EELS instrumentation for spectroscopy and imaging. Microsc. Microanalysis and Microstructures **2**, 315–332 (1991)

8.57 H. Liebl: The image aberration caused by the acceleration field between concentric spherical electodes. Optik **83**, 129–135 (1989)

8.58 L.M. Peng, J.M. Cowley: Thermal diffuse scattering and REM image contrast preservation. Ultramicroscopy **29**, 168–176 (1989)

8.59 A. Bleloch, A. Howie, R. Milne, M. Walls: Elastic and inelastic scattering effects in reflection electron microscopy. Ultramicroscopy **29**, 175–185 (1989)

8.60 G. Meyer-Ehmsen: Direct calculation of the dynamical reflectivity matrix for RHEED. Surf. Sci. **219**, 177–188 (1989)

8.61 L. Wang, J.M. Cowley: Electron channelling effects in filtered CBED RHEED. Ultramicroscopy **55**, 228–240 (1994)

8.62 N. Yamamoto, S. Muto: REM of multilayer. Jpn. J. Appl. Phys. **23**, L804–L807 (1984)

8.63 G. Ndubuisi, J. Liu, J. M. Cowley: REM observation of the prismatic faces of sapphire. *Proc. XIIth Int'l Congr. on Electron Microscopy*, ed. by R. Fisher (San Francisco Press, San Francisco 1990) Vol. 1, pp.330–331.

8.64 J. Liu, J.M. Cowley: Energy-filtered scanning reflection imaging of copper. *Proc. 42th Ann. Meeting of EMSA* (San Francisco Press, San Francisco 1989) pp.542–543.

8.65 J. Spence, J. Mayer: Zero loss Omega filtered REM and RHEED on the Zeiss 912. *Proc. 49th Ann. Meeting of EMSA*, (San Francisco Press, San Francisco 1991) pp.616–617.

Index

Springer Series in Optical Sciences

Editorial Board: A. L. Schawlow A. E. Siegman T. Tamir

Managing Editor: H. K. V. Lotsch